STATISTICS: Textbooks and Monographs

Recent Titles

Quality By Experimental Design

Third Edition

Thomas B. Barker

Professor

John D. Hromi Center for Quality and Applied Statistics

Rochester Institute of Technology

Rochester, New York

Chapman & Hall/CRC

Taylor & Francis Group

Boca Raton   London   New York   Singapore

This book was produced using Adobe® PageMaker® 7.0. The original first edition book was produced on a Xerox® Alto computer in 1985, making it the first desktop-published book in the world using electronic composition.

Portions of the input and output contained in this book are printed with permission of Minitab Inc.

About the cover: The triangle is one of the strongest structural configurations. The triangle of statistics, engineering, and management builds the strongest quality organization. The Deduction–Induction graphic is originally attributed to George Box.

Published in 2005 by
CRC Press
Taylor & Francis Group
6000 Broken Sound Parkway NW, Suite 300
Boca Raton, FL 33487-2742

Library of Congress Cataloging-in-Publication Data

Barker, Thomas B., 1941-
    Quality by experimental design / Thomas B. Barker.-- 3rd ed.
        p. cm.
    Includes bibliographical references and index.
    ISBN 0-8247-2309-0
    1. Quality control--Statistical methods. 2. Experimental design. I. Title.

TS156.B375 2005
658.5'62--dc22   2005043989

Taylor & Francis Group
is the Academic Division of T&F Informa plc.

Visit the Taylor & Francis Web site at
http://www.taylorandfrancis.com

and the CRC Press Web site at
http://www.crcpress.com

To Mason E. Wescott,
a teacher of teachers

"... the engine was a better engineer
than the engineers. It told us what
kind of piston rings it liked! We
just ran errands for it, bringing it
a variety to choose from."

–Charles F. Kettering
(on the development of the
diesel engine by empirical
methods)

Foreword

I have known and respected Tom Barker as teacher, practitioner, and colleague for well over ten years. Having heard him in seminars and delivering papers on the design and analysis of experiments, I was delighted that the same easy-to-understand style carried into the writing of his first book. This third edition takes the positive aspects of the first edition and adds to them. Some of the improvements include (1) highlighting key teaching points, (2) additional demonstrations and exercises that can be used to get specific points across to the reader or to a class, and (3) the addition of templates and worksheets for applying the information in a practical situation in your own company or class. I know these work because they have stood the test of years of teaching in academe and in industry.

Particularly important to any person trying to design an experiment are the discussions in the early chapters on the philosophy and technique for setting up experiments. In my experience, this is where most experiments fail. No amount of good analysis can extract reliable information from bad data or poorly conceived experiments; contrarily, ineffective or incorrect decisions can be made from these data.

The benefits of the approach espoused by Tom are (1) increased efficiency, e.g., one research chemist estimated that he saved his company eighteen months of lapsed time and over two hundred thousand dollars worth of experiments by designing the right series of experiments, (2) timely accomplishment of the goals of the experiment and appropriate follow-up experiment, (3) visualization through graphical and numerical presentations of the data rather than assumptive or subjective conclusions, and (4) control of the experimental process through careful planning, e.g., occasionally the collection of information in preparing an experiment precludes the need for even designing a statistical experiment.

FOREWORD

This is a well written book that can be used for your own reference or as a text. Good luck and good experimentation!

John T. Burr
Assistant Professor
Center for Quality and Applied Statistics
Rochester Institute of Technology

Preface to the Third Edition

"That DOE book reads like someone is talking to me. ... not like those other stuffy, mathematically oriented statistics books." I was so gratified when a colleague passed this comment along from a student who had use *Quality by Experimental Design* in a recent class. It is a sentiment I have often heard. When I set out more than 20 years ago to write a statistical experimental design book, with the intention that it would be for *students* new to the field and not necessarily for the seasoned faculty, I decided on the "conversational style" extolled by the above reader.

This third edition continues in that tradition. But, without destroying the charm and spirit of the original work, this revision is expanded to include new topics in inference, more realistic practice problems, examples utilizing modern computer solutions (from Minitab®), and a large dose of the philosophy and methods of *Robust Design* (sometimes known as the "Taguchi Approach").

When I first encountered Dr. Genichi Taguchi at Xerox in the autumn of 1982, I was polite but skeptical. His loss function was intriguing, and I thought, "This is a way to involve the management team in quality thinking." I liked the idea of using engineer terms to describe the inverse of this loss function and I had already been using the "Signal to Noise" in my explanation of the F test in ANOVA. I began to think that Taguchi and I were kindred spirits. However, when it came to his "orthogonal arrays" we were at odds. I was steeped in the use of the CCD and shunned any thoughts of making non-linear inferences from the 3-level designs that he advocated. But I had taken a bite from the Taguchi "apple" and while some of this fruit was bitter, there was still enough juice to be squeezed from it that I got past the first and second stages of encountering a new idea (1st stage "utter bunk!"; 2nd stage "true but trivial"). As evidenced by sections from the first edition of this book, I had reached the third stage ("this is good stuff – and I thought of it myself long ago").

At that point I made a decision and created another separate text, ***Engineering Quality by Design – Interpreting the Taguchi Approach***. This book joined a bandwagon of more than a dozen books on this topic that all marched to success for a number of years. But eventually the "Taguchi" craze fell to other new initiatives in quality and statistics. ISO with its promise of markets in foreign lands took the center stage only to be followed by the Jack Welsh/General Electric endorsed six sigma initiative.

Six sigma has been one of the most influential corporate activities in the field of quality and statistics since they were devised nearly a century ago. Six sigma has sparked an intensive interest in statistical methods, linking these methods to management goals of reducing cost. In the previous two editions of this book, I have always had a link to management goals. My definition of efficiency in statistical experimental design links required scientific information with the least expenditure of resources. Science, Engineering, and Management are linked with the structure of experimental design. This had been my fundamental philosophy long before I knew of Taguchi and long before six sigma was launched. Again, I had reached that third stage of accepting a new idea

If you are a new reader of this third edition, I welcome you to some of the most powerful ideas in scientific investigation and engineering understanding. For the seasoned "QED'ers" who have learned from the first or second editions, read the book again and appreciate new insights it offers. In doing so, you will also review the concepts you have most likely become a bit rusty with. Remember, most technical subjects have a mere six month half-life if you do not use them. I encourage you to experiment and keep the ideas of statistical experimental design alive and working in your endeavors!

I thank my wife and fellow faculty member, Anne, who has stood by me, and encouraged my writing. A special thanks is extended to Hank Altland who carefully read the manuscript and made many helpful suggestions in addition to his incredible encouragement.

Thomas B. Barker
Professor
Center for Quality & Applied Statistics
Kate Gleason College of Engineering
Rochester Institute of Technology
Rochester, NY
tbbeqa@mac.com

Preface to the Second Edition

In the nine years since *Quality by Experimental Design* (or the QED book as it has become known by my students) was introduced, there have been, as there always are in life, a number of high and not so high points. The excitement of looking at a first publication, the support of my readers in response to reviews, and the continuing volume of sales solidified my conviction that the approach I had taken to expand the subject of experimental design to include the human and engineering aspects, beyond the mathematical side, was the "right stuff" for such a book.

The first edition (*QEDI*) was written entirely while I was employed by Xerox. Soon after *QEDI* was in print, I "graduated" from what I consider to be the "University of Xerox." This was my additional undergraduate degree from the real world of experimental design. In my new position of professor at the Rochester Institute of Technology, I began to acquire new insights into this real world, for I now had the opportunity through external teaching assignments to meet and work with engineering and scientific professionals from almost every discipline and from a wide array of companies around the world. Over these nine years at RIT, I have worked with a variety of researchers: from "sheep-dip" chemists and mining engineers in Australia to microelectronics and silicon crystal growers in Arizona. If the problem is wobbly spindles in an automotive bearing assembly or cracked toilet bowls, the use of experimental design is a universal constant – constant and continual improvement through systematic investigation.

Experimental design is not just orthogonal structures and fancy analysis techniques. Experimental design is a mind-set and a structured procedure that must be integrated as a part of the engineering process... if I complete this train of thought, this preface would become a book!

Now, it is up to you, the reader, to jump on board the statistical experimental design express to survey the vistas, explore the terrain, and find how quality can emerge from experimental design.

How the Second Edition Is Different

Armed with a broader base of experimental applications and hundreds of presentations of experimental design classes, this book is expanded to include the insights gleaned from this exposure. To allow this expansion and not grow the book to a two-volume set, the computer programs included in *QEDI* have been left out. In the relatively few years between the two editions, statistical computing has become more user-friendly and expansive enough to include programs to cover experimental design. Therefore, my BASIC programs are no longer necessary.

The introductory chapters on the philosophy of experimental design contain more of the essential material that is necessary for organizing and implementing the first and most important phase of experimentation. Aids to organization, including sample meeting agendas and information gathering forms, are in these chapters.

A new chapter on the response variable has been added. In this chapter we find methods that turn subjective feelings into numerical values that can be analyzed using ordinary statistical methods. The concept of the indicator variable is introduced to help the experimenter get to the root cause, rather than simply treat the symptom. The chapters on factorial and fractional factorial designs have been expanded to show more examples of application, and fill-in-the-blank experimental design templates are provided in the Appendix of Chapter 5. With these templates, it is possible to set up an experimental structure without the help of a computer or to visualize what the computer-generated design is doing for you.

The wording in the chapters on multi-level designs, blocked designs, nested designs, and evolutionary operation (EVOP) has been sharpened. The chapters on analysis have been expanded to include more visual material to help solidify the concepts of separating the signal from the noise. This is often the hardest part of designed experiments for engineers to grasp and a special effort was made to clearly explain these concepts. The chapter on ANOVA now includes a conceptual and practical investigation of residuals and the power of the idea of tracing the sources of variation. The chapter on YATES ANOVA now fully explains the meaning of a half effect. This chapter has been expanded to explore modern methods of deconfounding which include the "minimum number of runs" technique to resolve confounding among 2-factor interactions and the powerful "fold-over" design.

The chapters on regression and its enabling chapter on matrix algebra have been untouched, but Chapter 18 on blocked and nested design analysis has a new section on split-plot analysis.

The final chapters on the derivation and utilization of empirical equations have been expanded to include a new example. These chapters integrate the Taguchi philosophy of robust product design. Since I became acquainted with Dr. Taguchi's ideas, I have found it difficult to speak or write on the subject of experimental design without including his philosophies and methods. I do not consider myself a member of any one "school" or philosophy of methods, but I would rather use a combination of both classical and "Taguchi" approaches – taking the best features from both sides.

I have evidence to convince me that the method we apply depends upon the situation that confronts us. Applying a classical method to a problem seeking a Taguchi solution will fail, much as forcing a Taguchi method on a problem crying for a classical solution will fall into oblivion. I firmly believe in fitting the correct method to the situation, rather than force-fitting the situation to the method. This book offers a number of situations and their matching methods to act as examples for the reader's application. Your job is to find the parallel between your specific situation and the correct technique of design and analysis. Each situation will be different. Each application will be just a bit different. The greatest fun in experimental design is knowing that you will never see the same problem twice! When you experiment, you will have fun.

I want to thank those people who helped in the production of this book. At Marcel Dekker, Inc., Maria Allegra, my editor, showed both patience and persistence with a project that seemed to be on-again off-again over what appeared to be an interminable time-frame. Andrew Berin, my production editor, echoed Maria's qualities and gave the book his meticulous attention. I believe this book will be so much better because of his efforts.

At RIT, a great number of students read the manuscript before it reached the copy editing process. Their insights and the fact that they are the customer helped me create a publication that reflects customer need. In particular, Cheng Loon and Bruce Gregory provided essential suggestions and encouragement on the chapters regarding factorial designs and the fractional factorial design templates.

Professor Daniel Lawrence, a fellow CQAS faculty member, must be especially thanked for his critical review of the chapters on t tests and ANOVA. He gave me the confidence that I had successfully explained these methods in everyday language. It is wonderful to have a colleague who understands that the purpose of this book is to explain experimental design while still making it correct mathematically.

I thank my wife and fellow faculty member, Anne, who has stood by me, encouraged my writing, and aided the content by suggesting changes to be sure the book is clear and correct. I am fortunate to have, in the same household, a colleague who teaches from the same book (although we do not always agree.)

Thomas B. Barker

Preface to the First Edition

There have been a great many textbooks and reference manuals on the subject of statistically designed experiments. The great work by Owen Davies is still considered a classic by this author. Why then a new book to crowd others on the shelves of the professional statistician? Mainly because this book is written to introduce the *nonstatistician* to the methods of experimental design as a philosophical way of life. It is intended to bridge the gap between the the experimenter and the analyst, and break down the aura of mystery created by so many statisticians. In short, this is a short book (as contrasted with Davies' encyclopedia) that lays the foundation for logical experimental thinking, the efficient means to implement the experiment, the statistical methods of sorting signal from noise, and the proper methods of utilizing the information derived in the experiment.

To present the reasoning behind each concept and yet allow the book to stand as a reference, a unique format has been created. For each chapter there are two parts. In the main text we present the entire story complete with examples. An extract is then made of the key concepts in the appendix for each chapter. This serves as a handy reference after the learning is over and the concepts are put to practical use. After all, experimental design is not an academic subject, but a down-to-earth, practical method of getting necessary information.

Throughout this book, we shall rely upon modern computer techniques to perform the tedious arithmetic sometimes associated with large scale experimentation. The appendix of each chapter also contains the BASIC language versions of the programs needed to complete the computing.

The method of presenting the material is based on the discovery and building block approach to learning. No new concepts are thrown out without a basis on previous information. When a new idea is presented, it is developed in such a way that the reader has the thrill of discovery. I believe that the learning will last longer if he or she is able to *develop* the concept rather than have it given from memory.

I would like to thank all the people who have been associated with the preparation of this book. At Xerox Corp., Doug Boike, my manager, saw the need for a practical treatment of experimental design and encouraged me to complete the manuscript. The Innovation Opportunity Program provided the resources to allow my scribblings to become the bits and bytes via the Alto work station. Karen Semmler and Mark Scheuer both worked on this transcription and showed me how to format and change these electronic images through the many revisions necessary before a "quality" manuscript was in hand. Loretta Lipowicz and Nuriel Samuel both helped translate the computer programs into the "PC" format from my TRS-80 originals. Without their help the project could never have been finished.

I would especially like to thank all of the many design-of-experiments students at Rochester Institute of Technology who found the "glitches" in the manuscript as they used it as "supplemental notes" for the courses I have taught over the past 15 years. These students also encouraged me to "finish your book soon" and helped me find the title through a *designed* experiment.

Of course, special thanks are extended to my wife Anne, who always has encouraged my writing and aided the content by making sure it was clear and correct. It is fortunate to have such a knowledgeable statistician in the same household. My children Audrey and Greg who grew up as I spent 9 years on this effort must be thanked for allowing me to close the den door and work on "daddy's project."

Thomas B. Barker

How to Use This Book

The chapters of this book have been arranged in a functional rather than chronological reading order. The concepts of experimental design methods and the philosophy behind these methods found in Parts I and II can be grasped without the formality of statistical analysis. Most other books on experimental design begin with foundations of math and fundamentals of complex statistics which leave ordinary readers in a state of confusion. They often do not get through the first chapter of such a book without asking, Is this experimental design? Do I have to do all this complicated math? Then they put the book aside and revert to their old ways.

By placing the motivational aspects of experimentation in the front and gradually building in a logical progression, the reader will get to at least Chapter 5 in this book before being motivated to ask the next question, How do I manage to make decisions with variation plaguing me? This is the time to make an excursion into that part of this book that shows how to separate the information (signal) from the residual error (noise.) It is suggested that this learning take place in the third part of the book as outlined in the following diagram.

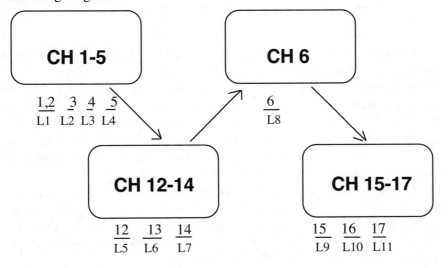

For use in an engineering curriculum, the previous outline would serve as a first- quarter course in experimental design with the appropriate reading assignments shown underlined. The appropriate lecture numbers (L) are shown below the chapters.

For a second-quarter course, a similar outline is found below. In this quarter, the "specialized" designs that are used when randomization is not possible and other topics like components of variance are explored. It is suggested that a class project be utilized in the last four weeks of the quarter to put the concepts of the entire two-quarter sequence to work.

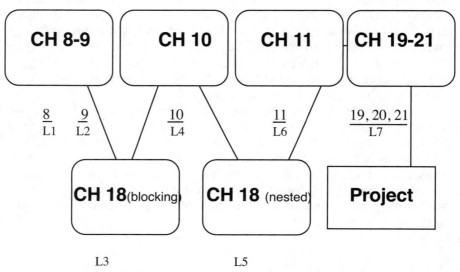

For a semester course, the outline for the first quarter plus the chapters on blocking and their analysis in Chapter 18, followed by the case history and Monte Carlo would fit into the 15-week format. A two-semester enriched course could follow the quarter outline and expand the time on least squares (Chapters 16-17) in the first semester while spending more time on examples and class projects in the second semester. Learning by doing is the best way to teach and learn experimental design. This is not my opinion, but the constant statement made by my students.

I would be pleased to share my teaching experience with other professors who wish to embark on using this book and the philosophies it promotes. An answer key using an interactive computer system (as I did for the book, *Engineering Quality by Design:Interpreting the Taguchi Approach*) is being prepared. My email is: TBBEQA@MAC.COM.

Contents

PART I
The Philosophy of Experimentation

PART II
Statistical Experimental Design

CONTENTS

CONTENTS

PART III
Sorting the Signal from the Noise

CONTENTS

PART IV
The Derivation of Empirical Equations from
Statistically Designed Experiments

PART V
Utilization of Empirical Equations

PART I

The Philosophy of Experimentation

1

Why Design Experiments?

In today's industrialized society, almost every product that eventually reaches the market has a long lineage of testing and modification to its design before it ever sees the light of day. In view of the success of such products, we may ask, Why upset the apple cart? What we are doing *now* gets the product out the door. Why change?

It is always difficult to argue with success, but those of us who are on the inside of industry know that this "success" is a most difficult commodity to come by, especially if the time frame imposed upon us is short. This time frame is often short and is usually based on a customer need or a competitive threat.

The customer is more informed today than ever before. We live in an age of "customerism" where an informed consumer is our best customer or our worst enemy. Improved products constantly raise the level of expectation for the next generation of products. In many cases, technology and invention are moving more rapidly than our ability to perform a good engineering design or create a manufacturing environment by the old "tried and true" methods. Quality products that perform as advertised and perform with little variation are now a part of the informed customer's expectation. In many cases, only through testing are we able to produce the information necessary to determine what the quality of a product or a service is.

It is our job as experimenters to find the most efficient schemes of testing and to apply such schemes to as broad a gamut of applications as possible to obtain the information required to make a successful product.

3

Uses of Experimental Design

Given the need for finding efficient methods of uncovering information about our products, processes, or services, we need to be more specific in the exact application and use of experimentation. The prime area of application and the area we shall emphasize in this book is the *characterization* of a process. We shall use the word "process" in this text rather broadly to mean *any* phenomenon that is worthy of investigation. A process could be the way we **assemble** a copy machine in a manufacturing environment. A process could be the way **banking** is practiced in the world of finance. In this very general definition of a process, we study the result of making purposeful changes in the way the process is operated and watch the way the results (or *responses*) of the process change.

Another application of experimental design utilizes experimental designs to *troubleshoot* a problem by interchanging components. In this way, we can induce a failure at will and understand the source of this failure. I discovered a faulty power cord on a projector by interchanging cords. A third use of experiments is to access *routine analytical errors* that occur in the measurements of our response variables. Instead of just watching the variation in the data, said to be the result of a so-called random process, we trace the sources of the changes in the values to gain **control** of the overall variance and improve the quality of our product, process, or service.

Table 1-1 is a summary of the three general uses of experimental design. Do not confuse these *uses* with the *reasons* we will see later. As we develop the concepts even further, we will see further uses within the context of these generalities. The common element in the application of experimental design is the purposeful change we exert on the factors. We don't wait for changes to take place, we make the changes and watch the results happen! Experimenters rely on changes made on purpose, not on changes made by fate.

TABLE 1-1
Uses of Experimental Design
1. Characterization of a process
2. Troubleshooting
3. Quantification of errors

While purposeful change is the important method of experimental design, efficiency is the added value provided by the *statistical* approach to experimentation.

Efficiency

A good experiment must be *efficient*. Efficiency is defined in the box that follows. This definition is the only aspect of experimentation that must be committed to memory, since every time we encounter an experimental situation, we must know if our approach is efficient.

> **An efficient experiment is an *experiment* that derives the *required information* with the least expenditure of *resources*.**

This is a very precise definition and forms the basis of all of the methods associated with **S**tatistical **E**xperimental **D**esign (**SED**). Since we need to memorize the definition, let's take it apart to understand just what it means.

First, Experiment

The first key word in the understanding of this definition is the concept of an *experiment* as opposed to an isolated test. A test merely looks at a problem as a "go" - "no-go" situation. A test is usually success oriented. A test does not ask why an event has occurred. A test only cares *if* the event has happened. If the event does not occur as hoped, the team working on the project will be very disappointed.

Such was the disappointment of the group testing the formulation of "crash proof" jet aircraft fuel called AMK (anti-misting kerosene). A jelly additive was supposed to keep the fuel from vaporizing and causing the conflagration that is the major cause of death in such disasters. Well, the team set up an elaborate "experiment" (as they called it) to *test* the additive's effect. This was done at great expense by purposely crashing a radio-controlled jet liner (a Boeing 720, the military version of a 707). The tanks were filled with the AMK fuel. The drone radio control pilot flew the plane over Edwards Air Force Base and then, on command of the ground crew, crashed the entire $7 million plane into the desert!

Unfortunately, the team's hopes were dashed on the desert floor. The plane burst into flames and was a charred cinder instead of the crumpled wreck of twisted metal that the testers had anticipated. The additive had failed to perform in its intended job function.

AMK is currently not in use because of this disaster of a test. In the postmortem that took place after the crash, the team equivocated that the plane had "crashed wrong." But who knows how a plane will crash?! Crash configuration is the kind of thing we can't predict but must anticipate. In the terminology of Genichi Taguchi, the crash configuration is an *outside noise*, which is beyond our engineering control.

An experiment goes far beyond the test by drawing out the reasons for the *"go"* or *"no-go"* in a *series* of ordered tests. While the testing approach is success oriented, the experiment approach is **information** oriented. An experiment gets to *why* something works, rather than just *if* it works. When we know *why* something happens, we are much more likely to make it work!

It is unfortunate that many of us spend so much of our time *testing* rather than **experimenting** and then try to let mathematics (in the form of multiple regression) do the organizing that should have been done before the testing but is left until after the fact. This commonly reduces our scientific endeavors to the not-so-scientific ground of "the lure of accumulated data."

All too often the tester laments the fact that after a number of years of work trying "a little of this and some of that," he or she is no further ahead than when the work had begun. Worse yet, this tester has little chance of reconstructing the data into anything that could guide subsequent endeavors.

While a set of designed experiments may not contain the exact results in the form of a product that can be sold, the information derived along the way will point in the correct direction to the proper conditions that will produce a final product that meets the customer needs. Think of the situation this way. If I have half a dozen factors that control the output of my product, and each factor has a range of possible set points, then there are a very large number of possible *combinations* of these settings. With just three settings per factor, there would be 729 possible combinations! Somewhere within those 729 is an optimum setting. I could search randomly as in the *testing* method, or I could search *systematically* and find the direction to the correct set of combinations. SED is the systematic approach. It is statistical, since it selects a sample from the population of possible configurations. From this sample and through statistical analysis, we find the direction to the best configuration without trying every possible combination in the entire population of configurations.

The first order of business as we strive for efficiency is a well-planned **experiment** and not just some disjointed testing.

> ## An experiment is a structured set of coherent tests that are analyzed as a whole to gain an understanding of the process

Second, Required Information

Before we can plan the tests that constitute our experiment, we need to decide firmly on what we are looking for. This may sound like a trivial task, but it is one of those deceptive essentials that usually gets by the experimenter. Motivated by a real need, he or she rushes off into the lab to get quick results (see Figure 1-1). In such a zeal to obtain information, the experimenter often misses the fundamental understanding that *lasting* results will be built upon.

While we need to define information before we begin our experiment, it is not necessary to strive to obtain every iota of information in our experiment. We need to define just what is required to accomplish the task. This is fortunate as well as unfortunate. It is fortunate since we need to isolate only a relative handful of components from the virtual bag of causes that exist on a typical problem. It is unfortunate, on the other hand, since many times the vital few components are quite difficult to characterize or even recognize. Since the investigator is human, he or she will often select the easiest or the most obvious factors from the trivial many, and by throwing these into the experiment, very little is learned that would lead to the essential fundamental understanding necessary to solve the problem. Often, the experiment is perfect from a procedural standpoint, but the basic factors chosen were not the correct factors.

This type of misapplication has in the past given statistical experimental design techniques a bad name since the experiment was perfect statistically, but the patient died anyway.

The region between what is easy to define and what is difficult or impossible to identify is the territory of required information. The exploration of this region requires people who are skilled in the disciplines of the process under study. Those trained only in the statistics of experimentation will fall into the many traps set along the way by "Mother Nature" or the physics of the process. Prior knowledge and understanding of the nature of the process under study is essential for a good experiment. In Chapter 2 we will show how to avoid such problems by proper organization before the experiment is begun.

Third, Resources

The final part of our definition of efficiency concerns the area of resources. Resources can be money, people, buildings, prototype machines, chemicals, and, in every experiment, *time*. In most industrial experimentation, management can provide enough of the material resources of an experiment. Time, however, is costly, and large quantities are usually unavailable at any price. Here is where SED shines, for it strikes a balance between what is defined as needed and the way to get there at the least cost. We will show in later chapters how statistical design is inherently more resource efficient.

When we realize that every piece of experimental data we gather is merely a statistic (that is, a sample subject to variability), then it becomes clear why we need a body of methods called *statistics* to treat these statistics. By using statistical methods, we will obtain the required information to do the job in a systematic and controlled manner.

A General Example

Unfortunately, most experimenters do not realize that a short-term investment in a good experiment is actually more cost effective than a series of shot-in-the-dark attempts to fix a problem. Figure 1-1 shows the usual method of testing and is contrasted with Figure 1-2, which shows the use of statistical experimental design. These examples are further enhanced by the assignment of chances of success based on actual experience. This general example shows that there is much better control and lower cost with the SED approach to the design and analysis of experiments.

In Figure 1-1 the chance of solving the problem is only about 20%. The chance of solving the right problem is only 30%. There is also a high probability that a new problem will be created as a result of the "solution" to the first problem! It takes a long time to get a reward for a good job in the first approach because we are constantly going back to a previous step. There is an old engineering saying that is still very true today: "There is never enough time to do the job right, but lots of time to do it over." This is the unfortunate problem with the approach in Figure 1-1; we do it over too often. A computer simulation of this "jump-on-the-problem" mode shows that there is a high cost (59 days) with a very high uncertainty (44 days standard deviation) using this faulty approach to problem solving.

In Figure 1-2, we need only to include a statistically designed experiment as the investigation methodology. Now the chances are much higher for solving the problem the first time. We are almost certain (95%) to solve the problem on only one try, and we will solve the **right** problem. The problem also stays solved without creating any new problems. This is because the statistical approach looks at the whole situation and not just a fragmented segment. Now the time to do the job is reduced to only 40 days, and more important, the variation is reduced to a standard deviation of only 8 days (better than one-fifth the variation of the first method).

An experimental design is more than an engineering tool. It is a necessary part of the management of the entire engineering process. Of course, as we do not restrict the meaning of the word *"process*," we should not restrict the meaning of the word *"engineering"* to only industrial/manufacturing organizations. The engineering concept can be expanded to include any process that needs to be improved to meet the quality requirements of the customer. It is possible, then, to engineer medicine, banking, baking, and airline service as well as the obvious manufacturing methods.

A Note on the Simulation

In Chapter 20 we will cover the concepts and methods of "Monte Carlo Simulation." That method was utilized to perform the simulations found in Figures 1-1 and 1-2. To see this method in action, let's follow the flow in Figure 1-1.

*We start with the recognition of the problem and get the orders to fix the problem. We next spend a half-day in a meeting and plan a simple test. The test takes 2 days and the analysis takes an additional day. So far we have spent 3 1/2 days on the project and ask if the problem is solved. If there is a solution, we can quit at this point, but the chances (only 20%; p=.2) are slim that we did it, and we have to go back to more testing and analysis. We continue until we leave this part of the loop and progress to the question "did the right problem get solved?" Again, the chances are rather slim (30%; p=.3) that the correct problem did get solved and we go back to the meeting, back through the test and back through the analysis. Even when we make it past this next loop, there is one more hurdle regarding a **new** problem resulting from our solution! This brings us back to the meeting and the subsequent reiterations. The last obstacle in our path to a reward is the solution forever question. Only when we have a high chance of passing all of the decision points will we be able to claim our reward. The reward is staying in business. SED is an enabling technology that helps take the chance out of experimenting and puts it on a more certain basis.*

SYSTEM FLOW DIAGRAM
EMPIRICAL INVESTIGATIONS—JUMP ON THE PROBLEM

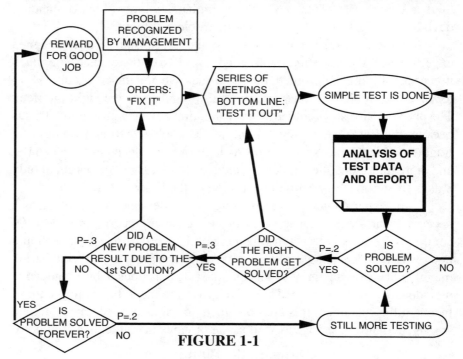

FIGURE 1-1

Figures 1-1 and 1-2 show a simulation of two approaches to solving a typical problem. In both cases, the orders come down from above (management) to fix a problem. In the usual case (Figure 1-1), the workers jump on the job without too much consideration of what they are doing! (probably from fear of management). In the case that employs SED (Figure 1-2), the predictability of the time necessary to accomplish the task and the overall cost are both more reasonable than in the "jump-on-the-problem" approach.

Going-in Assumptions for Simulation

Time per test = 2 days
Analysis time= 1 day
Meeting time = 1/2 day
Costs:$30/hr Management
          $16/hr Workers
We are looking at 4 factors

Results of Simulation

Figure 1-1 "Jump on the Problem"
Average time = 59 days; Std Dev = 44
Average cost = $8520; Std Dev = $6135

Figure 1-2 "SED" Approach
Average time = 40 days; Std Dev = 8
Average cost = $5320; Std Dev = $1035

SYSTEM FLOW DIAGRAM
EMPIRICAL INVESTIGATIONS – USING STATISTICAL DESIGN

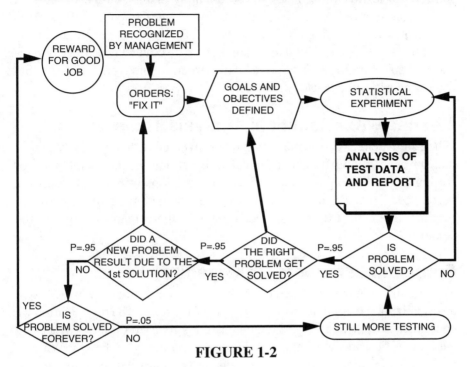

FIGURE 1-2

Reasons for Designed Experiments

It's probably quite clear why we need to utilize statistically designed experiments instead of the old fashioned way that we may have learned in the school of "hard knocks." Here, in Table 1-2, are four reasons that summarize our need.

TABLE 1-2
FOUR REASONS FOR EXPERIMENTAL DESIGN
1. A **structured** plan of attack
2. Meshes with **statistical analysis** tools
3. Forces experimenter to **organize**
4. Is more **efficient**

Structured Plan of Attack

Impromptu testing often leads to the unhappy results shown in the first simulation (Figure 1-1). By using the statistical (see Chapters 4-10) and the organizational (see Chapter 2) aspects of SED, we not only gain in the structure of the experimental test plan, but we discover a necessary organizational structure to the discipline. This structure uncovers a process that forces the experimenter to organize.

Meshes with Statistical Analysis Tools

When statistical experimental design was invented during the early part of the 20th century, both the experimental design structures and the methods for sorting the real changes from the random variations were developed for a perfect fit. This perfect fit means that when the data from the experiment becomes available, the analysis method will be there, ready to help sort the signal from the noise.

Forces Experimenter to Organize

The organization that is forced on the experimenter is not the type found in an organization chart that shows who reports to whom. The organization we encounter in statistical experimental design is focused on the engineering and scientific aspects of the problem or opportunity at hand. While a problem is a current encounter that is causing a concern and needs immediate attention and a solution, an opportunity is a problem that is headed off at the pass before it causes concern. Organization for experimentation helps identify likely problems and turns them into opportunities.

Working on opportunities is far less stressful than problem solving "under the gun." So, in a way, SED is a stress avoidance method that can contribute to a better quality of life for the investigators.

Efficiency

We have seen how efficiency is the keystone to constructing the correct experiment. The experiment is not just a matrix structure, but the underlying approach that is an extension of and an enhancer of the scientific/engineering endeavor. Because we do obtain the required information at the least expenditure of resources using statistical experimental design, we are inherently efficient and aimed at success.

Problems for Chapter 1

1. Define a problem that you have encountered in your work experience or any related experience that could or did require an experiment.

2. a) If an experiment was performed on the above-mentioned problem, determine if it was efficient.
    and/or
   b) Set down what is required information with regard to problem #1 and what resources are available to perform the experiment. Is this the most efficient approach?

3. a) Give an example of a test.
   b) Give an example of an experiment.

4. A manufacturer has been receiving complaints about a redesigned product. The redesign had been initiated to effect a cost saving. The savings amounted to nearly $100,000 over a 2-year period and were afforded by reducing the thickness of the sheet metal used in the mounting bracket of a mop. Comment on the mistake made in the selection of the response variable ($) used and how this fits into the system flow diagram (Figure 1-1).

5. Comment on how you could see statistical experimental design improving your quality of life by reducing the stress in the investigations you embark upon.

APPENDIX 1

Key concepts from this chapter

SOME USES OF EXPERIMENTAL DESIGN
1. Characterization of a process
2. Troubleshooting
3. Quantification of errors

FOUR REASONS FOR EXPERIMENTAL DESIGN
1. A **structured** plan of attack
2. Meshes with **statistical analysis** tools
3. Forces experimenter to **organize**
4. Is more **efficient**

EFFICIENCY
An efficient experiment is an *experiment* that derives the *required* information with the least expenditure of *resources*.

TEST
A one-shot, go or no-go activity. Usually success oriented (i.e., "make the thing work!").

EXPERIMENT
A structured set of coherent tests that are analyzed as a whole to gain an understanding of the process. Always information oriented.

REQUIRED INFORMATION
Information that is necessary and sufficient to accomplish our goals.

RESOURCES
The things we have to derive the required information: people, money, and time. Time is the most costly and valuable.

2

Organizing the Experiment

If a designed experiment does anything at all, it should force the experimenter to organize. In many situations, pressures to produce results cause ill-planned, premature experiments. In a great number of cases, there is not enough prior "lab bench" knowledge available to formulate the proper questions that need to be studied. Here we fall into the human trap of investigating the obvious. In doing so, we violate the prime directive of efficiency by not deriving the required information.

In a great many ill-planned experiments, we generate an impressive array of purposeful changes, but neglect to define the meaningful measure of the results of these changes. Without a good response variable, we again fail to obtain the required information.

The most flagrantly violated area involved in designing experiments is also the most trivial sounding. There are so many situations in which the experimenter has confused *goals* with *objectives*. In doing so, he or she has failed to identify the problem. It is a well-known fact among repair people (auto mechanics, appliance service people, etc.) that the identification of the problem is the major portion of the repair. Once the problem is identified, the fix is easy. We can learn from this experience. If the real problem can be identified, at least half the work is completed. Understanding the goal and objective of our effort is a first step in this identification process.

The organization of experimentation is the critical aspect of our efforts to assure successful implementation and execution of our investigations. In this chapter we will discover the critical elements of a good experiment and show how to assemble these elements, following a systematic but flexible procedure.

15

The Elements of a Good Experiment

The three essential items for a successful, efficient experiment are as follows:

       1. Knowledge of the process.
       2. A response variable.
       3. Clear goals and objectives.

We will look at each one of these elements and determine why that element is essential and how we can achieve it. Once again, the organizational aspects of our experiment will "make or break" our efforts. Scrimping on this front end of experimentation will most assuredly lead us to failure. Don't become so enamored with the statistics of experimentation that you forget the science and engineering that the statistical methods serve.

Prior Knowledge

The first requirement of an efficient experiment is sufficient prior knowledge to get the effort started. Now, at first it may seem contradictory that knowledge is necessary, since we are running the experiment to obtain knowledge. While this may appear to be circular thinking, logic dictates that we can't run an experiment if we don't know what we are doing!

Without sufficient prior knowledge of the process, we are completely halted in any *good* thinking. We tend to grasp onto the obvious. Prior knowledge requires some preliminary study and consideration of all the background material within the time frame available. Many investigators pride themselves on what they call "research." Often they are merely re-doing what has already been done. It is wasteful to do "re-search" by expending resources on problems that have already been solved by others. A major side benefit of the organizational aspect of SED is the discovery that an experiment is not even needed, since prior work has already been done and the answer is there, ready for our use. Of course, for information to be readily available, there must be reports on the prior efforts. Investigators are often lax in writing such reports, and when they do, the reports are often ponderous piles of technical detail that are not reader-friendly.

To help in the organizational aspects of the experiment, the appendix to this chapter has an information-gathering and organization form that has been very successful in the past. This form helps to determine if you are ready to begin the experiment. If any part of the form cannot be completed, you need to go back to the **Store of Knowledge** before proceeding.

Also in the appendix is a sample of a report that communicates. This sample shows the five elements of reporting: goal, objective, results, conclusions, and recommendations. A report written in this format can easily fit on a single sheet of paper, and more important, communicate the essentials of the experimental effort. Traditional reports that involve pages and pages of essentially hollow background material before reaching the "bottom line" do not communicate and are not usually read by the intended audience. The background material is important for those skilled in the art of the process under investigation, but should be included as an *appendix* to the one-page summary report that says everything necessary to bring about appropriate actions. The report adds to the **Store of Knowledge** (Figure 2-1).

So, besides previous research that has been documented in reports, where does prior knowledge come from? There are many sources, but the following are the primary ones available to the experimenter at the store:

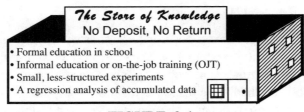

FIGURE 2-1

Most of us have obtained our prior knowledge in the formality of college and bring this knowledge to the job. While college has educated us in general thinking about a discipline, we learn to apply our knowledge while on the job and enrich our basic information with the particular nuances (and nuisances!) of the application under study. While working on a problem, we often try small experiments. These "lab bench" trials form the substance of our more structured, efficient experimentation. There is nothing wrong with a little fiddling around before the main event.

The final possible source of prior knowledge comes from the "lure of accumulated data." Often, unstructured efforts are used to solve a problem in a less than efficient manner. Records of this work may sometimes yield enough information to structure a proper investigation (SED). A number of excellent books on regression analysis describe the care required in using this method of "data mining" that does a numerical brainstorming.

We can see why those who scoff and are skeptical at the need for knowledge to build knowledge should not scorn this effort. Again, we call upon common sense to build on what we know, to gain more of what we need to know. A good experiment will often ask more probing questions about the process under investigation while it provides timely answers to the needs of the present problem or opportunity.

The Qualities of a Response

The response variable is unfortunately the most neglected variable in many experiments. The attitude is usually, "Oh, we can think of what to measure after the process has produced the product." Anyone who has a working knowledge of basic statistics knows the fallacy behind this statement, since, to specify how many samples from the process are sufficient to find an important difference, we need to know the variability of the response or responses (yes, there are many cases involving more than one dependent variable). We must be sure that the response variables are:

> **1. Quantitative**
> **2. Precise**
> **3. Meaningful**

Some examples of responses include such **quantitative** measures as percent contamination, customer satisfaction, product wear-out, and time to failure. Unlike quantitative responses, qualitative responses give us trouble since a "good" or "bad" as a response will not "compute."

If the response is not quantitative, it will be impossible to perform the arithmetic operations of the statistical analysis. There are arbitrary numerics that can be applied to qualitative data. In the case of a binomial response such as "good" or "bad," a "0" or "1" can be used. An ascending order counting scale can be used to quantify qualitative responses. These scales, of course, are arbitrary, and all attempts should be made to quantify a qualitative response on a physical basis. Quantification of subjective responses is the work of fundamental sciences. We shall embark on this work in Chapter 3.

To get sensitivity in our experiment, it is essential that our response is repeatable or **precise**. If it is not, we end up with so much variation (or noise) due to our measurements that nothing of value emerges from the experimental work. Knowing the precision of a response is part of the fundamentals of statistics and will also be covered in Chapter 3.

Finally, the response must have a **meaning** related to the subject under investigation. Studying an effect for its own sake adds nothing to the efficiency of our experiment There are a great number of responses that are quantitative and precise but don't relate to the problem. We should not run the experiment unless we have a meaningful response.

Goals and Objectives

It is surprising how many experimenters go off and try to solve a problem without first understanding the roots of the problem behind it. This mistake is a matter of confusing goals with objectives.

A **goal** is the ultimate end result of a task ("make it work").

An **objective** is a statement of how the task will be accomplished ("understand how it works").

For example, we may want to make a disabled car work. However, before we can fix the car we must present some hypothesis about what is broken. We then test this hypothesis, and if we are correct we can then accomplish the goal and apply the repair. Figure 2-2 shows this process in action. Sometimes it takes a number of cycles from the conjecture stages to the formal solution of the problem. These cycles form the scientific method of conjecture-deduction, information-inference. Knowledge builds with this process.

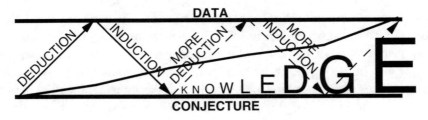

FIGURE 2-2

There is often a hierarchy of goals and objectives in an industrial organization that could look like this:

Goal	Objective
Corporate: $$$$$	Sell a product
Marketing: Sell a product	Have a good product
Manufacturing: Have a good product	Make it work
Development: Make it work	Test the hypothesis that a functional relationship exists between a response and some independent factors

What becomes obvious from this hierarchy is the fact that one organization's goal becomes another organization's objective. When we reach the experimental design stage, we must be sure that our goals and objectives are clearly defined so as to state the problem in scientific terms.

While the goal is stated correctly most of the time without any special form, the *objective* needs the help of the format that is now suggested. This format is a formal statement of the conjecture part of the concept shown in Figure 2-2. We state our conjecture by testing our idea or *hypothesis* that the response is a function of the factors we believe will influence it. This form of the goal and objective has proven to be an effective way to separate the means from the end in experimentation. This separation helps avoid confusion and reaches a solution to the problem through *understanding*.

Goal words include "fix," "improve," "make less costly," "make the darn thing work," etc. An ***objective*** searches for understanding by testing the hypothesis that the response changes as a function of the changes in the factors we take control of and change during the experiment.

To illustrate the goal and objective in action, let's look at an example. A Company is currently producing sandpaper and wants to make an improvement in the useful life of this product. Life testing in general is a major part of corporate R&D expense. Ultimately, the company would like to see that the sandpaper does its job without wearing through. The indicator responses that relate to the life of the sandpaper are how fast the paper gets a hole in it, how long the particles of grit stick on the paper, and how fast the grit particles are reduced to smaller particles. We consider the thickness of the paper, the amount of glue that holds the grit to the paper, and the particle size distribution of the grit as our controlling factors. The goal and objective are, then, for this example:

Goal: Improve the life of 100 grit sandpaper.

Objective: To test the hypothesis that paper wear, grit adhesion, and grit wear are functions of paper basis weight, glue thickness, and grit size distribution.

Observe how we have told the world what we want to accomplish (**Goal**) and how we are about to do so (**Objective**). In one paragraph-long sentence, the objective has made very clear what we are measuring and what we are changing.

Gathering Information

The experimental design form on page 28 is a useful way to organize the information needed to design your experiment. This form provides questions on all the controlling or independent variables, as well as the responses.

It is best to organize an experiment as a team effort and use the brainstorming technique to scope the entire problem. From the list of brainstormed controlling factors, the team selects those with the most probable impact on the process under study. This selection is based on the accumulated team experience. The team effort assures us that there is a lower likelihood that the key factors will be selected for the experiment. There is nothing worse than completing an experiment and finding out at the very end that there is more change due to unknowns than to the factors included in the study.

In one study involving the ultrasonic bonding of a fine wire to a microcircuit, the experimenter, who was a student in my experimental design class, did an outstanding job of applying the statistics but found in the final analysis that the "residual" or unexplained variation was greater than the changes due to the controlling factors in the experiment. I asked him who was on the brainstorming team that selected the factors. He replied, "Me and my buddy."

It was clear to me that the critical mass was missing in this effort. A team implies a group of multi-disciplined individuals. The team should have a positive, synergistic effect on the quality of the factor selection process, since more than one or two opinions are responsible for selecting the controlling factors and the response variables. The trap that the student fell into was not expanding the membership of the team to include the following critical mass elements:

1. The experts
2. The peripheral experts
3. The operators or technicians
4. The customers

Organizational Psychology

Let's examine the list of the players in the brainstorming session to see why they are present and what role they play in the experimental effort. First, we have the *experts*. It is certainly obvious why they are there. They know the answer to the problem. Unfortunately, the answers that the experts bring usually don't work, or if they do work, they add more expense to the final design. In one brainstorming effort involving a problem with a loose threaded connection to a fuel regulator device, the experts had decided that a chemical bonding agent would fix the problem and prevent the nut from unscrewing from the regulator body. This "fix" would involve an added step in the manufacturing process and additional material. Worse than the extra cost for each regulator built, the chemical was quite toxic and would require an exhaust and recovery system to meet EPA governmental health standards. Fortunately, this brainstorming session did include other team members. Let's talk about their role in the process.

The *peripheral experts* consist of knowledgeable people who can ask embarrassing questions of their colleagues, the experts. These people have a peripheral knowledge of the subject, but do not have direct design authority on this problem. They are at the edge of the effort, and from their questions, they are able to strip away the blinders that the experts often wear. This gives the experts peripheral vision to see beyond the obvious answer. From an organizational, human resources view, the encouragement of cross-partici-pation in problem solving enriches the quality of work life by expanding horizons beyond the monotony of a single project. This also builds synergy in an organization and promotes a camaraderie that is often lost in highly segmented design or manufacturing groups. Peripheral experts are at the periphery of the problem and give the experts peripheral vision.

The *operators* or *technicians* are intimate with the process. They actually see the inner workings of the device or system under investigation. In the case of the fuel nut problem mentioned above, the operator (a UAW union member) made only one contribution during the brainstorming session. He said, "Air pressure." Now all day long, this assembly person put a nut on the regulator and torqued it to 31.7 Newton meters of indicated force. He also noticed that the pressure gauge attached to the pneumatic torque wrench would fluctuate during the process. He *saw* the pressure change to a lower value as he did his job. As the experiment that was eventually completed showed, the most important factor was air pressure! The toxic chemical bonding agent was not even needed. A pressure regulator was attached to the system and the problem disappeared.

There is another reason to involve the technicians in the brainstorm-ing process. They the ones who actually run the experiment. Technicians are conscientious persons and want to do the best job possible. Often, the experiment will contain runs that seem to the operator to be capable of producing substandard quality products. Operators will often "fix" the run to give it the ability to "make a good part." Of course, the experiment must make some good stuff and some bad stuff to make knowledge. If all the runs make good stuff, we won't learn anything. Bringing the operators in at the beginning of the experiment and explaining that there will be some runs that will not make a good output will prevent them from making a "field modification" during the running of the experiment. Bringing the operators in at the beginning of the experiment will give them a sense of ownership of the experiment and make them feel a part of the process rather than a tool of the process. This will allow them to do the very best job possible.

The final part of the brainstorming team is the *customer*. From other areas of quality studies we should realize that the customer is the person or entity that gets our output. The customer is not necessarily the end product user. "Next in line" is another way of thinking of the customer. Modern automobiles no longer have axle rods to hold the wheels on the chassis. Rather, a spindle rotates in a hub that is attached to the chassis. The spindle is a forged part (for strength), and it is machined after the forging process. The lathe is the customer of the forge. In a U.S. automotive company the spindles were forged off-center, and because of this, the lathe tools that were supposed to give the spindle the bearing surface were breaking. The lathe operations personnel attended the brainstorming session for the solution to the forging problem. Who were better equipped to identify the meaningful response variables that were in this process? When the experiment was finished, these were the people who would need to accept the results. Instead of surprising them, they were involved from the very beginning. Acceptance of the results is thus guaranteed!

As you can see, we involve the critical players in the brainstorming session to glean their knowledge and involve them in the process to gain their acceptance of the approach and the final results.

The Brainstorming Process

While the technique of brainstorming is quite simple, there are a few enhancements that are applicable to the experimental design activity. After the critical mass of players have been assembled, you will have at least four people if each group has only one representative, and more likely there will be eight to twelve people participating in the event. You should schedule the first meeting to last only one hour and schedule it early in the morning.

Since there will be a second meeting that should be no more than 2 days after the first, Tuesday is a good day for the brainstorming to begin. During this meeting, the leader will explain the goal of the experiment and write it on the top of the first piece of "flip chart" paper. Then each participant will freely contribute *any* type of factor, even the most ridiculous, to the session. Three types of factors should be identified during the approximately 20 minutes of brainstorming:

1. **Controlling factors**
   (*under our design authority*)
2. **Noise factors**
   (*outside of our design authority*)
3. **Response variables**

It is important that all three types of factors are represented at this session. The first or *controlling* factors are the obvious ones and will make up about 60% of the list. The *noise* factors are important since these are the nasty factors that could distort our experiment if left to vary on their own. There is another reason to identify the noise factors that we will show as we progress in this book and investigate ***robust design***. While we may have a primary *response* variable, it is important to identify any response that is likely to change as a result of our experimentation. I have seen many experiments optimize the prime response and at the same time ruin another response. Trade-offs are often required between responses. If all responses are not measured, then we will not know where to make such trade-offs.

After the brainstorming is complete and the ideas have been captured on the flip chart paper and form a "wall paper" around the room, it is time to classify the types of factors into the categories shown above. With a red marking pen put a letter "N" for noise adjacent to each noise factor. With a green marking pen place the letter "R" next to each response variable. The remaining factors are then by default the controlling factors. Some factors can take on a dual role and should be categorized accordingly.

The brainstorming team leader can now take the pages of paper and transcribe the information into a more manageable form such as a spread-sheet using the format shown on page 30. This information from the brainstorming meeting must be in the hands of each of the team members by early afternoon of the day the brainstorming took place. Send the spreadsheet via e-mail with a reminder notice of the next meeting.

If we had a Tuesday brainstorming meeting, then the follow-up meeting would be no later than Thursday. By delivering the summary information from the brainstorming meeting by the afternoon of that first meeting, the individual team members are able to study the list, add ideas in the privacy of their own thoughts, and begin the *elevation* process. In a typical brainstorming effort, there will be more than 100 factors identified. Approximately 67% of the items will fall in the controlling factor category, about 20% will be noise factors, and the remainder will be responses. Since it is not humanly possible to execute an experiment with more than a dozen factors, the list of controlling factors will have to be pared. Rather than think of this process as an elimination effort, it is better to consider the job to be an elevation of the most critical factors to the experimental situation. If we eliminate a factor, we may never think of it as a potential candidate for future experiments. However, if we elevate a handful of factors to the actual experiment, we will still have the remainder in a priority order for future efforts. Each member votes on the importance of the factors using 1 to 5.

The second meeting in the brainstorming process is called the follow-up effort. You should allocate 4 hours to this consolidation process. Forming a consensus on the factors that will be included in the actual experiment can be easier than you might think *if* you use the votes provided by the individual team members. By computing the standard deviation of the votes, you can easily see where the lack of consensus is taking place. Negotiating skills are put to the test in this part of the second meeting as we get members to change their vote. Often misconceptions cause the original discrepant votes.

After the list of factors has been put together, the team must next decide on the values or *levels* of each of the elevated factors. For example, a factor like temperature would have levels or conditions of 100° and 200°. Level selection will take about 1/4 of the time. If the values of the levels of a factor cannot be determined from the prior knowledge available, then the team must engage in some "lab bench" work to establish the levels. In choosing levels, the team must be careful not to make too small or too large a change. If too small a change is put into an experiment, the variation from other sources will "swamp" the changes from the factor under study and ruin the chances of learning anything about that factor. If the changes made are very large, then superfluous nonadditive effects will arise and make the functions we are investigating more complex than is necessary. Again, prior knowledge is essential in level setting.

The third activity in the consolidation meeting is the identification of possible nonadditive effects such as curved (quadratic function) effects and synergistic or antagonistic "interactive" effects. We will discuss *interactions* in more detail in Chapter 5 and suggest methods to differentiate an interaction from a reaction. Armed with the effort from the follow-up meeting, it is possible to apply the statistical experimental design methods that we will learn about in subsequent chapters. Without this information, such SED methods are useless.

Experimental Phases

Once we have selected the factors for our experiment and written the goals and objectives, it is important that we embark on the experiment in phases or stages. If we were to put all our experimental resources into one basket and then discover a hole in our experimental basket near the end of the effort, we will have lost a great quantity of resources. Besides wasting the resources, we will have gained poor (if any) information and worse, probably will turn management against the use of experimental design again.

The trick is to plan the experiment in phases. Each phase builds on the prior information of the earlier phase. A general rule of thumb says that no more than 25% of the total effort should be spent in the first phase of the experiment. Figure 2-3 shows the phases of experimentation and indicates the possible experimental design methods used in each phase. We will devote the remainder of this book to the "how-to" aspect of designing experiments with the help of well-established statistical experimental design methods and statistical analysis approaches.

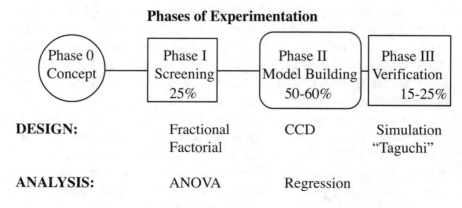

Phases of Experimentation

| Phase 0 Concept | Phase I Screening 25% | Phase II Model Building 50-60% | Phase III Verification 15-25% |

| **DESIGN:** | Fractional Factorial | CCD | Simulation "Taguchi" |
| **ANALYSIS:** | ANOVA | Regression | |

FIGURE 2-3

Problems for Chapter 2

1. Pick an experimental situation from your experience (or invent one) and state the goal and the objective.

2. List three quantitative response variables.

3. List three qualitative response variables. Attempt to quantify them.

APPENDIX 2

Key concepts from this chapter

Essential for an experiment:
>   1. Knowledge of the process
>   2. A response variable
>   3. Clear goals and objectives

A response is what we measure, and it must be:
>   1. Quantitative
>   2. Precise
>   3. Meaningful

Goals and Objectives:
>   A *goal* is the end result
>   An *objective* is a statement of how the task will be accomplished.

Example of Goals and Objectives:
Goal: Make the picture quality of the photographic film better.

Objective: To test the hypothesis that image quality is a function of metol, hydroquinone, sulfite, KBr, alkali, temperature, and speed.

The general form of the objective:
To test the hypothesis that (response(s)) is/are a function of the (controlling variables).

EXPERIMENTAL DESIGN ORGANIZER

Goal of Experiment _____

_____

Objective of Experiment _____

_____

_____

_____

Date experiment to START_____

END _____

A) List the controlling factors to be studied:

FACTOR   TYPE* DIMENSIONS   RANGE OF INTEREST   LEVELS

1._____

2._____

3._____

4._____

5._____

6._____

7._____

B) List the responses in order of importance:

| | | | IMPORTANT | | ESTIMATE OF |
| RESPONSE | TYPE* | DIMENSIONS | DIFFERENCE | RISK LEVEL | VARIATION |

1._____

2._____

3._____

4._____

C) How many runs do you want to make?_____ Maximum #?_____

D) How long for one run?_____

E) Can all the runs be randomized? _____ Which factors are a problem?

_____ _____

_____ _____

*Quantitative (N) or Qualitative (Q)

GUIDELINES FOR BRAINSTORMING

Team Make-up: The "experts"
The "semi"-experts or peripheral experts
Technicians or operators
The customers

Scheduling: Two meetings no more than 2 days apart. This makes a Tuesday and a Thursday ideal. The Tuesday meeting should require only 1 hour, while the Thursday meeting needs 4 hours.

Philosophy of Brainstorming: Suspend judgment, strive for quantity, generate wild ideas, build on the ideas of others. The *team leader* must be enthusiastic, capture *all* ideas, get the right skills mix, push for quantity, strictly enforce the rules, keep intensity high, and get everybody to participate.

Brainstorming Meeting
Tuesday, 8:30 AM to 9:30 AM
Agenda

Introduction and statement of problem	10 min
Brainstorming	20 min
Factor type identification	30 min

(Tools for brainstorming include a "flip chart,"
black, red, and green marking pens, and tape.)

Follow-up To
Brainstorming Meeting
Thursday, 8:00 AM to 12 Noon
Agenda

Review of goals	10 min
Review of brainstorming ideas	50 min
-any additions?	
-purge ridiculous ideas	
-find redundancies and consolidate	
Elevate the critical factors for consideration in the experiment	60 min
Select the factors for the experiment	30 min
Identify the levels for the factors	30 min
Determine possible interactions	45 min
Review action items and assign the timetable	15 min

Elevation of Factors Voting Form

FACTOR NAME	Type of Factor (C) Controlling (N) Noise (R) Response	Ranking of importance of the factors. 1: low rank; 5:high rank Team Member's Ranking						Average	Std. Dev.
		M1	M2	M3	M4	M5	M6		

SAMPLE REPORT

Note: *The following report is the result of an actual experiment run by the author involving his American Flyer electric train hobby.*

GOAL: To develop modifications capable of producing a super-power, low-speed steam engine for use on an S gauge electric train set.

OBJECTIVE: To test the hypothesis that efficiency, speed, power, and motor heat characteristics are functions of field winding wire gauge and winding volume and armature winding wire gauge and winding volume.

RESULTS:
1) Efficiency is most influenced by the armature windings. By increasing the gauge of the wire in the armature, the efficiency can be increased. This same conclusion holds for the field, but the influence is approximately 2/3 that of the effect of the armature for the same amount of change. There is an interaction (nonadditive effect) between the field and the armature, especially evident at low voltage values. Efficiency increases as the voltage increases.
2) The field has the greatest influence on speed. The effect of the armature is 1/3 that of the field for the same change in the wire gauge. Finer gauge windings for both the field and armature produce a slower motor.
3) Power is most influenced by the field gauge windings. The heavier the wire, the more powerful the motor. It is necessary to match the armature to the field to get the most power from the combination.
4) It is necessary to match the field wire gauge to the armature gauge to prevent overheating of either component.

CONCLUSIONS: To come closest to the goal of a higher power, lower speed device, use #22 gauge wire for the field and #24 gauge wire for the armature. This will not overheat, will deliver the greatest power, but is the least efficient and fastest of the designs. The original **Gilbert** *American Flyer* design with a #24 field and #26 armature is still a good choice for optimum speed, efficiency, and heat characteristics, but falls down on power.

RECOMMENDATION: For greatest power, use 100 turns of #22 enamel magnet wire for the field assembly and 40 turns per pole for the armature using #24 enamel magnet wire.

(Note: The remainder of the report is an appendix containing the experimental design configuration, the analysis tables, and the important graphical representations of these analyses. This page gives a sample of these elements.)

Experimental Design

		Wire Gauge		Windings	
tc	Field	Armature		#F	#A
(1)	22	24		100	40/pole
a	26	24		200	40/pole
b	22	28		100	80/pole
ab	26	28		200	80/pole
standard	24	26		200	75/pole

(Further note: There should now be a discussion of the reasons for the levels chosen and any physics involved with the design as well as the responses being measured. The tabulated data is then displayed and the analysis summarized in the following plots.)

Graphical Results

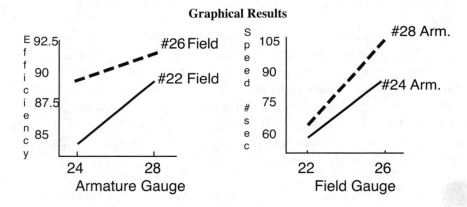

(Final note: A discussion of the graphical results should now give the reader the author's interpretation of the results and a basis for the conclusions.)

The above excerpt from the report on rewinding armatures and fields is offered as an example of the form used to effectively communicate the results of an experiment. The entire report was completed in only 4 pages. Keeping the report short and to the point encourages the customer to begin reading and to complete reading the entire report. By fitting *all* the graphs on one page, the number of pages is reduced *and* the entire experiment is seen in a single glance. Communication after experimentation is paramount to successful experimental efforts.

3

The Neglected Response Variable

In Chapter 2 we introduced the three elements of a good experiment. We learned that by having prior knowledge of the process, we would be able to identify the response variables that would change as a function of the controlling factors placed in the experiment. We also learned how to describe this process via the *goal* and *objective* of the experiment. Further, we learned that a response variable should be *quantitative, precise,* and *meaningful*.

In this chapter, we will explore these three qualities of the response and show how to assure that the responses we select meet these three qualities. It is unfortunate that in comparison to controlling variables, responses are often neglected and not given proper attention. This chapter gives the response variable the deserved attention necessary to build good experiments.

Quantitative

The first quality of a good response is its quantitative nature. As we will see in the third part of this book, we will be applying statistical analysis methods that are based of the requirement that the data is quantitative. If we are faced with responses that are only perceptions or feelings, we will be unable to complete the analysis no matter how well we designed the experiment. Examples abound where a crisis arises and an experiment is launched to "fix the darn thing." In one such situation, a company was working on the design and development of a new industrial mixer. The impellers of the mixer were attached to a shaft that was inserted into a chuck. One of the problems with this mixer was a defect called "fretting" that appeared at the end of the shaft that went into the chuck.

Figure 3-1 shows a drawing of the mixer, and Figure 3-2 shows the typical type of fretting due to chuck "working" on the shaft. An experiment was designed to look at the fretting response along with other responses. Fretting up to this time had been a judgmental factor, and the idea was to *eliminate* this cosmetic defect from the mixer. It would make no sense to go to NIST (National Institute of Standards and Technology, formerly NBS) and have a standard method for measuring fretting established since we want to make fretting disappear!

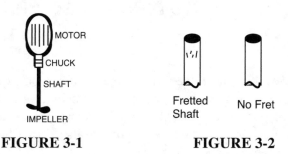

FIGURE 3-1   **FIGURE 3-2**

> *The first consideration for the quantification of a response is: "Is this a long-term response that will be around for the life of the product, or is this an 'elimination' variable that will only be measured during the design/development stages?"*

If the answer is long term, then a NIST type response needs to be developed. If, as in this case, the response is to be eliminated, we do not need to go into great detail in the quantification of the response. The simplest response could have been a rank order scale of the degree of fretting performed by an expert fret evaluator. However, the problem of a rank order statistic is the fact that the differences between ranks could be quite great or rather minor since a rank order treats any differences the same. Figure 3-3 shows some possible rank order differences with the same rankings. The difficulty is that ranks are not meaningful on a continuous basis. The response we need for statistical analysis needs to be a continuous equal-interval scale variable. Ranks are not equal in their interval.

An alternative approach that could produce an equal interval continuous scale utilizes a method called *successive categories*. The observer is given the task of *rating* the objects (like the fretted shafts) in such a way that the distance between ratings is the same. Now this can be a tough job for the evaluator, but if done well can overcome the problems with plain rank order.

Another advantage of the successive categories approach is the ability to evaluate more items than the number of ranking positions available. Let's say that we have a stack of 100 inkjet prints that are to be evaluated for image quality. If we were to utilize the rank order approach, we would need a work space large enough to place each print in a separate position. This would require a table over 850 inches long (assuming the width of each print is 8.5 inches)! With the successive category method, we restrict the number of positions to about 7 (plus or minus 2). Therefore, there are categories of prints that end up in, say, pile #4 (out of 7). All the prints in pile #4 appear to have about the same level of image quality. Figure 3-4 shows a diagram of the sorting concept behind successive categories.

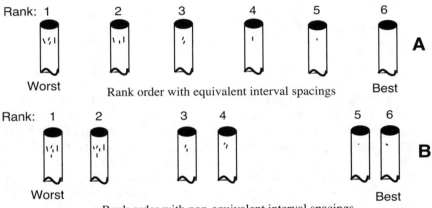

In **A**, the distance between the fret levels is about the same, so the ranks are equally spaced in the interval space. In **B**, there are three equivalent sets with a small interval between each of the two items in the set, but with a bigger distance between the sets. The **B** ranks do not have an equivalent interval value and are misleading.

FIGURE 3-3

In a successive category task, the rater is asked to place the objects into a fixed number of piles. In the case of the prints, we have used 5 categories or piles since there are only a few prints to be rated. The more objects in the rating process, the more categories may be used. With more categories, it is possible to stretch the resolution of the differences between categories. The observer is also asked to keep the relative differences between categories the same. As the number of objects becomes larger, the task becomes more tedious. Research at Xerox Corp. revealed that up to 144 copies could be sorted without significant increase in errors or noticeable fatigue on the part of the observers (1). Of course, sorting prints is a much different task than sorting ten-pound mixer shafts. In the case of the shafts, photographs were made of the fret areas and these photographs were then evaluated.

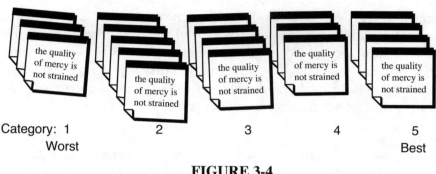

Category: 1   2   3   4   5
Worst   Best

FIGURE 3-4

Getting back to the print sort task, of the 19 prints, there are 3 in category #1, 5 in category #2, 4 in category #3, 3 in category #4, and 4 in category #5. It is not necessary to have the same number of objects in each category, and it is possible to have an empty category. However, if this "open category" happens, it is probably due to the fact that there are too many categories and the total number of categories should be reduced. Figure 3-5 shows the relationship between the number of categories and the number of objects to be rated for most situations.

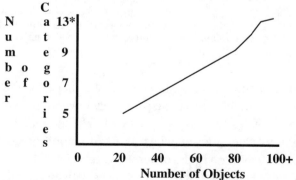

Empirical relationship between the number of categories and the number of objects being rated using the successive category method. *Uses a special method.*

FIGURE 3-5

Successive categories is not recommended for fewer than a dozen objects, since there is the distinct possibility of obtaining empty categories. Another caution in any rating system is to make sure that there *are* differences between the objects being rated. If all the objects are the same, then *any* rating system will simply become a random number reporting project with no possibility of actually sorting anything.

On this last point concerning the generation of random numbers as responses, we must overlap into the next quality of a good response momentarily. If we have a subjective response, we certainly do not want to base the experimental analysis on only one opinion or one rater. We usually gather many opinions. Since these are opinions, variations will occur between the raters. The simplest way to combine these variations is to average the rating of each object. Since we are working with interval *ratings* and not rank *ordering*, statistical methods allows us to compute the average. (Many times we go ahead and find averages despite their statistical validity anyway! Such practices sometimes lead to trouble, and other times we are able to get by due to the robustness of many statistical methods. It is good to know we are on firm ground with successive category data.)

Table 3-1 shows a typical set of data from a successive category sort for 10 of the 19 prints shown in Figure 3-4 for a portion of the raters interviewed.

TABLE 3-1
Print#

Rater	1	2	3	4	5	6	7	8	9	10...
Tom	5	1	3	3	2	4	5	1	2	4
Sue	4	1	4	3	1	5	5	1	4	3
Joe	5	1	3	4	2	4	4	2	3	3
Eric	4	1	2	3	1	3	4	1	2	3
Kelly	5	2	4	4	3	5	5	2	4	4
....										
Average	**4.6**	**1.2**	**3.2**	**3.4**	**1.8**	**4.2**	**4.6**	**1.4**	**3.0**	**3.4**
Std. Dev.	**.55**	**.45**	**.84**	**.55**	**.84**	**.84**	**.55**	**.55**	**1.0**	**.55**

Besides the simple average of the five observers, the standard deviation (Std. Dev.) is also reported in Table 3-1. We will look at the concept and methods of computing the standard deviation in the next section of this chapter where we investigate the precision of the response. But before we do that, we need to look at another method of assigning quantitative values to opinions using a more sophisticated psychometric method. This method will take care of the situations involving fewer than a dozen objects. We must also discuss the concept of standards of comparison or "anchors" that are used in rating systems.

In the successive category system, the observer gets to see all of the stimuli, form judgments, and revise those judgments during the sorting process. In another system called *magnitude estimation*, the observer gets to see the individual stimulus only once and must form a judgment at the time of that one and only observation. In such a system it is often helpful to have reference or standard objects that can act as "anchors" in order to make relative comparisons. The use of anchors can greatly reduce the variation between observers and within an observer and improves the precision of the responses. We will return to the magnitude estimation system in the last part of this section and show how it is a part of a hybrid approach that can be very helpful in rating situations that require different resolution in different parts of the rating spectrum. But first we will investigate a high-resolution self-anchoring system based on a very simple concept of showing all pairs of objects and making a comparison of the objects in each pair.

This method is called *pairs comparison*. It has the highest resolution of the methods we have discussed by virtue of the self-anchoring activity, but it is severely limited to fewer than a dozen objects since the number of pairs grows rapidly as the number of objects increases. Table 3-2 shows the number of pairs that are displayed to the subject as a function of the number of objects under review.

TABLE 3-2

# of Objects	# of Pairs
3	3
4	6
5	10
6	15
7	21
8	28
9	36
10	45
11	55
12	66
13	78

As the table shows, the number of pairs becomes large and probably unmanageable beyond 12 objects. The difficulty lies both in the judgment task of the observer as well as the presentation task by the giver of the test.

Fatigue can set in for both of these participants. To further understand how such fatigue can occur, let's look at the procedure in detail.

In the pairs comparison procedure, the test administrator picks up the two stimuli that form the pair to be shown. These are presented simultaneously to the observer, who then decides if there is a difference in the objects and assigns a value to the object to the observer's left. The value reported is the amount the left object obtains if there are 100 points to be divided between the two objects. So, if the observer sees no difference between the objects, the reported score is 50, or half of the total score. Of course, the right-hand object gets the remainder, which is, in this case of no difference, a value of 50. If the observer judges that the object in the left position deserves a higher score, then a value of 60 may be given. This leaves 40 for the right object. The degree of difference between the two objects dictates the actual score reported. This number may vary from 1 (for very low-scoring left objects) to 99 (for very high-scoring left objects).

To be sure that each observer follows the same rules and understands the task, a written instruction as shown below must be read to each participant before the task begins. The method we have just described is called the *"Comrey Constant Sum Pairs Comparison"* and is based on the work of the French psychometrician Comrey (2).

> You will be asked to judge the fretting of the following shafts. You will be presented with a pair of photographs of the area of the shaft with fret marks. Look at each pair and given that you have 100 points to divide between the two fret samples, give the largest share of the 100 points to the more highly fretted shaft. Report the score of the shaft on your left. So, for example, if the left shaft is more highly fretted, you might report a score of 65 (leaving 35 for the right hand shaft). If the right hand shaft appears more fretted, then you will report a score of, say, 25, which means that the right hand shaft was a 75. Remember, do not report the highest score, but the score that represents the value of the left hand position shaft.

Now we will learn how to compile the data from the constant sum method and complete the analysis. In the Appendix, there is a prescribed ordering of the presentation of objects based on a randomization / left-right trade-off established in the 1930s by another psychologist by the name of Ross (3). In Table 3-3 we see the results of the fret comparisons for one observer in the order of the Ross presentation. Table 3-4 is a data matrix that shows the relative scores of the column object to the row object. In Table 3-5, we have computed the ratio of the column object's value to the row object's value. A ratio calculation is appropriate, since in presenting the observer with the task of dividing 100 points between the two objects, we have in a subtle, painless way given the observer a ratio scaling task.

TABLE 3-3

Optimal Order For Pairs Comparison Of 4 Stimuli   Enter LEFT Stimulus score

Left	Right	
1	2	Left Score= 30
4	1	Left Score= 80
2	3	Left Score= 60
1	3	Left Score= 40
4	2	Left Score= 60
3	4	Left Score= 35

<table>
<tr><td colspan="5">

TABLE 3-4

Data Matrix
Stimulus Number

</td><td colspan="5">

TABLE 3-5

Ratio Matrix
Stimulus Number

</td></tr>
<tr><td></td><td>1</td><td>2</td><td>3</td><td>4</td><td></td><td>1</td><td>2</td><td>3</td><td>4</td></tr>
<tr><td>1</td><td>50</td><td>70</td><td>60</td><td>80</td><td>1</td><td>1.0</td><td>2.33</td><td>1.50</td><td>4.00</td></tr>
<tr><td>2</td><td>30</td><td>50</td><td>40</td><td>60</td><td>2</td><td>.43</td><td>1.0</td><td>.67</td><td>1.5</td></tr>
<tr><td>3</td><td>40</td><td>60</td><td>50</td><td>65</td><td>3</td><td>.67</td><td>1.5</td><td>1.0</td><td>1.86</td></tr>
<tr><td>4</td><td>20</td><td>40</td><td>35</td><td>50</td><td>4</td><td>.25</td><td>.66</td><td>.54</td><td>1.0</td></tr>
</table>

We now compute the ratio average for each column. Since we are operating with ratios, we cannot simply compute the arithmetic average, but we must compute the geometric average. This is the nth root of the product of the ratios, where n is the number of ratios in the product.

For Object #1:   Average Ratio $= \sqrt[4]{1 \times .43 \times .67 \times .25}$

Average Ratio $= .518$

For Object #2:   Average Ratio= $\sqrt[4]{1 \times 2.33 \times 1.5 \times .66}$

Average Ratio = 1.23

For Object #3:   Average Ratio= $\sqrt[4]{1 \times 1.5 \times .67 \times .54}$

Average Ratio = 0.858

For Object #4:   Average Ratio= $\sqrt[4]{1 \times 4 \times 1.5 \times 1.86}$

Average Ratio = 1.83

Summary: Fretting Level (higher values mean more fretting)

Least Fretting   Most Fretting

| Shaft: | 1 | 3 | 2 | 4 |
| Fret Scale: | 0.52 | 0.86 | 1.23 | 1.83 |

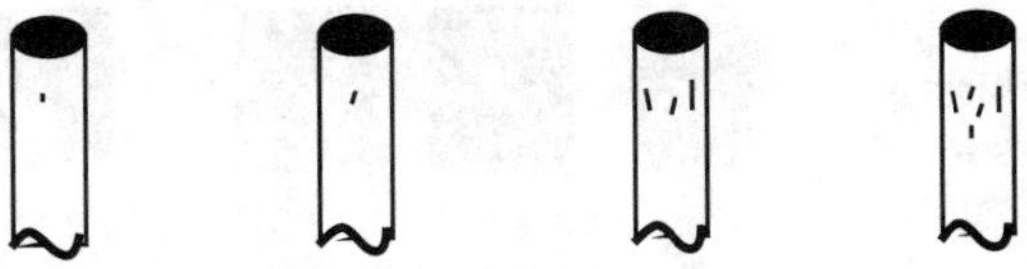

FIGURE 3-6

Figure 3-6 illustrates the ability of the pairs comparison method to find quantitative differences that reflect the degree of difference between rated objects. Using the constant sum analysis, the pairs comparison method allows ratio scales to be built. All ordinary statistical methods may be applied to ratio scaled values.

While the calculations for the constant sum scale are relatively easy, the use of a computer to complete these calculations, as with all such arithmetic operations, is highly recommended. We will use the constant sum method later in an example of a designed experiment. The appendix of this chapter has tables of optimal ordering for the pairs comparison method for up to 12 objects.

The last method of the many psychometric approaches to the quantification of perceptional information is a special application of the successive categories technique that also uses an anchor or reference object in conjunction with the magnitude estimation approach. Here's how it works. The observer is asked to sort the objects into three categories. Then each of these three categories are further subdivided into three more categories for a total of nine. Further subdivisions based on the triad sort are also possible to further refine the resolution of the rating system. Then a reference object is placed at the top end or the bottom end of the categories and a reference value assigned to it. The observer is asked to rate each of the categories with a magnitude value in reference to the anchor object's value. Figure 3-7 shows the method diagrammatically.

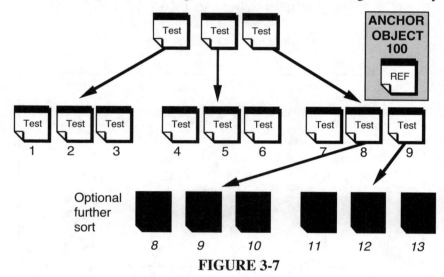

FIGURE 3-7

The advantage of this ***segmented triad sort*** is found in the reduction of variation between observers. By giving the observers a system, rather than have them invent one on their own, the task is made easier and more precise. In studies conducted at Xerox (1), it was found that half of the observers used some type of tiered sorting method when evaluating copies. These observers accomplished the task more rapidly and had better reproducibility than the others, who did not use such a systematic approach. By having everyone use the system of the segmented triad sort, it is possible to improve the precision.

The addition of the magnitude estimation after the sorting is completed removes the burden of trying to keep the distance between categories the same. Rather than try to do what is hard, we simply supply the distances after sorting.

To summarize, we have explored just a few of the powerful *psychometric methods* that have been developed by *experimental psychology* to *quantify* subjective judgments. Fundamental, physical based response variables that are inherently quantitative, such as temperature, tensile strength, miles per gallon, and optical density, do not need such manipulation, and fill the first requirement of a good response variable. However, there are many situations where the response is not physically defined and the methods we have just studied are needed to allow us to do the analysis of the experiments we will be building. We must not be neglectful of the **quantitative** nature of responses.

Precise

Even if our response is numerically based, but does not repeat well when measuring the same object, we will find that the experiment tells us very little since the variation in the response can overwhelm the changes induced by the controlling factors. We need to be sure that responses are precise or repeatable. Now, precision is not the same as accuracy. In fact, accuracy is related to the closeness to the actual, real value of the response, while precision is related to the repeatability of the response no matter where it is in relationship to the "true" value. This point may be illustrated by observing an archery match. The target has a center (bull's-eye), and the goal is to hit the center. If the archer can hit the center every time, the score will be large. If the archer never even hits the target, the score is zero. Let's look at a target with some situations that illustrate precision and accuracy.

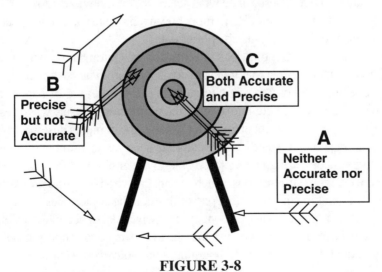

FIGURE 3-8

In Figure 3-8, the **A** situation with the arrows scattered all over the place represents poor precision, and since the target is missed every time, there is no accuracy. However, in situation **B**, the clustering of the arrows is almost perfect, but the bull's-eye is missed. Here we have precision, but we are biased since we do not hit the center of the target. This is an acceptable situation, since the responses that are used in SED will be looking for *relative* differences in most experimental situations. If the archery match we are watching has unfolded in Sherwood Forest, with *Friar Tuck* as the **A** archer, and *Little John* as the **B** shooter, then it must be *Robin Hood* himself who is the archer that has both accuracy and precision as the **C** marksman. This ideal situation comes closest to the target and does so repeatedly.

While we do not want to emulate the actions of Friar Tuck, it is possible to get by if we aspire to Little John's abilities with our responses. Being as good as Robin Hood is of course the best, but we sometimes do not know where the target lies (since the target might not yet be defined), and the only thing we can do is measure the repeatability of the response. Statistical methods that we will study in Chapter 12 provide for the quantification of the precision of a response.

Meaningful

The final quality of a response is how well it relates to the problem under study. Even if the response is quantitative and precise, it has no value in SED if it is meaningless. This is another situation, which parallels our study of the quantitative nature of a response, since what seems to be an obvious response is often the wrong thing to measure. Remember how a rank scale was obvious but wrong with respect to the interval needs of our data.

To illustrate this point, let's look at a copier. Getting the paper through the copier without a jam is a major design consideration. The customer complains if the machine breaks down due to jamming or misfeeding of paper. So, the number of jams or misfeeds is, of course, the response to measure when designing a new paper feeding mechanism. **WRONG!** If the copier is designed properly, it will rarely jam or misfeed. This means that a very large number of copies will have to be made during testing to be able to observe any differences between different design configurations. Such testing can wipe out large groves of trees that go into making the paper that passes through the copier. Besides, such "life" testing takes an extraordinary amount of time, which delays the introduction of the product in the market. An alternative to extensive testing is to simply put the product in the field and let the customers do the testing. Automotive companies are often forced into this mode of operation due to planned model-year introduction dates. Since customers resent being the quality control (QC) department, they will very often "walk away" from a company's products that behave poorly in the field. Most people agree that it is foolish to buy the first model year of a newly designed car.

However, back at the copier, we still need to find a meaningful response that will be *indicative* of good feeding characteristics. **Indicative** is the key concept. We must identify *indicator variables* that relate to the final product characteristic. The following is how a Japanese design team developed and used a meaningful response in the design of a new paper feeder for a copier.

Figure 3-9 is a side view of a paper tray and the spring-loaded arm that holds the picking mechanism that moves a single sheet of paper to the next position in the copier. If there is not enough pressure on the paper stack, there will be a misfeed. If there is too much pressure on the stack, there will be a multi-feed. We want the design configuration of the copier to have the fewest mis- or multi-feeds, but rather than count the numbers of such defects, we will change the spring tension and *force* the undesired events to take place. We measure the differences in the spring tension latitude as the **indicator variable**. The greater the spring tension latitude, the more likely the design is able to perform the job and deliver flawless feeding. Instead of testing the feeder by observing how many jams occur, we find the design configuration that gives the largest range of spring tension settings between no feed and multi-feeds. Table 3-6 shows some possible outcomes for this concept.

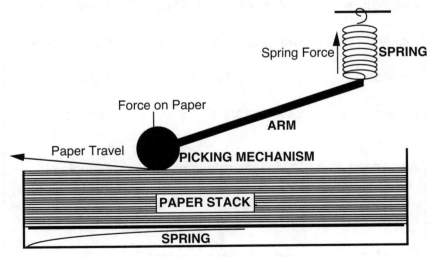

FIGURE 3-9

TABLE 3-6

CONFIGURATION	PRESSURE TO FIRST FEED	PRESSURE TO MULTI-FEEDS	FEEDING LATITUDE
PROTO A	10	15	5
PROTO B	8	13	5
PROTO C	12	20	8
PROTO D	6	21	15

Table 3-7 shows some more examples of meaningful response variables along with their obvious but meaningless counterparts. We can see that the response variable needs to have much thought behind it if we are to succeed with our experiments. If we measure the right thing, we will be much more successful in our experiments.

TABLE 3-7

Situation	Meaningful Response	Counterpart
Automotive	Piston ring wear rate	Engine failure
Lighting	Crimp force on filament	Bulb burn-out
Plastics	Size of void or flash	Yield
Photographic	Color balance	Number of reprintings
Aerospace	Valve leak rate in cc/min	Complaints from customer
Electronics	Size of solder bridges	Board failures
Chemical	Purity, molecular weight	Yield

Problems for Chapter 3

1. Comment on the following response variables with regard to the three attributes of quantitative, precise, and meaningful.

>   a) Maximum temperature of different types of firewood
>   b) Aroma of coffee
>   c) Surface finish to relate to internal voids in a casting
>   d) Number of solder defects per circuit board

2. Over a 2-month time frame, make a copy of the same document on any copier available to you. Note the brand of the copier and the model number as well as the time the copy was made. When you have 50 copies, use the successive category sort method described in this chapter to rate these copies. After the task is completed, consider any difficulties you had performing the sort. Write your comments as well as a summary of the results.

3. Select 4 copies from the sort in question #2 that represent poor to good copy quality and set up a pairs comparison rating investigation. Compute the constant sum values and compare them with the ratings from the successive category sort. Comment on your findings.

4. Compute the constant sum values from the following pairs comparison data.

Pair#	Object Numbers		Left Score
	Left	Right	
1	1	2	10
2	3	5	55
3	4	1	80
4	2	3	70
5	5	4	25
6	1	3	40
7	4	2	45
8	5	1	50
9	3	4	45
10	2	5	85

APPENDIX 3

Key concepts from this chapter

The types of psychometric scales:

Type of Scale	Method of Scaling	Comments
Nominal Scale	Attach a number to an object for the simple identification of that object.	Do not use in experimental design. Used to name an object rather than evaluate its level.
Ordinal Scale	Rank the objects in order of the increasing or decreasing value of the object.	Not effective as a method for building good responses. Treats all differences as the same.
Interval Scale	Use successive categories or magnitude estimation.	Good method for building responses.
Ratio Scale	Use *Comrey Constant Sum* with pairs comparison presentation.	Excellent, but limited to 12 objects.

Precision is the quality of a response that reports the repeatability of the value. This is measured with the standard deviation(s).

Accuracy is the quality of the response that reports how close the observation comes to the true value of the characteristic.

Indicator variables do not measure the ultimate product requirement, but show how that characteristic will change as a function of the experiment. Such a response variable is quantitative and continuous. So instead of looking at yield or failure rate, we measure what causes the failure or reduces the yield.

Optimal Order for Pairs Comparison Based on Ross Algorithm (1934) for 4 to 8 objects. For computerized generation of these orders, see Ref. 4.

For 4 Objects:		
Pair #	Left	Right
1	1	2
2	4	1
3	2	3
4	1	3
5	4	2
6	3	4

For 5 Objects:		
Pair #	Left	Right
1	1	2
2	3	5
3	4	1
4	2	3
5	5	4
6	1	3
7	4	2
8	5	1
9	3	4
10	2	5

For 6 Objects:		
Pair #	Left	Right
1	1	2
2	4	6
3	5	1
4	2	3
5	6	5
6	1	3
7	4	2
8	6	1
9	3	4
10	2	5
11	1	4
12	5	3
13	6	2
14	4	5
15	3	6

For 7 Objects:		
Pair #	Left	Right
1	1	2
2	3	7
3	4	6
4	5	1
5	2	3
6	7	4
7	6	5
8	1	3
9	4	2
10	5	7
11	6	1
12	3	4
13	2	5
14	7	6
15	1	4
16	5	3
17	6	2
18	7	1
19	4	5
20	3	6
21	2	7

Optimal Order for Pairs Comparison Based on Ross Algorithm (1934) for 4 to 8 objects. Continued.

For 8 Objects:			For 9 Objects:		
Pair #	Left	Right	Pair #	Left	Right
1	1	2	1	1	2
2	4	8	2	3	9
3	5	7	3	4	8
4	6	1	4	5	7
5	2	3	5	6	1
6	8	5	6	2	3
7	7	6	7	9	4
8	1	3	8	8	5
9	4	2	9	7	6
10	6	8	10	1	3
11	7	1	11	4	2
12	3	4	12	5	9
13	2	5	13	6	8
14	8	7	14	7	1
15	1	4	15	3	4
16	5	3	16	2	5
17	6	2	17	9	6
18	8	1	18	8	7
19	4	5	19	1	4
20	3	6	20	5	3
21	2	7	21	6	2
22	1	5	22	7	9
23	6	4	23	8	1
24	7	3	24	4	5
25	8	2	25	3	6
26	5	6	26	2	7
27	4	7	27	9	8
28	3	8	28	1	5
			29	6	4
			30	7	3
			31	8	2
			32	9	1
			33	5	6
			34	4	7
			35	3	8
			36	2	9

Optimal Order for Pairs Comparison Based on Ross Algorithm (1934) for 4 to 8 objects. Continued.

For 10 Objects:

Pair #	Left	Right	Pair #	Left	Right
1	1	2	33	5	6
2	4	10	34	4	7
3	5	9	35	3	8
4	6	8	36	2	9
5	7	1	37	1	6
6	2	3	38	7	5
7	10	5	39	8	4
8	9	6	40	9	3
9	8	7	41	10	2
10	1	3	42	6	7
11	4	2	43	5	8
12	6	10	44	4	9
13	7	9	45	3	10
14	8	1			
15	3	4			
16	2	5			
17	10	7			
18	9	8			
19	1	4			
20	5	3			
21	6	2			
22	8	10			
23	9	1			
24	4	5			
25	3	6			
26	2	7			
27	10	9			
28	1	5			
29	6	4			
30	7	3			
31	8	2			
32	10	1			

Optimal Order for Pairs Comparison Based on Ross Algorithm (1934) for 4 to 8 objects. Continued.

For 11 Objects:

Pair #	Left	Right	Pair #	Left	Right
1	1	2	33	10	9
2	3	11	34	1	5
3	4	10	35	6	4
4	5	9	36	7	3
5	6	8	37	8	2
6	7	1	38	9	11
7	2	3	39	10	1
8	11	4	40	5	6
9	10	5	41	4	7
10	9	6	42	3	8
11	8	7	43	2	9
12	1	3	44	11	10
13	4	2	45	1	6
14	5	11	46	7	5
15	6	10	47	8	4
16	7	9	48	9	3
17	8	1	49	10	2
18	3	4	50	11	1
19	2	5	51	6	7
20	11	6	52	5	8
21	10	7	53	4	9
22	9	8	54	3	10
23	1	4	55	2	11
24	5	3			
25	6	2			
26	7	11			
27	8	10			
28	9	1			
29	4	5			
30	3	6			
31	2	7			
32	11	8			

Optimal Order for Pairs Comparison Based on Ross Algorithm (1934) for 4 to 8 objects. Continued.

For 12 Objects:

Pair #	Left	Right	Pair #	Left	Right
1	1	2	34	1	5
2	4	12	35	6	4
3	5	11	36	7	3
4	6	10	37	8	2
5	7	9	38	10	12
6	8	1	39	11	1
7	2	3	40	5	6
8	12	5	41	4	7
9	11	6	42	3	8
10	10	7	43	2	9
11	9	8	44	12	11
12	1	3	45	1	6
13	4	2	46	7	5
14	6	12	47	8	4
15	7	11	48	9	3
16	8	10	49	10	2
17	9	1	50	12	1
18	3	4	51	6	7
19	2	5	52	5	8
20	12	7	53	4	9
21	11	8	54	3	10
22	10	9	55	2	11
23	1	4	56	1	7
24	5	3	57	8	6
25	6	2	58	9	5
26	8	12	59	10	4
27	9	11	60	11	3
28	10	1	61	12	2
29	4	5	62	7	8
30	3	6	63	6	9
31	2	7	64	5	10
32	12	9	65	4	11
33	11	10	66	3	12

REFERENCES

1. Nester, C. E., and T. B. Barker (1980). **Customer Preference Model (CPM)**, *Xerox Internal Report*, Rochester, NY.

2. Kling, J. W., and L. A. Riggs (1971). **Experimental Psychology (3rd ed.)** Holt, Rinehart & Winston, Inc., p. 71.

3. Ross, R. T. (1934). Optimum Orders for the Presentation of Pairs in the Method of Paired Comparisons, **Journal of Educational Psychology**, 25: 375-382.

4. Barker, T. B. (1993). **QED Programs for MSBASIC**. Rochester Institute of Technology, Rochester, New York.

PART II

Statistical Experimental Design

The Factorial 2-Level Design
General Factorial Designs

So far, all we have said about an experiment is that it should be efficient and well thought out before execution. Since most people will buy this concept, where does statistical experimental design have any advantage over the basic, "normal" way experimenters have been doing business for centuries? Let's look at a long established method that is recommended in physics books (1) and is usually called the "classical, one-factor-at-a-time" (1-F.A.A.T.) technique. We will investigate a 1-F.A.A.T. to see if it does get the required information with the least expenditure of resources.

Let's look at a chemical process that needs to be studied to improve the contamination in the end product. Our **goal** is to reduce contaminants. The team has elevated three factors or items that we will change in this investigation. These factors - pressure, temperature and reaction time - are expected contributors to the contamination response of the process. Table 4-1 shows the factors and the levels or conditions they will take on in the investigation.

TABLE 4-1

Factor	Low Level	High Level
Pressure (P)	100 psi (P1)	200 psi (P2)
Temperature (T)	70 degrees (T1)	90 degrees (T2)
Reaction Time (t)	10 minutes (t1)	20 minutes (t2)

The **objective** is to test the hypothesis that percent contamination is a function of pressure, temperature, and time.

In running the individual tests in the experiment, a 1-F.A.A.T. would follow this line of thinking. First, get a base line reading on the low level values of each of the three factors and then make systematic changes to each of the factors, one at a time. By taking the difference between the base line and each change, we can access the effect of each factor on the response. Let's watch that process in action and observe the results for this investigation.

The first test is the base line where all three factors are at their low settings. The statistical jargon for this would say "run at their low *levels*." **The word *level* means the value at which you set each factor in the experiment.** When all the factors are set at their levels in an experimental run, we call this a *treatment* or *treatment combination.* The abbreviation for treatment combination is **tc.** This term comes from the farmers who first used experimental design to *treat* the *combination* of the crop variety, fertilization, and other agricultural factors.

TABLE 4-2

Treatments				**Trial #1**	**Trial #2**	**Trial #3**	**Trial #4**	**Average**
P1	T1	t1	#1	2.2	2.8	3.2	3.6	2.95
P2	T1	t1	#2	3.4	3.9	4.3	4.7	4.07
P1	*T2*	t1	#3	4.6	5.0	5.4	5.8	5.20
P1	T1	*t2*	#4	3.7	4.1	4.5	4.9	4.30

The effect of pressure is $4.07 - 2.95 = 1.12$ i.e., #2 - #1
The effect of temperature is $5.20 - 2.95 = 2.25$ i.e., #3 - #1
The effect of reaction time is $4.30 - 2.95 = 1.35$ i.e., #4 - #1

The next treatment combination we encounter would change one of the factors, say pressure, to its high level (while the others are at their original, low levels) and observe the contamination. We make similar one-at-a-time changes in the remaining two factors and end up with four different tc's, which when analyzed in a rather simple manner can give us the answer we are looking for.

However, we know that this process has exhibited a high level of instability in the past, and to be sure of the results, we repeat the runs (4 times) to get a better idea of the true values and take the average of four trials. A basic concept of statistical methods has led us to perform these repeats and find the average, which is a better estimate of the true value of the contamination than a single isolated point.

The only statistics we have used so far is a simple average—nothing too complicated and certainly quite intuitive. There is a more underlying question, which is: have we designed our experiment efficiently? Did we get the required information? At this point we really can't judge the information part of the efficiency definition, since we haven't tried the results in a production situation. However, it looks like we are pretty close to learning how to control the process. It appears that we should stay at the low levels of each of the factors to keep the contamination at a low level since increasing each factor increased contamination. What about the "least cost" aspect of efficiency? The least cost would have been to run only four treatment combinations and take the differences between single value responses. However, we notice that a strange thing is happening in this process. As we run more trials, the contamination builds up. Notice that the base case trial #4 is more than 1.5 times higher in contamination as the first run (i.e., 3.6 versus 2.2).

Checking into the actual records of the experiment, we find that the reaction container was too large and difficult to clean out between runs, and thus the material that stuck to the walls of the container became part of the next batch that was run. This type of chemical reaction has the tendency to spill contamination over into the next batch. What we are observing is a systematic build-up of contamination merely caused by starting with a dirty reactor. Since we ran our experiment in a systematic order of one complete trial, we have mixed up (or, in the statistical jargon, "confounded") the effect of cleaning (or not cleaning, as in this case) with the factors under study. If cleaning has a large effect, as it seems to have here, the signal we wish to observe could be obliterated by the noise generated by the order of running the experiment. What's worse is the fact that the factor "time" is always last in the order of testing, and this is the condition when the worst of the carry-over contamination is present. Therefore, the effect of time on the contamination will be made up of any effect that reaction time really has and the accumulated effect of not cleaning the reaction container. Now do you still think we have the correct, required information?

To draw that conclusion, we shall have to try another method of experimentation. Let us say for now that the classical one-factor-at-a-time approach, even with replicated trials, is probably not the most efficient method of experimentation. We will have to continue our search for efficiency.

In the above 1- F.A.A.T. experiment, the analysis is performed on each result independently of the other results. This is an intuitive and logical approach to the assessment of changes in our responses as a result of the changes we make to the controlling factors we are studying. While this is ostensively logical, we still wonder if this 1-F.A.A.T. experimental approach obtains the required information.

The class of experimental designs we are about to investigate may at first look more complex in comparison with the 1- F.A.A.T.'s, but we shall see that they are a simple extension of basic testing concepts. Let's look at the structure of these 2-level factorials and compare them with the 1- F.A.A.T. designs.

Orthogonality

A set of experimental designs that can look at k factors in n observations, with each factor at two levels, is called the 2-level factorial design. The name *factorial* comes from the fact that we are studying *factors*. In such a design, the observations are not analyzed separately but as an experimental unit to provide mathematically independent (or "orthogonal") assessments of the effects of each of the factors under study. The number of observations (tc's) in such an experiment is determined by taking the number of levels (2) to the power of the number of factors (k):

$$tc = 2^k \qquad\qquad (4\text{-}1)$$

Thus for a 3-factor experiment, there would be 2^3 or 8 observations, or in statistical terms, there would be 8 treatment combinations (tc). For 5 factors there would be 32 treatment combinations. In these designs we look at all possible combinations of the two levels.

Besides the formalities in naming the components of the experiments (factors, levels, treatment combinations, etc.), some conventions are widely accepted for describing the experimental runs or treatments. Previously, we used the very simple notation of a subscript to indicate the low (we used 1) and the high (we used 2) levels for the factors under study. Since we wish to obtain an analysis that will provide mathematically independent assessments of the effects of each of the factors and we would also like this analysis to be fairly easy to perform, we now introduce the concept of the orthogonal design. By properly coding the levels of our experiment into what we call "design units" (or standard units) we can accomplish our goals of having independent assessments and a far easier analysis. The design units are mathematically coded values related directly to the physical values of the factors under study.

The coding method is not difficult. What we require is an experimental design in which the covariance of any pair of columns is zero. Covariance is a measure of the relationship between two variables. The larger the covariance, the more the variables are related to each other. If there is independence between a set of variables, then we observe zero covariance between the variables. This does not at all mean that the factors under study are physically independent. It only means that the mathematical analysis will produce results that are independent for each variable. The covariance looks much like the variance in its definition, but it involves two variables rather than just one as with the variance. Look at the formula for the variance and then compare it with the formula for the covariance. Notice the similarities. Problem #1 has a hint for this comparison.

$$\text{Variance:} \qquad s^2 = \frac{\sum (x - \bar{x})^2}{n - 1} \qquad \qquad (4\text{-}2)$$

$$n = \text{\# of x's}$$

$$\text{Covariance:} \qquad s_{xy} = \frac{\sum (x - \bar{x})(y - \bar{y})}{n - 1} \qquad (4\text{-}3)$$

$$n = \text{\# of pairs of y's and x's}$$
$$\bar{x} \text{ is the average x}$$
$$\bar{y} \text{ is the average y}$$

The above are defining formulas. There are calculation formulas found in the appendix.

Design Units

The method used to code the levels produces a set of design unit values consisting of - and + signs. These really represent the values -1 and + 1 and are devised by taking the physical values of the factors under study and transforming them using the following method:

$$X_{DU} = \frac{X - \bar{X}}{(X_{hi} - X_{lo})/2} \qquad \qquad (4\text{-}4)$$

where X is the value of the level, X_{hi} is the high level, X_{lo} is the low level, $\bar{X}$ is the average of X_{hi} and X_{lo}, and X_{DU} is the design unit level.

By using Expression 4-4 we can change the physical values of the factor under study to mathematically independent design unit values.

This coding works every time, and as long as you have only two levels, the results will always be the same. You will get a -1 for the low level and a +1 for the high level of X. Let's illustrate the method with the factors in the experiment on percent contamination.

For the pressure factor, there were two levels, 100 and 200 psi. The average of these is 150, and the difference between the high and the low levels is 100. Therefore if we substitute these values into Expression 4-4:

For the low level: $$X_{DU} = \frac{100 - 150}{(200 - 100)/2}$$

$$X_{DU} = \frac{-50}{50} = -1$$

For the high level: $$X_{DU} = \frac{200 - 150}{(200 - 100)/2}$$

$$X_{DU} = \frac{50}{50} = 1$$

This coding process begins to become trivial as soon as we realize that the values will always become +1 and -1, but upon analysis (as covered in Chapter 14) you will see that this coding will help us obtain that easy analysis we desire.

From the coding we have just developed, the convention of using the minus and plus signs has emerged. We can code (see Problems) the other two variables similarly and thus produce an experimental design matrix. The matrix (Table 4-3) begins at first to look much like the design of a one-factor-at-a-time experiment. We note that there is a treatment combination with all the factors at their low levels (indicated by a row of minus signs). Next we change the first factor to plus while leaving the others at minus. Then we change the second factor to plus with the others at minus. But the next treatment combination is different! Here we change two factors at the same time. An engineer once remarked to me that everyone who knows anything about engineering realizes that in a test you don't change two things at the same time, since you won't be able to discover which one caused the change.

How close to the truth and far from discovery was this poor chap. The truth is that in any single isolated test, my engineer friend is correct, but in an orthogonal, experimental design, we are interested not in the results of one single treatment combination but in the average change in the response over a number of treatment combinations. Remember: An experiment is a set of coherent tests that are analyzed as a **whole** to gain understanding of the process.

TABLE 4-3

A Pressure	B Temperature	C Time
-	-	-
+	-	-
-	+	-
+	+	-
-	-	+
+	-	+
-	+	+
+	+	+

Note the pattern in the columns of Table 4-3. The first column varies the plus and minus alternately, while the second column varies them in pairs and the third in groups of four. In general, we can reduce the pattern of - and + signs in any column to a formula:

The number of like signs in a set = 2^{n-1} (4-5)

Where n is the column number in the matrix

$$\text{for } n = 1, \ 2^{1-1} = 2^0 = 1$$
$$\text{for } n = 2, \ 2^{2-1} = 2^1 = 2$$
$$\text{for } n = 3, \ 2^{3-1} = 2^2 = 4$$
$$\text{for } n = 4, \ 2^{4-1} = 2^3 = 8$$
$$\text{etc.}$$

Yates Order

The convention of alternating -'s and +'s produces an order in the experimental design, which is useful in the design and analysis. It is called Yates order after the British statistician, Frank Yates. There are other conventions in such designs, which we shall study. The next convention helps us label each run or treatment combination. We can identify each of the treatment combinations with a unique letter or combination of letters (Table 4-4). To do so, we name each column by an uppercase letter of the alphabet. Whenever there is a plus sign appearing in a row under a letter, we bring that letter over to the left to create a new column called the treatment combination (tc) identity column. To illustrate, for the first treatment combination (first row), there are no + signs, and just to show that we did not miss this column, we put a (1) in the tc identifier column. The next row has a + in the A column, so an "a" (lowercase by convention) goes in the tc column. The next entry is a "b," while the fourth row is both "a" and "b" to give "ab." The remainder of the entries follow this same pattern.

TABLE 4 -4

tc	A	B	C	Row
(1)	-	-	-	1
a	+	-	-	2
b	-	+	-	3
ab	+	+	-	4
c	-	-	+	5
ac	+	-	+	6
bc	-	+	+	7
abc	+	+	+	8

By using the logic behind the construction of the tc identifier column in reverse, we can produce the plus and minus table when given just the identifiers in the proper Yates order. The Yates order for the tc's can also be generated without the plus and minus matrix. To do so, first place a "1" at the top of a column. Now multiply the first letter of the alphabet by this "1" to produce the second entry. The scheme repeats by multiplying each new entry (letter of the alphabet) by the existing entries. The next letter of the alphabet (b) is multiplied by 1 to get "b" and then by "a" to get "ab." We now have (1), a, b, ab, which when multiplied by the next letter, "c," produces c, ac, bc, abc, and this then completes the tc column of Table 4-4.

In many publications (Ref. 2) on 2-level factorial designs, the treatment combination identifiers are the only pieces of information given, and the designs are completed by the user, who places plus signs in the columns of the letters that appear in the tc identifier for that particular row of design, and minus signs in all other positions. Therefore, if I have " acd" for my identifier, there will be a + in the A, C, D columns for that treatment combination, and a - in the B column.

	A	B	C	D
acd	+	-	+	+

The treatment combination identifiers have a practical, mundane aspect also. They uniquely label the experimental run you are making. Therefore, in labeling the results of the run (the label on the sample bottle, the marking on the test copy, etc.), you should use the tc identifier rather than a sequence number (which is not necessarily unique) or the full description of the run (which is cumbersome). An entire experiment worth half a million dollars was almost lost by using sequence numbers as identifiers. The experimenters changed the sequence, and a lot of scrambling was required to reconstruct the true identity of the tests! Unfortunately, computing software such as Minitab® uses sequence numbers to identify runs. We will now show how to add the tc ID labels to a Minitab constructed experiment.

Using Minitab

Building factorial designs is a clerical job that can be subject to mistakes. Using a statistical computing software package will take much of the tedium out of this effort and, if done correctly, will prevent mistakes from entering the process. Let's build the design for the three factors - temperature, pressure and reaction time - using Minitab software.

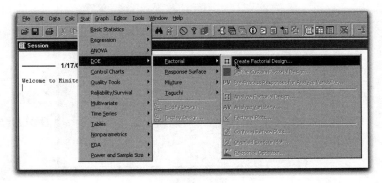

FIGURE 4-1
*After launching Minitab, go to the **Stat** dropdown and select **DOE**, then **Factorial**, then **Create Factorial Design**.*

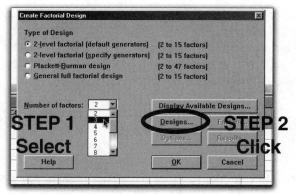

FIGURE 4-2

*A dialog box appears. We need to indicate the number of factors. Scroll to 3, then click on **Designs**.*

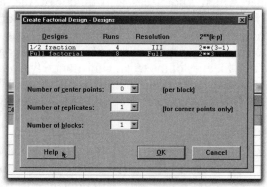

FIGURE 4-3

*Choose **Full factorial**. We will learn more about fractional factorials and center points in Chapter 5 Click **OK**. If you ever need to know what a certain command or dialog box means, simply click **Help**.*

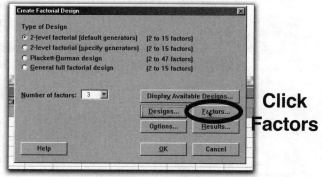

FIGURE 4-4

*Having chosen the design, we have more options with regard to the Factors, Options, and Results. Click **Factors**.*

After launching Minitab, we enter the drop-down menu as shown in Figure 4-1 and navigate to **DOE** > **Factorial** > **Create Factorial Design**. This sequence opens a dialog box shown in Figure 4-2. There are 6 items in this box that we need to look at. First, we need to decide on the type of factorial design we are selecting. The default at the top of the list is for 2-level factorial {default generators}, which is what we want for this exercise. In subsequent chapters, we will learn about the other three options. Notice that the maximum number of factors is 15.

My "rule of thumb" for the maximum number of factors is a bit smaller than this and is 9. The number 9 is based on the difficulty of the logistics of handling more than 9 factors. Think about it. By the time you have the last factor set at its level, the first one may have drifted to a different setting and you need to keep chasing your tail to get all the factors properly set. I once did an experiment with 12 factors and while it was finally finished and produced results that were very beneficial for Xerox, the experiment took nearly a year of daily experimentation. One set of conditions required more than 3 days of constant tweaking to get them where they needed to be!

Continuing with the computer exercise, in the main dialog box, we click on **Designs**, which brings up a sub dialog box shown in Figure 4-3. Minitab shows all the possible designs that we could use including fractional factorial designs that are the subject of the next chapter. We choose the **Full Factorial**. At this point, we are not intending to use a design enhancement called "Center Points" and we will do the experiment only once for each treatment combination, so we choose one (1) as the **Number of Replicates**. Blocking is an advanced topic (Chapter 8), and for now we will utilize only one **Block**. Notice that there is a Help button in every dialog box. If you get stuck or do not understand something in that box, simply click on **Help** and Minitab will come to your rescue! To leave any dialog box after making your selection, simply click on **OK**.

This takes us back to the main dialog box where we click on **Factors**. This is where we add the physical names to the design and specify the actual levels we will be using. Unless we add this specific information to the design, Minitab will simply produce an experimental structure with the "-1" and "+1" levels, which we have learned are called "Design Unit" values and what Minitab calls "Standard Units."

When we click on Factors, a new sub dialog box comes up (**Figure 4-5**). Start by clicking in the second entry where factor A is shown and type in the physical name of the A factor, which is Pressure. You can tab to the levels

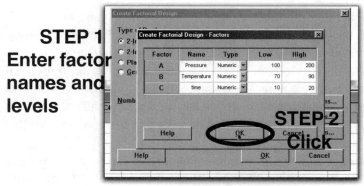

FIGURE 4-5

This dialog box is used to place names on the factors and assign levels. If you do not fill it in as we have done above, the experiment will look just like the design unit configuration shown in Table 4-4, p. 64. Click OK.

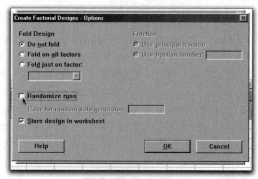

FIGURE 4-6

*Next click Options, which produces the above dialog box. Ignore the question on Folding. The Randomize runs and Store design in worksheet are default options. Un-click Randomize runs so we can see the design in standard order. Later we will randomize the experiment. Click **OK** and return to the primary dialog box. Click OK to get:*

↓	C1	C2	C3	C4	C5	C6	C7
	StdOrder	RunOrder	CenterPt	Blocks	Pressure	Temperature	time
1	1	1	1	1	100	70	10
2	2	2	1	1	200	70	10
3	3	3	1	1	100	90	10
4	4	4	1	1	200	90	10
5	5	5	1	1	100	70	20
6	6	6	1	1	200	70	20
7	7	7	1	1	100	90	20
8	8	8	1	1	200	90	20

FIGURE 4-7

*Here is the design in standard (or YATES) order. We will redo this design in random order later in Chapter 14 and add the contamination response. But first, we need to add the **tc ID's** that label the treatments.*

and fill them in as we have done in Figure 4-5. Click **OK** and we return to the main dialog box (not shown in the figures) where we click on **Options**. Again, Minitab gives an advanced option (Folding) that we will cover in Chapter 14. Go with the default (**Do not fold**) as well as the default **Store design in worksheet**. However, for this exercise only, turn off the default option to **Randomize runs**. For this example in the book, I want you to see what the design looks like in Yates (or as Minitab calls it, standard order). We should always randomize an experiment to provide protection from unknown, unwanted other sources of variation that could bias the results. Click **OK**, return to the main dialog box, and skip going to the **Results** button for this exercise. Click **OK** in the main dialog box and the design will appear in the worksheet as shown in Figure 4-7.

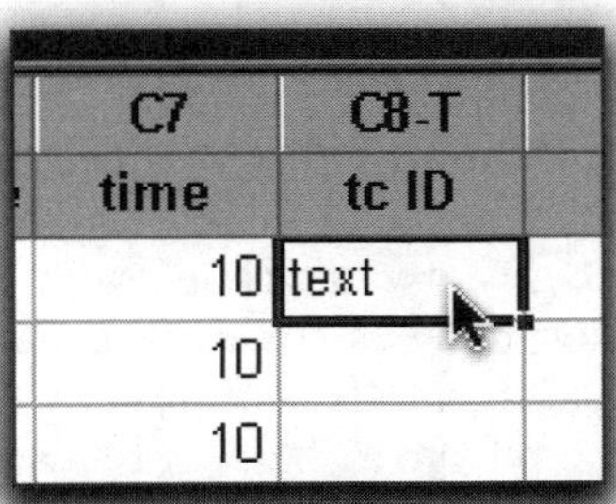

C7	C8-T	
time	tc ID	
10	text	
10		
10		

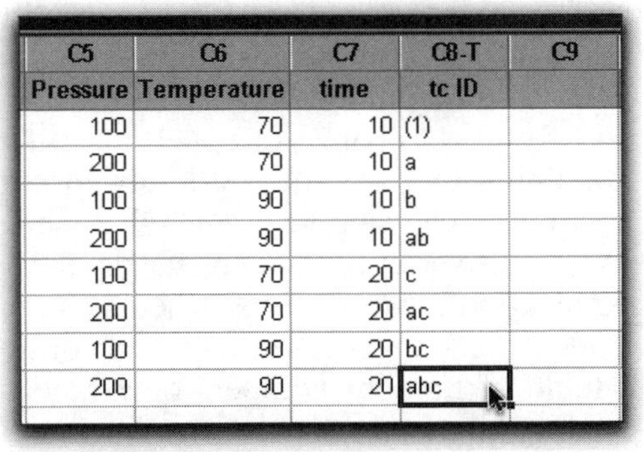

	C5	C6	C7	C8-T	C9
	Pressure	Temperature	time	tc ID	
	100	70	10	(1)	
	200	70	10	a	
	100	90	10	b	
	200	90	10	ab	
	100	70	20	c	
	200	70	20	ac	
	100	90	20	bc	
	200	90	20	abc	

FIGURE 4-8

To add the treatment combination identifiers (tc ID) to the Minitab worksheet, select the column right after the final column (C8). Since you will be inserting text to this column, click on the first entry and type in the word "text" to make this a "T" type column. Name this column "tc ID" and manually insert the identifiers as described earlier in this chapter. More complex identifiers are in the appendix of Chapter 5.

Now we will add a column that Minitab neglects. Notice that in the worksheet (Figure 4-7) column 1 (**C1**) has been designated by Minitab as **StdOrder** and column 2 (**C2**) is designated as **RunOrder**. Unfortunately, both of these orders use the same numerical labels and can become confused. We have seen how the treatment combination (tc) identifiers are a descriptive, nonambiguous labeling system for the runs. We add these tc ID's as described in Figure 4-8. I recommend strongly that you add this to the Minitab worksheet and use these tc ID's to label your physical runs during the execution of the experiment. Add the tc ID's to the last column of the worksheet, to avoid interfering with the columns Minitab has created.

Thus far, we have merely shown the mechanics of the construction of the 2-level factorial designs. It certainly looks like more work to build and understand these designs than the simple one-factor-at-a-time types shown before. What is the gain in efficiency in using these designs?

The first thing you may notice as a difference between the design in Table 4-3 and the design in Table 4-2 is that there are four more treatment combinations in the factorial (Table 4-3) than in the one-at-a-time design (Table 4-2). Certainly, from a resources viewpoint the factorial appears to be less efficient since it has four more runs. However, we have questioned the efficiency of the one-at-a-time method for its ability to get the required information. Let's see if the four extra resource units in the more "expensive" factorial really get us any more information over the simpler method.

If you recall, there was a problem with residual contamination buildup in the chemical reaction example we are studying. Therefore, when we perform this factorial experiment we will not run the set of treatments in the sequence order of the design but will randomize the running of the experiment. This randomization will assure that each treatment combination has an equally likely chance of appearing first in the running order and that the remaining tc's are equally likely to appear in any subsequent run order positions. The random order will assure us that any unwanted and unknown sources of variation will appear randomly rather than systematically. Randomization can be accomplished by simply creating a lottery by writing the tc identifiers on pieces of paper and drawing them from a "hat," or by using a table of random numbers or a random order computer program. By randomizing, we now have the original tc's in the order shown in Table 4-5 for the actual running of the experiment.

To run the experiment, read the conditions left to right, across each row. For Run #1 (the c condition), we would have low pressure (A), low temperature (B), and high time (C). (See Table 4-4 to confirm these levels.)

TABLE 4-5

Random Order	tc	% Contamination
1	c	4. 1
2	(1)	2. 6
3	a	3. 9
4	abc	3. 2
5	ac	1. 7
6	b	4.4
7	ab	8.0
8	b c	1. 8

After the actual running of the experiment, we must put the results back into Yates order for our analysis. We shall go through a simple analysis at this point, and in Chapter 14 we show a more formal approach to separating the signal from the noise.

For this analysis we shall write (in Table 4- 6) the entire matrix of plus and minus signs, which will include all of the effects we are able to draw conclusions about from the design. You will easily recognize the first three factor columns, but the last four are new. In *setting up* a design, we do not write any columns beyond the single factors. The "new" columns are only needed for the analysis.

Let's look at the first three single-effects columns, since these produce a direct comparison with the results in the one-at-a-time-design. First notice that there is more than only one change from the "base case" for each of the factors. Factor C, the time, is found at its high level four times and at its low level four times! Looking a little closer, we see that the pattern of plus and minus signs for factors A and B is exactly the same for the four low C's (-) and the four high C's (+). If we assume that the effects of A and B are exactly the same over the two levels of C and that C is mathematically independent of factors A and B, then we can "sum over" factors A and B (treat them as if there were no variation induced by them) and then find the average difference between the low level of C and its high level. The first assumption depends upon the physical reality of the situation. For the moment we shall make this assumption and return to a verification later. The second assumption is easy to check and the results are general for all the 2-level factorial designs (even the fractional factorials in the next chapter). Recall from page 61 that mathematically independent factors demonstrate this quality by having zero covariance. By using Equation 4-3 we can show that there is zero covariance between all the effects (see Problem 11).

TABLE 4-6

Random Order	tc	SINGLE EFFECTS			INTERACTION EFFECTS				% Contamination
		A	B	C	AB	AC	BC	ABC	
1	1	-	-	-	+	+	+	-	2.6
2	a	+	-	-	-	-	+	+	3.9
3	b	-	+	-	-	+	-	+	4.4
4	ab	+	+	-	+	-	-	-	8.0
5	c	-	-	+	+	-	-	+	4.1
6	ac	+	-	+	-	+	-	-	1.7
7	bc	-	+	+	-	-	+	-	7.8
8	abc	+	+	+	+	+	+	+	3.2

The same logic concerning independence that we just applied to factor C applies to factors A and B, so now we can find the average differences for the three single factors by summing the responses at the low levels (-'s) and the high levels (+'s) and take the difference of the averages. To illustrate, we will continue using factor C since, from this design, it is the most clear-cut example.

	Time Effect (C)	
	C High (+)	C Low (-)
	4.1	2.6
	1.7	3.9
	7.8	4.4
	3.2	8.0
Total =	16.8	18.9
Average =	4.20	4.73 difference= -0.53

Differences between averages are as follows:

For factor C (time) the difference is -0.53

For factor A (pressure) the difference is -0.53

For factor B (temperature) the difference is 2.11

These results are not the same as the results we calculated in the one-at- a-time design. The effects of pressure and time are reversed, and the effect of temperature is just a bit greater. But there is even more to look at. Now is the time to inspect the "interaction" columns, which contain the remaining required information in this experiment

Plotting Interactions

We can compute effects for interactions in the same way we compute effects for single factors. Simply add up the responses at the plus levels and at the minus levels for the interactions, find the averages, and take the differences of these averages. Since we already have a clue that there seems to be something unusual in the way factors A and C work together, let's take the AC interaction column and find the effect of the AC interaction.

This AC interaction column is generated by taking the algebraic product of the sign in column A times the sign in column C for each row. The first entry in the AC interaction column is a "+," which is the product of the "-" in the A column times the "-" in the C column. The remaining seven signs of AC are created in the same way. The sum of the responses at the plus values for the AC interaction is 11.9, and the sum of the responses at the minus values is 23.8. This results in an average difference of 2.98, which is an even greater effect than the factor B effect, which was the largest of the single factors. Indeed, there is something going on with the combination or "interaction" of factors A and C.

The easiest way to understand an interaction is to plot the results in an interaction graph, which is an iso-plot of the response variable for one of the factors over changes in the other factor. To plot Figure 4-9 we build a two-way table (Table 4-7) for the interaction factors.

The values in the boxes are the averages of the cells, which are made up of the responses corresponding to the combination of factors A and C. For instance, the entry in Table 4-7 with the average of 2.45 (high A & high C) comes from the responses in Table 4-6. The value 1.7 is the "ac" treatment and corresponds to the event when A and C are both at their + levels. The 3.2 value is the second time in this experiment that A and C are both + together. The interaction graph can be plotted with either factor as the iso-level factor, and in many cases it is informative to plot two interaction graphs as shown in Figure 4-9 using each factor as the iso-level.

TABLE 4-7

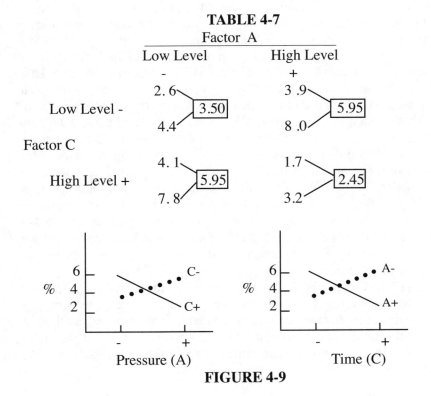

FIGURE 4-9

In this case, the plots are quite similar in shape. Both have been plotted to show the method. More important than the method of plotting is the result of this experiment. The discovery made in this experiment is graphically displayed in the interaction plots (Figure 4-9). The level of contamination varies with both pressure and the time of the reaction. Increasing one factor by itself while the other stays at a low level gives a higher level of contamination than if both factors are increased. The type of interaction or superadditivity shown is common in physical and especially chemical reactions and is a very important piece of information. Since we never ran the treatment combinations, which involved the changing of two factors at the same time in the one-at-a-time design, the opportunity to make this discovery was passed by. *Thus, the one-at-a-time experiment did not produce the required information.*

Cost

There is still the other half of the definition of efficiency related to cost. The factorial design in this particular experiment gives the average effects of a factor. This average is made up of four observations in this particular 3-factor investigation. If we wanted to get the same degree of certainty in the results from the one-factor-at- a-time experiment, it would be necessary to replicate the experiment four times; then, when we looked at average differences, we would have four observations involved in the average. In the 1-F.A.A.T. we did just that, and it took 16 treatment combinations to get the job done. That is, there were four different set-ups (the base case and the one change for each of the 3 factors), which had to be repeated four times to get the averages. In the factorial design we make more efficient use of the tc's by using the runs more than once in our computations. We may do this because the factors are mathematically orthogonal (independent) and balanced. Based on the criterion of resources alone, the work in the one-at-a-time experiment needs 16 units of work, while the work in the factorial requires only 8 units. This means that the factorial is twice as efficient as the 1-F.A.A.T. In general, as the number of factors increases, the efficiency factor goes up. Table 4-8 shows efficiencies for up to seven factors, which is about as many that we can handle in one experiment without resorting to the other methods that we shall study in Chapter 5.

The 2-level factorial design is a very efficient method of getting the information we require. It gets the required information at the least expenditure of resources. Besides standing on its own as an excellent method of experimental organization, it forms the backbone of many of the more advanced methods of efficient experimentation, which we shall study in the next chapters.

TABLE 4-8

# Factors	#tc's	Factorial # in Average	One-at-a-time # tc's Times # in Average	Factorial Efficiency
3	8	4	4 x 4 = 16	2
4	16	8	5 x 8 = 40	2.5
5	32	16	6 x 16 = 96	3
6	64	32	7 x 32 = 224	3.5
7	128	64	8 x 64 = 512	4

General Factorial Designs

The 2^k factorial designs are a very important subset of the more general crossed factorial design configurations. However, there are situations when two levels are insufficient to obtain the required information. Often an experimenter needs to investigate classification factors (qualitative factors) such as types of fruit or various vendors. In these cases, there is no simple, quantitative measure of the levels of these factors and a name must be used instead. If there are only two names (levels), then the 2^k factorial works perfectly, but if there are three or more names, we need to construct a design matrix with more than two levels.

Such designs are called General Factorial Designs. Let's look at an example of such a design that incorporates both classification and quantitative factors. Since all types of experimental designs had their roots in the field of agriculture, we will use an example from apple growing and preserving. Apple growing is an important upstate New York industry. The farmers harvest the crop in the late summer and early fall. However, people like to eat apples all year round. While there is an abundance of apples at low prices in the fall, the winter and spring bring higher and higher prices to the consumer. Besides being a supply-and-demand situation, one price-inflating factor is storage. There are different kinds of apples grown and there are different types of storage methods. In this problem, a large apple co-op is experimenting to determine optimum storage conditions. The response to be measured is the time to spoilage as measured in months. Table 4-9 shows the design matrix for this investigation. We will add the spoilage data in Chapter 13 and illustrate how a method called "ANOVA" can help determine differences between the levels of the factors. For now let's just build the design.

The general factorial design is almost intuitive. Build a matrix with i columns and j rows where there are i levels of the first factor and j levels of the second factor. So, for this apple-storage example with 5 levels of temperature and 3 levels of apple type, we have a 5 by 3 matrix as shown in Table 4-9. There will be 2 replicates per treatment combination, for a total of 30 observations.

TABLE 4-9

Storage Condition

Apple Types		36°	38°	40°	42°	44°
	Ida Red					
	McIntosh					
	Delicious					

There are five storage conditions (refrigerator temperatures of 36°, 38° ,40° , 42° , and 44° F) and three different types of apples (Ida Red, McIntosh, and Delicious). There will be two replicates per treatment combination for a total of 30 observations.

Using Minitab

We can construct the general factorial design much in the same way we construct a 2^k factorial using the DOE option in Minitab. The only difference is in the choice of the type of design in the main dialog box.

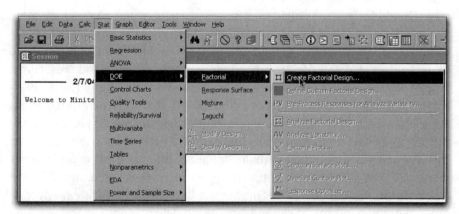

FIGURE 4-9

*We begin in the usual manner after launching Minitab with a double click from the Windows desktop. From the Minitab dropdown, go to **Stat > DOE> Factorial > Create Factorial Design**. This brings up the main dialog box as shown in Figure 4-10.*

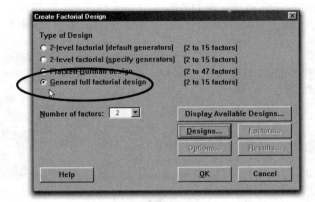

FIGURE 4-10

*The key to building a general factorial is to click the **General full factorial design** selection.*

Figure 4-9 shows the usual dropdown sequence from the statistics (**Stat**) over to the **Create Factorial Design** selection. When you release the mouse, the main **Create Factorial Design** dialog box appears. The default selection is **2-level factorial [default generators]**. By clicking on **General full factorial design**, we are able to build designs with up to 15 factors with as many reasonable levels as we wish. Select two factors for this example and click on **Designs**. This brings up a sub dialog box shown in Figure 4-11,

FIGURE 4-11

*In this sub-dialog box, type in the names of the factors and select from the scroll bar the number of levels. In the **Number of replicates** scroll bar section select 2 for this example. Click OK to move on.*

where we name our factors and indicate the number of levels. We also indicate the number of replicates and if blocking (a topic for later) is needed. When we click **OK** we go back to the main dialog box, where we click

Factors; this brings up the next sub dialog box shown in Figure 4-12. Here, we are able to assign both classification and quantitative levels to the factors.

a
b

FIGURE 4-12

In this sub-dialog box (b), type in the levels of the factors. Select the type of factor from the scroll bar. Be sure to scroll sideways for factors with more than four levels. Click OK to move on.

For this example we enter the numeric factor A (temperature) levels of 36, 38, 40, 42, and 44. For the B factor (apple type), we need to scroll down to indicate that this is a **Text** or classification factor. Then we may use the names of the three types of apples we are investigating, Ida Red, McIntosh, and Delicious. When we finish entering the levels, we click **OK** and go back to the main dialog box, where we click on **Options** as shown in Figure 4-13.

a
b

FIGURE 4-13

*Click on "Options" from the main dialog box (a). Leave the defaults of **Randomize runs** and **Store design in worksheet** clicked. It is prudent to enter a **base** for the randomizer. If you need to make any changes to the design at a later time, the same random order will be in effect by using the same base.*

The default options shown in Figure 4-13b are exactly what we want for a good experimental design. They include automatic randomization, and of course we want the design accessible in the worksheet so we may add the responses and then let Minitab do the analysis for us (see Chapter 13 for the data collection and analysis). Adding a seed or **base** number allows you to make subsequent changes to the design (such as factor level names or conditions, but not factors or number of levels) and still retain the same random order. This is especially helpful if you have already begun to execute the experiment and a change is needed due to some physical problems encountered during the data collection process. Click **OK** and we return to the main dialog box where we click on **Results**; this brings up the sub dialog box shown in Figure 4-14.

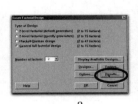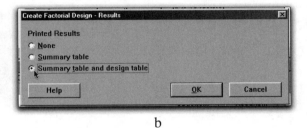

a

b

FIGURE 4-14

*From the main dialog box (a), click **Results**. In the Results sub-dialog box, change the default "**Summary table**" selection to "**Summary table and design table**." Click OK to return to the main dialog box for the last time.*

In the Results sub dialog box, we will change the default selection to obtain both the summary of our design that is printed in the session window of Minitab as well as the design table. The design table shows the design unit levels of the experiment (with levels at 1, 2, 3, 4, etc.). This can be helpful in understanding the fundamental design structure. If you want to use the design table to do this, you might want to uncheck the randomize option in the Options sub dialog box since Minitab will give the design table in random order. If you do this, you may also want to uncheck the **Store design in worksheet** for this unrandomized information and then go back and recheck the **Randomize** and **Store in worksheet** after getting the design table. This will reduce the number of worksheets produced by Minitab. Click **OK** and

 go to the main dialog box (Figure 4-15) for the last time. Click **OK** in this main dialog box, and the design parameters that you selected will be accomplished. The design will be built in the work-sheet (Figure 4-16), and the summary table and design table will be displayed in the session window.

FIGURE 4-15

↓	C1	C2	C3	C4	C5	C6-T	C7
	StdOrder	RunOrder	PtType	Blocks	Temperature	Apple Type	Months
1	2	1	1	1	36	McIntosh	
2	30	2	1	1	44	Delicious	
3	28	3	1	1	44	Ida Red	
4	19	4	1	1	38	Ida Red	
5	17	5	1	1	36	McIntosh	
6	1	6	1	1	36	Ida Red	
7	26	7	1	1	42	McIntosh	
8	29	8	1	1	44	McIntosh	
9	25	9	1	1	42	Ida Red	
10	15	10	1	1	44	Delicious	
11	8	11	1	1	40	McIntosh	
12	23	12	1	1	40	McIntosh	
13	13	13	1	1	44	Ida Red	
14	3	14	1	1	36	Delicious	
15	24	15	1	1	40	Delicious	
16	12	16	1	1	42	Delicious	
17	18	17	1	1	36	Delicious	
18	6	18	1	1	38	Delicious	
19	16	19	1	1	36	Ida Red	
20	14	20	1	1	44	McIntosh	
21	10	21	1	1	42	Ida Red	
22	7	22	1	1	40	Ida Red	
23	21	23	1	1	38	Delicious	
24	9	24	1	1	40	Delicious	
25	20	25	1	1	38	McIntosh	
26	27	26	1	1	42	Delicious	
27	5	27	1	1	38	McIntosh	
28	11	28	1	1	42	McIntosh	
29	4	29	1	1	38	Ida Red	
30	22	30	1	1	40	Ida Red	

FIGURE 4-16

This is the design for the five levels of temperature and three levels of apple type. This has been produced in random order and is ready to be used to gather the spoilage information in column C7, which was added manually before saving the project.

Problems for Chapter 4

1. Equations 4-2 and 4-3 show the definition of the variance and the covariance. Look at the formula for the variance and then compare it with the formula for the covariance. Notice the similarities. Here is a hint. Expand the square in the variance formula to the more fundamental expression of $(x - \bar{x}) * (x - \bar{x})$, which is equivalent to $(x - \bar{x})^2$.

2. Compute the covariance for the following sets of X and Y values:

SET #1		SET #2		SET #3	
X	Y	X	Y	X	Y
-1.0	2	5	3	1	7
-0.5	4	6	1	2	6
0.0	6	8	4	3	5
0.5	8	9	5	4	4
1.0	10	1	6	5	3
				6	2
				7	1

3. Build a table showing the number of treatment combinations for 2-level experiments with 3 to 10 factors.

4. Set up the coding for plus and minus signs using Expression 4-4 for the following two factors (i.e., change the physical units to design units):

	Low Level	High Level
Temperature	70°	90°
Time	10 min	20 min

5. Set up an experiment in design units for the following problem:

Temperature	(A) 90°	200°
Speed	(B) 70 ft/min	120 ft/min
Chemical	(C) 0.5 grams/liter	40 grams/liter

6. From the following treatment combination identifiers, set up the plus-and-minus sign table for this 2-level design:
   1,a, b,ab, c,ac, bc, abc, d, ad, bd, abd, cd, acd, bcd, abcd

7. Put the following tc's in Yates order: ab, b, c, a, (1), abc, bc, ac.

8. Use the results in Table 4-6 to compute the effects of the four interactions.

9. Plot the interaction graph for the BC interaction from Table 4-6.

10. Compute the covariance between all single effects in a 2^3 factorial design.

11. Compute the covariance between the AC interaction and the BC interaction in a 2^3 factorial design.

12. Design a generalized full factorial for five levels of runout (A through E) and three levels of speed (5, 10, 15 inches per second). Replicate twice.

APPENDIX 4

Definitions:

FACTOR: An item to which you make purposeful changes and observe a change in the response variable.

LEVEL: The value assigned to the changes in the factor.

TREATMENT
COMBINATION: The levels of all of the factors at which a test run is made or simply the set of conditions for a test in an experiment. Abbreviation: tc.

ORTHOGONAL: Independent mathematically

ORTHOGONAL
DESIGN: An experimental design constructed to allow independent analysis of each single factor and the interactions
between all factors.

CODING: A mathematical technique used to force the treatment
combinations in a design into orthogonality.

COVARIANCE: The co-variation of two factors as defined by the average
product of the differences from the mean.

DESIGN UNITS: Mathematical representations of the levels of the factors
in an orthogonal design. These are selected to create the
orthogonality. Minitab calls them "Standard Units."

For the 2-level design, the coding to give the design units is:

$$\frac{X - \overline{X}}{(X_{hi} - X_{lo}) / 2}$$

INTERACTION: A result of the nonadditivity of two or more factors on the
response variable. The situation where the change in a
response (its slope) differs depending on the level of the
other factor or factors.

SINGLE EFFECT: The influence of the factor on the response all by itself.
This is often called the "Main Effect." However, in the
common vernacular, the word "Main" implies big or
important and this may not be the case, so the idea of
single is more appropriate.

Formulas:

COVARIANCE: The co-variation of two factors as defined by the average product of the differences from the mean. The calculating formula is:

$$c_{xy} = \frac{\Sigma X_i Y_i \; - \; \frac{\Sigma X_i \Sigma Y_i}{n}}{n-1}$$

NUMBER OF
RUNS: There are tc treatment combinations in a 2-level factorial design equal to:

$$tc = 2^k$$

where k is the number of factors.

ASSIGNMENT
OF LEVELS: We use a minus sign for the low level of a factor and a plus sign for the high level of a factor.

To build the columns of plus and minus signs, alternate these signs in sets with:

The # of like signs in a set $= 2^{n-1}$
where n is the column number.

REFERENCES

1. Taffel, A. (1981). **Physics, Its Methods and Meanings (4th ed.)**, Allyn & Bacon, New York.

2. Davies, O. L., ed. (1956). **Design and Analysis of Industrial Experiments (2nd ed.)**, Hafner Publishing Co., New York.

## 5

Fractional Factorials
at Two-Levels

In Chapter 4 we studied a method of extracting the required information in the form of the single (main) effects of the factors under study and the interactions of these factors. We will now raise a question concerning the cost aspect from our definition of an efficient experiment. We must always keep in mind the balance that must be maintained between information and cost. To develop this point, we shall examine a rather trivial example and then expand into a real, extensive experiment. Before discussing these examples, we will look at the information content of 2^k factorials in general.

If we were to study only three factors, then there would be three single effects and four interactions possible. With four factors, there are four single effects and eleven interactions. Table 5-l shows how the number of treatment combinations and information increases as the number of factors increases.

TABLE 5-1

k Factors	2^k	Single Effects	Interactions 2-Factor	3-Factor	4-Factor	5-Factor	6-Factor
5	32	5	10	10	5	1	-
6	64	6	15	20	15	6	1
7	128	7	21	35	35	21	7
8	256	8	28	56	70	56	28

In general,you find the number of interactions by using the combinations formula:

$$C_h^k = \frac{k!}{(k - h)!(h!)} \qquad\qquad (5\text{-}1)$$

Where: k = # of factors
          h = # of factors in interaction

Thus for a 10-factor experiment the number of 4-factor interactions is given by:

$$C_4^{10} = \frac{10!}{(10 - 4)!(4!)} = 210 \qquad\qquad (5\text{-}2)$$

Now that is a lot of information on 4-factor interactions! However, 4-factor interactions are quite rare. In such a 10-factor experiment, we need to spend 210 degrees of freedom (each degree of freedom is a measure of work and information) in determining what is most likely zero information! Now, that is not efficient. The most likely interactions expected to take place are some of the 45, 2-factor or first-order interactions. If we add in the 10 single effects, we find that all we need are 55 degrees of freedom to describe the required information in this situation. The 2^{10} experiment has 1024 treatment combinations or more than 18 times the volume of information we need to do the job efficiently.

About Interactions

While interactions are important, they do not abound. In some disciplines, such as mechanical assembly, interactions do not even exist. If I take four parts and simply put them together as shown in Figure 5-1, I have an additive effect of the dimensions of the individual parts. Since an interaction is the result of nonadditivity, this mechanical assembly process is interaction free.

On the other hand, if I am working with a chemical process, it is very possible that one chemical's influence on the response very much depends on the setting of another part of the chemical process. Figure 5-2 shows such a nonadditivity for the antioxidants BHA and BHT, which are often used as food preservatives. Two-factor interactions such as this are very important

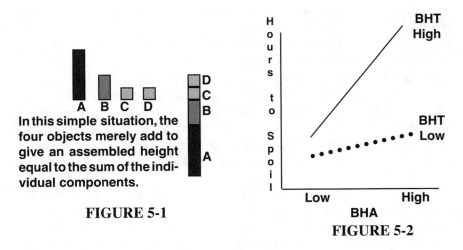

In this simple situation, the
four objects merely add to
give an assembled height
equal to the sum of the individual components.

FIGURE 5-1

FIGURE 5-2

and one of the reasons that we run factorial experiments. However, interactions that are more complex than those involving two factors are rare. Often researchers confuse an interaction with a reaction. They say that in the presence of a catalyst, two chemicals produce a desirable product. If that desirable product was more than just the result of the additivity of the reaction, we would have a legitimate interaction. However, the result is usually the simple linear addition of the components, much like the mechanical additive example shown in Figure 5-1.

In an experiment run at Xerox Corp., the initial analysis indicated that there was the possibility of a *4-factor* interaction. Upon discussing the results with the technical personnel responsible for the experiment, it was discovered that they made changes to other parts of the experiment above and beyond the prescribed changes called for in the experimental design matrix to "assure good product"! What at first appeared to be a 4-factor interaction was the result of the "fixing" of the experiment by well-meaning but ignorant people. Of course, there was no 4-factor interaction, and a valuable lesson can be learned from this botched experiment. Often, people who do not understand the purpose of experimentation will impose their own ideas of its purpose. These ideas are often rooted in early forays into the science found in chemistry or physics labs in high school or college. In these labs we were told that we were doing "experiments." However, these labs were allowing us to perform demonstrations of previously proven ideas. But if we did not demonstrate the idea perfectly (or nearly so), we got a poor grade. Thus, we often have the mistaken concept that an experiment must make a good result (like a product that can be sold). True experiments make good *information*. The good information leads us to make good products.

In another experiment that was a two-level fractional factorial with 16 tc's, the technician made a blunder in setting up one of the runs. Instead of using thin wall stock, he used thick wall stock. Upon analysis, the experiment proclaimed a three-factor interaction! Again this was a false indication, due to the mistake in the execution of the experiment. When the faulty run was redone, the three-factor interaction disappeared. Besides the absence of the higher order interaction, this example warns us to be very careful in *running* fractional factorial designs. Since we are taking a very small sample from a rather large population of possible runs, a mistake in the set-up can lead to rather erroneous results. So, we must be careful in the way we conduct our experiments.

Based on results like these, we are led to the conclusion that the two-level factorial is not efficient when the number of factors increases above five or six because the amount of information becomes excessive and the work becomes oppressive. Since many of our problems are likely to contain more than five factors, we need a method to get at the required information with the least cost. If we can come up with a method that uses a fraction of the treatment combinations, we can accomplish that goal. We want to decrease the volume of work and excess information while still obtaining the required information.

A Simple Example

Let us take a very simple example and then expand upon it to the point where we can understand and observe the value of the method. Suppose that we wish to find the weight of three objects A, B, and C. In any normal weighing operation, we would place the objects on the balance one at a time to find the weights. In a full factorial design, however, we would need 2^3 or 8 treatment combinations to accomplish the same goal. However, think about the information that will emerge from the full factorial beyond the "effect" of placing the objects on the scale. We will be able to find the interactions of AB, AC, BC, and ABC. Consider the fact that in weighing an object, the weight of that object will not change if it is alone on the balance or if it has other objects with it. Therefore, there can be no interactions in this experiment. If we should run the full experiment, any effects for the interactions can only be attributed to random error, since the effects of these interactions will be zero according to physical laws. If these interactions are zero and we have a fairly errorless system for weighing, what good are these pieces of information? They are actually worthless. It appears that a full factorial is not efficient since we have empty information and the cost is excessive. So how do we become efficient? First, we determine the number of treatment combinations necessary to obtain the required information. In this case we need to find three weights with a degree of freedom for each weight. Since in every experiment we need one degree of freedom for the grand mean, the least amount of work to run this experiment is four treatments. That just happens to be half the number of weighings needed by the 2^3 full factorial design. We need a 1/2 fractional factorial design. To construct such a design, we revert to a smaller basic building-block full factorial design. Table 5-2 shows the matrix for a 2^2 design for factors A and B. We also show the AB interaction column in this matrix.

TABLE 5-2

tc	A	B	AB
(1)	-	-	+
a	+	-	-
b	-	+	-
ab	+	+	+

Now this full factorial still shows a two-factor interaction (AB). Since we have ruled out the possibility of all interactions, we still have excess information. Let's turn that excess AB interaction information into something we can use. Since the expected value of the effect of the AB interaction is zero, we can superimpose the effect of the C factor on this *empty* interaction. That is, we will make factor C utilize the signs of the AB interaction column to describe the presence or absence of object C on the balance just as we assign the presence or absence of the objects in the A and B columns. Table 5-3 shows the new design.

TABLE 5-3

tc	A	B	C
(c)	-	-	+
a	+	-	-
b	-	+	-
ab(c)	+	+	+

Note that the "tc" labels have modified. Instead of the usual "1" in the first entry, we have "(c)" since C is at its high level. The use of parentheses is a convention to indicate that factor C is not a part of the full factorial base design but a part of the fractional design. This convention distinguishes a fractional factorial from a full factorial and in no way means that factor C is inferior to the other factors with respect to its information detection ability. The information detection ability of *all* the factors is equivalent in a fractional factorial design.

We will now introduce an algebraic system that will help to identify the information in the experiment. Recall how the fractional factorial was created. We made the factor C effect the same as the AB interaction effect. That is, we have mathematically combined the two effects. The statistical word for this combining is **confounding**. We can express this in writing this with a special symbol,"≈ ":

$$C \approx AB \tag{5-3}$$

This special symbol indicates that the measure of the effects are confounded but not necessarily equivalent or equal. In fact, in some cases it is possible for one effect to be negative and the other positive, and the measured result can be zero, since the measured effect is the algebraic sum or the *linear combination* of the two effects. The effects remain independent physically, but the fractional factorial design has mathematically combined or confounded them in such a way that it is impossible to assess these effects independently. Of course, in the weighing experiment, there are no physical interactions, and thus we are able to extract the physical interpretation from the confounded mathematical result.

The system that defines all the confounding in a fractional factorial design is based on a modulus (base 2) algebra system. The base is 2 since we have a 2-level design. With other designs with different levels, we use the number of levels to determine the base for the modulus algebra system. Relationship 5-3 forms the basis of the fractional factorial or what we shall call the **generator** of the design. From this generator, we derive a relationship call the **defining contrast** or **defining relationship**. We now need to introduce some new ideas and rules to make the 2^{k-p} system work.

The first idea is that of the identity element. The number 1 is the identity since multiplication or division by 1 does not change the identity of the number. The second concept is modulus algebra. Modulus algebra is a counting method that looks at the remainder of a number after the modulus or its multiple has been removed from the number. So, in the $\text{Mod}_{\text{Base 2}}$ system used for 2^{k-p} designs, the number 1 is a 1, but 2 (when 2 is removed) becomes a 0. The number 3 has one 2 removed with a remainder of 1, so the modulus of 3, Base 2 is 1.

To summarize then:

Counting in $\text{Mod}_{\text{Base 2}}$ 1=1, 2=0, 3=1, 4=0, 5=1, 6=0, etc.

An example will illustrate the use of these ideas in fractional factorials:

Start with the generator, $\qquad C \approx AB$

multiply the generator through on both sides by C :
$$CC \approx ABC$$
$$C^2 \approx ABC$$

apply the $\text{Mod}_{\text{Base 2}}$ to the exponent of C: $\qquad C^0 \approx ABC$

since anything to the 0 power is 1, we get: $\qquad 1 \approx ABC \qquad$ (5-4)

Expression 5-4 is the defining contrast that is used in determining the complete set of confounded effects in the design. We multiply the defining contrast through by each single factor (A, B, C) and apply the modulus algebra and identity rules to both the left and right side of the confounded sign.

$$A1 \approx A^{0}_{2} BC \text{ which is } A \approx BC$$
$$B1 \approx A B^{0}_{2}C \text{ which is } B \approx AC$$
$$C1 \approx ABC^{0}_{2} \text{ which is } C \approx AB$$

Let us now examine what the defining contrast has told us. In Table 5-4, we observe *all* the interaction columns that have been obtained by multiplying, row by row, the appropriate columns. So, for the AC interaction column, we go to the first row of the matrix and multiply the minus in the A column by the plus in the C column to get a minus in the resulting AC column. We follow this procedure for the remaining three rows of the matrix and then repeat the procedure for the BC column.

TABLE 5-4

tc	**A**	**B**	**C$\approx$AB**	AC	BC
(c)	-	-	+	-	-
a	+	-	-	-	+
b	-	+	-	+	-
ab(c)	+	+	+	+	+

Now notice that the pattern of the minus and plus signs in the AC column is exactly the same as the pattern of signs in the B column. Our defining contrast told us that B$\approx$AC and now we can see what this means in a physical way. If I find the effect of B, I also find the effect of AC since they have the same signs! I don't know if the B factor is causing the change to take place or if it is the AC interaction that is making the change.

Or, it could be a combination of the two effects causing the change. Here are some possible numerical results from confounded effects and the consequences of such confounding.

Single Effect	Interaction Effect	Observed Result	Consequence
+5	-5	0	No action for either effect. Factors raise their ugly heads later.
+5	+5	+10	If a +10 change is important, but +5 is not important, then unnecessary action will be taken.

In the weighing experiment, all the single effects are confounded with the 2-factor interactions. In any other experimental situation this would be undesirable, since two-factor interactions are likely to exist. (Of course in such a weighing experiment, the interactions are physically impossible.) We now need a set of rules to determine priorities about the confounding in fractional factorial designs.

Rules for Confounding
1. Never confound single effects with each other.
2. Do not confound single effects with 2-factor interactions.
3. Do not confound 2-factor interactions with each other.

Rule #1 should *never* be violated. Since both the single effects and 2-factor interactions are likely to occur, we should avoid violating Rule #2, but in certain *exploratory* efforts, we may break the rule. Rule #3 is often broken, because while interactions are important, they are not prolific (see the Topics of Minimum Number of Runs and Fold-over in Chapter 14 for more on this important concept).

Fractionalization Element

Before getting into the larger and more realistic fractional factorial designs, let us present the remaining conventions involved with designating fractional factorials. The name given to this type of design is a 2^{k-p}, where k is the number of factors and p is the fractionalization element of the design. The result of subtracting p from k produces n, the exponent of the number of levels. The numerical value of 2^{k-p} (or 2^n) gives the number of treatment combinations in the design. Since we know the number of factors and the maximum number of tc's we can economically afford, we are able to determine the value of p. In the case of the weighing experiment p=1, our design was a 2^{3-1}. We use the p to determine the fraction of the whole factorial we are using by computing $1/2^p$.

If we have a 2^{6-2} then this is a $1/2^2$ or a 1/4 fractional factorial, with 2^4 (16) treatment combinations. The value of p also determines the number of generators for the design, and the number of terms or words (counting "1") in the defining contrast is simply 2^p. These last two pieces of information are important in designing fractions smaller than the very simple 1/2 fractional factorial where there is only one generator and two terms in the defining contrast.

> There are p generators.
> The fraction is $1/2^p$.
> There are 2^p terms in the defining contrast.

More Factors—Smaller Fractions

The weighing example represents the fewest number of factors and the smallest fractional design possible. For real experimental work in the presence of interactions, this design is not at all good, since it violates the second rule of confounding. For most situations, a general rule of thumb says that the smallest fractional design that is possible is one that has 8 treatment combinations. The base full factorial for such a design would be a 2^3 with factors A, B, C as shown in Table 5-5.

TABLE 5-5

tc	A	B	C	ABC=D
(1)	-	-	-	-
a(d)	+	-	-	+
b(d)	-	+	-	+
ab	+	+	-	-
c(d)	-	-	+	+
ac	+	-	+	-
bc	-	+	+	-
abc(d)	+	+	+	+

Let us define a new factor D that will be confounded with the ABC interaction. We choose the ABC interaction, since it is the highest order interaction in the 2^3 base design. The generator of this fractional factorial design is then:

$$D \approx ABC \tag{5-5}$$

and the defining contrast is:

$$DD \approx ABCD$$

$$\overset{0}{D^2} \approx ABCD$$

$$1 \approx ABCD \tag{5-6}$$

In such a design, by using the modulus algebra, we find that:

$$
\begin{array}{ll}
A \approx BCD & AB \approx CD \\
B \approx ACD & AC \approx BD \\
C \approx ABD & AD \approx BC \\
D \approx ABC &
\end{array}
$$

These are all the effects we may find in this experiment, since there are 8 treatment combinations and thus 7 degrees of freedom for determining the effects. Note that the single factors are confounded with three-factor interactions, but the two-factor interactions are confounded with each other. This is the smallest design we may use to determine single effects and is usually reserved for first cut or *screening experiments*. We can go one step further and add still another factor to this design to create the matrix in Table 5-6.

TABLE 5-6

tc	A	B	C	ABC≈D	BC≈E
(e)	-	-	-	-	+
a(de)	+	-	-	+	+
b(d)	-	+	-	+	-
ab	+	+	-	-	-
c(d)	-	-	+	+	-
ac	+	-	+	-	-
bc(e)	-	+	+	-	+
abc(de)	+	+	+	+	+

This is a 2^{5-2} fractional factorial. Thus with p of 2, the fraction is a $1/2^2$ or a 1/4 fraction. We can observe in Table 5-6 that there are two generators in this design:

$$D \approx ABC$$
$$E \approx BC$$

These generators were chosen with little freedom, since the 2^3 base design has only 8 runs and is so small that there is very little to choose from with regard to the interactions available. Besides the limited choice in generators, the modulus algebra needed to determine the defining contrasts becomes a bit more complex.

In general, we will need 2^p terms or words in the defining contrast. For this design we will then need $2^2 = 4$ words. (Since these terms have the letters of the alphabet in them, we will call them *"words"* from now on.) To obtain these words, we begin as before by multiplying both sides of the generator by the letter on the left side of the generator:

$$D \approx ABC \qquad\qquad E \approx BC \quad \text{(generators)}$$

$$DD \approx ABCD \qquad\qquad EE \approx BCE$$

$$D^{\overset{0}{2}} \approx ABCD \qquad\qquad E^{\overset{0}{2}} \approx BCE \quad \text{(a 2 in mod}_2 \text{ becomes a 0)}$$
$$1 \approx ABCD \qquad\qquad 1 \approx BCE$$

Now, we write this as a continuous string expression separating the words with commas:

$$1 \approx ABCD, BCE$$
$$\text{Word: (1)} \quad\quad \text{(2)} \quad\quad \text{(3)}$$

When we count the number of words, we come up one word short! The last word is obtained by multiplying the first found words together. So:

$$(1)(ABCD)(BCE) \approx A B^{\overset{0}{2}} C^{\overset{0}{2}} DE \approx ADE$$

And now the defining contrast has all four words:

$$1 \approx ABCD, BCE, ADE$$
$$\text{Word: (1)} \quad\quad \text{(2)} \quad\quad \text{(3)} \quad\quad \text{(4)}$$

Before we examine this defining contrast to determine the pattern of confounding in this experiment, we need to note that the algorithm we have just used to determine the words in the defining contrast is based on the more fundamental mathematical system called *group theory* . We do not need to learn or understand this form of math but need only to be able to utilize the results that are found in the 2^{k-p} designs. The defining contrast is a very important part of the process of designing fractional factorials. It does two things for us. It determines the quality of the design with respect to the confounding rules and then determines the pattern of confounding in the design. If we look at the defining contrast for this 2^{5-2}, we can see that there will be confounding between the single effects and the two-factor interactions since there are 2 three-letter words.

$$1 \approx ABCD, BCE, ADE \qquad\qquad (5\text{-}7)$$

$$\mathbf{A} \approx BCD, ABCE, \mathbf{DE}$$
$$\mathbf{B} \approx ACD, \mathbf{CE}, ABDE$$
$$\mathbf{C} \approx ABD, \mathbf{BE}, ACDE$$
$$\mathbf{D} \approx ABC, BCDE, \mathbf{AE}$$

Such a design would only be appropriate if we were in a "screening" situation and merely attempting to identify factors for *further* experimentation. In such cases the consideration is that the number of factors will collapse to a smaller subset, and therefore with fewer factors, there will be fewer possibilities for interactions. (This is called the sparsity-of-effects justification for fractional factorials.)

If we are investigating a situation with the possibility of interactions, the 2^{5-2} found in Table 5-6 is inadequate, and we must search for a design that can remain efficient on both aspects of minimal resource allocation *and* required information. If we were to continue to design and check for efficiency in the sort of groping manner that we have been using, we could spend a considerable amount of effort and time finding the "ideal experiment." [Of course, if we can define what is "ideal," we can have a computer search for us. This has been done under criteria such as d-optimal (determinant optimal) and other optimality searches. We shall not pursue these approaches, but rather we will do things logically.]

The Logical Approach (Information Analysis)

There is a short-cut that can guide us into the proper area and then after a check-out, give us the answer we desire. The short-cut makes use of Table 5-1, where we analyzed the information content of various designs. Let us say that we need to characterize a process in which we expect two-factor interactions to be present along with the usual single effects.

Information Analysis

There are seven factors that we determine to be important in this process. Table 5-1 shows that a full factorial design would require 128 treatments to get all of the information. We only require a fraction of this information, so we go to a fractional design. The question is what fraction? A half, a quarter, an eighth? We may calculate the amount of information as shown below and confirm what we observe in Table 5-1.

Effects	Number (in general)	Number (this case)	df
Single	k	7	7
Two-factor interactions	$\dfrac{k(k-1)}{2}$	$\dfrac{7(7-1)}{2} = 21$	21
Total			28

The total required df for this experiment comes to 28. That means we need a design that has at least 29 treatment combinations rather than 128. The closest design we have in integer powers of 2 is a 2^5 or 32 treatments. Thus, the "p" fractionalizing element of the design is computed as follows: $7 - 5 = 2$. The value of p tells us that we have a 1/4 fractional factorial with 2^2 terms in the defining contrast, and two generators. The choice of the generators is important. Table 5-7 shows the base design and some possible generators. If we selected the generators: $F \approx ABCDE$ and $G \approx BCDE$, then the defining contrast will be:

$$FF \approx ABCDEF \qquad\qquad GG \approx BCDEG$$

$$\overset{0}{F^2} \approx ABCDEF \qquad\qquad \overset{0}{G^2} \approx BCDEG$$

$$1 \approx ABCDEF, BCDEG$$
$$ABCDEF * BCDEG \approx AB^{\overset{0}{2}}C^{\overset{0}{2}}D^{\overset{0}{2}}E^{\overset{0}{2}}FG$$

$$1 \approx ABCDEF, BCDEG, \textbf{AFG} \qquad\qquad (5\text{-}8)$$

We immediately see that in the "AFG" part of the defining contrast, there is single factor and two-factor interaction confounding.

$$A \approx FG$$
$$F \approx AG$$
$$G \approx AF$$

This generator set is not appropriate for the information we seek. There are adequate degrees of freedom to get the single effects free of two-factor interactions. The trick is to use the correct parts of the design to attain this promise.

Let us try a different set of generators. Let $F \approx ABCD$ and $G \approx BCDE$. Now the defining contrast becomes:

$$FF \approx ABCDF \qquad GG \approx BCDEG$$

$$F^2 \approx ABCDF \qquad G^2 \approx BCDEG$$

$$1 \approx ABCDF, BCDEG$$

$$ABCDF * BCDEG \approx AB^2C^2D^2EFG$$

$$1 \approx ABCDF, BCDEG, AFEG \qquad\qquad (5\text{-}9)$$

This design seems to do the job. The only problem may be with the "AEFG" word. In it we have two-factor interactions confounded together. There will be 6 interactions that are confounded:

$$AF \approx EG$$
$$AE \approx FG$$
$$AG \approx FE$$

The remaining 15 two-factor interactions will be free of each other. In most physical systems not all the two-factor interactions are likely to occur. Those factors that are unlikely to interact are assigned to the confounded pairs. In this example, factors A,E,F,G would be chosen as nonpossible interacting pairs. Even if it is not possible to make such an assignment, there are techniques of analysis that are able to separate confounded two-factor interactions (see Chapter 14).

TABLE 5-7

A	B	C	D	E	ABCDE	ABCD	BCDE
-	-	-	-	-	-	+	+
+	-	-	-	-	+	-	+
-	+	-	-	-	+	-	-
+	+	-	-	-	-	+	-
-	-	+	-	-	+	-	-
+	-	+	-	-	-	+	-
-	+	+	-	-	-	+	+
+	+	+	-	-	+	-	+
-	-	-	+	-	+	-	-
+	-	-	+	-	-	+	-
-	+	-	+	-	-	+	+
+	+	-	+	-	+	-	+
-	-	+	+	-	-	+	+
+	-	+	+	-	+	-	+
-	+	+	+	-	+	-	-
+	+	+	+	-	-	+	-
-	-	-	-	+	+	+	+
+	-	-	-	+	-	-	-
-	+	-	-	+	-	-	+
+	+	-	-	+	+	+	+
-	-	+	-	+	-	-	+
+	-	+	-	+	+	+	+
-	+	+	-	+	+	+	-
+	+	+	-	+	-	-	-
-	-	-	+	+	-	-	+
+	-	-	+	+	+	+	+
-	+	-	+	+	+	+	-
+	+	-	+	+	-	-	-
-	-	+	+	+	+	+	-
+	-	+	+	+	-	-	-
-	+	+	+	+	-	-	+
+	+	+	+	+	+	+	+

The type of reasoning used in the example above should be applied to all 2-level fractional factorial designs where single effects and two-factor interactions are the required information.

- Define required information (k single effects(S), k*(k -1)/2 interactions(I))
- Determine minimum number of degrees of freedom (#df=S+I)
- The number of runs in the base 2-level design is $\geq$ #df and is 2^n where n=k-p
- Compute "p," the fractionalization element (p = k - n)
- Select the p generators (see *Statistical Tables* at the end of this book)
- Calculate defining contrast
- Check if required information is attained

Other Designs

Over the years, a number of alternative fractional factorial design configurations have been developed by various mathematicians. These design configurations are often named for their authors. One such set of designs is known as the Plackett-Burman (P-B) design. It is interesting to note the history of these design configurations. The paper (5) published by R. L. Plackett and J. P. Burman appeared in 1946, just after World War II. According to Richard A. Freund (a noted experimenter who worked for Eastman Kodak Co. and a past president of ASQC) the motivation for the P-B designs was to be able to accomplish empirical tolerance stack-up analyses of mechanical systems to improve the quality and reliability of war materials. The work done by Plackett and Burman was classified as a "war secret" for fear that if it fell into the hands of the Axis Powers, this enemy of the free world would gain an advantage by the use of such experimental methods to improve the quality of their war machines.

With this important bit of history, we now should be able to see why the P-B designs are not suited to investigations where interactions are present. The task presented to Plackett and Burman was to provide a set of small, 2-level experimental designs that would operate in completely non-interacting systems (mechanical parts tolerance stack-up). Therefore, the P-B designs do not provide information on interactions. This is a major short-fall and one reason to warn against the use of the P-B designs. Another reason to avoid the P-B designs lies in the fact that the pattern of confounding in P-B designs (6) is not easily discerned as it is with the classical 2^{k-p} designs since there are no defining contrasts available for these designs.

With that warning, let us now look at the structure of the P-B designs. First, these designs are very easy to construct, and second, they allow non-integer powers of 2 number of treatment combinations. So instead of forcing the number of runs to be 4 or 8 or 16 or 32, etc., we may select configurations with 12 or 20 or 24, and so on runs. To build the design matrix, simply take the "generator" column (found in Table 5-8) and cycle the column as shown in Table 5-9a. Here are the rules:

1. Write down the first column (generator column).
2. Shift the column cyclically one place N-1 times.
   N is the number of factors being investigated.
3. Add a final row of minus signs.

TABLE 5-8
Generator columns for some Plackett-Burman designs

#tc's	Generators
8	+ + + - + - -
12	+ + - + + + - - - + -
16	+ + + + - + - + + - - + - - -
20	+ + - - + + + + - + - + - - - - + + -
24	+ + + + + - + - + + - - + + - - + - + - - - -

We will show the building of the P-B matrix using the 12 tc generator.

TABLE 5-9a

Write the generator column down and cycle it for up to 11 factors.

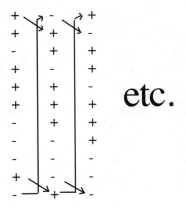

etc.

TABLE 5-9 b
Plackett-Burman 12 for 9 factors

A	B	C	D	E	F	G	H	I	ABC
+	-	+	-	-	-	+	+	+	-
+	+	-	+	-	-	-	+	+	-
-	+	+	-	+	-	-	-	+	-
+	-	+	+	-	+	-	-	-	-
+	+	-	+	+	-	+	-	-	-
+	+	+	-	+	+	-	+	-	+
-	+	+	+	-	+	+	-	+	-
-	-	+	+	+	-	+	+	-	+
-	-	-	+	+	+	-	+	+	-
+	-	-	-	+	+	+	-	+	+
-	+	-	-	-	+	+	+	-	+
-	-	-	-	-	-	-	-	-	-

Another problem with some of the P-B designs like the P-B 12 above is the fact that the three-factor interactions are unbalanced. Notice that there are 8 minus conditions for ABC and only 4 plus conditions. This can lead to incorrect information with regard to any effects confounded with the three-factor interactions. For this reason, I do not recommend the P-B designs.

The final different design we will mention, only as a warning, is called the *random balance* (**RB**) design. "The only problem with a random balance design is random balance." This a direct quote from a paper (7) written by George Box and pretty much summarizes the state of this unhappy design. The idea is to design an experiment by randomly choosing the levels for each column by a coin toss. We show such a "design" in Table 5-10. This approach is about the same as trying to achieve a Shakespearian play by putting 1 million monkeys at a word processor and letting them randomly type away!

TABLE 5-10

Coin toss outcome			A	B	C
H	T	T	+	-	-
T	H	H	-	+	+
H	H	H	+	+	+
T	H	T	-	+	-
H	T	H	+	-	+
T	T	T	-	-	-

. . . . etc.

Putting It All to Use

Having learned the concepts of fractional factorial designs, let's use them to build a real experiment with physical factors rather than just A's, B's, and C's.

An engineering team has brainstormed a machining process that makes spindles for automobile wheel bearings. Figure 5-3 is a drawing of such a spindle.

Spindle (side view)
FIGURE 5-3

Table 5-11 lists the seven factors and their possible physical interactions. This is a typical example of the type of information that emerges from the brainstorming sessions that were described in Chapter 2. The appendix to this chapter has a selection of fractional factorial design templates that may be used to help in the focusing of this brainstormed information to construct the proper experiments. We have a choice of two templates, the **T8-7** and the **T16-7.**

To help make this choice, we must perform an information analysis. With seven factors, we have the need for seven single effects and 21 possible two-factor interactions. However, we have identified (in the brainstorming session) that there are only 7 of these possible interactions that are likely to take place. On the templates, there is a section designated to help in the identification of such likely interactions. To illustrate which interactions are likely, the team works to construct a hypothetical, qualitative interaction plot. In the spindle example, the team identified a likely interaction between the speed and the feed left. This has been designated in Table 5-11 by drawing lines connecting the speed factor to the feed left factor and placing a dot at the point of intersection. To determine if this is a real interaction and not just a reaction, we go through the following procedure.

- Draw an x-y graph axis with the response on y and factor X_1 on x.
- Hold a pen or pencil on the graph at the origin and parallel to the x axis.
- Work with the team to determine the slope and intercept of the line for the second x factor (X_2) at its low level.
- Repeat the above step for the high level of the X_2 factor.
- If the lines are parallel, there is only a reaction and **no** possible interaction. If the lines are nonparallel, then there is a possible interaction.

Figure 5-4 will serve to show this procedure in action for the speed-feed interaction. We first draw the graph axis. Then place the marker parallel to the axis. Now move the marker up and determine the likely slope for the first level of X_2 . We repeat the procedure and observe for this set of factors, the lines are non-parallel.

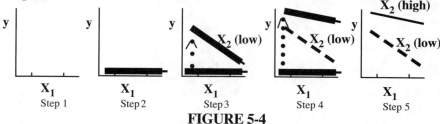

FIGURE 5-4

Now we are ready to select the correct design matrix to meet our needs. We require 7 df for the single effects and we have identified the likelihood of seven two-factor interactions, so we need another 7df to study these effects. The **T16-7** will deliver 15 df, which is sufficient for this problem. The **T8-7** lacks 7df and is inadequate for this problem. It is interesting to note that the **T8-7** is the same design as the Taguchi "**L8**" configuration that has been used incorrectly for situations like this one since the interactions have not been taken into consideration.

TABLE 5-11

Factor	Working Range	Levels		Possible Interaction
Material	N/A	Soft	Hard	
Speed	150-250 rpm	175	225	
Feed Left	.2 - .4 ips	.25	.35	
Feed Right	.3 - .5 ips	.35	.45	
Clearance	5% - 15%	6%	12%	
Heat Cycle	4 - 10 sec.	5.5	8.5	
Heat Power	50% - 100%	75%	85%	

We will now combine the physical knowledge of the possible interactions and the knowledge of the confounding pattern in the **T16-7** to produce an efficient experiment that gets the required information at the least expenditure of resources.

On the information page of this template, we see the generators as well as the defining contrast. The information page shows the complete pattern of confounding and for our current efforts we also see the pattern of confounding

TABLE 5-12

Generators: E≈ABC F≈BCD G≈ACD
Defining Contrast:
1≈ABCE,BCDF,ACDG
   ADEF,BDEG,ABFG,CEFG
Complete confounding pattern:

A≈BCE,ABCDF,CDG,DEF,ABDEG,BFG,ACEFG	AB≈CE,FG
B≈ACE,CDF,ABCDG,ABDEF,DEG,AFG,BCEFG	AC≈BE,DG
C≈ABE,BDF,ADG,ACDEF,BCDEG,ABCFG,EFG	AD≈EF,CG
D≈ABCDE,BCF,ACG,AEF,BEG,ABDFG,CDEFG	AE≈DF,BC
E≈ABC,BCDEF,ACDEG,ADF,BDG,ABEFG,CEF	AF≈DE,BG
F≈ABCEF,BCD,ACDFG,ADE,BDEFG,ABG,CEG	AG≈BF,CD
G≈ABCEG,BCDFG,ACD,ADEFG,BDE,ABF,CEF	BD≈CF,EG

among the 21 possible two-factor interactions. This information is reproduced in Table 5-12. We now need to place the physical factors in the appropriate columns to avoid unnecessary confounding among the effects. Since we have built the table of confounded interactions in alphabetical order, the "A" factor shows up in the first six positions of this table. We could decide to place the factor that does not interact with anything in the A column of the experiment, thus freeing the remaining confounded interactions, or as in this particularly difficult case (we need 14df to study the effects and have only 15df to do so) we will assign the material (which interacts with many factors) to the A column.

TABLE 5-13

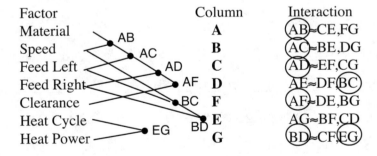

Factor	Column	Interaction
Material	A	AB≈CE,FG
Speed	B	AC≈BE,DG
Feed Left	C	AD≈EF,CG
Feed Right	D	AE≈DF,BC
Clearance	F	AF≈DE,BG
Heat Cycle	E	AG≈BF,CD
Heat Power	G	BD≈CF,EG

The remaining factors stay in alphabetical order except for clearance and heat cycle, which are switched out of order. This arrangement produces interactions that are free from confounding except for one set, BD and EG. Fortunately, in this experiment, the BD (speed-feed) interaction influences the chip rating of the machining process while the EG (heat cycle-heat power) influences the hardness of the part. Since these are two physically independent responses, there is no problem with confounding this pair of interactions. The experiment is efficient! Table 5-14 shows the final experiment in the format of the template. Note that there are two zero treatments, one for the soft steel and one for the hard steel. These should be replicated two or three times each.

TABLE 5-14

Random Run Order	tc	A: Matrl	B: Speed	C: Lfed	D: Rfed	E: HTcyc	F: Clear	G: Htpwr	Response (Y)
6	(1)	(-)Soft	(-)175	(-).25	(-).35	(-)5.5	(-)6%	(-)75%	
2	a(eg)	(+)Hard	(-)175	(-).25	(-).35	(+)8.5	(-)6%	(+)85%	
18	b(ef)	(-)Soft	(+)225	(-).25	(-).35	(+)8.5	(+)12%	(-)75%	
17	ab(fg)	(+)Hard	(+)225	(-).25	(-).35	(-)5.5	(+)12%	(+)85%	
13	c(efg)	(-)Soft	(-)175	(+).35	(-).35	(+)8.5	(+)12%	(+)85%	
15	ac(f)	(+)Hard	(-)175	(+).35	(-).35	(-)5.5	(+)12%	(-)75%	
5	bc(g)	(-)Soft	(+)225	(+).35	(-).35	(-)5.5	(-)6%	(+)85%	
8	abc(e)	(+)Hard	(+)225	(+).35	(-).35	(+)8.5	(-)6%	(-)75%	
16	d(fg)	(-)Soft	(-)175	(-).25	(+).45	(-)5.5	(+)12%	(+)85%	
3	ad(ef)	(+)Hard	(-)175	(-).25	(+).45	(+)8.5	(+)12%	(-)75%	
9	bd(eg)	(-)Soft	(+)225	(-).25	(+).45	(+)8.5	(-)6%	(+)85%	
1	abd	(+)Hard	(+)225	(-).25	(+).45	(-)5.5	(-)6%	(-)75%	
14	cd(e)	(-)Soft	(-)175	(+).35	(+).45	(+)8.5	(-)6%	(-)75%	
11	acd(g)	(+)Hard	(-)175	(+).35	(+).45	(+)8.5	(-)6%	(+)85%	
10	bcd(f)	(-)Soft	(+)225	(+).35	(+).45	(-)5.5	(+)12%	(-)75%	
7	abcd(efg)	(+)Hard	(+)225	(+).35	(+).45	(+)8.5	(+)12%	(+)85%	
4	zero$_S$	Soft	(0)200	(0).30	(0).40	(0)7.0	(0)9%	(0)80%	
12	zero$_H$	Hard	(0)200	(0).30	(0).40	(0)7.0	(0)9%	(0)80%	

Using Minitab

We will now create the experimental design shown in Table 5-14 using Minitab software. We begin much like we did in Chapter 4. As shown in Figure 5-4, select **Stat> Factorial> Create Factorial Design** from the dropdown menu. This brings the main dialog box as shown in Figure 5-5, where we scroll to 7 factors and click on "Display Available Designs."

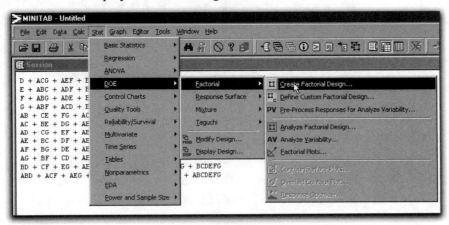

FIGURE 5-4

*After launching Minitab, go to the **Stat** dropdown and select **DOE**, then **Factorial**, then **Create Factorial Design**.*

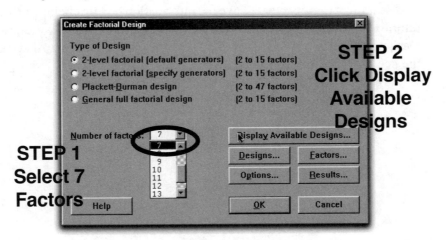

FIGURE 5-5

*A dialog box appears. We need to indicate the number of factors. Scroll to 7, then click on **Display Available Designs**.*

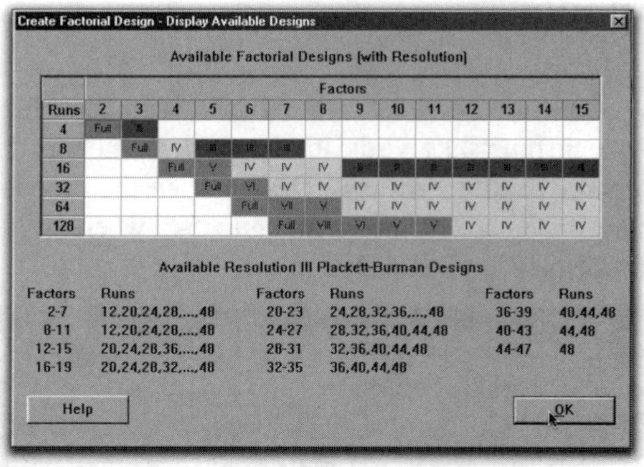

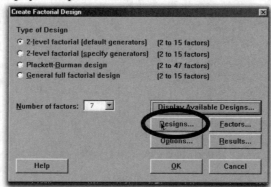

FIGURE 5-6

While this table does not allow you to choose a design, it lists those available in this software package. The concept of resolution will be discussed in the next section of the book. Click OK to return to the main dialog box.

Figure 5-6 merely shows the available designs and, with a color coding, makes some recommendations. The green code (medium gray above) indicates designs with little confounding. A yellow code (light gray above) is a slightly more confounded. The red code (dark gray) is highly confounded and should be used with caution. We will discuss the idea of resolution indicated by the Roman Numeral designations in this table in the next section of this chapter. Leave the information box by clicking **OK** and return to the main dialog box shown in Figure 5-7. Here we click on **Designs**, which takes us to Figure 5-8 where we choose the design that meets the needs of the physical problem.

FIGURE 5-7

Now click on "Designs" to select the design appropriate for this experimental situation.

We select a 1/8 fractional factorial design from the available designs shown in the dialog box in Figure 5-8. This design matches the previous configuration we selected from the design templates in the appendix of this chapter. We apply these same considerations to pick the 2^{7-3} when we use Minitab to create the design. Minitab does not make the selection for you, nor does Minitab have any "expert system" that gives you the best design. You need to do the thinking. Minitab just does the "statistical word processing" for you. This same dialog box has the place to specify the number of center points, and if we can afford them, the number of replicates in the factorial part of the design (designated as "corner points" in Minitab language). We can also specify the number of blocks. Blocking will be covered in Chapter 8. After making the selections, click **OK** to move back to the main dialog box and click on **Factors**.

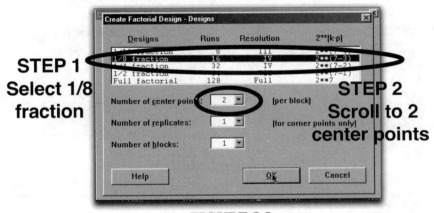

FIGURE 5-8

To match the design we created using the template, we choose a 1/8 fraction 2^{7-3} from the available designs. We will add two centerpoints and no replicates to the points in the factorial part of the design (designated "corner points" by Minitab). The design will be run in one block (blocking is a topic for Chapter 8). Click OK.

This brings up a sub dialog box where we name the factors with their physical designations and add the physical levels. If a factor is a classification ("Text" in Minitab language), we scroll to **Text** and then are allowed to use verbal identifications for the levels. This is necessary for the first factor, which is the hardness of the steel that is being machined. Figure 5-9 shows the entry of the names and levels for all the factors in this experiment. Units such as degrees or pounds cannot be used since doing so would make them "Text" types. It is best to have numeric factors (the default) since later, when we do the analysis of the experiment, we may interpolate. It is not possible to interpolate on a classification. After completing all the information on the factor names and levels, we click **OK** and return to the main dialog box and click **Options**. Figure 5-10 shows this step; and the **Options** sub dialog box. We will un-click the "Randomize runs" default for this example so we may compare the manually generated design with the computer generated design. We will come back to this option later and re-click Randomize runs. We bring up the last sub dialog box by clicking on **Options** in the main dialog box as shown in Figure 5-11(a).

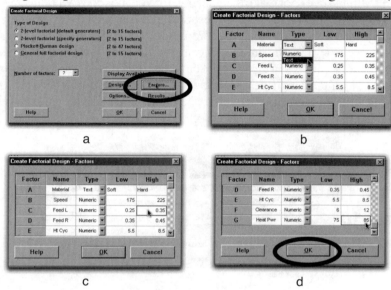

FIGURE 5-9

In the main dialog box (a), we click on "Factors," which brings up a sub-dialog box (b) where we add the physical names of the factors and their levels. The first factor (b) is non-numeric, and we select "Text" as the factor type. All the other factors are numeric and we simply fill in the name and the levels (c-d). Click OK (d) to return to the main dialog box.

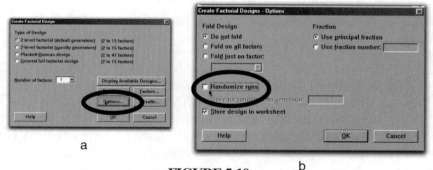

FIGURE 5-10

In the main dialog box (a), we click on "Options," which brings up a sub-dialog box (b). For this illustration, we will unclick "Randomize runs" so that the displayed experiment is shown in standard Yates order (Table 5-15). This will allow us to scrutinize the individual treatment combination and judge if they are reasonable. Being reasonable does not mean if they will produce good product, but merely if they will produce a measurable product. By re-clicking "Randomize runs," a completely randomized version of the design is shown in Table 15-16.

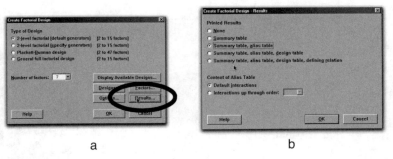

a   b

FIGURE 5-11

In the main dialog box (a), we click on "Results," which brings up a sub-dialog box (b). The default options are sufficient and we click OK.

The **Options** sub dialog box establishes the information that is displayed in the session window. Recall that in Minitab, there are two default windows displayed simultaneously. The **session window** is in text format and is placed at the top of the split screen. It contains a summary of the commands that are generated by the selections in the dialog boxes as well as the information generated such as the confounding pattern (called "Alias"), design table (in standard units), and other optional material we select from the results box. The **worksheet** contains the design and the data we will later analyze. Figure 5-12(a) shows a section of the session window and worksheet. Back to the sub dialog box, **Results** in Figure 5-11(b), we will simply use the default options since much of the additional material is redundant or may be found in the design templates in the appendix of this chapter. It is best to actually use the design templates to select the design that accommodates the problem and then have Minitab do the clerical work of filling out the table, which is what we do by clicking **OK** in the main dialog box in Figure 5-12. This produces the worksheet in Figure 5-13 and the design shown in Table 5-15. Table 5-16 is the design in random order produced by clicking **Randomize runs** in the **Options** sub dialog box.

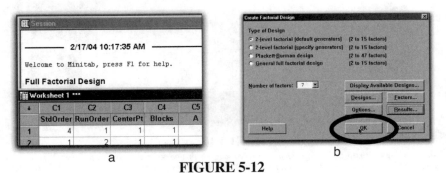

a   b

FIGURE 5-12

In the main dialog box (b), we click OK and the experimental design is produced as shown in Figures 5-13 and 5-4.

	C1 StdOrder	C2 RunOrder	C3 CenterPt	C4 Blocks	C5-T Material	C6 Speed	C7 Feed L	C8 Feed R	C9 Ht Cyc	C10 Clearance	C11 Heat Pwr
1	1	1	1	1	Soft	175	0.25	0.35	5.5	6	75
2	2	2	1	1	Hard	175	0.25	0.35	8.5	6	85
3	3	3	1	1	Soft	225	0.25	0.35	8.5	12	75
4	4	4	1	1	Hard	225	0.25	0.35	5.5	12	85
5	5	5	1	1	Soft	175	0.35	0.35	8.5	12	85
6	6	6	1	1	Hard	175	0.35	0.35	5.5	12	85
7	7	7	1	1	Soft	225	0.35	0.35	5.5	6	75
8	8	8	1	1	Hard	225	0.35	0.35	8.5	6	85
9	9	9	1	1	Soft	175	0.25	0.45	5.5	12	75
10	10	10	1	1	Hard	175	0.25	0.45	8.5	12	85
11	11	11	1	1	Soft	225	0.25	0.45	8.5	6	75
12	12	12	1	1	Hard	225	0.25	0.45	5.5	6	75
13	13	13	1	1	Soft	175	0.35	0.45	5.5	6	75
14	14	14	1	1	Hard	175	0.35	0.45	8.5	6	85
15	15	15	1	1	Soft	225	0.35	0.45	5.5	12	75
16	16	16	1	1	Hard	225	0.35	0.45	8.5	12	85
17	17	17	0	1	Soft	200	0.30	0.40	7.0	9	80
18	18	18	0	1	Hard	200	0.30	0.40	7.0	9	80
19	19	19	0	1	Soft	200	0.30	0.40	7.0	9	80
20	20	20	0	1	Hard	200	0.30	0.40	7.0	9	80

FIGURE 5-13

Since we went with the default option to store the design in the worksheet, we have the entire 20 runs. While we asked for 2 replicates of the center points, Minitab creates two replicates for each of the two levels of the classification factor (Material). This is similar to the two Zero Point runs in manually produced design (Table 5-14).

TABLE 5-15
Standard Order

Sd	Rn	CP	Blk	Mtrl	Spd	Fd L	Fd R	HtCyc	Clr	Ht Pwr
1	1	1	1	Soft	175	0.25	0.35	5.5	6	75
2	2	1	1	Hard	175	0.25	0.35	8.5	6	85
3	3	1	1	Soft	225	0.25	0.35	8.5	12	75
4	4	1	1	Hard	225	0.25	0.35	5.5	12	85
5	5	1	1	Soft	175	0.35	0.35	8.5	12	85
6	6	1	1	Hard	175	0.35	0.35	5.5	12	75
7	7	1	1	Soft	225	0.35	0.35	5.5	6	85
8	8	1	1	Hard	225	0.35	0.35	8.5	6	75
9	9	1	1	Soft	175	0.25	0.45	5.5	12	85
10	10	1	1	Hard	175	0.25	0.45	8.5	12	75
11	11	1	1	Soft	225	0.25	0.45	8.5	6	85
12	12	1	1	Hard	225	0.25	0.45	5.5	6	75
13	13	1	1	Soft	175	0.35	0.45	8.5	6	75
14	14	1	1	Hard	175	0.35	0.45	5.5	6	85
15	15	1	1	Soft	225	0.35	0.45	5.5	12	75
16	16	1	1	Hard	225	0.35	0.45	8.5	12	85
17	17	0	1	Soft	200	0.30	0.40	7.0	9	80
18	18	0	1	Hard	200	0.30	0.40	7.0	9	80
19	19	0	1	Soft	200	0.30	0.40	7.0	9	80
20	20	0	1	Hard	200	0.30	0.40	7.0	9	80

TABLE 5-16
Random Order

Sd	Rn	CP	Blk	Mtrl	Spd	Fd L	Fd R	HtCyc	Clr	Ht Pwr
6	1	1	1	Hard	175	0.35	0.35	5.5	12	75
2	2	1	1	Hard	175	0.25	0.35	8.5	6	85
20	3	0	1	Hard	200	0.30	0.40	7.0	9	80
18	4	0	1	Hard	200	0.30	0.40	7.0	9	80
13	5	1	1	Soft	175	0.35	0.45	8.5	6	75
11	6	1	1	Soft	225	0.25	0.45	8.5	6	85
1	7	1	1	Soft	175	0.25	0.35	5.5	6	75
17	8	0	1	Soft	200	0.30	0.40	7.0	9	80
19	9	0	1	Soft	200	0.30	0.40	7.0	9	80
16	10	1	1	Hard	225	0.35	0.45	8.5	12	85
9	11	1	1	Soft	175	0.25	0.45	5.5	12	85
5	12	1	1	Soft	175	0.35	0.35	8.5	12	85
14	13	1	1	Hard	175	0.35	0.45	5.5	6	85
8	14	1	1	Hard	225	0.35	0.35	8.5	6	75
3	15	1	1	Soft	225	0.25	0.35	8.5	12	75
15	16	1	1	Soft	225	0.35	0.45	5.5	12	75
7	17	1	1	Soft	225	0.35	0.35	5.5	6	85
12	18	1	1	Hard	225	0.25	0.45	5.5	6	75
4	19	1	1	Hard	225	0.25	0.35	5.5	12	85
10	20	1	1	Hard	175	0.25	0.45	8.5	12	75

Resolution

George Box (3) has introduced a handy nomenclature for the 2^{k-p} designs. This descriptive label gives an immediate indication of the information capacity of the design. The term "resolution" has been applied to the number of individual letters (or elements) in the shortest word of the defining contrast. The lower the resolution value, the less information is produced.

For example, a Resolution IV (Four) has in its defining contrast at least one word with only four letters in it. This is exactly like Expression 5-9 and the defining contrast in template **T16-7**. A Resolution III (Three) design has at least one word with only three letters in it and will confound single effects with two-factor interactions. According to the rules of confounding, we would *ideally* use only Resolution IV and above designs.

Final Thoughts

While the experimental design templates and the Minitab software provide a convenient method for setting up the experimental design, remember that the entire design consists of gathering the prior knowledge and using the statistical experimental design methods to focus this prior knowledge. Simply filling in a table is not experimental design, but only a part of the entire process.

We have looked at three types of fractional factorial designs. The classical 2^{k-p} is the recommended technique to use because of the short-falls of the random balance (RB) and the Plackett-Burman (P-B) designs. However, these designs are often recommended as "screening" designs. The concept behind a screening design is the fact that we have a large number of factors and only a few are expected to cause a reaction to take place. If a *reaction* (simple additivity) is all that is happening, then such screening designs are appropriate. Classical 2^{k-p} designs can be used as screening designs (a 2^{7-3} is typical of such a screening design) and is actually better than a RB or P-B, since we *know* the confounding pattern. When interactions are present, the RB and P-B designs are inappropriate since an interaction can "wipe out" a single effect and give us a biased answer, and we will never know it!

We must not lose sight of the fact that when we run the experiments, we must strive to prevent other factors that are not a part of our investigation from systematically introducing their influence on the response. We are able to do this by randomizing the runs in the experiment as described in Chapter 4. Remember that the templates are presented in a systematic order to allow an easily applied analysis, but they should be randomized before any actual data is gathered. Without randomization the results could be biased.

Problems for Chapter 5

1. Extend Table 5-1 to 12 factors using Relationship 5-1.

2. If we are interested in only single factors and two-factor (first order) interactions, how much excess information in terms of degrees of freedom is there in a full factorial 2^{12} experiment?

3. Show the design matrix for a full 2^4 factorial.

4. Using the 2^4 factorial from #3, add a fifth factor (without adding any more rows).
     a) What fraction do you have?
     b) What is the generator?
     c) What is the defining contrast?
     d) Is this an efficient design if we need to investigate both single effects and first-order (two-factor) interactions?
     e) Justify your answer to d.

5. Using the 2^4 factorial in #3 as the base design, add two more factors: E,F (without adding any more rows).
     a) What fraction do you have?
     b) What are the generators?
     c) What is the defining contrast?
     d) Is this an efficient design if we need to investigate both single effects and two-factor interactions?
     e) Justify your answer to d.

6. A chemical process is being investigated for yield increases. The team has come up with the following factors, levels and likely 2-factor interactions. Use the appropriate design template to construct an efficient experiment for this problem.

Factor	Likely interaction	Levels	
Temperature		$100°$	$200°$
Pressure		14psi	28psi
Catalyst Conc.	This is not a 3 factor interaction	.5%	.75%
Batch Size		1 Liter	10 Liters
Time		5 min.	15 min.
Raw Material		Bag A	Bag B

7. Show the design for a 2^{8-3} fractional factorial. Include the generators and defining contrast.

8. An engineer has defined a problem that includes eight factors. He would like to determine the effect of each factor and if any two-factor interactions exist. He also has a resource constraint of 16 weeks. It is possible to run two treatment combinations per week. What would your advice be to this person if a two-level design is sufficient to solve the problem?

9. Using Table 5-10, attempt to assign the factors to different columns and still keep the confounding to a minimum.

10. Try using two different generators than the ones used in Table 5-6 to see if you can get a better resolution design than the one found in the text.

11. Explain in words the interaction between BHA and BHT shown in Figure 5-2.

APPENDIX 5

Degrees of Freedom (df):

A counting method devised to assess and define the amount of mathematically independent information in an experiment. In general, each time we compute an average value, we deduct one degree of freedom from the total available. Thus in a three-level experiment, upon computing the mean across the three levels (which is done in each analysis), we deduct 1 df and have 2 left. In a two-level design, we have only one degree of freedom for each factor. While it is customary to say "we lose a df" it is more conceptually correct to say we have **converted** one df into the summary information (the average). You never lose df's but you can look at the information they contain in a different form (such as the average). Thus the idea of "conversion."

Interaction degrees of freedom are determined by the product of the degrees of freedom of the single factors that make up the interaction. Thus, if A has 2 df and B has 3 df, then the AB interaction has 6df.

Confounding: The linear combination of two or more effects. Factors that are confounded have to share the same degree of freedom and are not independent of or orthogonal to each other.

If AB≈CD and the effect found is 10, this could mean that either AB or CD has the effect or the combined effect of both is equal to 10.

Some possible combinations of AB and CD are:

AB	CD	Observed Result
-9	19	10
10	0	10
5	5	10
-20	30	10

Generator: Usually a high-order interaction column with which a factor is equated. There are p generators in the fractional designs.

Defining Contrast: A relationship used to show the confounded sets of factors in a fractional design. There are 2^p terms in the defining contrast, including the identity term (1).

2^{k-p} Defining Contrast Algorithm and Modulus Algebra Rules

1. The number 1 is the identity. The identity does not change the value of a number when multiplying or dividing. The identity appears only on the left side of the confounded sign.

2. In modulus algebra, the base or any multiple of the base is removed from the number in question and the remainder is the modulus of that number. For $Mod_{Base\ 2}$, a 0=0, 1=1, but 2=0. A 3=1, and 4=0, 5=1, 6=0, etc.

3. To construct the defining contrast, multiply both sides of each generator by the single effect in the generator. Apply the modulus algebra rules to the resulting exponents in the expression to create the identity.

4. If there is only one generator, the process stops after step #3. If there is more than one generator, the remaining words in the defining contrast are derived by multiplying the first found words together first in pairs, then in triples, etc. up to the number of generators used to build the design.

2-LEVEL FRACTIONAL FACTORIAL DESIGN TEMPLATES

The following 2-level fractional factorial design templates have been prepared for your use. They cover a selection of possible situations for up to seven factors in 8 runs.

Table 11 in the *Statistical Tables* Appendix of this book covers situations for up to 11 factors in as many as 128 runs. Computer programs (4) are available to build such designs from these tables.

For 3 factors in 4 runs:

Resolution III

Generator: C≈AB

Defining Contrast: 1≈ABC

Confounding Pattern:

    A≈BC

    B≈AC

    C≈AB

tc	A	B	C
(c)	-	-	+
a	+	-	-
b	-	+	-
abc	+	+	+

Also Measures: **BC   AC   AB**

Description: All single effects are confounded with 2-factor interactions. This design configuration is good only if there are *no* possible 2-factor interactions.

TEMPLATE: **T4-3** (3 factors in 4 runs)
FACTOR NAMES

Random Run Order	tc	A:_____	B:_____	C:_____	Response (Y)
_____	(c)	(-)_____	(-)_____	(+)_____	_____
_____	a	(+)_____	(-)_____	(-)_____	_____
_____	b	(-)_____	(+)_____	(-)_____	_____
_____	abc	(+)_____	(+)_____	(+)_____	_____
_____	zero	(0)_____	(0)_____	(0)_____	_____

2-LEVEL FRACTIONAL FACTORIAL DESIGN TEMPLATES

For 4 factors in 8 runs:	tc	**A**	**B**	**C**	**D**	**AB**	**AC**	**AD**
Resolution IV	(1)	-	-	-	-	+	+	+
Generator: D≈ABC	a(d)	+	-	-	+	-	-	+
Defining Contrast: 1≈ABCD	b(d)	-	+	-	+	-	+	-
Complete confounding pattern:	ab	+	+	-	-	+	-	-
A≈BCD AB≈CD	c(d)	-	-	+	+	+	-	-
B≈ACD AC≈BD	ac	+	-	+	-	-	+	-
C≈ABD AD≈BC	bc	-	+	+	-	-	-	+
	abc(d)	+	+	+	+	+	+	+

Also measures: **BCD ACD ABD ABC CD BD BC**

Description: All single effects are confounded with 3-factor interactions. All the 2-factor interactions are confounded with each other. This design configuration is good only if solid prior knowledge exists about the 2-factor interactions. **Helpful strategy:** *If one factor is not expected to interact at all, then place it in the A column, thus allowing the BC, BD, and CD interactions to be free of confounding.*

> Likely physical interactions:
> **A**:_____
> **B**:_____
> **C**:_____
> **D**:_____

TEMPLATE: **T8-4** (4 factors in 8 runs)
FACTOR NAMES

Random
Run Order tc   **A**:_____ **B**:_____ **C**:_____ **D**:_____ **Response (Y)**

_____ (1)   (-)_____ (-)_____ (-)_____ (-)_____ _____

_____ a(d) (+)_____ (-)_____ (-)_____ (+)_____ _____

_____ b(d) (-)_____ (+)_____ (-)_____ (+)_____ _____

_____ ab   (+)_____ (+)_____ (-)_____ (-)_____ _____

_____ c(d) (-)_____ (-)_____ (+)_____ (+)_____ _____

_____ ac   (+)_____ (-)_____ (+)_____ (-)_____ _____

_____ bc   (-)_____ (+)_____ (+)_____ (-)_____ _____

_____ abc(d) (+)_____ (+)_____ (+)_____ (+)_____ _____

_____ zero (0)_____ (0)_____ (0)_____ (0)_____ _____

2-LEVEL FRACTIONAL FACTORIAL DESIGN TEMPLATES

For 5 factors in 8 runs:

Resolution III

Generators: D≈AB E≈AC
Defining Contrast: 1≈ABD,ACE,BCDE
Complete confounding pattern:

A≈BD,CE,ABCDE
B≈AD,ABCE,CDE
C≈ABCD,AE,BDE
D≈AB,ACDE,BCE
E≈ABDE,AC,BCD

tc	A	B	C	D	E	BC	ABC
(de)	-	-	-	+	+	+	-
a	+	-	-	-	-	+	+
b(e)	-	+	-	-	+	-	+
ab(d)	+	+	-	+	-	-	-
c(d)	-	-	+	+	-	-	+
ac(e)	+	-	+	-	+	-	-
bc	-	+	+	-	-	+	-
abc(de)	+	+	+	+	+	+	+

Also measures:	BD	AD	AE	AB	AC	DE	CD
	CE						BE

Description: All single effects are confounded with 2-factor interactions. All the 2-factor interactions are confounded with each other. This design configuration is good only if the 2-factor interactions do not exist. *This is a dangerous design to use without solid prior knowledge!*

```
Likely physical interactions:
A:_____
B:_____
C:_____
D:_____
E:_____
```

TEMPLATE: **T8-5** (5 factors in 8 runs)
FACTOR NAMES

Random Run Order tc	A:_____	B:_____	C:_____	D:_____	E:_____	Response (Y)
_____ (de)	(-)_____	(-)_____	(-)_____	(+)_____	(+)_____	_____
_____ a	(+)_____	(-)_____	(-)_____	(-)_____	(-)_____	_____
_____ b(e)	(-)_____	(+)_____	(-)_____	(-)_____	(+)_____	_____
_____ ab(d)	(+)_____	(+)_____	(-)_____	(+)_____	(-)_____	_____
_____ c(d)	(-)_____	(-)_____	(+)_____	(+)_____	(-)_____	_____
_____ ac(e)	(+)_____	(-)_____	(+)_____	(-)_____	(+)_____	_____
_____ bc	(-)_____	(+)_____	(+)_____	(-)_____	(-)_____	_____
_____ abc(de)	(+)_____	(+)_____	(+)_____	(+)_____	(+)_____	_____
_____ zero	(0)_____	(0)_____	(0)_____	(0)_____	(0)_____	_____

2-LEVEL FRACTIONAL FACTORIAL DESIGN TEMPLATES
For 6 factors in 8 runs:

tc	A	B	C	D	E	F	ABC
(def)	-	-	-	+	+	+	-
a(f)	+	-	-	-	-	+	+
b(e)	-	+	-	-	+	-	+
ab(d)	+	+	-	+	-	-	-
c(d)	-	-	+	+	-	-	+
ac(e)	+	-	+	-	+	-	-
bc(f)	-	+	+	-	-	+	-
abc(def)	+	+	+	+	+	+	+

Resolution III

Generators: D≈AB E≈AC F≈BC

Defining Contrast:

1≈ABD,ACE,BCF

BCDE,ACDF,ABEF,DEF

Complete confounding pattern:

A≈BD,CE,ABCF,ABCDE,CDF,BEF,ADEF

B≈AD,ABCE,CF,CDE,ABCDF,AEF,BDEF

C≈ABCD,AE,BF,BDE,ADF,ABCEF,CDEF

D≈AB,ACDE,BCDF,BCE,ACF,ABDEF,EF

E≈ABDE,AC,BCEF,BCD,ACDEF,ABF,DF

F≈ABDF,ACEF,BC,BCDEF,ACD,ABE,DE

Also measures:	BD	AD	AE	AB	AC	BC	AF
	CE	CF	BF	EF	DF	DE	BE
							CD

Description: All single effects are confounded with 2-factor interactions. All the 2-factor interactions are confounded with each other. This design configuration is good only if the 2-factor interactions do not exist. *This is a very dangerous design to use without solid prior knowledge!*

> Likely physical interactions:
> A:_____
> B:_____
> C:_____
> D:_____
> E:_____
> F:_____

TEMPLATE: **T8-6** (6 factors in 8 runs)
FACTOR NAMES

Random Run Order	tc	A:_____	B:_____	C:_____	D:_____	E:_____	F:_____	Response (Y)
_____	(def)	(-)_____	(-)_____	(-)_____	(+)_____	(+)_____	(+)_____	_____
_____	a(f)	(+)_____	(-)_____	(-)_____	(-)_____	(-)_____	(+)_____	_____
_____	b(e)	(-)_____	(+)_____	(-)_____	(-)_____	(+)_____	(-)_____	_____
_____	ab(d)	(+)_____	(+)_____	(-)_____	(+)_____	(-)_____	(-)_____	_____
_____	c(d)	(-)_____	(-)_____	(+)_____	(+)_____	(-)_____	(-)_____	_____
_____	ac(e)	(+)_____	(-)_____	(+)_____	(-)_____	(+)_____	(-)_____	_____
_____	bc(f)	(-)_____	(+)_____	(+)_____	(-)_____	(-)_____	(+)_____	_____
_____	abc(def)	(+)_____	(+)_____	(+)_____	(+)_____	(+)_____	(+)_____	_____
_____	zero	(0)_____	(0)_____	(0)_____	(0)_____	(0)_____	(0)_____	_____

2-LEVEL FRACTIONAL FACTORIAL DESIGN TEMPLATES

For 7 factors in 8 runs:

Resolution III

Generators: D≈AB E≈AC F≈BC G≈ABC
Defining Contrast:
1≈ABD,ACE,BCF,ABCG
 BCDE,ACDF,CDG,ABEF,BEG,AFG
 DEF,ADEG,BDFG,CEFG,ABCDEFG
Confounding pattern (2-factor int. only):
A≈BD,CE,FG
B≈AD,CF,EG
C≈AE,BF,DG
D≈AB,EF,CG
E≈AC,DF,BG
F≈BC,DE,AG
G≈AF,BE,CD

There are 12 other higher order interactions that are confounded with each of these effects.

tc	A	B	C	D	E	F	G
(def)	-	-	-	+	+	+	-
a(f)	+	-	-	-	-	+	+
b(e)	-	+	-	-	+	-	+
ab(d)	+	+	-	+	-	-	-
c(d)	-	-	+	+	-	-	+
ac(e)	+	-	+	-	+	-	-
bc(f)	-	+	+	-	-	+	-
abc(def)	+	+	+	+	+	+	+

Also measures:
BD	AD	AE	AB	AC	BC	AF
CE	CF	BF	EF	DF	DE	BE
FG	EG	DG	CG	BG	AG	CD

Description: All single effects are confounded with 2-factor interactions. All the 2-factor interactions are confounded with each other. This design configuration is good only if the 2-factor interactions do not exist. *This is an extremely dangerous design to use without solid prior knowledge!* The T16-7 is better.

TEMPLATE: **T8-7** (7 factors in 8 runs)
FACTOR NAMES

> Since this design is so confounded, a follow-up design will be necessary to resolve any interactions.

	A:____	B:____	C:____	D:____	E:____	F:____	G:____	Response (Y)
Random Run Order / tc								
____ (def)	(-)____	(-)____	(-)____	(+)____	(+)____	(+)____	(-)____	_____
____ a(fg)	(+)____	(-)____	(-)____	(-)____	(-)____	(+)____	(+)____	_____
____ b(eg)	(-)____	(+)____	(-)____	(-)____	(+)____	(-)____	(+)____	_____
____ ab(d)	(+)____	(+)____	(-)____	(+)____	(-)____	(-)____	(-)____	_____
____ c(dg)	(-)____	(-)____	(+)____	(+)____	(-)____	(-)____	(+)____	_____
____ ac(e)	(+)____	(-)____	(+)____	(-)____	(+)____	(-)____	(-)____	_____
____ bc(f)	(-)____	(+)____	(+)____	(-)____	(-)____	(+)____	(-)____	_____
____ abc(defg)	(+)____	(+)____	(+)____	(+)____	(+)____	(+)____	(+)____	_____
____ zero	(0)____	(0)____	(0)____	(0)____	(0)____	(0)____	(0)____	_____

Sixteen Treatment Combination
Design Matrix Templates

This set of experimental design templates will be the most useful for a majority of experimental situations. While the previous eight treatment combination templates are useful for a mere handful of factors, the sixteen run experiments turn out to be to be "workhorses" of experimental design. There is a psychological barrier to experiments with more than 20 runs and while you may aspire to attempt a 32tc experiment, you will most likely never complete such an endeavor. Many times, a new problem arises and the large experiment is put on the "back burner" for completion another day. That other day often never happens, and you are left with a half-baked experiment.

The strategy is to never bite off more than you can chew and by using a reasonable 16 tc design from the beginning, you can do an efficient job and *complete* the work. These templates for from 5 to 9 factors fill the needs of most experimenters.

2-LEVEL FRACTIONAL FACTORIAL DESIGN TEMPLATES

For 5 factors in 16 runs:

Resolution V

Generator: E≈ABCD
Defining Contrast:
1≈ABCDE

Likely physical interactions:
A:_____
B:_____
C:_____
D:_____
E:_____

Complete confounding pattern:
A≈BCDE
B≈ACDE
C≈ABDE
D≈ABCE
E≈ABCD

Description: All single effects are confounded with 4-factor interactions. All the 2-factor interactions are confounded with 3-factor interactions. This design configuration is an excellent choice for 5 factors. It is a *perfectly* efficient design for situations requiring single effects and 2-factor interactions, since the required amount of information is equal to 15 df. This design delivers exactly 15df.

tc	A	B	C	D	E	AB	AC	AD	BC	BD	CD	ABC	ABD	ACD	BCD
(e)	-	-	-	-	+	+	+	+	+	+	+	-	-	-	-
a	+	-	-	-	-	-	-	-	+	+	+	+	+	+	-
b	-	+	-	-	-	-	+	+	-	-	+	+	+	-	+
ab(e)	+	+	-	-	+	+	-	-	-	-	+	-	-	+	+
c	-	-	+	-	-	+	-	+	-	+	-	+	-	+	+
ac(e)	+	-	+	-	+	-	+	-	-	+	-	-	+	-	+
bc(e)	-	+	+	-	+	-	-	+	+	-	-	-	+	+	-
abc	+	+	+	-	-	+	+	-	+	-	-	+	-	-	-
d	-	-	-	+	-	+	+	-	+	-	-	+	+	+	+
ad(e)	+	-	-	+	+	-	-	+	+	-	-	+	-	-	+
bd(e)	-	+	-	+	+	-	+	-	-	+	-	+	-	+	-
abd	+	+	-	+	-	+	-	+	-	+	-	-	+	-	-
cd(e)	-	-	+	+	+	+	-	-	-	-	+	+	+	-	-
acd	+	-	+	+	-	-	+	+	-	-	+	-	-	+	-
bcd	-	+	+	+	-	-	-	-	+	+	+	-	-	-	+
abcd(e)	+	+	+	+	+	+	+	+	+	+	+	+	+	+	+
zero	0	0	0	0	0	0	0	0	0	0	0	0	0	0	0

Also measures: BCDE ACDE ABDE ABCE ABCD CDE BDE BCE ADE ACE ABE DE CE BE AE

TEMPLATE: **T16-5** (5 factors in 16 runs)
FACTOR NAMES

Random Run Order tc	A:____	B:____	C:____	D:____	E:____	Response (Y)
_____ (e)	(-)____	(-)____	(-)____	(-)____	(+)____	_____
_____ a	(+)____	(-)____	(-)____	(-)____	(-)____	_____
_____ b	(-)____	(+)____	(-)____	(-)____	(-)____	_____
_____ ab(e)	(+)____	(+)____	(-)____	(-)____	(+)____	_____
_____ c	(-)____	(-)____	(+)____	(-)____	(-)____	_____
_____ ac(e)	(+)____	(-)____	(+)____	(-)____	(+)____	_____
_____ bc(e)	(-)____	(+)____	(+)____	(-)____	(+)____	_____
_____ abc	(+)____	(+)____	(+)____	(-)____	(-)____	_____
_____ d	(-)____	(-)____	(-)____	(+)____	(-)____	_____
_____ ad(e)	(+)____	(-)____	(-)____	(+)____	(+)____	_____
_____ bd(e)	(-)____	(+)____	(-)____	(+)____	(+)____	_____
_____ abd	(+)____	(+)____	(-)____	(+)____	(-)____	_____
_____ cd(e)	(-)____	(-)____	(+)____	(+)____	(+)____	_____
_____ acd	(+)____	(-)____	(+)____	(+)____	(-)____	_____
_____ bcd	(-)____	(+)____	(+)____	(+)____	(-)____	_____
_____ abcd(e)	(+)____	(+)____	(+)____	(+)____	(+)____	_____
_____ zero	(0)____	(0)____	(0)____	(0)____	(0)____	_____

2-LEVEL FRACTIONAL FACTORIAL DESIGN TEMPLATES

For 6 factors in 16 runs:

Resolution IV
Generators: E≈ABC F≈BCD
Defining Contrast:
1≈ABCE,BCDF,ADEF

Complete confounding pattern:

A≈BCE,ABCDF,DEF	AB≈CE
B≈ACE,CDF,ABDEF	AC≈BE
C≈ABE,BDF,ACDEF	AD≈EF
D≈ABCDE,BCF,AEF	AE≈DF,BC
E≈ABC,BCDEF,ADF	AF≈DE
F≈ABCEF,BCD,ADE	BD≈CF
	BF≈CD

Likely physical interactions:
A:_____
B:_____
C:_____
D:_____
E:_____
F:_____

Description: All single effects are confounded with 3-factor interactions. All the 2-factor interactions are confounded with each other. This design configuration is good when prior knowledge of the likelihood of 2-factor interactions exists. **Helpful strategy:** *If one factor is not expected to interact at all, then place it in the A column, thus allowing the CE, BE, EF, and DE interactions to be free of confounding. If the DF interaction is not expected, then the BC interaction is free of confounding also.*

tc	A	B	C	D	ABCD	AB	AC	AD	BC	BD	CD	E	ABD	ACD	F
(1)	-	-	-	-	+	+	+	+	+	+	+	-	-	-	-
a(e)	+	-	-	-	-	-	-	-	+	+	+	+	+	+	-
b(ef)	-	+	-	-	-	-	+	+	-	-	+	+	+	-	+
ab(f)	+	+	-	-	+	+	-	-	-	-	+	-	-	+	+
c(ef)	-	-	+	-	-	+	-	+	-	+	-	+	-	+	+
ac(f)	+	-	+	-	+	-	+	-	-	+	-	-	+	-	+
bc	-	+	+	-	+	-	-	+	+	-	-	-	+	+	-
abc(e)	+	+	+	-	-	+	+	-	+	-	-	+	-	-	-
d(f)	-	-	-	+	-	+	+	-	+	-	-	-	+	+	+
ad(ef)	+	-	-	+	+	-	-	+	+	-	-	+	-	-	+
bd(e)	-	+	-	+	+	-	+	-	-	+	-	+	-	+	-
abd	+	+	-	+	-	+	-	+	-	+	-	-	+	-	-
cd(e)	-	-	+	+	+	+	-	-	-	-	+	+	+	-	-
acd	+	-	+	+	-	-	+	+	-	-	+	-	-	+	-
bcd(f)	-	+	+	+	-	-	-	-	+	+	+	-	-	-	+
abcd(ef)	+	+	+	+	+	+	+	+	+	+	+	+	+	+	+
zero	0	0	0	0	0	0	0	0	0	0	0	0	0	0	0
Also measures:					DE	CE	BE	EF	AE	CF	BF				
					AF				DF						

TEMPLATE: **T16-6** (6 factors in 16 runs)
FACTOR NAMES

Random Run Order tc		A:_____	B:_____	C:_____	D:_____	E:_____	F:_____	Response (Y)
_____	(1)	(-)_____	(-)_____	(-)_____	(-)_____	(-)_____	(-)_____	_____
_____	a(e)	(+)_____	(-)_____	(-)_____	(-)_____	(+)_____	(-)_____	_____
_____	b(ef)	(-)_____	(+)_____	(-)_____	(-)_____	(+)_____	(+)_____	_____
_____	ab(f)	(+)_____	(+)_____	(-)_____	(-)_____	(-)_____	(+)_____	_____
_____	c(ef)	(-)_____	(-)_____	(+)_____	(-)_____	(+)_____	(+)_____	_____
_____	ac(f)	(+)_____	(-)_____	(+)_____	(-)_____	(-)_____	(+)_____	_____
_____	bc	(-)_____	(+)_____	(+)_____	(-)_____	(-)_____	(-)_____	_____
_____	abc(e)	(+)_____	(+)_____	(+)_____	(-)_____	(+)_____	(-)_____	_____
_____	d(f)	(-)_____	(-)_____	(-)_____	(+)_____	(-)_____	(+)_____	_____
_____	ad(ef)	(+)_____	(-)_____	(-)_____	(+)_____	(+)_____	(+)_____	_____
_____	bd(e)	(-)_____	(+)_____	(-)_____	(+)_____	(+)_____	(-)_____	_____
_____	abd	(+)_____	(+)_____	(-)_____	(+)_____	(-)_____	(-)_____	_____
_____	cd(e)	(-)_____	(-)_____	(+)_____	(+)_____	(+)_____	(-)_____	_____
_____	acd	(+)_____	(-)_____	(+)_____	(+)_____	(-)_____	(-)_____	_____
_____	bcd(f)	(-)_____	(+)_____	(+)_____	(+)_____	(-)_____	(+)_____	_____
_____	abcd(ef)	(+)_____	(+)_____	(+)_____	(+)_____	(+)_____	(+)_____	_____
_____	zero	(0)_____	(0)_____	(0)_____	(0)_____	(0)_____	(0)_____	_____

2-LEVEL FRACTIONAL FACTORIAL DESIGN TEMPLATES

For 7 factors in 16 runs:

Resolution IV

Generators: E≈ABC F≈BCD G≈ACD

Defining Contrast:

1≈ABCE,BCDF,ACDG
   ADEF,BDEG,ABFG,CEFG

Likely physical interactions:
A:_____
B:_____
C:_____
D:_____
E:_____
F:_____
G:_____

Complete confounding pattern:

A≈BCE,ABCDF,CDG,DEF,ABDEG,BFG,ACEFG

B≈ACE,CDF,ABCDG,ABDEF,DEG,AFG,BCEFG

C≈ABE,BDF,ADG,ACDEF,BCDEG,ABCFG,EFG

D≈ABCDE,BCF,ACG,AEF,BEG,ABDFG,CDEFG

E≈ABC,BCDEF,ACDEG,ADF,BDG,ABEFG,CEF

F≈ABCEF,BCD,ACDFG,ADE,BDEFG,ABG,CEG

G≈ABCEG,BCDFG,ACD,ADEFG,BDE,ABF,CEF

AB≈CE,FG

AC≈BE,DG

AD≈EF,CG

AE≈DF,BC

AF≈DE,BG

AG≈BF,CD

BD≈CF,EG

Description: All single effects are confounded with 3-factor interactions. All the 2-factor interactions are confounded with each other. This design configuration is good when prior knowledge of the likelihood of 2-factor interactions exists. **Helpful strategy:** *Place factors that are not expected to interact at all in the A and B columns, thus allowing the DG, DF, DE, and CD interactions to be free of confounding. This still leaves the CE, FG; the EF,CG; and the CF,EG interactions confounded in pairs. A good amount of prior knowledge is required to use this design.*

tc	A	B	C	D	ABCD	AB	AC	AD	BC	BD	CD	E	ABD	G	F
(1)	-	-	-	-	+	+	+	+	+	+	+	-	-	-	-
a(eg)	+	-	-	-	-	-	-	-	+	+	+	+	+	+	-
b(ef)	-	+	-	-	-	-	+	+	-	-	+	+	+	-	+
ab(fg)	+	+	-	-	+	+	-	-	-	-	+	-	-	+	+
c(efg)	-	-	+	-	-	+	-	+	-	+	-	+	-	+	+
ac(f)	+	-	+	-	+	-	+	-	-	+	-	-	+	-	+
bc(g)	-	+	+	-	+	-	-	+	+	-	-	-	+	+	-
abc(e)	+	+	+	-	-	+	+	-	+	-	-	+	-	-	-
d(fg)	-	-	-	+	-	+	+	-	+	-	-	-	+	+	+
ad(ef)	+	-	-	+	+	-	-	+	+	-	-	+	-	-	+
bd(eg)	-	+	-	+	+	-	+	-	-	+	-	+	-	+	-
abd	+	+	-	+	-	+	-	+	-	+	-	-	+	-	-
cd(e)	-	-	+	+	+	+	-	-	-	-	+	+	+	-	-
acd(g)	+	-	+	+	-	-	+	+	-	-	+	-	-	+	-
bcd(f)	-	+	+	+	-	-	-	-	+	+	+	-	-	-	+
abcd(efg)	+	+	+	+	+	+	+	+	+	+	+	+	+	+	+
zero	0	0	0	0	0	0	0	0	0	0	0	0	0	0	0
Also measures:					DE	CE	BE	EF	AE	CF	BF				
					AF	FG	DG	CG	DF	EG	AG				
					BG										

TEMPLATE: **T16-7** (7 factors in 16 runs)
FACTOR NAMES

Random Run Order	tc	A:___	B:___	C:___	D:___	E:___	F:___	G:___	Response (Y)
_____	(1)	(-)___	(-)___	(-)___	(-)___	(-)___	(-)___	(-)___	_____
_____	a(eg)	(+)___	(-)___	(-)___	(-)___	(+)___	(-)___	(+)___	_____
_____	b(ef)	(-)___	(+)___	(-)___	(-)___	(+)___	(+)___	(-)___	_____
_____	ab(fg)	(+)___	(+)___	(-)___	(-)___	(-)___	(+)___	(+)___	_____
_____	c(efg)	(-)___	(-)___	(+)___	(-)___	(+)___	(+)___	(+)___	_____
_____	ac(f)	(+)___	(-)___	(+)___	(-)___	(-)___	(+)___	(-)___	_____
_____	bc(g)	(-)___	(+)___	(+)___	(-)___	(-)___	(-)___	(+)___	_____
_____	abc(e)	(+)___	(+)___	(+)___	(-)___	(+)___	(-)___	(-)___	_____
_____	d(fg)	(-)___	(-)___	(-)___	(+)___	(-)___	(+)___	(+)___	_____
_____	ad(ef)	(+)___	(-)___	(-)___	(+)___	(+)___	(+)___	(-)___	_____
_____	bd(eg)	(-)___	(+)___	(-)___	(+)___	(+)___	(-)___	(+)___	_____
_____	abd	(+)___	(+)___	(-)___	(+)___	(-)___	(-)___	(-)___	_____
_____	cd(e)	(-)___	(-)___	(+)___	(+)___	(+)___	(-)___	(-)___	_____
_____	acd(g)	(+)___	(-)___	(+)___	(+)___	(+)___	(-)___	(+)___	_____
_____	bcd(f)	(-)___	(+)___	(+)___	(+)___	(-)___	(+)___	(-)___	_____
_____	abcd(efg)	(+)___	(+)___	(+)___	(+)___	(+)___	(+)___	(+)___	_____
_____	zero	(0)___	(0)___	(0)___	(0)___	(0)___	(0)___	(0)___	_____

2-LEVEL FRACTIONAL FACTORIAL DESIGN TEMPLATES

For 8 factors in 16 runs:

Resolution IV

Generators: E≈BCD F≈ACD G≈ABC H≈ABD

Defining Contrast:

   1≈BCDE,ACDF,ABCG,ABDH
      ABEF,ADEG,ACEH,BDFG,BCFH,CDGH
      CEFG,DEFH,BEGH,AFGH,ABCDEFGH

Two-factor interaction confounding pattern:

   AB≈CG,DH,EF
   AC≈DF,BG,EH
   AD≈CF,BH,EG
   AE≈BF,DG,CH
   AF≈CD,BE,GH
   AG≈BC,DE,FH
   AH≈BD,CE,FG

Likely physical interactions:
A:_____
B:_____
C:_____
D:_____
E:_____
F:_____
G:_____
H:_____

Description: All single effects are confounded with 3-factor interactions. All the 2-factor interactions are confounded with each other. This design configuration is good when prior knowledge of the likelihood of 2-factor interactions exists and/ or when further experimentation is planned.

tc	A	B	C	D	ABCD	AB	AC	AD	BC	BD	CD	G	H	F	E
(1)	-	-	-	-	+	+	+	+	+	+	+	-	-	-	-
a(fgh)	+	-	-	-	-	-	-	-	+	+	+	+	+	+	-
b(egh)	-	+	-	-	-	-	+	+	-	-	+	+	+	-	+
ab(ef)	+	+	-	-	+	+	-	-	-	-	+	-	-	+	+
c(efg)	-	-	+	-	-	+	-	+	-	+	-	+	-	+	+
ac(eh)	+	-	+	-	+	-	+	-	-	+	-	-	+	-	+
bc(fh)	-	+	+	-	+	-	-	+	+	-	-	-	+	+	-
abc(g)	+	+	+	-	-	+	+	-	+	-	-	+	-	-	-
d(efh)	-	-	-	+	-	+	+	-	+	-	-	-	+	+	+
ad(eg)	+	-	-	+	+	-	-	+	+	-	-	+	-	-	+
bd(fg)	-	+	-	+	+	-	+	-	-	+	-	+	-	+	-
abd(h)	+	+	-	+	-	+	-	+	-	+	-	-	+	-	-
cd(gh)	-	-	+	+	+	+	-	-	-	-	+	+	+	-	-
acd(f)	+	-	+	+	-	-	+	+	-	-	+	-	-	+	-
bcd(e)	-	+	+	+	-	-	-	-	+	+	+	-	-	-	+
abcd(efgh)	+	+	+	+	+	+	+	+	+	+	+	+	+	+	+
zero	0	0	0	0	0	0	0	0	0	0	0	0	0	0	0
Also measures:					AE	CG	DF	CF	AG	AH	AF				
					BF	DH	BG	BH	DE	CE	BE				
					DG	EF	EH	EG	FH	FG	GH				
					CH										

TEMPLATE: **T16-8** (8 factors in 16 runs)
FACTOR NAMES

Random Run Order	tc	A:____	B:____	C:____	D:____	E:____	F:____	G:____	H:____	Response (Y)
____	(1)	(-)____	(-)____	(-)____	(-)____	(-)____	(-)____	(-)____	(-)____	_____
____	a(fgh)	(+)____	(-)____	(-)____	(-)____	(-)____	(+)____	(+)____	(+)____	_____
____	b(egh)	(-)____	(+)____	(-)____	(-)____	(+)____	(-)____	(+)____	(+)____	_____
____	ab(ef)	(+)____	(+)____	(-)____	(-)____	(+)____	(+)____	(-)____	(-)____	_____
____	c(efg)	(-)____	(-)____	(+)____	(-)____	(+)____	(+)____	(+)____	(-)____	_____
____	ac(eh)	(+)____	(-)____	(+)____	(-)____	(+)____	(-)____	(-)____	(+)____	_____
____	bc(fh)	(-)____	(+)____	(+)____	(-)____	(-)____	(+)____	(-)____	(+)____	_____
____	abc(g)	(+)____	(+)____	(+)____	(-)____	(-)____	(-)____	(+)____	(-)____	_____
____	d(efh)	(-)____	(-)____	(-)____	(+)____	(+)____	(+)____	(-)____	(+)____	_____
____	ad(eg)	(+)____	(-)____	(-)____	(+)____	(+)____	(-)____	(+)____	(-)____	_____
____	bd(fg)	(-)____	(+)____	(-)____	(+)____	(-)____	(+)____	(+)____	(-)____	_____
____	abd(h)	(+)____	(+)____	(-)____	(+)____	(-)____	(-)____	(-)____	(+)____	_____
____	cd(gh)	(-)____	(-)____	(+)____	(+)____	(-)____	(-)____	(+)____	(+)____	_____
____	acd(f)	(+)____	(-)____	(+)____	(+)____	(-)____	(+)____	(-)____	(-)____	_____
____	bcd(e)	(-)____	(+)____	(+)____	(+)____	(+)____	(-)____	(-)____	(-)____	_____
____	abcd(efgh)	(+)____	(+)____	(+)____	(+)____	(+)____	(+)____	(+)____	(+)____	_____
____	zero	(0)____	(0)____	(0)____	(0)____	(0)____	(0)____	(0)____	(0)____	_____

2-LEVEL FRACTIONAL FACTORIAL DESIGN TEMPLATES

For 9 factors in 16 runs:

Resolution III

Generators: E≈ABC F≈BCD G≈ACD
H≈ABD J≈ABCD

Defining Contrast:

1≈ABCE,BCDF,ACDG,ABDH,ABCDJ
ADEF,BDEG,CDEH,DEJ,ABFG,ACFH
AFJ,BCGH,BGJ,CHJ,CEFG,BEFH,BCEFJ
AEGH,ACEGJ,ABEHJ,DFGH,CDFGJ,BDFHJ
ADGHJ,ABCDEFGH,ABDEFGJ,ACDEFHJ,BCDEGHJ,ABCFGHJ,EFGHJ

Likely physical interactions:
A:_____
B:_____
C:_____
D:_____
E:_____
F:_____
G:_____
H:_____
J:_____

Two factor interaction confounding pattern:

AB≈CE,DH,FG	AJ≈**F**
AC≈BE,DG,FH	BJ≈**G**
AD≈CG,BH,EF	CJ≈**H**
AE≈BC,DF,GH	DJ≈**E**
J≈AF,DE,BG,CH	EJ≈**D**
AG≈CD,BF,EH	FJ≈**A**
AH≈BD,CF,EG	GJ≈**B**
	HJ≈**C**

tc	A	B	C	D	J	AB	AC	AD	BC	BD	CD	E	H	G	F
(j)	-	-	-	-	+	+	+	+	+	+	+	-	-	-	-
a(egh)	+	-	-	-	-	-	-	-	+	+	+	+	+	+	-
b(efh)	-	+	-	-	-	-	+	+	-	-	+	+	+	-	+
ab(fgj)	+	+	-	-	+	+	-	-	-	-	+	-	+	+	+
c(efg)	-	-	+	-	-	+	-	+	-	+	-	+	+	+	+
ac(fhj)	+	-	+	-	+	-	+	-	-	+	-	-	+	-	+
bc(ghj)	-	+	+	-	+	-	-	+	+	-	-	-	+	+	-
abc(e)	+	+	+	-	-	+	+	-	+	-	-	+	-	-	-
d(fgh)	-	-	-	+	-	+	+	-	+	-	-	-	+	+	+
ad(efj)	+	-	-	+	+	-	-	+	+	-	-	+	-	-	+
bd(egj)	-	+	-	+	+	-	+	-	-	+	-	+	-	+	-
abd(h)	+	+	-	+	-	+	-	+	-	+	-	-	+	-	-
cd(ehj)	-	-	+	+	+	+	-	-	-	-	+	+	+	-	-
acd(g)	+	-	+	+	-	-	+	+	-	-	+	-	-	+	-
bcd(f)	-	+	+	+	-	-	-	-	+	+	+	-	-	-	+
abcd(efghj)	+	+	+	+	+	+	+	+	+	+	+	+	+	+	+
zero	0	0	0	0	0	0	0	0	0	0	0	0	0	0	0
Also measures:	FJ	GJ	HJ	EJ	AF	CE	BE	CG	AE	CF	BF	DJ	CJ	BJ	AJ
					DE	DH	DG	BH	DF	EG	AG				
					BG	FG	FH	EF	GH	AH	EH				
					CH										

Description: All single effects are confounded with one 2-factor interaction. All the 2-factor interactions are confounded with each other. This design configuration is good when **solid** prior knowledge of the likelihood of 2-factor interactions exists and/or when further experimentation is planned. **Helpful strategy:** *Let J be the not-likely-to-interact factor to prevent confounding among the single effects.*

TEMPLATE: **T16-9** (9 factors in 16 runs)
FACTOR NAMES

Random Run Order	tc	A:___	B:___	C:___	D:___	E:___	F:___	G:___	H:___	J:___	Response (Y)
_____	(j)	(-)___	(-)___	(-)___	(-)___	(-)___	(-)___	(-)___	(-)___	(+)___	_____
_____	a(egh)	(+)___	(-)___	(-)___	(-)___	(+)___	(-)___	(+)___	(+)___	(-)___	_____
_____	b(efh)	(-)___	(+)___	(-)___	(-)___	(+)___	(+)___	(-)___	(+)___	(-)___	_____
_____	ab(fgj)	(+)___	(+)___	(-)___	(-)___	(-)___	(+)___	(+)___	(-)___	(+)___	_____
_____	c(efg)	(-)___	(-)___	(+)___	(-)___	(+)___	(+)___	(+)___	(-)___	(-)___	_____
_____	ac(fhj)	(+)___	(-)___	(+)___	(-)___	(-)___	(+)___	(-)___	(+)___	(+)___	_____
_____	bc(ghj)	(-)___	(+)___	(+)___	(-)___	(-)___	(+)___	(+)___	(+)___	(+)___	_____
_____	abc(e)	(+)___	(+)___	(+)___	(-)___	(+)___	(-)___	(-)___	(-)___	(-)___	_____
_____	d(fgh)	(-)___	(-)___	(-)___	(+)___	(-)___	(+)___	(+)___	(+)___	(-)___	_____
_____	ad(efj)	(+)___	(-)___	(-)___	(+)___	(+)___	(+)___	(-)___	(-)___	(+)___	_____
_____	bd(egj)	(-)___	(+)___	(-)___	(+)___	(+)___	(-)___	(+)___	(-)___	(+)___	_____
_____	abd(h)	(+)___	(+)___	(-)___	(+)___	(-)___	(-)___	(-)___	(+)___	(-)___	_____
_____	cd(ehj)	(-)___	(-)___	(+)___	(+)___	(+)___	(-)___	(-)___	(+)___	(+)___	_____
_____	acd(g)	(+)___	(-)___	(+)___	(+)___	(-)___	(-)___	(+)___	(-)___	(-)___	_____
_____	bcd(f)	(-)___	(+)___	(+)___	(+)___	(-)___	(+)___	(-)___	(-)___	(-)___	_____
_____	abcd(efghj)	(+)___	(+)___	(+)___	(+)___	(+)___	(+)___	(+)___	(+)___	(+)___	_____
_____	zero	(0)___	(0)___	(0)___	(0)___	(0)___	(0)___	(0)___	(0)___	(0)___	_____

REFERENCES

1. National Bureau of Standards (1957). **Fractional Factorial Experiments Designs for Factors at Two Levels**. U.S. Department of Commerce.

2. Davies, O.L. (1956). **Design and Analysis of Industrial Experiments, 2nd ed.** Hafner Publishing Co., New York.

3. Box, G. E. P., W. G. Hunter, and J. S. Hunter (1978). **Statistics for Experimenters.** Wiley, New York.

4. Barker, T. B. (1993). **QED Programs for MS BASIC**. Rochester Institute of Technology, Rochester, New York.

5. Plackett, R. L., and Burman, J. P. (1946). The Design of Optimum Multifactorial Experiments. **Biometrika**, 33, 305-325.

6. Draper, N. R., and Stoneman, D. M. (1966). Alias Relationships for Two-Level Plackett and Burman Designs. **Technical Report No. 96**. Department of Statistics, The University of Wisconsin, Madison, Wisconsin.

7. Box, G. E. P. (1959). Comments on the Random Balance Design. **Technometrics**, v.1, n.2, May 1959.

8. Box, G. E. P., and Hunter, J. S. (1961). The 2^{k-p} Fractional Factorial Designs, **Technometrics**, 3, 311, 449.

6

Multi-Level Designs

While the factorial and fractional factorial designs that we have explored in the past chapters are considered the "workhorses" of modern empirical investigations, there is still a lingering question: Are these designs efficient? Remember that, to be efficient, an experiment must get the required information at the least expenditure of resources. The factorials certainly get the required information in the form of single effects and the interactions *and* do so at the least expenditure of resources, but what if there is a *nonlinear* relationship between the response we are measuring and the factors we are changing in the design? In that case we could be misled. A classic example is shown in Figure 6-1. Here the levels of the experiment have been set at two extremes, and a straight line between them shows that the slope is zero.

However, the solid line is the form of the real underlying relationship, which is quadratic in nature with the maximum midway between the two

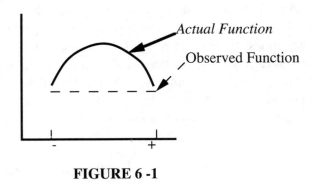

FIGURE 6 -1

levels set in the experiment. Here indeed, we did not get the required information and in fact by using the two-level design probably discarded the factor under investigation since its effect (over 2 levels) was zero!

Does this condemn the two-level type design to oblivion? No, of course not. This is where prior knowledge of the process comes into play. Remember that this knowledge is the main ingredient when any experiment is undertaken. If the factor we were looking at was temperature and the process was a photographic developer investigation, then we should have known from our prior experience that there is a peak temperature level where a response such as quality will be optimized. Therefore, if we do not see a relationship as a result of the two-level experiment, we should not throw away prior information but follow on with more extensive investigations. This is one reason for the "25% rule," which states that for the first part of your experimental investigations, you should expend only 25% of the total resources available. This assures that there is enough time and money to complete the work. Remember that a good experiment will probably pose more questions that will lead to more experiments that will of course eventually lead to the final answer. Experimentation is a process, not an answer.

One last word of caution before we get into the methods of designing efficient experiments that look at more than two levels: Sometimes an experienced worker will be so convinced of an outcome, he or she will keep testing until *by chance* the results match that preconceived idea. This kind of narrow thinking will only impede progress. We should be ready and willing to modify our convictions in the light of evidence. Without such a philosophy, we will not advance.

The usual, intuitive approach to investigate a curved relationship is to design a 3-level experiment. Such a design is able to investigate up to a quadratic polynomial relationship (i.e., a curve as shown in Figure 6-1). In general, we may determine a polynomial up to one *order* less than the number of levels in the factor. So, in the case of a two-level design, we may investigate first-order or linear relationships. For a three-level design, we may make inferences about a second-order or quadratic relationship. Now in this case, "second order" does not mean "not important" as some researchers imply, but describes the power of the exponent applied to the factor in a polynomial equation. Mathematically this means that in a simple linear relationship with a straight line fit, the equation is:

$$Y = b * X_1 \qquad \text{where b is the slope} \qquad (6\text{-}1)$$
$$X_1 \text{ is the control factor}$$
$$Y \text{ is the response}$$

With a second order, the equation is:
$$Y = b_1 * X_1 + b_{11} * X_1^2 \qquad\qquad (6\text{-}2)$$

Notice that the lower-order term (linear) is retained in the second-order polynomial. This inclusion is dictated by the polynomial expansion rules for a "proper polynomial." Also, notice the subscript on the second "slope" (b_{11}). This is not read as eleven, but as "one, one." It signifies the fact that X is multiplied by itself (i.e., X^2 is equivalent to $X*X$). If, in fact, Equation 6-2 is true, then the second-order nature of this relationship is very important and should not be shrugged off as "only a second-order effect." It is indeed the mathematical form of the relationship found in Figure 6-1. So certain second-order and above effects can be important, and there must be ways of investigating these relationships.

A simple three-level design for two factors and three levels is designated by 3^2, where the "3" refers to the number of levels and the "2" is the number of factors. This same notation holds for bigger experiments also. A 5^2 would have five levels and 2 factors under investigation. To find the number of treatment combinations, we simply solve the power relationship. So, a 3^2 design would have 9 treatment combinations. A 5^2 would have 25 treatment combinations.

Now what do we get for all this experimental work? We measure the information in an experiment by degrees of freedom as described in the last chapter. So for a 3^2 design, we have 9 treatment combinations from which we lose 1 degree of freedom (df) due to finding the grand mean of the data (this corresponds to the intercept term in regression). This leaves us with 8 df to estimate the effects in the process we are investigating.

The allocation of these 8 df are as follows:

TABLE 6-1

Measurement of	df Used Up	Math Form
Two single effects (linear)	2	X_1, X_2
Two quadratic effects	2	X_1^2, X_2^2
Linear interaction	1	$X_1 * X_2$
• •		
Linear/quadratic interactions	2	$X_1 * X_2^2 ; X_1^2 * X_2$
Quadratic interaction	1	$X_1^2 * X_2^2$
Total	8	

A dotted line is drawn under the linear interaction, since the quadratic forms of the interactions do not usually happen. Therefore, we have 3 degrees of freedom that will estimate effects that are equal to zero. This is not a big problem and does not violate our quest for efficiency. However, in the next-sized design (a 3^3 Table 6-2) with three factors, we see less payback for the experimental resources spent.

TABLE 6-2

Measurement of	df Used Up	Math Form
Three main effects (linear)	3	X_1, X_2, X_3
Three quadratic effects	3	X_1^2, X_2^2, X_3^2
Linear interactions	4	$X_1 * X_2, X_1 * X_3$ $X_2 * X_3, X_1 * X_2 * X_3$
Linear/quadratic interactions and quadratic interactions	16	complex!

Of the 26 df in this experiment, 16 tc's or 16 df of the expensive information will be squandered on the so-called higher-order effects that inefficient experiments will seek to discover and only waste resources doing so.

So how can we produce efficient experiments on multi-level problems with more than three factors? This is a most likely area of investigation. To answer that question, let us perform a sleight-of-hand trick with the 3^2 experiment we examined just a moment ago. The geometric form of this experiment is:

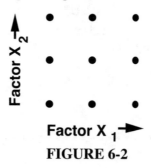

FIGURE 6-2

Now if we twist the design in Figure 6-2 45° while leaving the factor axis alone, the pattern becomes that which is shown in Figure 6-3:

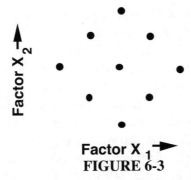

Factor X$_1$ ➤
FIGURE 6-3

To take the design out of the abstract and into the real world, we add names to the factors and put levels along the axis. At the same time, we should recognize a familiar geometric pattern in the middle of all this. It is the 2^k (in this case 2^2) factorial!

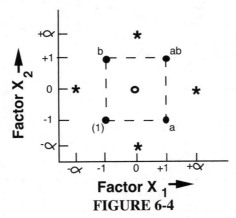

Factor X$_1$ ➤
FIGURE 6-4

The Central Composite Design (CCD)

This design is made up of a 2^k design that allow us to investigate the required interactions, plus a center point (at the zero, zero position) and two extreme points for each factor, which are shown in Figure 6-4 as asterisks. Counting up the number of levels in the new design we see that instead of the original three levels, we now have five levels for each factor. This will allow curvilinear investigations up to fourth-order or a quartic relationship. While quartics are not too common, the real value of the design is the increased scope of the experiment. We have expanded the "volume" of the design by expanding the levels over a wider inference space. So now if we see a curvilinear relationship (second order), we have not used up all the degrees of freedom in fitting this equation that is necessary in the three-level design. Now we can't get something for nothing. What is the price in this design? After all, we have 5 levels for 2 factors. In the usual factorial design (which will be balanced), there are 25 treatment combinations (5^2). The above design has only 9! If we compare the information in the 5^2 from a "degrees-of-freedom" view to our reduced version, we see that *all* is not lost. We actually only lose the non-essential information that we would have thrown away anyway. Table 6-3 shows that only 7 df are needed to capture the required information from a 5^2 design. This amount of information is more closely matched by the design shown in Figure 6-4, which produces 8 df worth of understanding.

TABLE 6-3
Analysis of 5^2 Information Content

Measurement of	df Used Up	Math Form
Two linear single effects	2	X_1, X_2
Two quadratic single effects	2	X_1^2, X_2^2
One linear interaction	1	X_1*X_2
Two cubic single effects	2	X_1^3, X_2^3
Two quartic single effects	2	X_1^4, X_2^4
• •		
Various quadratic & cubic interactions that are unlikely	15	complex!

The design shown in Figure 6-4 is called a **Central Composite Design** (CCD). It is made up of a 2-level factorial or fractional factorial, a center point (thus *central*), and a one-factor-at-a-time! Now of course, we might be suspicious of the inclusion of a 1-F.A.A.T. in any of our experiments since it is unable to detect interactions. However, in the CCD, the factorial portion gets the interaction information and the 1-F.A.A.T. expands the inference space to allow the detection of curvilinear effects. The CCD obtains all the information above the dotted line in Table 6-3. Once again, we have achieved efficiency, since we have obtained the required information and have done so at the least expenditure of resources. The CCD captures *exactly* the required information and does so with far fewer runs than the fully crossed factorial would require. This particular CCD does the work of investigating two factors at five levels with about 1/3 the effort. Central composite designs are easy to construct since they consist of 2^{k-p} factorial designs and the one-factor designs to round out the levels plus a center point. The general designation is:

$$
\text{\# of tc's} = 2^{k-p} + (2*k) + \text{\# of center points}
$$
$$
\text{where } k = \text{\# of factors}
$$
$$
p = \text{the fractionalization element}
$$

Table 6-4 gives the number of treatment combinations and the distance from center to the "star" positions for various values of k and p. These distances are different for each design and will allow the design to be orthogonal (or independent) in design units and also assures that the error in prediction remains constant (rotatable) around the design. These "alpha star points" will be covered in more detail in Chapter 16 when we learn to apply the appropriate analysis for the CCD design.

TABLE 6-4

Some Orthogonal Designs for Three to Eight Factors

Number of factors:	2	3	4	5	5	6	6	7	8
Size of factorial:	full	full	full	full	1/2	full	1/2	1/2	1/4
# of points in 2^{k-p}:	4	8	16	32	16	64	32	64	64
# of star points:	4	6	8	10	10	12	12	14	16
Length of star points from center: (alpha α)	1.414	1.682	2.0	2.378	2.0	2.828	2.378	2.828	2.282

While each of these designs can get by with a single center point, we will see in the analysis that the center point plays a pivotal role in deriving our equations. Therefore to assure the integrity of the data from this treatment (the zero point), it is prudent to replicate it between 3 and 5 times. This replication also provides a measure experimental error. We may consider the 3 to 5 replicates as cheap insurance against experimental hazards.

Table 6-4 will probably be sufficient to satisfy most of the design requirements encountered in normal experimental situations. If it is necessary to build larger designs or work with different fractions than the ones shown, the length of the distance from the center of the design to the star points may be calculated according to Expression 6-3.

$$\text{alpha star} = \sqrt[4]{2^{(k-p)}} \quad \text{or alpha star} = 2^{(k-p)/4} \qquad (6-3)$$

To use these alpha distances, we simply extend the design from the two- level factorial's -1 and +1 distances in proportion to the size of the alpha star lengths. So if we decide that three factors will be studied in this experiment, the design in *design* units will be as follows:

TABLE 6-5

tc	A	B	C
(1)	-	-	-
a	+	-	-
b	-	+	-
ab	+	+	-
c	-	-	+
ac	+	-	+
bc	-	+	+
abc	+	+	+
$-\alpha_a$	-1.682	0	0
$+\alpha_a$	+ 1.682	0	0
$-\alpha_b$	0	-1.682	0
$+\alpha_b$	0	+ 1.682	0
$-\alpha_c$	0	0	-1.682
$+\alpha_c$	0	0	+ 1.682
zero	0	0	0

Since we usually can't work in design units in a physical world, the design shown in Table 6-5 will have to be converted to physical units. So we'll add some substance to the problem and talk about the factors we are working with.

Consider a study of a polymer process where the yield is measured by taking the ratio of the product produced to the raw materials put into the reactor. The closer this ratio comes to 1, the higher the yield. The three control factors under study are temperature, pressure, and time of reaction. Therefore, the **goal** of this project is to increase yield and the **objective** is to test the hypothesis that yield is a function of temperature, pressure, and time of reaction. Table 6-6 shows the working ranges for the factors under study.

TABLE 6-6

Factor	Range of Interest
Temperature	100 to 200°
Pressure	20 to 50 psi
Time	10 to 30 min

If we pick the extremes of the ranges for the -1 and +1 levels of our factorial portion of the CCD, then there is no room to move to the alpha star points. So in this case, if we wish to utilize the entire range, we will assign the end points of the ranges to the alpha distances. This means that the -1 and + 1 levels of the factorial must be moved in proportionally from the 1.682 distances as prescribed in Table 6-5.

To do this we set up a proportion of the distance from the center point to the star (call this the delta star) to the distance from the center point to the -1 and +1 positions (call this the delta factorial). On a number line it looks like this:

TABLE 6-7

	-1.682	-1	0	+1	+1.682
Temperature	100		150		200
Pressure	20		35		50
Time	10		20		30

The proportion for temperature is:

$$\frac{50 \text{ (delta star)}}{1.682} = \frac{\text{x (delta factorial)}}{1}$$

$$1.682 \text{ x} = 50$$

$$\text{x} = 29.73$$

This 29.73 value is the distance from the center point to the +1 and -1 temperature point settings.

so the low level of temperature is: 150 - 29.73 = 120.27
and the high level of temperature is: 150 + 29.73 = 179.73

The completion of the calculation of the levels for pressure and time is left as an exercise (see Problem 1). Figure 6-5 shows the geometry of this design.

While the exact values of temperature have been computed above, it would be silly to try to set them at the level of precision suggested by the design. It would be better from an engineering operations point of view to set the temperature at 120° and 180°. This slight modification will not upset the operational integrity of the design.

A Note on Choosing Levels

The choice of levels in an experiment is critical to the success, execution, and analysis of the treatment combinations. Should we pick levels so close together that the noise of the system swamps the information we are seeking, then we will get nothing for our efforts. On the other hand, if we spread the levels so wide across the possible range of investigation, we will most certainly get measurable effects, but these effects could be an artifact of the choice of levels and not an indication of the scientific impact of the factor under study. Another problem in choosing levels that are too far apart is the fact that, when in combination with other widely spaced levels of other factors, we may have a treatment combination that utterly fails to produce any product at all! In the case of our polymer process, this usually is termed an experimental treatment that has produced "glop."

The CCD is able to help us choose levels that will be mathematically independent with errors that will not be inflated by the addition of higher-order polynomial terms, but is not able to tell us how far to set these levels apart. This is where prior knowledge of the process (a necessary ingredient for any successful experiment) must be used. For this reason, the CCD should not be used until there is sufficient information to assure us that the levels we pick will not lead us into the dangers outlined above. It is very tempting to apply the CCD on a new problem right at the beginning without the up-front work of fractional factorial screening designs. As we outlined in the first chapter, it is important to build on prior experimental knowledge before putting all our eggs in one experimental basket that may have holes in it.

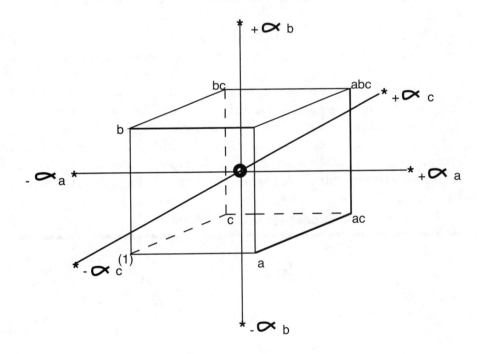

FIGURE 6-5
Geometric diagram of a CCD for 3 Factors

Strategies of Using a CCD

One of the attractive aspects of a CCD is the fact that it may be part of a sequential knowledge-building process. Recall the concept of knowledge building illustrated in Figure 2-2 of Chapter 2 as shown below. Figure 6-7 makes this generic concept more concrete by showing the actual experimental designs utilized to build knowledge. Notice how a fractional factorial with a set of centerpoints is our first attack on the problem or opportunity we encounter.

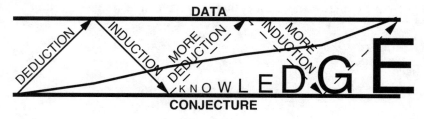

FIGURE 6-6

Generic Concept of Knowledge Building (from Chapter 2)

From the analysis of the data obtained via this preliminary experimental investigation, we are able to ascertain if further information is needed to build the knowledge base necessary to achieve our goal. If there is an indication of possible curvature from the preliminary 2-level experiment (details of such an analysis are covered in Chapter 17), then we will **add** sufficient experimental trials to **augment** the 2-level factorial into a CCD. Let us watch this process in action.

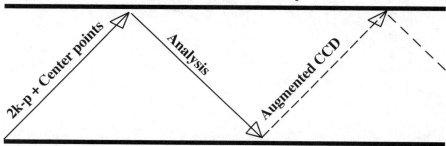

FIGURE 6-7

Sequential Experimentation

We will expand the design from Chapter 5 that investigated the machining process that makes spindles for automobile wheel bearings. Recall the factors, working ranges, levels, and column assignments from this investigation as shown in Table 6-8. The designs we built in either the "manual mode" or with Minitab included center points. While we will expand on the details of the analysis in Chapter 17, let's just simplify the analysis at this point and say that there is an indication of likely curvature.

TABLE 6-8

Factor	Working Range	Levels		Column in Design
Material	N/A	Soft	Hard	A
Speed	150-250 rpm	175	225	B
Feed Left	.2 - .4 ips	.25	.35	C
Feed Right	.3 - .5 ips	.35	.45	D
Clearance	5% - 15%	6%	12%	F
Heat Cycle	4 - 10 sec.	5.5	8.5	E
Heat Power	50% - 100%	75%	85%	G

Fortunately, in building the fractional factorial, we had not utilized the entire working range of the factors in setting the low and high levels in the fractional factorial 2^{7-3}. This gives us "room" to expand the factorial to the alpha points. Also, notice that the distances between the levels of the factors over the entire working range are exactly as prescribed by the design level distances for a 16 tc factorial design! None of these occurrences are merely "fortunate." They are a result of planning and good prior knowledge of the physics involved in the engineering aspects of this process. It is prudent to anticipate the need for a curvilinear design like the CCD when beginning the process of selecting levels for the preliminary fractional factorial design. If we had used the entire working range in selecting the levels for the fractional factorial design, there would be no "room" left if we needed to augment the design to a CCD. If this were the case, then we would need to redesign the entire experiment all over again and include a new factorial that would be inefficient since that new factorial part of the CCD would be redundant with the original design.

So, let's augment the design we built in Chapter 5 (Table 5-14) to build the CCD. There is a slight difficulty with the physics of this process, since there is a classification factor (Material) that does not have continuous levels.

Again, we will use newly gained knowledge from the first segment of our investigation and utilize the level of the material that was determined to be best from the analysis of the fractional factorial. Having the information on this classification factor from the first segment is a very strong reason to do such experimentation sequentially. If we had tried to accomplish the entire investigation in a complete CCD from the beginning, the axial points would need to be run with both the soft and hard material, which would have doubled the number of runs in this part of the design. Table 6-9 shows the axial runs that would be added to the original fractional factorial to complete the CCD.

Notice that there is no change in the "Material" column since we selected the best material from the first part of the investigation. Also note that there are additional center point replicates. If there has been a change between the two segments (fractional factorial and subsequent axial points) of the investigational process, these added centerpoints when compared with the centerpoints of the fractional factorial can identify this change. In Chapter 17, which covers analysis, we will delve deeper into this procedure and suggest remedies for shifts between segments. But for now let's look at the additional trials that expand the automobile wheel bearing investigation into a CCD. The top of Table 6-9 in italics shows the 16 treatment combinations from the original 2^{7-3} fractional factorial in a single spaced display. Of course, here we were looking at all 7 factors. The analysis of this portion of the experiment has shown that the hard material is superior, so we will run all the axial points with the hard material. The analysis also shows that there is an indication of a curvilinear effect. So we add axial points for the quantitative factors to augment the factorial portion and expand it to a CCD. The double spaced portion of Table 6-9 in normal type contains the added treatment combinations for this CCD. These levels represent the extreme low and high values listed in Table 6-8 for each factor. We will add three additional centerpoints made with the hard material. Of course these additional runs will be randomized as indicated to the left of each treatment combination.

Using Minitab to Build a CCD

We will now build the 3 factor CCD based on the chemical reaction described in Table 6-6. This will also serve as an answer to Problem 2. If you have access to Minitab, you should use this example as a practice problem. We will also expand the wheel-bearing example from a fractional factorial to an augmented CCD, although this is not as easy to do in Minitab due to limitations in the choice of fractional factorials in this program.

TABLE 6-9

		A	B	C	D	E	F	G
		Matrl	Speed	Lfed	Rfed	HTcyc	Clear	Htpwr
	(1)	Soft	175	.25	.35	5.5	6%	75%
	a(eg)	Hard	175	.25	.35	8.5	6%	85%
	b(ef)	Soft	225	.25	.35	8.5	12%	75%
	ab(fg)	Hard	225	.25	.35	5.5	12%	85%
	c(efg)	Soft	175	.35	.35	8.5	12%	85%
	ac(f)	Hard	175	.35	.35	5.5	12%	75%
	bc(g)	Soft	225	.35	.35	5.5	6%	85%
	abc(e)	Hard	225	.35	.35	8.5	6%	75%
	d(fg)	Soft	175	.25	.45	5.5	12%	85%
	ad(ef)	Hard	175	.25	.45	8.5	12%	75%
	bd(eg)	Soft	225	.25	.45	8.5	6%	85%
	abd	Hard	225	.25	.45	5.5	6%	75%
	cd(e)	Soft	175	.35	.45	8.5	6%	75%
	acd(g)	Hard	175	.35	.45	8.5	6%	85%
	bcd(f)	Soft	225	.35	.45	5.5	12%	75%
	abcd(efg)	Hard	225	.35	.45	8.5	12%	85%
	zero	Soft	200	.30	.40	7.0	9%	80%
	zero	Hard	200	.30	.40	7.0	9%	80%
6	$-\alpha_B$	Hard	150	.30	.40	7.0	9%	80%
2	$+\alpha_B$	Hard	250	.30	.40	7.0	9%	80%
13	$-\alpha_C$	Hard	200	.20	.40	7.0	9%	80%
15	$+\alpha_C$	Hard	200	.40	.40	7.0	9%	80%
5	$-\alpha_D$	Hard	200	.30	.30	7.0	9%	80%
8	$+\alpha_D$	Hard	200	.30	.50	7.0	9%	80%
3	$-\alpha_E$	Hard	200	.30	.40	4.0	9%	80%
9	$+\alpha_E$	Hard	200	.30	.40	10.0	9%	80%
1	$-\alpha_F$	Hard	200	.30	.40	7.0	5%	80%
14	$+\alpha_F$	Hard	200	.30	.40	7.0	15%	80%
11	$-\alpha_G$	Hard	200	.30	.40	7.0	9%	50%
10	$+\alpha_G$	Hard	200	.30	.40	7.0	9%	100%
7	zero	Hard	200	.30	.40	7.0	9%	80%
4	zero	Hard	200	.30	.40	7.0	9%	80%
12	zero	Hard	200	.30	.40	7.0	9%	80%

To begin the Minitab process, we use the dropdown menu for **Stat=>**
DOE=> Response Surface=>Create Response Surface Design as shown
in Figure 6-8.

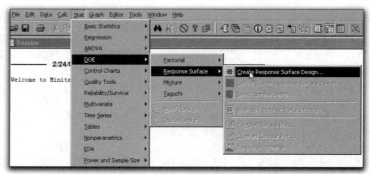

FIGURE 6-8

A dialog box appears (Figure 6-9) in which we choose the default Central
composite type of design and scroll down to choose 3 factors. Click on **Display**
Available Designs which brings up an information box shown in Figure 6-10. Ignore
the **Blocked** option since we have not yet covered blocking. Note the relative
inflexibility of Minitab to allow much fractionalization in CCD's. Only 1/2 and 1/4
fractions are supported. We'll see how to work around this with our 1/8 fraction wheel-
bearing design. For our simple CCD we can use the 3 factor unblocked; after checking
OK we are returned to the main dialog box where we click **Designs** and go to the dialog
box shown in Figure 6-12 to make our choice - the first selection "Full Factorial, 1
Block." Keep all other default selections for this example. Click **OK** and return

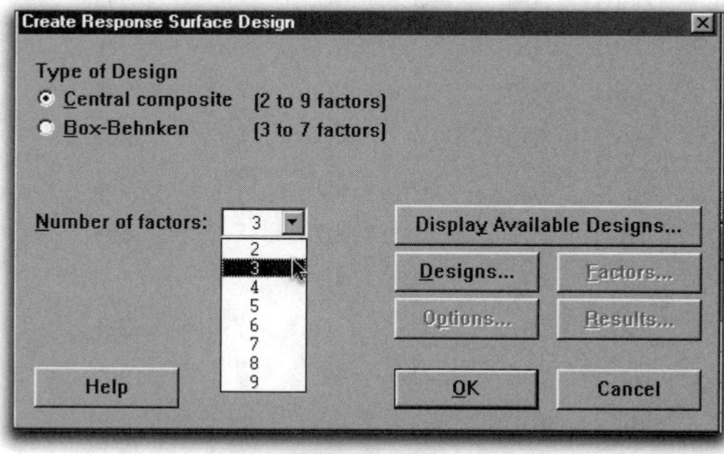

FIGURE 6-9

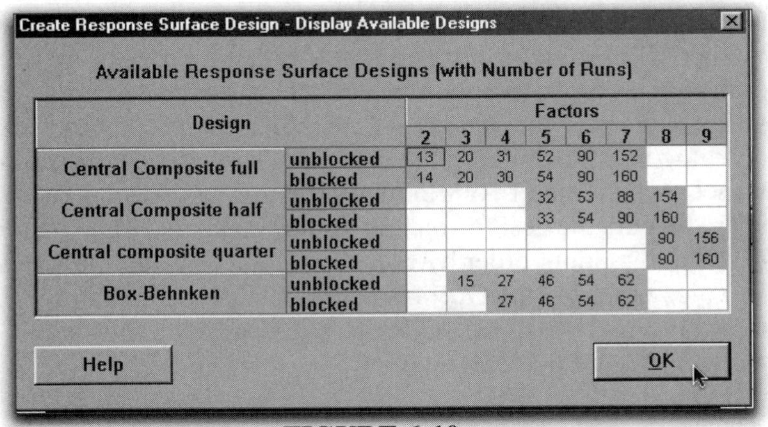

Create Response Surface Design - Display Available Designs

Available Response Surface Designs (with Number of Runs)

Design		Factors							
		2	3	4	5	6	7	8	9
Central Composite full	unblocked	13	20	31	52	90	152		
	blocked	14	20	30	54	90	160		
Central Composite half	unblocked				32	53	88	154	
	blocked				33	54	90	160	
Central composite quarter	unblocked							90	156
	blocked							90	160
Box-Behnken	unblocked		15	27	46	54	62		
	blocked			27	46	54	62		

Help   OK

FIGURE 6-10

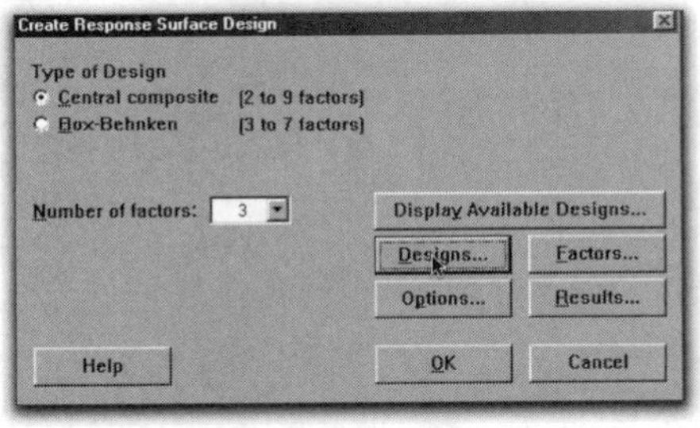

Create Response Surface Design

Type of Design
○ Central composite   [2 to 9 factors]
○ Box-Behnken   [3 to 7 factors]

Number of factors: 3 ▾   Display Available Designs...
                                Designs...   Factors...
                                Options...   Results...

Help   OK   Cancel

FIGURE 6-11

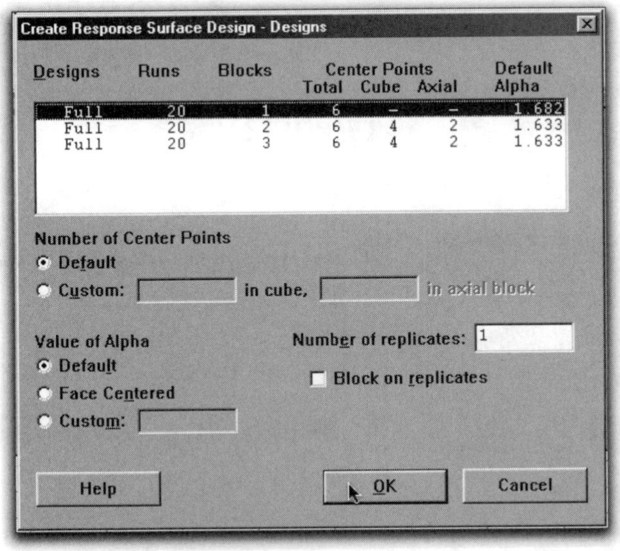

Create Response Surface Design - Designs

Designs	Runs	Blocks	Center Points Total Cube Axial			Default Alpha
Full	20	1	6	–	–	1.682
Full	20	2	6	4	2	1.633
Full	20	3	6	4	2	1.633

Number of Center Points
○ Default
○ Custom: _____ in cube, _____ in axial block

Value of Alpha   Number of replicates: 1
○ Default
○ Face Centered   □ Block on replicates
○ Custom: _____

Help   OK   Cancel

FIGURE 6-12

to the main dialog box as shown in Figure 6-13 (a) and click **Factors**. Just as in other Minitab dialog boxes for factors that we have seen before, we enter the names of the factors; for the CCD, we enter the entire working range as shown in Figure 6-13 (b). Click **OK**, which takes us back to the main dialog box shown in Figure 6-14 (a) where we select **Options**. We will unclick the default **Randomize runs** for this example to allow a comparison with the design you build in Problem #2. We will also show this design in random order by going back and clicking on **Randomize runs**. Click **OK** from the **Options** to return to the main dialog box again as shown in Figure 6-15 (a). Click **Results** and keep the default option as shown in Figure 6-15 (b).

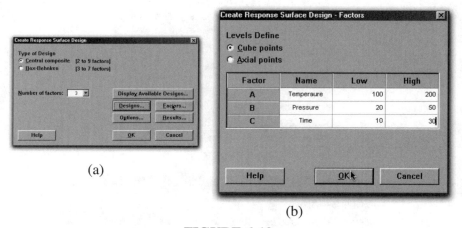

(a)

(b)

FIGURE 6-13

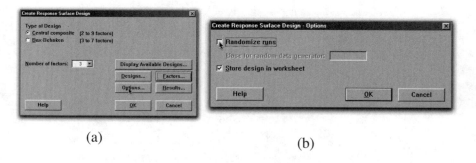

(a)

(b)

FIGURE 6-14

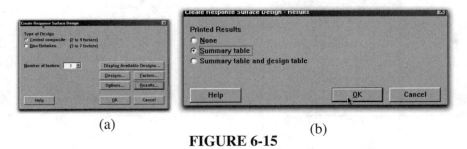

(a)

(b)

FIGURE 6-15

The last click of **OK** in the main dialog box produces the design that is placed in the worksheet. The design in Figure 6-16 is in standard order. However, the number of significant digits displayed are far more than could be held in a practical experiment. Worse, an operator would laugh at settings like "99.999" for a temperature. To fix this programming artifact, we need to find the offending numbers and change them to reasonable values. Figure

↓	C1	C2	C3	C4	C5	C6	C7	C8
	StdOrder	RunOrder	PtType	Blocks	Temperaure	Pressure	Time	
1	1	1	1	1	099.999	19.9999	09.9999	
2	2	2	1	1	199.999	19.9999	09.9999	
3	3	3	1	1	099.999	49.9999	09.9999	
4	4	4	1	1	199.999	49.9999	09.9999	
5	5	5	1	1	099.999	19.9999	29.9999	
6	6	6	1	1	199.999	19.9999	29.9999	
7	7	7	1	1	099.999	49.9999	29.9999	
8	8	8	1	1	199.999	49.9999	29.9999	
9	9	9	-1	1	65.910	34.9999	19.9999	
10	10	10	-1	1	234.089	34.9999	19.9999	
11	11	11	-1	1	149.999	9.7731	19.9999	
12	12	12	-1	1	149.999	60.2268	19.9999	
13	13	13	-1	1	149.999	34.9999	3.1821	
14	14	14	-1	1	149.999	34.9999	36.8178	
15	15	15	0	1	149.999	34.9999	19.9999	
16	16	16	0	1	149.999	34.9999	19.9999	
17	17	17	0	1	149.999	34.9999	19.9999	
18	18	18	0	1	149.999	34.9999	19.9999	
19	19	19	0	1	149.999	34.9999	19.9999	
20	20	20	0	1	149.999	34.9999	19.9999	

FIGURE 6-16

Temperaure	Pressure	Time
100.0	19.9999	29.9999
200.0	19.9999	29.9999
100.0	49.9999	29.9999
200.0	49.9999	29.9999
65.9	34.9999	19.9999
234.0	34.9999	19.9999
150.0	9.773103	19.9999
150.0	60.2268	19.9999
150.0	34.9999	3.1821
150.0	34.9999	36.8178
150.0	34.9999	19.9999

(a)

Temperaure	Pressure	Time
100.0	20.0	29.9999
200.0	20.0	29.9999
100.0	50.0	29.9999
200.0	50.0	29.9999
65.9	35.0	19.9999
234.0	35.0	19.9999
150.0	9.8	19.9999
150.0	60.2	19.9999
150.0	35.0	3.1821
150.0	35.0	36.8178
150.0	35.0	19.9999

(b)

FIGURE 6-17

16-16 shows the offending numbers (circled), which happen to be the alpha star values. Figure 16-17 (a & b) show that by changing the number of significant digits from a very long string that Minitab calculates, to (for this case) only one decimal place for both alpha points, the reasonable values for all levels are displayed. You need to determine the number of significant places for each experiment you build. Figure 6-18 shows the design with reasonable, working levels. Figure 6-19 is the same experiment in random order.

↓	C1 StdOrder	C2 RunOrder	C3 PtType	C4 Blocks	C5 Temperaure	C6 Pressure	C7 Time	C8
1	1	1	1	1	100	20.0	10.0	
2	2	2	1	1	200	20.0	10.0	
3	3	3	1	1	100	50.0	10.0	
4	4	4	1	1	200	50.0	10.0	
5	5	5	1	1	100	20.0	30.0	
6	6	6	1	1	200	20.0	30.0	
7	7	7	1	1	100	50.0	30.0	
8	8	8	1	1	200	50.0	30.0	
9	9	9	-1	1	66	35.0	20.0	
10	10	10	-1	1	234	35.0	20.0	
11	11	11	-1	1	150	9.8	20.0	
12	12	12	-1	1	150	60.2	20.0	
13	13	13	-1	1	150	35.0	3.2	
14	14	14	-1	1	150	35.0	36.8	
15	15	15	0	1	150	35.0	20.0	
16	16	16	0	1	150	35.0	20.0	
17	17	17	0	1	150	35.0	20.0	
18	18	18	0	1	150	35.0	20.0	
19	19	19	0	1	150	35.0	20.0	
20	20	20	0	1	150	35.0	20.0	

FIGURE 6-18

↓	C1	C2	C3	C4	C5	C6	C7	C8
	StdOrder	RunOrder	PtType	Blocks	Temperaure	Pressure	Time	
1	19	1	0	1	150	35.0	20.0	
2	6	2	1	1	200	20.0	30.0	
3	3	3	1	1	100	50.0	10.0	
4	4	4	1	1	200	50.0	10.0	
5	20	5	0	1	150	35.0	20.0	
6	16	6	0	1	150	35.0	20.0	
7	11	7	-1	1	150	9.8	20.0	
8	10	8	-1	1	234	35.0	20.0	
9	5	9	1	1	100	20.0	30.0	
10	7	10	1	1	100	50.0	30.0	
11	2	11	1	1	200	20.0	10.0	
12	8	12	1	1	200	50.0	30.0	
13	12	13	-1	1	150	60.2	20.0	
14	14	14	-1	1	150	35.0	36.8	
15	13	15	-1	1	150	35.0	3.2	
16	1	16	1	1	100	20.0	10.0	
17	9	17	-1	1	66	35.0	20.0	
18	18	18	0	1	150	35.0	20.0	
19	15	19	0	1	150	35.0	20.0	
20	17	20	0	1	150	35.0	20.0	

FIGURE 6-19

Comments on the Minitab CCD

We noted earlier the relative inflexibility of Minitab to allow much fractionalization in CCD's when we looked at the **Available Designs** information box (Figure 6-10). Only 1/2 and 1/4 fractions are supported and such fractions are only for larger numbers of factors. The next example will show how to overcome this. The other difference from conventional wisdom and common practice in building the CCD is the relatively large number of center point replicates that Minitab includes as a default. Problem 6 asks the student to find the reasoning behind the default number of center-points in a CCD. We can select a more reasonable number of center points as a custom feature as shown in Figure 6-20.

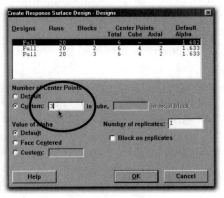

FIGURE 6-20

Building a Custom CCD in Minitab

A number strategies can be utilized to establish a sequential approach to the investigation of a response surface. All require the proper brainstorming and team approach to factor/level selection as shown in Chapter 2. Remember to incorporate as much prior knowledge in this planning phase because: *the more you know, the less you will need to do experimentally*. On the "flip" side of expression: *the less you know, the more you will need to do experimentally.*

- Build a Fractional Factorial with center points over the entire range of levels. If there is indications of a curve, build a CCD from scratch repeating the factorial portion.

- Build a Fractional Factorial with center points allowing for axial points, given that there is curvature. This means that the levels in the factorial portion of the experiment are not as widely spaced as in the first approach. Augment to complete the CCD.

- Build a CCD and run only the factorial portion first. Do the analysis and complete the experiment if there is curvature.

The first approach will require the most experimental resources but will investigate the entire range of the factors in the factorial portion of the effort. This can be important if interactions are relatively weak since by using a wider range of levels, interactions will manifest their influence more strongly.

The second approach is an ideal way to build knowledge and in most situations is the recommended approach. However, if interactions are weak and/or there is considerable error, the manifestation of these interactions will be less using such an approach. The third approach accomplishes the same end result as the second, but is easier to accomplish in Minitab since we do not need to customize the design. However, remember that Minitab does not offer the flexibility of all useful fractional factorials in the CCD set of designs.

We will now use Minitab and customize the fractional factorial on the wheel-bearing problem from Chapter 5 that we previously modified manually and used to produce the CCD in Table 6-9. First we open the previously saved fractional factorial file we had saved when we built the 2^{7-3} design. Figures 6-21 and 6-22 show this process.

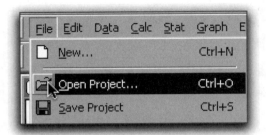

FIGURE 6-21

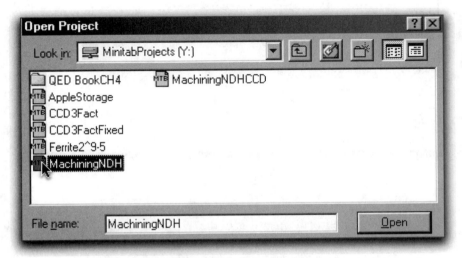

FIGURE 6-22

The modifications are made to the worksheet as shown in Figure 6-23. While it is possible to use editing commands in Minitab, such as "patterned data," adding the points manually to expand the 2^{7-3} design to a CCD is probably just as efficient and might even be a bit quicker. Remember that all the runs in the axial part of this CCD will use the hard material, so the first thing to do is type in "Hard" in the material column. You can speed up this process by using a copy-and-paste command from the edit dropdown (Figure 6-24). Next, enter the lowest value for the RPM (150) and then the highest value of the RPM (250). Follow the remainder of this column with the mid value of the RPM (200). Since there are six factors that are included with alpha points, we need to add 12 more rows for the lowest and highest values of these factors. We will also include three more center point replicates, so there will be a total of 15 new rows added. Continue to add the alpha points as shown in Figure 6-23.

Soft	225	0.35	0.45	5.5	12	75
Hard	225	0.35	0.45	8.5	12	85
Soft	200	0.30	0.40	7.0	9	80
Hard	200	0.30	0.40	7.0	9	80
Soft	200	0.30	0.40	7.0	9	80
Hard	200	0.30	0.40	7.0	9	80
Hard	150	0.30	0.40	7.0	9	
Hard	250	0.30	0.40	7.0	9	
Hard	200	0.20	0.40	7.0	9	
Hard	200	0.40	0.40	7.0	9	
Hard	200	0.30	0.30	7.0	9	
Hard	200	0.30	0.50	7.0	9	
Hard	200	0.30	0.40	4.0	9	
Hard	200	0.30	0.40	10.0	9	
Hard	200	0.30	0.40	7.0	5	
Hard	200	0.30	0.40	7.0	15	
Hard	200	0.30	0.40	7.0		

FIGURE 6-23

After completing the insertion of the additional alpha runs as well as the new center points (the completed design is shown in Figure 6-25), the next step is to make our modified design a "Custom Response Surface" so that Minitab can accept our response variable(s) and do the analysis (which we will do in Chapter 17). From the dropdown heading **Stat**, go to **Response Surface** and then to **Custom Response Surface**. A dialog box (Figure 6-26) will appear and all we need to do is select the factors we want to include in the design. Highlight the 6 factors as shown in Figure 6-26 and click **Select**. Click on **Low/High**. Minitab scans your input from the worksheet and puts the low an d high values into

FIGURE 6-24

C1	C2	C3	C4	C5-T	C6	C7	C8	C9	C10	C11
StdOrder	RunOrder	CenterPt	Blocks	Material	Speed	Feed(Left)	Feed(Right)	HtCyTime	Punch/Die	HtPwr
1	1	1	1	Soft	175	0.25	0.35	5.5	6	75
2	2	1	1	Hard	175	0.25	0.35	8.5	6	85
3	3	1	1	Soft	225	0.25	0.35	8.5	12	75
4	4	1	1	Hard	225	0.25	0.35	5.5	12	85
5	5	1	1	Soft	175	0.35	0.35	8.5	12	85
6	6	1	1	Hard	175	0.35	0.35	5.5	12	75
7	7	1	1	Soft	225	0.35	0.35	5.5	6	85
8	8	1	1	Hard	225	0.35	0.35	8.5	6	75
9	9	1	1	Soft	175	0.25	0.45	5.5	12	85
10	10	1	1	Hard	175	0.25	0.45	8.5	12	75
11	11	1	1	Soft	225	0.25	0.45	8.5	6	85
12	12	1	1	Hard	225	0.25	0.45	5.5	6	75
13	13	1	1	Soft	175	0.35	0.45	8.5	6	75
14	14	1	1	Hard	175	0.35	0.45	5.5	6	85
15	15	1	1	Soft	225	0.35	0.45	5.5	12	75
16	16	1	1	Hard	225	0.35	0.45	8.5	12	85
17	17	0	1	Soft	200	0.30	0.40	7.0	9	80
18	18	0	1	Hard	200	0.30	0.40	7.0	9	80
19	19	0	1	Soft	200	0.30	0.40	7.0	9	80
20	20	0	1	Hard	200	0.30	0.40	7.0	9	80
21	21	1	1	Hard	150	0.30	0.40	7.0	9	80
22	22	1	1	Hard	250	0.30	0.40	7.0	9	80
23	23	1	1	Hard	200	0.20	0.40	7.0	9	80
24	24	1	1	Hard	200	0.40	0.40	7.0	9	80
25	25	1	1	Hard	200	0.30	0.30	7.0	9	80
26	26	1	1	Hard	200	0.30	0.50	7.0	9	80
27	27	1	1	Hard	200	0.30	0.40	4.0	9	80
28	28	1	1	Hard	200	0.30	0.40	10.0	9	80

FIGURE 6-25

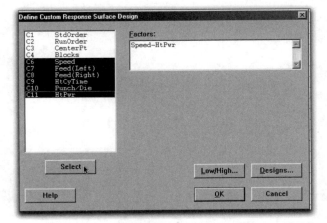

FIGURE 6-26

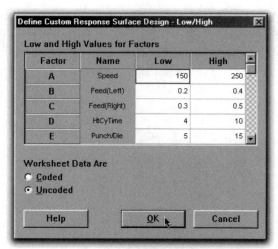

FIGURE 6-27

the proper places in the sub dialog box as shown in Figure 6-27. You must go
through this step of clicking on **Low/High** in order to click on **OK** in the main
dialog box, even though Minitab has already found your low and high values.
Click **OK** and the design is ready for use as shown in Figure 6-28.

C5-T	C6	C7	C8	C9	C10	C11	C12	C13	C14	C15
Material	Speed	Feed(Left)	Feed(Right)	HtCyTime	Punch/Die	HtPwr	StdOrder_1	RunOrder_1	Blocks_1	PtType
Hard	175	0.25	0.35	8.5	6	85	2	2	1	1
Soft	225	0.25	0.35	8.5	12	75	3	3	1	1
Hard	225	0.25	0.35	5.5	12	85	4	4	1	1
Soft	175	0.35	0.35	8.5	12	85	5	5	1	1
Hard	175	0.35	0.35	5.5	12	75	6	6	1	1
Soft	225	0.35	0.35	5.5	6	85	7	7	1	1
Hard	225	0.35	0.35	8.5	6	75	8	8	1	1
Soft	175	0.25	0.45	5.5	12	85	9	9	1	1
Hard	175	0.25	0.45	8.5	12	75	10	10	1	1
Soft	225	0.25	0.45	8.5	6	85	11	11	1	1
Hard	225	0.25	0.45	5.5	6	75	12	12	1	1
Soft	175	0.35	0.45	8.5	6	75	13	13	1	1
Hard	175	0.35	0.45	5.5	6	85	14	14	1	1
Soft	225	0.35	0.45	5.5	12	75	15	15	1	1
Hard	225	0.35	0.45	8.5	12	85	16	16	1	1
Soft	200	0.30	0.40	7.0	9	80	17	17	1	1
Hard	200	0.30	0.40	7.0	9	80	18	18	1	1
Soft	200	0.30	0.40	7.0	9	80	19	19	1	1
Hard	200	0.30	0.40	7.0	9	80	20	20	1	1
Hard	150	0.30	0.40	7.0	9	80	21	21	1	1
Hard	250	0.30	0.40	7.0	9	80	22	22	1	1
Hard	200	0.20	0.40	7.0	9	80	23	23	1	1
Hard	200	0.40	0.40	7.0	9	80	24	24	1	1
Hard	200	0.30	0.30	7.0	9	80	25	25	1	1
Hard	200	0.30	0.50	7.0	9	80	26	26	1	1

FIGURE 6-28

We have just made a modification of an original factorial design to add the axial star points which expands it into a CCD. While in this example, we accomplished this augmentation by manually inserting the new rows, it is possible to automate this process in Minitab by selecting from the DOE dropdown the **Modify Design** option (Figure 6-29). This method may be used only if all the factors are numeric. Simply click on the option, **Add axial points** in the modify dialog box. Click **Put in new worksheet**. Then click **Specify** to select the type of design (usually the CCD) and click **OK** to put the new design into the new worksheet.

FIGURE 6-29

Final thoughts

The CCD is one of the "modern miracles" of experimental design. It is as much a fractional factorial design as any of the 2^{k-p} designs, if not more so. Consider a five-level full factorial experiment with two factors. Figure 6-30 (a) shows the geometry of this design matrix. There are 25 total treatment combinations in this design. Now look at the CCD that covers this same sample space in Figure 6-30 (b) where (with an unreplicated center point) there are only 9 treatment combinations. Problem 7 asks you to justify the efficiency of the CCD based on a comparison of these two alternative design approaches.

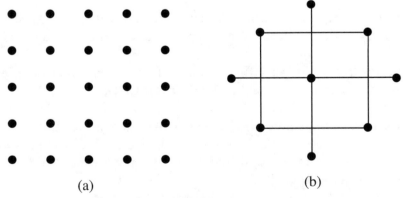

(a)   (b)

FIGURE 6-30

Problems for Chapter 6

1. Complete the calculation of the physical levels for the problem outlined in the text using the working ranges found in Table 6-6. Compute the levels for pressure and time that are set for the -1 and + 1 design unit positions.

2. With the results from Problem 1, build the randomized treatment combination table that would be used to run the experiment with temperature, pressure, and time as a CCD. Compare with the Minitab produced design in Figures 6-18 and 6-19.

3. Compute the alpha star distances for the following designs:

No. of Factors	Type of Factorial
5	1/4 fractional
6	1/4 fractional
9	FULL
12	1/64 fractional

4. Set up a CCD for the following problem:

Factor	Working Range
Screw Speed	10 to 25 ips
Temperature	3000 to 4000°
Holding Time	30 to 90 sec
Feed Particle Size	0.5 to 5 mm

5. How many extra runs would have been necessary if the CCD had been designed without the sequential approach as described on page 148.

6. Search the literature and find the reason for the large number of center points assigned as default by Minitab in the CCD.

7. Using the information shown in Figure 6-30, and the definition of efficiency, write a justification for using a CCD instead of a full factorial.
   a) What is the fraction of the full factorial if you use the CCD?
   b) Write a paragraph justifying the use of the CCD.
   c) Expand this to a 3 factor situation and repeat a and b above.

APPENDIX 6

Philosophy: There is a balance between skepticism and learning. We should always be skeptical of results that are counterintuitive, but we must also be willing to modify our convictions in the light of experimental evidence. So if a two-level experiment seems to indicate no difference in a factor under study, we may still want to study this factor further in a multi-level design.

Some Functional Relationships and Their Polynomial Forms:

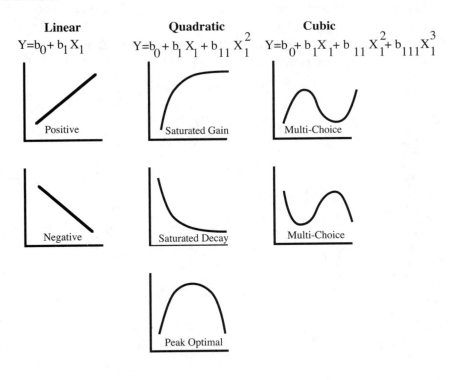

Linear
$$Y = b_0 + b_1 X_1$$
Positive

Negative

Quadratic
$$Y = b_0 + b_1 X_1 + b_{11} X_1^2$$
Saturated Gain

Saturated Decay

Peak Optimal

Cubic
$$Y = b_0 + b_1 X_1 + b_{11} X_1^2 + b_{111} X_1^3$$
Multi-Choice

Multi-Choice

CCD: Center Composite Design
- An efflcient design that allows investigation of interactions (two-level) and curvilinear main effects.
- The number of treatment combinations in a CCD
#tc's $= 2^{k-p} + 2*(k) +$ # of center points, where k = # of factors, p is the fractionalization element.

The center point is usually replicated to obtain an estimate of experimental error. The number of replicates is dependent on the cost of experimentation, and should be at least 3, but does not need to be greater than 5 or 6.

REFERENCES

1. Box, G.E.P. and Wilson, K.B. On the experimental attainment of optimum conditions. *J. Roy. Statist. Soc. B*, 13 (1951), 1-45.

2. Box, G.E.P. The exploration and exploitation of response surfaces: Some general considerations and examples. *Biometrics* 10 (1954), 16-60.

3. Box, G.E.P. and Hunter, J. S. Multi-factor experimental designs for exploring response surfaces. *Ann. Math. Statist.* 28 (1957), 195-241.

7

3-Level Designs

In Chapter 6 we showed how 3-level designs are inherently inefficient due to the low volume of information generated at a relatively high cost. There are some modifications that can be made to the basic 3-level designs that can make them efficient by producing the required information at a low cost. The 3^2 design with its three levels and only two factors fits into this category as we showed in Chapter 6. We also showed that for 3-level designs with more than two factors, the efficiency degraded rapidly. The solution to this problem was to use the central composite design, which has five levels.

Sometimes it is not possible to obtain the required five levels for the CCD. In such a circumstance we could run a "modified" CCD with most of the factors at five levels and a few at three levels. Such a design would not suffer terribly in the analysis. Of course, we would not be able to derive functions beyond a quadratic for the factors that have only three levels. There could also be an increase in confounding among the factors that are actually quadratic if only three levels are run. This is due to the fact that when we run a center point, we are changing all the factors at the same time. An outline of such a design for four factors using a 1/2 fractional factorial as the base would be as follows, given that only three of the factors could be run at five levels as shown in Table 7-1.

In this design, we have simply left out the alpha star positions for factor C. This omission of the alpha star levels could be taken to the extreme by leaving out all the alpha star positions for all the factors and simply running the center point with a 2^{k-p} factorial. This is a good thought at first but leads to confusion

TABLE 7-1

tc	A	B	C	D
(1)	-	-	-	-
a(d)	+	-	-	+
b(d)	-	+	-	+
ab	+	+	-	-
c(d)	-	-	+	+
abc(d)	+	+	+	+
$-\alpha_a$	−1.682	0	0	0
$+\alpha_a$	+1.682	0	0	0
$-\alpha_b$	0	−1.68 2	0	0
$+\alpha_b$	0	+1.682	0	0
$-\alpha_d$	0	0	0	−1.682
$+\alpha_d$	0	0	0	+1 .682
zero	0	0	0	0

in the analysis since we don't know which center point of which factor is causing the curvilinearity to take place. We have seen this confusion, called "confounding," in fractional factorial designs, which is common in designs that lack completeness. We shall avoid such incomplete designs. While the modified CCD is a viable alternative to investigating a reduced set of multi-levels, it can be improved upon by using a special type of three-level designs.

This type of design that will investigate three levels with a minimum of work is called the Box-Behnken (B-B) design (1). The concept makes an intelligent use of the 2^k factorial. To build a B-B design, take subsets of the factors in pairs and build 2^2 factorials for all possible pairs while holding the other factors at a center point for each 2^2 factorial. So, for the example of the four-factor experiment that we used as the modified CCD in Table 7-1, a B-B 3-level design is shown in Table 7-2.

In comparing the design in Table 7-1 (the CCD) with the design in Table 7-2 (the B-B), we can see that there is a greater expenditure of resources in the B-B design. The B-B requires 47% more work than the CCD, and the CCD operates at five levels while the B-B operates only at three. However, the B-B design is still more efficient then the comparable 3^4 full factorial that would require 81 treatment combinations.

To determine how many "sets" of 2-level subdesigns in a B-B, we need to find the number of pairs of factors in the problem at hand. This is based on the combinations formula that we introduced in Chapter 4. In this case, we will take

Box-Behnken

TABLE 7-2

A	B	C	D	
-	-	0	0	
+	-	0	0	A&B Together in a 2^2
-	+	0	0	
+	+	0	0	
-	0	-	0	
+	0	-	0	A&C Together in a 2^2
-	0	+	0	
+	0	+	0	
-	0	0	-	
+	0	0	-	A&D Together in a 2^2
-	0	0	+	
+	0	0	+	
0	-	-	0	
0	+	-	0	B&C Together in a 2^2
0	-	+	0	
0	+	+	0	
0	-	0	-	
0	+	0	-	B&D Together in a 2^2
0	-	0	+	
0	+	0	+	
0	0	-	-	
0	0	+	-	C&D Together in a 2^2
0	0	-	+	
0	0	+	+	
0	0	0	0	Zero Point

k factors, two at a time, which reduces to the special case for pairs:

$$\text{\# Of Pairs} = \frac{k(k-1)}{2} \qquad (7\text{-}1)$$

Therefore the number of treatment combinations in a B-B design will be equal to the number of pairings of factors times the number of treatments in the subdesign (which will always be four) plus the zero or center point.

$$\text{\# tc's} = 4 * \left(\frac{k(k-1)}{2} \right) + 1 \qquad (7\text{-}2)$$

As in any experimental design, the tc's are randomized and run in random order. While the B-B designs are not as efficient as the CCD, they fill a gap between the 2^{k-p} and the larger, multi-level full factorial designs and certainly are far more efficient than the full 3^k factorials.

Once again, from a degrees of freedom analysis, we can see that the B-B designs are smaller than 3^k designs, they use fewer df, and thus can't find some information. This information is in the form of the higher-order interactions that are not likely to occur. We do not obtain information on three-factor or greater interactions, or quadratic interactions from a B-B design. We do, however, get information on single linear effects and single quadratic effects along with the two-factor linear interactions.

3^{k-p} Designs

While the Box-Behnken design is an appropriate, clever, and useful approach to solving the problem of investigating nonlinear effects, some researchers and engineers (especially those who follow the so-called *Taguchi Methods*) prefer to use fractions of the 3^k factorial designs. These 3-level designs are completely balanced with all levels of all factors found in combination with all other factors' levels. In contrast, the CCD has its extreme low and high levels "orphaned" off the axes of the main part of the design and is *not* in balance. The B-B design does not combine more than two factors at a time in a factorial manner (beyond the zero point). While we have shown how these "flaws" in the CCD and B-B designs are really virtues from an information theory perspective, there is still a controversy over the merits of the alternative approach (3^{k-p}) to the investigation of curvilinear functions. The CCD and B-B designs were specifically engineered to fit the criteria for multi-level designs specified in the paper by Box and Wilson (1951). It has been said that the 3-level factorial and fractional factorial designs are "methods searching for an application."

We shall see some possible uses of these controversial 3-level fractional factorials in the chapters on applications and Taguchi Methods. For now, we will simply study this class of experiments to understand how they are constructed. As we present the methods for building the 3^{k-p} fractions, we will also see some of the problems encountered in the application of such designs to building response surfaces and writing equations on an unknown process.

Generating a 3^{k-p} Design

To build three-level fractional factorials we will use some of the same methods that were used in Chapter 5 to construct the 2^{k-p} designs. We shall see that the 3^{k-p} are more complex to build and the interpretation of the physical nature of the interactions is somewhat obscure in these 3^{k-p} designs. Also, to handle the 3-level designs, it is necessary to change the notation from the -1, +1 coded levels of the 2-level designs to a more general system using 0,1,2 notation designating the low, mid, and high levels of the 3^k factorials. We will also need to expand the modulus algebra of the 2^{k-p} from base 2 to base 3. In base 3 modulus algebra, any modulus of 3 becomes zero. Therefore, $3 = 0, 6 = 0, 9 = 0$, etc. The "remainder" of any modulus of 3 keeps its value. This gives the following mod 3 table:

MOD 3

$0 = 0$	$4 = 1$
$1 = 1$	$5 = 2$
$2 = 2$	$6 = 0$
$3 = 0$	etc.

Let's look at a simple 3^2 design with two factors at three levels and partition the degrees of freedom. This design has nine treatment combinations, and after accounting for the grand mean, we have 8 degrees of freedom to measure the effects.

df	Effect
2	Linear (L) Factor A
	Quadratic (Q) Factor A
2	Linear (L) Factor B
	Quadratic (Q) Factor B
4	Interaction of Factors A & B
TOTAL 8	

TABLE 7-3

Row	A	B	AB	AAB≈A^2B	ABB≈AB^2	AABB≈A^2B^2
1	0	0	0	0	0	0
2	1	0	1	2	1	2
3	2	0	2	1	2	1
4	0	1	1	1	2	2
5	1	1	2	0	0	1
6	2	1	0	2	1	0
7	0	2	2	2	1	1
8	1	2	0	1	2	0
9	2	2	1	0	0	2
	Ll	L2	L1+L2	L1+L1+L2	L1+L2+L2	L1+L1+L2+L2
	2df	2df	1dft	1df	1df	1df

In Table 7-3 we have labeled the top of the columns with both conventional and unconventional notation. The usual notation for effects is seen in the first three columns, but the quadratic interaction effects have been noted first in the unconventional manner (i.e., AAB) of multiplying out the quadratic function and then in the conventional manner (i.e., A^2B) of showing the exponent. This notation is intended to help demonstrate how the name of each effect is related with the linear combinations that create the entries in each column. At the base of each column, we have defined the linear combinations (L) in terms of the basic Ll and L2 columns, which represent the A and B effects, respectively.

Let's use the mod 3 algebra and derive values for the fourth column (AAB or A^2B). We see at the bottom of this column that the linear combination rule is to take $(Ll + Ll + L2)_{mod3}$. For the first row, the values of Ll and L2 are both zero, so the entry in the fourth column, first row is also a zero. In the second row, we add two Ll's, which give a 2 plus L2, which is still zero, so we end up with a 2 in the second row, fourth column. We can continue down the column, row by row in a similar manner. The sixth row is interesting. We take Ll twice, which gives 4 and add on L2 to give 5, which in mod 3 produces a 2, which is the entry in column 4, row 6. Besides the unconventional column-heading notation for the interactions, we have also represented the components of the interaction unconventionally again for learning purposes. The 4 df for the general AB interaction have been split into 4 parts instead of 2 parts with 2 df each, which is the usual approach. We have gone just one step further to observe how a notation convention evolved.

Let us combine the interaction columns in pairs. Since there are four columns, we can combine them in pairs in six different ways. Table 7-4 shows the 6 linear combinations possible.

TABLE 7-4

I AB+ AAB	II AB+ ABB	III AB + AABB	IV AAB+ ABB	V AAB+ AABB	VI ABB + AABB
0	0	0	0	0	0
0	2	0	0	1	0
0	1	0	0	2	0
2	0	0	0	0	1
2	2	0	0	1	1
2	1	0	0	2	1
1	0	0	0	0	2
1	2	0	0	1	2
1	1	0	0	2	2

Columns III and IV come up "empty," that is, with no pattern of variation. We call this "emptiness" the Identity (I or 1). This is the same identity that we used in the 2^{k-p} designs of Chapter 5. Column III is the linear combination of AB and AABB, which leads to A^3B^3. In mod 3, this turns into I since any power that is a modulus of 3 becomes zero, and any term to the zero power is one. Further, I x I = I. By stretching this point further, and realizing the underlying concept of the identity, we conclude that the AB interaction is identical to the A^2B^2 interaction (note how ***identi*** cal and ***identi*** ty come from the same root). This leads to the convention that the 4 df available for 2-factor interactions in 3^k designs will break down into only *two* components with 2 df each. The reason that no further breakdown of the df is possible is the redundancy that we have seen in columns III and IV of Table 7-4.

The convention in deciding which part of the 2 df is used as standard notation is somewhat arbitrary, but the conventional choice results in a neater notation than its alternative. The convention requires that the first term (or letter) of the interaction "word" is raised to the first power only. Therefore A^2B^2 and A^2B are eliminated as viable choices for notation, not because they can't exist physically, but because they are redundant with AB and AB^2.

A further inspection of Table 7-4 reveals that the pattern of levels of columns V and VI are the same as A and B, respectively. If we look at the column headings we see that the linear combination of V is $A^2B+A^2B^2$, which in mod 3 reduces to A (i.e., A^4B^3). Similarly column VI produces B (i.e., A^3B^4). Thus the pattern of "confounding" is predicted by the linear combination.

The first two columns (I and II) present at first a more obscure relationship, but the linear combination gives us a clue. Column I's linear combination is A^3B^2, which reduces to B^2. Now B^2 is found from L2 +L2, which is exactly the pattern in Column I. Column II follows the same concepts (see Problem 5).

Information and Resources

All of the above insights derived from Tables 7-3 and 7-4 help in the sorting of information available in the 3-level designs. We are also able to balance our resources with knowledge of the effects that can be expected in a particular experimental situation. We will now regroup the 3^2 experimental table into the four 2 df columns that are not redundant and follow the column labeling conventions for such designs.

TABLE 7-5

A	B	AB	AB^2
0	0	0	0
1	0	1	1
2	0	2	2
0	1	1	2
1	1	2	0
2	1	0	1
0	2	2	1
1	2	0	2
2	2	1	0

If we do not expect any interactions, then we may use one of the interaction columns to introduce a third factor, much as we did in 2-level fractional factorials. With this, we step into the realm of the 3^{k-p} fractional factorial. The **k** refers to the number of factors and the **p** is a fractionalization element as before. Since each 3-level factor requires 2 df, we may add only two factors when we have 9 treatments. Let's add at first a third factor C and confound it with the AB column. In doing so, we have established a confounding pattern that can be determined using defining contrasts similarly to the 2^{k-p} designs. But because of the extra information content of the 3^{k-p} designs, the rules for their construction are far more complex.

3^{k-p} Design Rules

1. There is a fractionalization element $p = k - n$ where k is the number of factors under study and n is the nth root equal to 3 of the number of treatment combinations in the base design.
2. There will be p generators chosen from the available 2 df positions of the interactions in the base design.
3. The fraction of the full design is $1/3^p$.
4. There are $(3^p-1)/2$ terms in the fundamental defining contrast (FDC). If p is greater than 1, then the FDC is generated by finding all the nonredundant (in mod 3) multiples of the first-found terms Groups) of the FDC using up to a square power in these group (**G**) multiplications.
5. Use the FDC and its square to determine the pattern of confounding in mod 3.

Let's continue the example from Table 7-5 and apply the above rules to complete the generation of the 3^{k-p} design and use the resulting defining contrast.

1. Our base design has 9 tc's and because k = 3 (factors **A,B,C**), and we compute $\sqrt[n]{9} = 3$, n = 2, and therefore p = 1.
2. We now have 1 generator, the **AB** interaction.
3. The fraction is $1/3^1 = 1/3$.
4. There will be $(3^1-1)/2 = 1$ term in the FDC, which is $I \approx ABC^2$ (don't count the identity term, I).
5. $I \approx ABC^2$ (FDC) and A^2B^2C (FDC)^2

Note: The symbol "=>" means "goes to this conventional form."
    The symbol "$\approx$" means "is confounded with."

$$I \approx ABC^2, \qquad\qquad A^2B^2C \qquad\qquad (7-3)$$

$A \approx A^2 BC^2 => AB^2C,$ $A^3B^2C => B^2C => BC^2$

$B \approx AB^2C^2,$ $A^2B^3C => A^2C => AC^2$

$C \approx ABC^3 => AB,$ $A^2B^2C^2 => ABC$

$AC \approx A^2BC^3 => A^2B => AB^2,$ $A^3B^2C^2 => B^2C^2 => BC$

Summary of Confounding:

$$\mathbf{A} \approx AB^2C, BC^2$$
$$\mathbf{B} \approx AB^2C^2, AC^2 \tag{7-4}$$
$$\mathbf{C} \approx AB, ABC$$
$$\mathbf{AC} \approx AB^2, BC$$

The implication of the confounding in this design is that the two-factor interactions are linearly combined with each other and with factor C (because of the generator). Even if we had picked the alternative generator ($C \approx AB^2$), we would have run into this same difficulty (see Problem 6).

Because of the extent of the confounding, this 9 tc, 1/3 fraction can be considered only as a single effects design. The AB interaction that is confounded with C certainly is not allowable since we would not be able to distinguish which effect (the single effect or the interaction) was causing the change in our response. This is not a recommended design configuration.

Larger 3^{k-p} Designs

Since powers of 3 grow very rapidly, there are only a small number of 3^{k-p} base designs that are useful. The 9 tc (3^2) has a limited application when a curved effect is suspected, but it is unable to detect interactions. The next base design is a 27 tc (3^3). We will look at this design in detail. The 81 tc (3^4) is much too large for most budgets and fails as an efficient design because of its excessive cost. So we really only have a 9 tc or a 27 tc base designs to fractionalize. Since the confounding of single effects and two factor interactions occurs in the 9 tc design, there is really only one useful 3-level fractional factorial design (the 27 tc). Of course, depending on the degree of fractionalization, a number of different 27 tc configurations are possible. We will now look at a 3^{5-2}, which is one of those configurations.

In the base 3^3 design there are 26 df. Table 7-6 shows the allocation of these 26 degrees of freedom broken down into the 2 df units for the single effects, the 2-factor interactions, and the 3-factor interactions.

While there are a large number of possible combinations of generators for the 3^{5-2}, the best from a confounding standpoint is to use the three-factor interactions as follows:

$$D \approx AB^2C^2 \tag{7-5}$$
$$E \approx AB^2C \tag{7-6}$$

TABLE 7-6

Row	A	B	C	AB	AB²	AC	AC²	BC	BC²	AB²C	AB²C²	ABC²	ABC
1	0	0	0	0	0	0	0	0	0	0	0	0	0
2	1	0	0	1	1	1	1	0	0	1	1	1	1
3	2	0	0	2	2	2	2	0	0	2	2	2	2
4	0	1	0	1	2	0	0	1	1	2	2	1	1
5	1	1	0	2	0	1	1	1	1	0	0	2	2
6	2	1	0	0	1	2	2	1	1	1	1	0	0
7	0	2	0	2	1	0	0	2	2	1	1	2	2
8	1	2	0	0	2	1	1	2	2	2	2	0	0
9	2	2	0	1	0	2	2	2	2	0	0	1	1
10	0	0	1	0	0	1	2	1	2	1	2	2	1
11	1	0	1	1	1	2	0	1	2	2	0	0	2
12	2	0	1	2	2	0	1	1	2	0	1	1	0
13	0	1	1	1	2	1	2	2	0	0	1	0	2
14	1	1	1	2	0	2	0	2	0	1	2	1	0
15	2	1	1	0	1	0	1	2	0	2	0	2	1
16	0	2	1	2	1	1	2	0	1	2	0	1	0
17	1	2	1	0	2	2	0	0	1	0	1	2	1
18	2	2	1	1	0	0	1	0	1	1	2	0	2
19	0	0	2	0	0	2	1	2	1	2	1	1	2
20	1	0	2	1	1	0	2	2	1	0	2	2	0
21	2	0	2	2	2	1	0	2	1	1	0	0	1
22	0	1	2	1	2	2	1	0	2	1	0	2	0
23	1	1	2	2	0	0	2	0	2	2	1	0	1
24	2	1	2	0	1	1	0	0	2	0	2	1	2
25	0	2	2	2	1	2	1	1	0	0	2	0	1
26	1	2	2	0	2	0	2	1	0	1	0	1	2
27	2	2	2	1	0	1	0	1	0	2	1	2	0

L_1 L_2 L_3 $L_4 = L_1 + L_2$ $\quad$ $L_6 = L_1 + L_3$ $\quad$ $L_8 = L_2 + L_3$ $\quad$ $L_{10} = L_1 + 2L_2 + L_3$ $\quad$ $L_{12} = L_1 + L_2 + 2L_3$

$L_5 = L_1 + 2L_2$ $\quad$ $L_7 = L_1 + 2L_3$ $\quad$ $L_9 = L_2 + 2L_3$ $\quad$ $L_{11} = L_1 + 2L_2 + 2L_3$ $\quad$ $L_{13} = L_1 + L_2 + L_3$

The FDC is found by clearing the generators in the usual way:

$$I \approx AB^2C^2D^2, AB^2CE^2 \tag{7-7}$$
$$\textbf{(G1)} \quad \textbf{(G2)}$$

We now take the two groups that have been produced in the first-found words (terms) of the defining contrast and multiply G1(Group 1) times G2 (Group 2) and G1 times (G2)^2 following rule 4 of the algorithm. This gives all the possible nonredundant multiples of Gl and G2 up to the second power. If there had been a G3, we would have had to apply the multiplication to all *pairs* and *triples* of the first-found groups. So applying this concept to the items in Expression 7-7, we obtain the following:

$$\textbf{G1 x G2}$$
$$AB^2C^2D^2 x AB^2CE^2 \Rightarrow A^2B^4C^3D^2E^2 \Rightarrow AB^2DE \tag{7-8}$$

and

$$\textbf{G1 x (G2)}^2$$
$$AB^2C^2D^2 x A^2BC^2E \Rightarrow A^3B^3C^4D^2E \Rightarrow CD^2E \tag{7-9}$$

Therefore the FDC is:

$$I \approx AB^2C^2D^2, AB^2CE^2, AB^2DE, CD^2E \tag{7-10}$$

And the square of this FDC is:

$$I^2 \approx A^2BCD, A^2BC^2E, A^2BD^2E^2, C^2DE^2 \tag{7-11}$$

Now we can use the FDC and its square to determine the pattern of confounding in our 3^{5-2} design. For the single effects and some selected 2-factor interactions, we can see the results in Table 7-7. Only factors A and B are clear of 2-factor interactions, and the 2-factor interactions are confounded among themselves. This design could be used only when prior knowledge of likely and unlikely interactions exists.

TABLE 7-7

FDC $(FDC)^2$

$I \approx AB^2C^2D^2, AB^2CE^2, AB^2DE, CD^2E; \ A^2BCD, A^2BC^2E, A^2BD^2E^2, C^2DE^2$

$A \approx ABCD, ABC^2E, ABD^2E^2 \ ACD^2E, BCD, BC^2E, BD^2E^2, AC^2DE^2$

$B \approx AC^2D^2, ACE^2, ADE, BCD^2E, ABC^2D^2, ABCE^2, ABDE, BC^2DE^2$

$C \approx AB^2D^2, AB^2C^2E^2, AB^2CDE, CDE^2, AB^2CD^2, AB^2E^2, AB^2C^2DE, DE^2$

$D \approx AB^2C^2, AB^2CDE^2, AB^2D^2E, AB^2C^2D, AB^2CDE^2, AB^2E, CDE$

$E \approx AB^2C^2D^2E, AB^2C, AB^2DE^2, CD^2E^2, AB^2C^2D^2E^2, AB^2CD^2E^2, AB^2DCD^2$

$AB \approx ACD, AC^2E, AD^2E^2, ABCD^2E, BC^2D^2, BCE^2, BDE, ABC^2DE^2$

$CD \approx AB^2, AB^2C^2DE^2, AB^2CD^2E, CE^2, AB^2CD, AB^2D^2E^2, AB^2C^2E, DE$

A Comment on 3-Level Fractions

It is appropriate to end this chapter with a comparison of the efficiency of 3^{k-p} versus the 2^{k-p} designs. A direct comparison of a 5 factor, 2-level design with the above 5 factor, 3-level design shows that a 1/2 (2^{5-1}) fraction with 16 tc's will give the single effects and the 2-factor interactions with no trouble at all. The defining contrast is $I \approx ABCDE$.

However, the 2-level design can't get at curved effects. The 3^{k-p} design can get the curved effects, but can't get the 2-factor interactions. This conflict between information and resources and the balance required for efficiency drives home an important axiom that *no experiment can be undertaken without sufficient prior knowledge of the process*. Because interactions are often so important in the processes we study, we may conclude that the inability of 3^{k-p} designs to detect interactions makes them a class of designs searching for an application. In Chapter 20 we will discuss an application that just may be the answer to that search.

Problems for Chapter 7

1. Set up a B-B design for the following problem:

	Factor	Working Ranges
	Screw speed:	10 to 28 ips
	Temperature:	300 to 400°
	Holding time:	30 to 90 sec.
	Feed particle size:	0.5 to 5mm

2. Compare the design from Problem 1 to the CCD design you developed as an answer to Problem 4 in Chapter 6. Comment on the efficiency. What if 3 levels were the maximum available?

3. How many tc's in a B-B for 6 factors? Compare this with a fractional factorial-based CCD that will obtain all single effects and all two-factor interactions.

4. A food engineer wishes to study the effect of changing the amount of ingredients in a pancake mix. He expects curvilinearity and wants to use a 3-level design. Resource constraints will hold the number of treatment combinations to 30 or fewer. Design a 3^{k-p} (Not a B- B) for this engineer.

Factor	Range
Flour	3/4 to 1 cup
Sugar	1-1/2 to 2-1/2 T
Salt	1/4 to 3/4 t
Baking Powder	1-1/2 to 2-1/2 t
Milk	1 to 1-1/2 cups
Eggs	1 to 3
Shortening	1-1/2 to 2-1/2 T

   a) Be sure to write a goal and objective and show the design in random order.
   b) Show the confounding pattern in this design.

5. By taking the proper linear combination, show that Column II of Table 7-4 is indeed A^2. (Hint: Make the comparison using the 0, 1, 2 levels.)

6. Design a 3^{3-1} using the generator $C \approx AB^2$. Work out the summary of confounding and comment on the precautions necessary in the application of this design.

7. Complete the computation of the defining contrast (I and I^2) for the following 3^{5-2} design given the following generators. Comment on the confounding pattern.

$$D \approx ABC^2 \qquad E \approx AB^2C$$

8. Answer Problem 7 again using the following generators:

$$D \approx AB^2C^2 \qquad E \approx ABC$$

9. Use the defining contrast generated in Problem 8 to determine the confounding for the main effects (A, B, C, D, E) and these interactions: AB, AD, DE.

APPENDIX 7

Box-Behnken Design (B-B):

A 3-level fractional factorial that is built on subassembly 2^2 factorials, while all other factors are held at a mid-point. There will be as many subassemblies in the B-B design as there are pairs of factors. The number of treatment combinations is equal to:

$$\#\text{tc's} = \left(\left(\frac{k^*(k-1)}{2} \right) * 4 \right) + 1$$

where k = # of factors.

A Box-Behnken for 3 factors would appear as follows:

A	B	C
-	-	0
+	-	0
-	+	0
+	+	0
-	0	-
+	0	-
-	0	+
+	0	+
0	-	-
0	+	-
0	-	+
0	+	+
0	0	0

3^{k-p} Design Rules:

1. There is a fractionalization element $p = k - n$ where k is the number of factors under study and n is the nth root equal to 3 of the number of treatment combinations in the base design.

2. There will be p generators chosen from the available 2 df positions of the interactions in the base design.

3. The fraction of the full design is $1/3^p$.

4. There are $(3^p-1)/2$ terms in the fundamental defining contrast (FDC). If p is greater than 1, then the FDC is generated by finding all the nonredundant (in mod 3) multiples of the first-found terms (Groups) of the FDC using up to a square power in these group (**G**) multiplications.

5. Use the FDC and its square to determine the pattern of confounding in mod 3.

Using Minitab

We have seen how using computer software reduces the tedium associated with the application of the rules for building experimental designs and the clerical task of making the tables for these designs. Let's reconstruct the B-B design we created in Table 7-2 using Minitab. We follow the same path that we used to build the CCD in Chapter 6. Go to **Stat=> DOE=> Response Surface=> Create Response Surface Design** (Figure 7-1). However, now we will choose the Box-Behnken instead of the default CCD. Scroll down to indicate 4 factors and then click on **Designs** as shown in Figure 7-2.

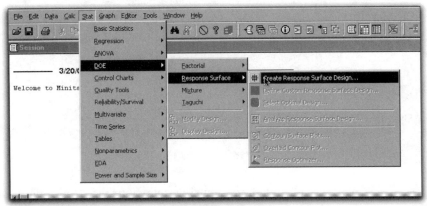

FIGURE 7-1

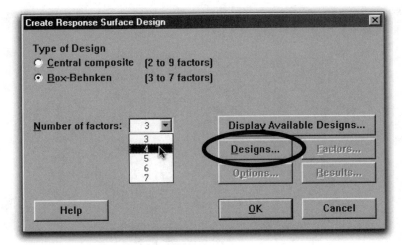

FIGURE 7-2

In the design sub dialog box, the default number of replicates of the centerpoint is 3. In order to match the design we have already constructed, we will un-click this default by clicking on **Custom** and inserting the number 1 as shown in Figure 7-3. Click **OK** and return to the main dialog box.

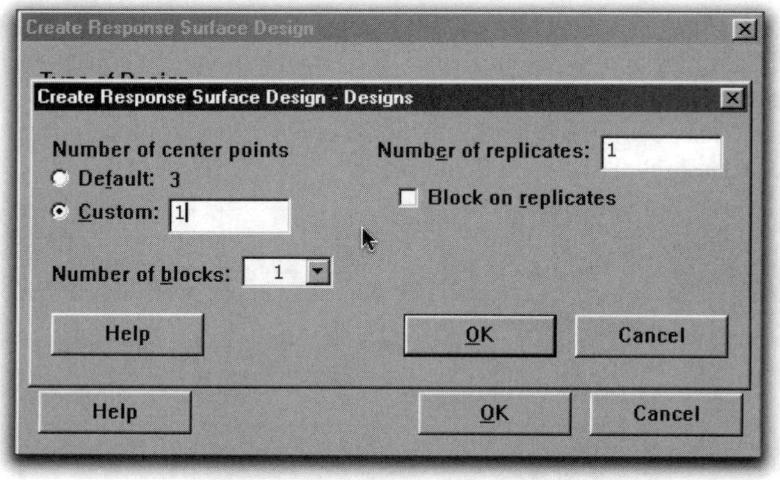

FIGURE 7-3

Since we had not added any physical names to the design we have already constructed, we simply click **OK** in the main dialog box as shown in Figure 7-4, and the experimental design matrix in standard or design units is displayed in the worksheet as illustrated in Figure 7-5. It is the same basic design that is shown in Table 7-2.

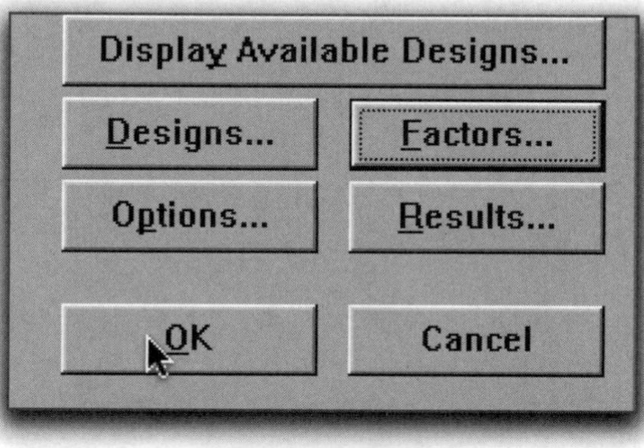

FIGURE 7-4

↓	C1	C2	C3	C4	C5	C6	C7	C8	C9
	StdOrder	RunOrder	PtType	Blocks	A	B	C	D	
1	1	1	2	1	-1	-1	0	0	
2	2	2	2	1	1	-1	0	0	
3	3	3	2	1	-1	1	0	0	
4	4	4	2	1	1	1	0	0	
5	5	5	2	1	0	0	-1	-1	
6	6	6	2	1	0	0	1	-1	
7	7	7	2	1	0	0	-1	1	
8	8	8	2	1	0	0	1	1	
9	9	9	2	1	-1	0	-1	0	
10	10	10	2	1	1	0	-1	0	
11	11	11	2	1	-1	0	1	0	
12	12	12	2	1	1	0	1	0	
13	13	13	2	1	0	-1	0	-1	
14	14	14	2	1	0	1	0	-1	
15	15	15	2	1	0	-1	0	1	
16	16	16	2	1	0	1	0	1	
17	17	17	2	1	-1	0	0	-1	
18	18	18	2	1	1	0	0	-1	
19	19	19	2	1	-1	0	0	1	
20	20	20	2	1	1	0	0	1	
21	21	21	2	1	0	-1	-1	0	
22	22	22	2	1	0	1	-1	0	
23	23	23	2	1	0	-1	1	0	
24	24	24	2	1	0	1	1	0	
25	25	25	0	1	0	0	0	0	
26									

FIGURE 7-5

3^{k-p} in Minitab

Minitab is not programmed to build 3^{k-p} fractional factorial designs but as we will see in the chapter on the application of Robust Design principles (Taguchi Quality Engineering), there are Minitab routines to harness designs at three levels that Minitab unfortunately calls "Taguchi Designs." While the use of these designs was popularized by Taguchi, they are in truth a subset of the 3^{k-p} fractional factorials. Granted that 3^{k-p} fractional factorials have very complex confounding structures, Taguchi's simplified catalog of these designs lacks the in-depth knowledge of confounding that is conveyed by the methods for building the 3^{k-p} fractional factorials shown in this chapter.

REFERENCE

1. Box, G.E.P. and K.B. Wilson (1951). On the experimental attainment of
optimal conditions, *J. Royal Statistical Society, Series B*, **13**,1.

8

Blocking in
Factorial Designs

During the early development of experimental design and analysis techniques, it was recognized that some of the requirements (sometimes called assumptions) necessary to perform a proper analysis of data from a designed experiment could not be fulfilled. The principal requirement that was in jeopardy was that of independence of the error variation throughout the experiment If the error is dependent upon, say, the order of running the experiment due to a learning effect or a chemical that becomes exhausted, then this systematic error could build up and superimpose itself on the factors we study and taint their actual effects! We would have confounded the error with the factors we have under study. One way to assure this independence and to prevent this type of confounding is to randomize the order of the experimental runs. By randomizing the runs, we are able to make any systematic but unknown variation *appear* as random variation.

Early applications of statistical experimental design took place in the field of agriculture. Many of the terms applied to that science have found their way into the words statisticians use to describe the procedures in use today. One such term is "blocking."

If we were to reconstruct those early days (in the 1920's and 1930's) with an experiment, we see the logic behind the methods used to remove some confusion from the results of an experiment constructed to study the effect of fertilizer on plant growth. Table 8-1 shows the factors and the levels for such a study.

TABLE 8-1

Factors Under Study	Levels	
Nitrogen fertilizer	10%	20%
Phosphorus fertilizer	5%	10%
Potash fertilizer	5%	10%
Plant type	Corn	Tomatoes

We will use a 2^4 full factorial design to study this set of factors. There will be 16 treatment combinations, and to obtain a good, precise measure of the response (crop yield) it is necessary to plant at least 1/4 acre per treatment combination. Thus we will need 4 acres of land to complete the experiment.

Table 8-2 shows the 16 treatment combinations for this experiment. Good experimental procedure dictates that we give some consideration to the assignment of the 4 acres of land. This assignment should be done at random as we mentioned before, since randomization will protect us against unknown systematic nuisance factors that could spoil our conclusions. But what can happen if we choose not to randomize? We will compare the base case randomized approach with some systematic alternatives:

Assign Acreage: a) At random (base case)
                        b) First 4 tc's to acre # 1, second 4 tc's to acre # 2, etc.
                        c) Some other systematic manner

Let's look at the implications of each of these proposed methods of assigning the acreage in the experiment. In Table 8-2, a random run order has been used to assign the acres at random. Random runs numbered 1 through 4 have been assigned to acre # 1; runs 5 through 8 are with acre #2; runs 9 through 12 are assigned acre #3; and runs 13 through 16 go with acre #4. To get an even better idea of the physical implications of these assignments, Figure 8-1 shows how the plantings and fertilizations will look if viewed from an aerial photograph.

Assigning the plants and the fertilizer at random is probably not a bad idea, for if there are any acre-to-acre differences, these differences are not confused or confounded with the factors under study as in the design found in Figure 8-2 and Table 8-3. Let's see what is improper or bad about the design using a systematic assignment of the acres.

TABLE 8-2

RUN#	tc	NITRO%	PHOS%	POTSH%	PLANT	ACRE#
12	(1)	10	5	5	Corn	3
2	a	20	5	5	Corn	1
13	b	10	10	5	Corn	4
8	ab	20	10	5	Corn	2
14	c	10	5	10	Corn	4
4	ac	20	5	10	Corn	1
1	bc	10	10	10	Corn	1
9	abc	20	10	10	Corn	3
11	d	10	5	5	Tomato	3
3	ad	20	5	5	Tomato	1
10	bd	10	10	5	Tomato	3
15	abd	20	10	5	Tomato	4
16	cd	10	5	10	Tomato	4
6	acd	20	5	10	Tomato	2
5	bcd	10	10	10	Tomato	2
7	abcd	20	10	10	Tomato	2

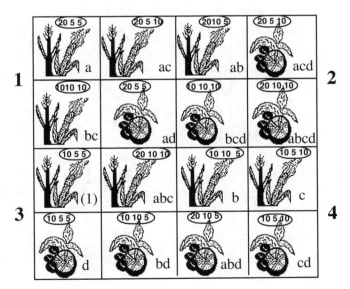

FIGURE 8-1
Fertilizers and Plantings over a Four-Acre Plot (Randomly Assigned)

TABLE 8-3

RUN#	tc	NITRO%	PHOS%	POTSH%	PLANT	ACRE#
12	(1)	10	5	5	Corn	1
2	a	20	5	5	Corn	1
13	b	10	10	5	Corn	1
8	ab	20	10	5	Corn	1
14	c	10	5	10	Corn	3
4	ac	20	5	10	Corn	3
1	bc	10	10	10	Corn	3
9	abc	20	10	10	Corn	3
11	d	10	5	5	Tomato	2
3	ad	20	5	5	Tomato	2
10	bd	10	10	5	Tomato	2
15	abd	20	10	5	Tomato	2
16	cd	10	5	10	Tomato	4
6	acd	20	5	10	Tomato	4
5	bcd	10	10	10	Tomato	4
7	abcd	20	10	10	Tomato	4

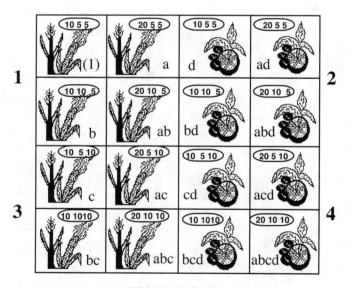

FIGURE 8-2
Fertilizers and Plantings over a Four-Acre Plot (Systematically Assigned)

In the Table 8-3 design, all the corn is systematically planted in acres designated as # 1 and #3, while the tomatoes are systematically planted in the #2 and #4 acres. On top of that, notice that acres #1 and #2 all have 5% potash while acres #3 and #4 have 10% potash. We can't tell if it is the natural fertility of the land or the fertilizer that is inducing the change in the growth of the plants in this confounded design.

We used the word "confounded" in Chapter 5. Recall that one of the rules of confounding was to not confound single effects with each other. We have done exactly that in the design of Table 8-3. We have in fact confounded both the type of plant and potash concentration with the land fertility. So to use the systematic approach of assigning the 4 acres of this example is not good practice and can lead to incorrect conclusions.

But what about the randomized design in Table 8-2? If there are acre-to-acre differences, then these differences will not show up as systematic effects, but as random effects. In our analysis, any differences between acres will be a part of the error (our ability to repeat the measurements). This could be as undesirable as the confounding among the single effects, since increasing the experimental error would decrease the sensitivity of the experiment and impair our ability to find differences between levels of the factors under study. We will expand on this concept in Chapter 18 where we will look at the analysis of these blocked experiments.

Basis of Blocking

The solution to this dilemma is to systematically assign the acres as "blocks" in the experiment. However, the assignment is not made in the simple manner first tried in Table 8-3 where single effects are confounded with the acres, but in a more clever manner. Recall how in Chapter 5 we were able to confound a new single effect with a high-order interaction and create the fractional factorial designs. The same thinking takes place in blocked factorial designs. We will confound the blocked "factor" with a higher-order interaction. (We use quotes on the word *factor* since we are not really studying the acres as a factor.) It is a nuisance that we can't run the whole experiment on one homogeneous piece of land. Therefore, we call the variation between acres a "nuisance factor" which we do not *study*, but rather **remove** from the experiment.

A blocked experiment will need to take place in cases where there is a potential or real source of variation creeping into our experiment. We do not wish to study this source of variation, but we must remove it from the results so that the random error is not inflated, nor is the nuisance factor confounded with other single factors. There are other reasons for blocking and other methods, but for now let's solve this agriculture problem and take a look at the general method of approaching blocking in 2^k factorials.

The basic idea is to confound the blocks (acres of land in this case) with higher-order interactions rather than single effects as we have tried and failed to do before. To construct the design, we need to know the number of blocks, and from this number we compute the number of primary or generating blocked effects. This is a concept similar to the fractionalization element of fractional factorial designs. We will use the symbol "p" to indicate the number of primary blocks. "p" is computed by finding the pth root of **b** (**b** = the number of blocks) that will produce the result of 2.

$$\sqrt[p]{\mathbf{b}} = 2 \qquad\qquad (8\text{-}1)$$

or,

$$p = \frac{\ln(\mathbf{b})}{\ln(2)} \qquad\qquad (8\text{-}2)$$

Where: b represents the number of blocks
           p represents the number of primary blocks
           ln is the natural log function

How does a primary block differ from an ordinary block? In choosing the interactions to be confounded with the blocks, we have a limited free choice. Once we have decided upon the p, primary blocks, the remaining blocks are immediately determined by this first choice. We do not have a free choice beyond the choice of the primary confounding patterns. This is similar to the resulting words in the defining contrast found in fractional factorial designs.

We must choose the primary blocks with great care so that the secondary or resulting blocks are not single effects or two-factor interactions. Let's return to our example and see what all this theory means.

Choice of Primary Blocks

In the 2^4 design we are working on, we have a four-factor interaction, **ABCD** that seems to be a likely candidate as an interaction to be confounded with the blocks. According to Expression 8-1 we will need 2 primary blocks, i.e.:

$$\sqrt[2]{4} = 2$$

with 4 blocks representing the 4 acres of the land in this experiment We need to pick another interaction to confound with the acres. We decide on a three-factor interaction and select ACD from the 4 three-factor interactions that are available.

The resulting secondary block is determined by using the same modus (2) algebra system we used in Chapter 5. We multiple the two primary blocks together, and all even power terms become the identity element and drop out of the expression.

$$ABCD * ACD = A^2BC^2D^2 = B \qquad\qquad (8\text{-}3)$$

This result is not at all acceptable. It says that one of the blocks will be confounded with a single effect (B, which is the nitrogen concentration). It is clear that our initial choice of primary blocks was not wise. Let's explore the criteria for selecting the correct primary blocks.

In general, we should choose primary blocks with the least number of common letters among them. The choice of the ABCD interaction was not in keeping with this concept. Let's examine all the higher-order interactions from this 2^4 design and see what we have to work with.

3-factor interactions:
    A B C
    A B   D
    A   C D
       B C D

4-factor interaction:

    A B C D

Setting the interactions down in the above staggered form is often a help in identifying the overlap and the resulting secondary blocks. From the above layout we can see that if we use the 4-factor interaction, we will always produce confounding with a single effect if we use any of the 3-factor interactions. Also, if we choose the 4-factor interaction and any 2-factor interaction, we will produce another 2-factor interaction as a secondary block. Therefore this is not a good idea.

The best we can do with this 16 tc design is to select two 3-factor interactions (say ABC and BCD), which will produce the 2-factor resultant of AD in the secondary block.

Now we have sufficient information to complete the blocked design. Table 8-4 shows the original design with the appropriate signs of the two primary blocks added at the end. The combination of these signs shows us which block (or acre of land in this case) goes with each treatment combination. You will notice that we do not need to use the secondary block in the determination of the arrangement of block numbers. The combination of the signs of the two interactions at two levels gives us exactly the four blocks we need. If we were to add the secondary block, AD, as a column, we see it would add no further information and only be redundant.

To use the primary block columns in determining the proper block for each treatment, we observe the combination of signs. There are four unique combinations among the 16 tc's. There is a "- -", a "+ -", a "+ +", and a "- +". We will assign a tc to each of these unique sign pair combinations. For example, block #1 is assigned to tc (1), bc, abd, and acd, since each of these treatments has a "- -". Similar assignments are made for the other 3 blocks as shown in Table 8-4.

While we have shown the logic behind the mechanics of building the blocked factorials, there are still a few loose ends left. We have not discussed degrees of freedom in these designs. In the original design, there are 15 df. By blocking, we are essentially adding another "factor." We do not want to study this "factor," but only want to remove its effect from other sources of influence in the analysis. Our scheme is to keep the unwanted factor from inflating the experimental error, while not getting mixed up with the single effects under study. If this factor (or any factor, for that matter) has four levels, then we need three degrees of freedom to estimate its effect. The four blocks correspond to the four levels and consume three degrees of freedom. We get these 3 degrees of freedom from the interactions ABC, BCD, and AD.

In the analysis of this experiment (found in Chapter 18), any effects found with the ABC, BCD, or AD interactions will not be these interactions alone but are the linear combination of the blocked effect and the interaction effect. Since we don't know and can't know each component's contribution to this linear combination, we are unable to draw any conclusion about the "effect" of blocks or the effect of the interactions. In combining the blocks with these interactions, we give up our ability to draw inferences about either. We lose our degrees of freedom for the analysis in this part of the experiment. However, the remainder of the experiment is clear of any other confounding, and we may draw conclusions about all the single effects and the remaining interactions.

Figure 8-3 shows the schematic diagram of the final experiment. Now we can see where the term "**blocked**" comes from. The 4 acres of land have been

TABLE 8-4

RUN#	tc	NITRO%	PHOS%	POTSH%	PLANT	ABC	BCD	BLOCK# (ACRE#)
12	(1)	10	5	5	Corn	-	-	1
2	a	20	5	5	Corn	+	-	2
13	b	10	10	5	Corn	+	+	3
8	ab	20	10	5	Corn	-	+	4
14	c	10	5	10	Corn	+	+	3
4	ac	20	5	10	Corn	-	+	4
1	bc	10	10	10	Corn	-	-	1
9	abc	20	10	10	Corn	+	-	2
11	d	10	5	5	Tomato	-	+	4
3	ad	20	5	5	Tomato	+	+	3
10	bd	10	10	5	Tomato	+	-	2
15	abd	20	10	5	Tomato	-	-	1
16	cd	10	5	10	Tomato	+	-	2
6	acd	20	5	10	Tomato	-	-	1
5	bcd	10	10	10	Tomato	-	+	4
7	abcd	20	10	10	Tomato	+	+	3

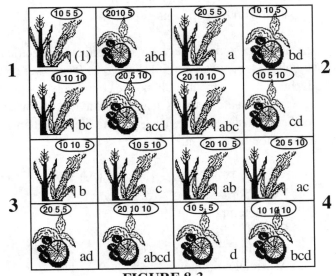

FIGURE 8-3

Fertilizers and Plantings over a Four-Acre Plot Using a Blocked Factorial

divided up into uniform "blocks" of land and the appropriate treatment combinations have been assigned to each block. Within each block we should assign the four treatment combinations at random. We can see this by looking at Figure 8-3. The boundaries we have set up are by measurement. Nature will impart her own differences in fertility across the land. There could be a diagonal swath of barren land running from the upper left corner to the lower right corner. While any such pattern of fertility could exist, our design can cope with it and will be able to remove its influence from the factors under study and still give us the sensitivity we need to complete the analysis.

Problems for Chapter 8

1. Compute the number of primary blocks in the following designs:
   a) a 2^6 with 4 treatments per block
   b) a 2^5 with 8 treatments per block
   c) a 2^4 with 4 treatments per block
   d) a 2^6 with 8 treatments per block

2. Select the primary blocks and compute the secondary blocks for the following designs:
   a) a 2^4 with 8 treatments per block
   b) a 2^6 with 16 treatments per block
   c) a 2^6 with 8 treatments pepblock

3. An experiment was designed to determine the effects of electrical coil breakdown voltage (i.e., when the coil shorts out). There were 6 factors under study, but the experiment could not be completely randomized due to changing environmental conditions. It is possible to make 8 runs under the same conditions, so the experiment was run in 8 blocks. Set up a reasonable design for this problem.

FACTORS	LEVELS	
1) Firing furnace	# 1 or	# 3
2) Firing temperature	1650°	1700°
3) Gas humidification	Present	Absent
4) Coil outer diameter	0.0300"	0.0305"
5) Artificial chipping	Yes	No
6) Sleeve	#1	# 2

4. A manufacturing engineer is investigating the effect of the following
   factors on the output of a lathe operation. There are two shifts that
   are involved in the study. Build a 32 tc design that will study the factors
   and remove the effect of the shifts as a random source of variation.

Factor	Levels
Speed	400 rpm – 600 rpm
Feed rate	5 ipm –10 ipm
Metal type	soft– hard
Tool angle	35° – 55°
Tool shape	chisel – stylus

5. Redo Problem 4 with only 16 tc's.

6. Confirm that the blocked design in Figure 8-3 is the same as the Minitab
   version shown in Figure 8-9.

APPENDIX 8

Blocking: A design method that allows the experimenter to remove an
                     unwanted, known source of variation from the data.

Blocked Effect: The source of variation being removed from the experiment.
                     We do not "study" this effect, but rather determine its
                     quantitative contribution to the variation and remove it from
                     the noise or error.

Primary Block(s):   Higher-order interaction effects that we have free choice in
                     determining. We will choose p primary blocks:

$$p = \frac{\ln (\mathbf{b})}{\ln(2)}$$

where ln is the natural log and **b** is the # of blocks.

Number of Blocks: There will be **b** blocks in a blocked 2^k design, where **b** will be an integer power of 2 and determined by the experimental conditions. If there are 32 treatments and it is possible to run only 8 treatments under uniform conditions, then there will be 4 blocks in the design.

Secondary Blocks: Resulting interactions derived by multiplying the primary blocks together using the modulo 2 algebraic system. So, if we selected 2 primary blocks (ABC, BCDE) from a 2^5 with 4 blocks, then the secondary block would be

$$ABC \times BCDE = AB^2C^2DE = ADE.$$

Degrees of Freedom: For blocks, the degrees of freedom will be one less than the total number of blocks. (df = **b** - 1 where **b** is the number of blocks)

Blocking with Minitab

Blocking is an option in the Minitab routines for generating 2^k and 2^{k-p} designs. We will rebuild the blocked design found in Figure 8-3. From the **Stat** dropdown we navigate to **DOE=> Factorial=> Create Factorial Design**, as shown in Figure 8-4. The main dialog box appears as shown in Figure 8-5. Choose four (4) factors from the scroll selector.

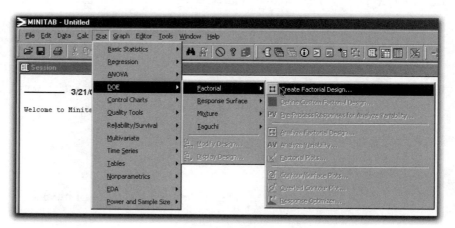

FIGURE 8-4

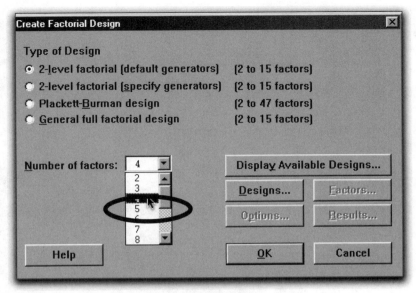

FIGURE 8-5

Click on **Designs**, which brings up the sub dialog box (Figure 8-6) where we choose the Full Factorial and 4 blocks. We will not include center points and replicate only once. Click **OK**, which takes us back to the main dialog box where we click on **Factors** shown in Figure 8-7.

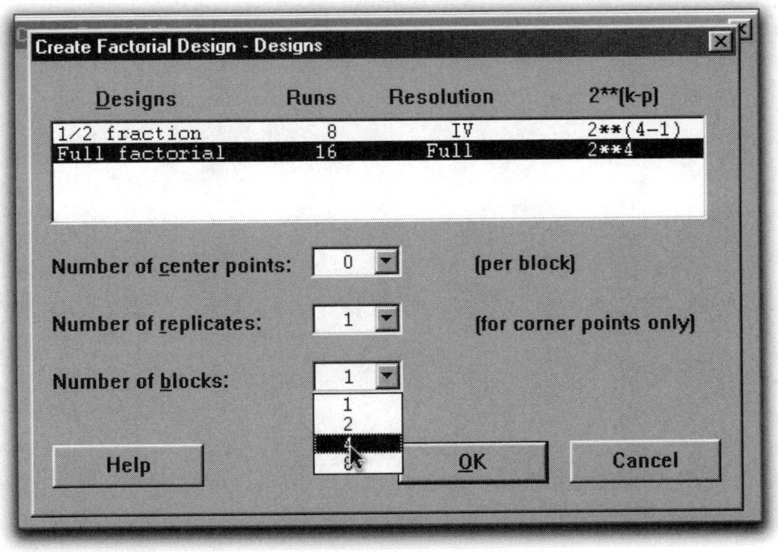

FIGURE 8-6

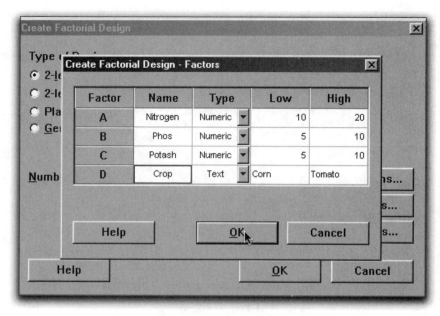

FIGURE 8-7

Enter the names and levels of the factors as shown in Figure 8-7. Be sure to make the crop factor a **Text** type. Click **OK** to return to the main dialog box. We will un-click the default **Randomize** in Options and stay with the defaults under Results. Click **OK** in the main dialog box (Figure 8-8) to produce the design shown in Figure 8-9.

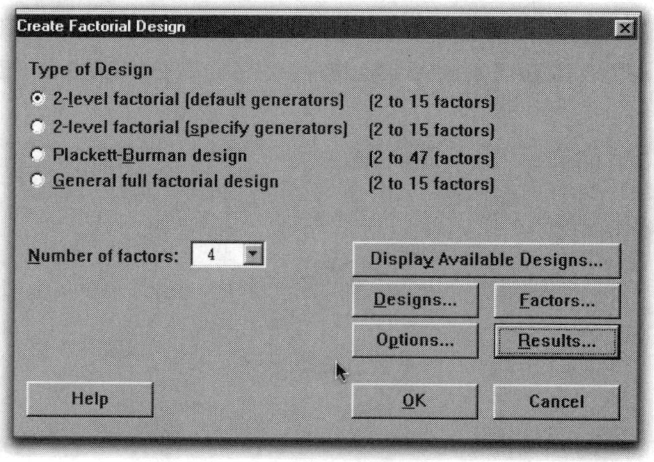

FIGURE 8-8

↓	C1 StdOrder	C2 RunOrder	C3 CenterPt	C4 Blocks	C5 Nitrogen	C6 Phos	C7 Potash	C8-T Crop
1	1	1	1	1	20	10	5	Corn
2	2	2	1	1	10	5	10	Corn
3	3	3	1	1	10	10	5	Tomato
4	4	4	1	1	20	5	10	Tomato
5	5	5	1	2	10	5	5	Corn
6	6	6	1	2	20	10	10	Corn
7	7	7	1	2	20	5	5	Tomato
8	8	8	1	2	10	10	10	Tomato
9	9	9	1	3	10	10	5	Corn
10	10	10	1	3	20	5	10	Corn
11	11	11	1	3	20	10	5	Tomato
12	12	12	1	3	10	5	10	Tomato
13	13	13	1	4	20	5	5	Corn
14	14	14	1	4	10	10	10	Corn
15	15	15	1	4	10	5	5	Tomato
16	16	16	1	4	20	10	10	Tomato

FIGURE 8-9

Minitab sorts the design runs by block rather than Yates order. Problem 6 asks that the reader confirm that the design as shown above in Figure 8-9 is the same as in Figure 8-3.

Blocked Fractional Factorials

Fractional factorials may be blocked. However, a whole new level of complexity is encountered when we block in 2^{k-p} designs. This is where a computer program is most useful. To illustrate this process, we will now redo the fertilizer example with a half fraction. Since Minitab remembers the previous inputs (factor names and levels, etc.), we only need to make one change in the **Designs** sub dialog box as shown in Figure 8-10 where we select "1/2 fraction" from the two choices available. In the **Options** sub dialog box (Figure 8-11), we request all the information. This is displayed in the Session Window. Figure 8-12 shows the new design. The session windows show the fractional factorial design generator (D ≈ ABC) and the block generators (Session Window 8-1) and the confounding pattern as well as the design table (Session Window 8-2).

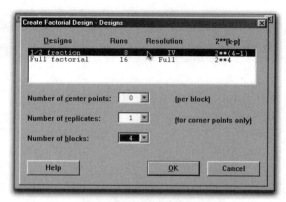

FIGURE 8-10

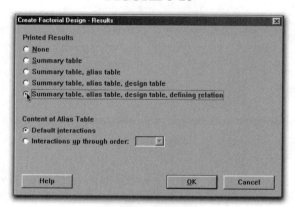

FIGURE 8-11

↓	C1	C2	C3	C4	C5	C6	C7	C8-T
	StdOrder	RunOrder	CenterPt	Blocks	Nitrogen	Phos	Potash	Crop
1	1	1	1	1	20	5	5	Tomato
2	2	2	1	1	10	10	10	Corn
3	3	3	1	2	20	10	5	Corn
4	4	4	1	2	10	5	10	Tomato
5	5	5	1	3	10	10	5	Tomato
6	6	6	1	3	20	5	10	Corn
7	7	7	1	4	10	5	5	Corn
8	8	8	1	4	20	10	10	Tomato
9								

FIGURE 8-12

Fractional Factorial Design

Factors: 4   Base Design: 4, 8   Resolution with blocks: III
Runs: 8   Replicates: 1 Fraction: 1/2
Blocks: 4   Center pts (total): 0

* NOTE * Blocks are confounded with two-way interactions.
Design Generators: D = ABC
Block Generators: AB, AC

SESSION WINDOW 8-1

Defining Relation: I = ABCD
Alias Structure
I + ABCD

Blk 1 = AB + CD
Blk 2 = AC + BD
Blk 3 = AD + BC

A + BCD
B + ACD
C + ABD
D + ABC

Design Table

Run	Block	A	B	C	D
1	1	+	-	-	+
2	1	-	+	+	-
3	2	+	+	-	-
4	2	-	-	+	+
5	3	-	+	-	+
6	3	+	-	+	-
7	4	-	-	-	-
8	4	+	+	+	+

SESSION WINDOW 8-2

Comments on Blocked Fractional Factorials

When we block a fractional factorial, the block influence requires degrees of freedom that would be assigned to interaction effects in the non-blocked version of the design. Thus, information on interactions is no longer available since the interactions are confounded with the blocks. In the above design there are 7 df available and with 4 factor; 4 of these df are used for the single effects leaving only 3 df. These 3 df are then used for the 4 blocks, leaving *no* df to investigate interactions of any degree. It is usually not wise to block in small fractional factorial designs since these blocked designs often become single effect experiments.

Randomized Block and Latin Square

Although the blocked 2^k factorial design is a common approach to compensate for non-uniformities in the experimental conditions, there are many other variations of the blocked experimental design concept. While the underlying motivation remains the same in all blocked designs, particular restrictions on randomization have resulted in some very clever schemes and treatment combination patterns to overcome the problem of not being able to randomize completely. Picking the proper design almost becomes intuitive and automatic once the basic concept is understood. The analysis and interpretation is where the difficulties arise. In this chapter, we shall look at some of the other types of blocked designs. Chapter 18 will then show how to properly analyze the data from these designs.

It is interesting that this particular aspect of experimental design (blocking) has attracted much attention in other books and in many papers, while it is used probably less than 20% of the time in industrial experimentation! However, if the experimental design team has identified known, unwanted nuisance factors, it is mandatory that blocking methods be used if we want valid results.

Complete Randomized Block

In this family of designs, all of the possible treatment combinations are present. The blocked 2^k designs belong to this group. The design is quite simple, but the analysis requires care. In such designs, it is possible to block on only one effect while investigating another factor. The usual type of nuisance effects that we block on include day-to-day variation, operator-to-operator differences, lot-to-lot raw material changes, and other sources of variance that we do not wish to study, but must be considered as potential sources of systematic error.

The following example shows the effect of day-to-day variation on the density of a black and white photographic film. The factor under study in this experiment is the type of developer solution. The objective is to test the hypothesis that optical density (how black the image gets) is a function of the type of developer. From some preliminary information, it is determined that 8 samples from each type of developer are necessary to control the risks of making the incorrect conclusions in the experiment. Because of the time necessary to process the film, only 8 runs may be made in a single day. It is also well known that there can be day-to-day differences in developing film, since developers age and become weaker. If we were to use only developer A on the first day and only developer B on the second day, we would not know if we were observing a developer deterioration or a day-to-day change. In statistical words, we would have confounded days with developers. To block the effect of the day-to-day variation out of the experiment, we run half of the tc's with developer A on day 1 and the other half on day 2. We do the same for developer B. Now, we have 4 runs from each type of developer within a single day. Since there could be some variation induced during a day, we randomize the order of running the 8 treatments within each day. Our design is shown in Table 9-1. (The values in parentheses are the responses, the optical density readings.)

TABLE 9-1

Developer A	Developer B	
tc #3 (1.1)	tc #6 (1.2)	
tc #7 (1.0)	tc #2 (1.3)	Day # 1
tc #1 (1.1)	tc #8 (1.2)	
tc #4 (1.3)	tc #5 (1.1)	
$\overline{X}= 1.125$	$\overline{X}= 1.20$	
tc #4 (1.2)	tc #7 (1.4)	
tc #8 (1.4)	tc #1 (1.5)	Day # 2
tc #5 (1.3)	tc #6 (1.4)	
tc #2 (1.2)	tc #3 (1.5)	
$\overline{X}= 1.275$	$\overline{X}= 1.45$	

We can see from the averages ($\overline{X}$'s) that there is a difference between days and also between developers. We shall see in a more extensive analysis in Chapter

18 how the day-to-day differences could have clouded the results if not analyzed as a blocked design. The blocked design allows us to remove a source of known nuisance variation from the error while still running a minimum of treatment combinations. We shall complete this analysis in Chapter 18 via analysis of variance (ANOVA).

Generalization of Results

Up to this point, we have worked with blocked experiments to remove the influence of unwanted sources of variation from the error. There is another reason to block in experimental design. The following example illustrates that concept.

A major energy company wants to test the effectiveness of an additive on automobile gasoline efficiency. The pilot plant prepares a quantity of unleaded gasoline and adds 10% alcohol to half of the batch. The objective of this experiment is to test the hypothesis that miles per gallon (mpg) is a function of type of gasoline. Since the results of the experiment could be used in a national advertising campaign, the test engineer wants to be sure that there is sufficient evidence to support a claim that gas-o-hol increases mpg. Considering the diversity of automobiles on the road, she decides to run the experiment not with just one automobile, but with a variety of automobiles. By doing this, the inference space is not restricted to only one make and model of vehicle, but is opened up to include the range of cars found on the road.

By looking at sales figures of domestic and foreign cars, the engineer picks the top five models that make up almost 65% of automobiles being driven, and adds five other models selected for their inherent efficiency. We now have a sufficient inference space capable of supporting any claims about the fuels. Table 9-2 shows the 10 cars and the results of fuel performance using each type of gasoline in all automobiles. We could look at the average of the mpg over the 10 cars for each type of fuel and do a simple "t" test of the difference between two means as outlined in Chapter 12. However, since there are 10 different types of cars in each average, with a wide range of inherent mpg capabilities, the variation between the cars could lead us to conclude that there is no significant difference between the fuels. We can get around this problem of inflated variance by a simple trick in the analysis. Instead of trying to draw an inference about the difference between the two means, we will find the differences between each pair of fuelings for each car and test to see if these differences, when averaged over the 10 cars, produce a value significantly different from zero. This method, which is the very simplest of blocked designs, is called the "paired comparison" method or sometimes the "correlated pair."

The experiment in Table 9-2 illustrates again that it is not so much the design but the analysis that is key in the understanding of information from experiments that are blocked. Chapter 18 will complete the analysis of this randomized block experiment to see if the gas-o-hol is indeed significantly better than the plain, unleaded fuel.

TABLE 9-2

Car Type	Plain Unleaded	Gas-o-hol	Difference
Civic	31	34	3
Focus	33	37	4
Corolla	39	38	-1
Allero	28	30	2
PT Cruiser	26	27	1
VW "Bug"	39	43	4
Coup deVille	18	21	3
Regal	23	26	3
Impala	14	18	4
Taurus	18	20	2
	$\overline{X} = 26.9$	$\overline{X} = 29.4$	$\overline{D} = 2.5$
	s= 8.75	s= 8.46	s_D= 1.58

Misconceptions in Using Blocked Designs

There is a popular expression often quoted about the conduct of statistical experimental designs that says, "Randomize if you can and block if you must." Most uninformed experimenters take this to mean that if there is some difficulty in randomizing, then a blocked design will somehow allow the runs to be done in an easier, systematic order. Remember, the reason we block is to prevent a known, unwanted source of variation from inflating the experimental error. We never *study or draw inferences* on the blocked source of variation.

A typical misconception in the use of the term *blocking* is illustrated by the following example. A chemical engineer is investigating, among other factors, the effect of temperature. It is actually quite inconvenient to randomize the levels of the temperature and instead the engineer "blocks on temperature." That is, a level of temperature is set and all of the treatment combinations requiring that level are completed before going on to the next level of temperature. Figure 9-1 shows the run order for this approach.

This approach is exactly what was warned about in Chapter 4, and the same consequences that this lack of randomization had then still hold even with the statistical jargon of "blocking" applied to this incorrect method. The heart of the problem is the simple fact that we are not blocking on temperature, since temperature is a factor we want to study. Granted, it is difficult to keep changing the temperature and the experiment might never be run if the imposition of randomization is enforced. What is actually happening in the experiment illustrated in Figure 9-1?

tc	Temperature	Time	Concentration	Run Order
bc	100	10	1.0	1
(1)	100	5	0.5	2
b	100	10	0.5	3
c	100	5	1.0	4
ac	200	5	1.0	5
ab	200	10	0.5	6
a	200	5	0.5	7
abc	200	10	1.0	8

While convenient to run, this arrangement will lead to possible inclusion of unwanted, unknown effects that will "taint" the actual physical results and lead to incorrect conclusions.
FIGURE 9-1

Rather than use the word "block" in reference to the arrangement of the experiment in Figure 9-1, look at the way the experiment has been run in two segments. There is the 100 degree segment and the 200 degree segment. We have split the experiment into two pieces. There are two ways we may treat this segmenting. The first and statistical approach is to consider this easier to run, segmented design as a "Split-Plot" experiment where the lack of randomization is accounted for in the way the data are treated in the analysis. Chapter 18 covers this type of analysis.

The other, physically based approach makes use of prior engineering knowledge of the unwanted influences that could taint the results if they were to change during the course of the execution of the experiment. Remember, we randomize to prevent the unknown, unwanted sources of variation from biasing the results. However, if we *know* these unwanted sources of variation, and, more important, can *control* them, we can with more assurance run the experiment in a convenient, segmented order. So the proper term to apply to such designs is **segmented design**, not blocked design.

Latin Squares

We have seen that blocking in experimental design can remove an unwanted source of nuisance variation and can also help to expand the inference space of our experiment and thus allow a more generalized conclusion. In all of the examples so far, we have blocked on only one effect. It is possible to block on two effects in a single experiment by using a design called the Latin Square. This design has attracted a lot of attention in the field of combinatorial mathematics, and it has been studied by mathematicians for its pure structure, for there are only a finite set of Latin Squares that may be constructed and the challenge is to find them *all*. However, while blocked designs are used only 20% of the time, the Latin Square is used even less than that (possibly only 5%), and this design, although an important part of the experimental design methodology, does not deserve this extensive publicity.

As the name implies, this design must be a square. That is, there are as many levels in the factor we are investigating as there are items in the two blocked "factors." The quotation marks around the word factor are important since we cannot study a blocked factor but only remove its effect from the analysis. This is a very important concept that is sometimes violated in a misapplication of the Latin Square.

Since there must be an equal number of levels among the factor under study and the blocks, it is sometimes difficult to apply the Latin Square to ordinary problems. This is one of the reasons that the use of this type of design is limited.

To build a Latin Square, we first usually identify the number of levels in the factor under study. Then we must match the number of levels in the "factors" we are blocking on to this number. Traditionally, the blocks are assigned to the rows and columns of the square and the factor under study is put into the body of the matrix. To create such a three-dimensional matrix in only two dimensions, we create the illusion of the third dimension by superimposing letters of the ordinary alphabet (Latin alphabet, thus *Latin Square*) on the two-dimensional matrix. Table 9-3 shows a 5 by 5 Latin Square design.

Block #1 runs across the top and block #2 runs along the side. The factor under study is identified by the letters A to E in the body of the experiment

TABLE 9-3

BLOCK #1

		1	2	3	4	5
	1	A	B	C	D	E
BLOCK #2	**2**	B	C	D	E	A
	3	C	D	E	A	B
	4	D	E	A	B	C
	5	E	A	B	C	D

In superimposing the letters on the two-dimensional matrix, we have cycled them so that no row or column has the same letter in the intersection more than once.

In running the experiment we follow the conditions set by the row and column block levels, and the letter identifies the level of the factor under study.

Now let's look at an example of a Latin Square with six levels in the factor under study. We need to investigate the quality of photographs taken on 6 different color films that are currently on the market for a competitive evaluation. Since we are interested in consumer opinion as a part of this assessment of quality, we will go directly to the customers and ask their opinion of the pictures. We carefully control the lighting, camera, and subject movement conditions in the creation of the pictures so these factors do not influence the judgments. We do, however, use six different picture content scenes, since it is a well-known fact that picture content can influence the opinion of an observer. By including the various picture contents, we can expand our inference space beyond only one type of picture. We also know that different observers could inflate the error in our determinations of picture quality and render our experiment useless. Therefore, we will block this nuisance factor out of the analysis. The design for this doubly-blocked experiment is found in Table 9-4.

Since we are not aware of any other possible nuisance factors, but suspect that they could bias the results of the experiment, we will randomize this experiment like any other experiment. There are two possible approaches to randomization of the Latin Square design. The simpler approach is to assign random ordering to each treatment combination and execute the runs in that order. Another approach is to assign a random number to each column and reorder the columns, and then do the same thing to the rows. To illustrate this approach, we will reorder Table 9-4 first by column and then by row to obtain the final random order. Table 9-5 reorders the columns and Table 9-6 reorders the rows.

Before we begin gathering data in the above experiment, we should review the reasons for running such a study. Many times, in such a detailed experiment, we become so involved with the design, that we often forget the original reason for the work! To bring us back to reality, we need to understand the goal and the objective in this study of film types.

The goal is to find the best film for the purposes we have identified and the applications we specified. An experiment of this nature could be the basis of a report in a consumer magazine. The goal in this instance would be to make a valid recommendation on film type.

The objective of the experiment is to test the hypothesis that judged image quality is a function of film type. Notice that we are not interested in studying differences in picture content or differences between observers. There is only one factor under study, the film type. The picture content is included in the design to expand the inference space, and the observer "factor" is merely a method of replicating the work to give it greater statistical validity.

Now it's time to gather some data. We start in the upper left corner of the design with Observer #3. Actually in selecting observers, we have "natural" randomization, since the selection process is a random process. However, given that we have the six observers in the room, we will take them in the order the design dictates. While there are a number of methods of attaching a numerical value to a psychological response, we will use the psychometric method called "magnitude estimation" as developed by S. S. Stevens. In this method the observer reports a numerical value that he or she associates with the "goodness" of the picture.

TABLE 9-4
Block #1 - Picture Content (for Generality)

		Rural	Portrait	Child	Old Man	Model	Food
	1	Fuji	Kodak	Fotomat	K Mart	Ilford	Agfa
Block	2	Kodak	Fotomat	K Mart	Ilford	Agfa	Fuji
#2	3	Fotomat	K Mart	Ilford	Agfa	Fuji	Kodak
Observer	4	K Mart	Ilford	Agfa	Fuji	Kodak	Fotomat
(nuisance)	5	Ilford	Agfa	Fuji	Kodak	Fotomat	K Mart
	6	Agfa	Fuji	Kodak	Fotomat	K Mart	Ilford

Take Table 9-4 and arrange the picture content columns in random order to produce Table 9-5. The original columns are in the following order:3,5,1,2,4,6.

TABLE 9-5
Block #1 - Picture Content *[randomized by column]*

		Child	Model	Rural	Portrait	Old Man	Food
	1	Fotomat	Ilford	Fuji	Kodak	K Mart	Agfa
Block	2	K Mart	Agfa	Kodak	Fotomat	Ilford	Fuji
#2	3	Ilford	Fuji	Fotomat	K Mart	Agfa	Kodak
Observer	4	Agfa	Kodak	K Mart	Ilford	Fuji	Fotomat
	5	Fiji	Fotomat	Ilford	Agfa	Kodak	K Mart
	6	Kodak	K Mart	Agfa	Fuji	Fotomat	Ilford

Finally take Table 9-5 and randomize the observers (rows). The row order is now: 3,6,1,4,5,2.

TABLE 9-6
Block #1 - Picture Content *[randomized by row]*

		Child	Model	Rural	Portrait	Old Man	Food
	3	Ilford	Fuji	Fotomat	K Mart	Agfa	Kodak
Block	6	Kodak	K Mart	Agfa	Fuji	Fotomat	Ilford
#2	1	Fotomat	Ilford	Fuji	Kodak	K Mart	Agfa
Observer	4	Agfa	Kodak	K Mart	Ilford	Fuji	Fotomat
	5	Fuji	Fotomat	Ilford	Agfa	Kodak	K Mart
	2	K Mart	Agfa	Kodak	Fotomat	Ilford	Fuji

In this case, the scale runs from 0 (poor) to 100 (good). We will present the pictures one at a time to the observer and obtain the responses as shown in Table 9-7.

TABLE 9-7
Block #1 - Picture Content

		Child	Model	Rural	Portrait	Old Man	Food
	3	75(I)	85(Fj)	80(Ft)	50(KM)	65(A)	88(K)
Block	6	97(K)	43(KM)	73(A)	85(Fj)	60(Ft)	65(I)
#2	1	82(Ft)	67(I)	87(Fj)	97(K)	37(KM)	75(A)
Observer	4	82(A)	87(K)	42(KM)	67(I)	72(Fj)	65(Ft)
	5	95(Fj)	70(Ft)	65(I)	80(A)	80(K)	43(KM)
	2	57(KM)	77(A)	92(K)	77(Ft)	57(I)	85(Fj)

We can find the average values for the six types of films from the body of the table. We can also find the average values of the picture content by finding the averages of the columns. The row averages give the differences between observers. These last two "factors" show a difference, but since these were added merely to increase the inference space of the experiment, we don't read any meaning into the differences. The only factor under study in this experiment is the film.

TABLE 9-8
Film Brand Averages

Ilford	Fuji	Fotomat	K Mart	Agfa	Kodak
66	80.7	70.7	45.3	75.3	90.2

In Chapter 18 we will carry the detailed analysis of this Latin Square to completion.

The Mis-Use of the Latin Square

We have emphasized the fact that in a Latin Square design, only one source of variation can be studied as a true factor. The other two "factors" are merely removed from the analysis to reduce the noise portion of the variation. Some experimenters do not realize that in blocked designs there is an underlying assumption that there is no possibility of any interactions taking place between the nuisance factors and the factor under study. Since the Latin Square design is capable of handling a multi-level situation, there is an appeal to use it as a three-factor multi-level fractional factorial design. Such a use would be contrary to the requirement of no possible interactions, and we shall show in the following example that this is a complete mis-application of the Latin Square.

While the following example is a simple 2-level design, the consequences can be expanded to the multi-level situation. Table 9-9 shows a Latin Square for the study of three factors. We have coded the levels with the conventional "-" and "+" signs used in Chapters 4 and 5 to indicate the low and high levels of the factors.

TABLE 9-9

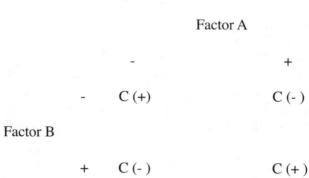

Factor A

		-	+
	-	C (+)	C (-)
Factor B			
	+	C (-)	C (+)

Since this is a 2-level design, we can put it in YATES order in a different tabular form that conforms to the conventions shown in Chapters 4 and 5. Reconstructing Table 9-9, we get Table 9-10.

TABLE 9-10

A	B	C	AB	AC	BC
-	-	+	+	-	-
+	-	-	-	-	+
-	+	-	-	+	-
+	+	+	+	+	+

By observing the matching patterns of the signs in the single effects and the interactions in Table 9-10, we can readily see why interactions are not allowed in a Latin Square design. The single effects are completely confounded with the interactions! Another way of showing that the information in a Latin Square is limited in scope is to look at the degrees of freedom. As we showed in Chapter 5, this is a quantitative way of measuring the volume of information in an experiment.

In the above design, there are 3 degrees of freedom (one less than the total number of observations). We require one degree of freedom for each single effect since there are two levels in each factor. We "use up" all the available (3) degrees of freedom immediately with the single effects. There are no degrees of freedom left for the interactions. From an information content we cannot investigate interactions in a Latin Square; therefore, this design cannot be used as a factorial design where the intent is to be able to obtain information on interactions.

The concept of information content as measured by degrees of freedom can be applied to even larger Latin Squares. For example, a 5 by 5 Latin Square has 24 degrees of freedom. The three single effects require 12 degrees of freedom, which leaves 12 degrees of freedom remaining. However, each two-factor interaction uses 16 degrees of freedom ($4 * 4$). There are three, two-factor interactions among the three factors, and if we were to measure them, we would need 48 degrees of freedom, which is 36 more than we have left after measuring the single effects! Since degrees of freedom cannot be created out of thin air, we can see that, again, interactions are not possible in a Latin Square. What is worse is the fact that, if interactions exist between the factors under study, their effects will be confounded with the single effects, which will bias the conclusions.

The solution to the multi-level fractional factorial is the central composite design (CCD) that we studied in Chapter 6. A three-factor central composite design requires even less work than a 5 by 5 Latin Square (there are

15 treatment combinations in the CCD with 3 factors). It is the allocation of the degrees of freedom in the CCD that makes the difference between the good information it obtains and the mis-information conferred upon us by a Latin Square when used as a factorial.

Problems for Chapter 9

1. An assembly line process is conducted in two plants, and there is a suspicion that the two plants do not produce the same level of quality. Because of the nature of the response variable, it is necessary to have a relatively large sample to draw the conclusions with low risk. The required sample size is 250 units per line. However, the assembly process is slow, and each line can make only 10 units per day. If there is evidence from the control charts on the lines that there is a hefty amount of day-to-day variation, then how would you design such an experiment to see if the lines differ? State the goal and objective for this experiment and propose a design that will handle the nuisance of day-to-day variation.

2. There is a suspicion, but no evidence, that two methods of analysis produce different results, although they are supposed to be identical. Propose an experiment that will settle the argument.

3. If I have four different copiers and wish to study the image quality produced by these devices, how many observers and test patterns will I need if I want to generalize the results with respect to test patterns and remove the effect of the observers?

4. Show the design for Problem 3 above. Put it in random order in readiness for running.

5. Consider an example of a situation from your experience that would require a blocked experimental design. Show the general design for this situation.

6. Think of an experiment you may have run that was called a blocked design. Based on the true concept of blocking (removal of an unwanted, known source of variation from the error), was your design a real blocked design or did you use the term *blocking* to cover up a lack of randomization?

APPENDIX 9

Blocking: A technique used to:
1) Remove an unwanted, known source of nuisance variation from the error in an experiment.
2) To expand the inference space of an experiment by generalizing over a set of conditions.

Randomized Block: One of the types of blocked experiments in which a block or condition is identified and within which the treatment combinations are randomized as if there were a "mini-experiment" taking place in that block alone. However, there will be more than one block in the whole experiment, which is considered in the total analysis.

Paired Comparison or Correlated Pairs: The simplest of the randomized block experiments where there are two levels of the factor under study. A "t" test on the average of the differences between individual paired observations is used to remove the unwanted source of variation.

Latin Square: A randomized block design that blocks on two sources of nuisance variation (or two generalizing variables, or one of each). In the Latin Square there must be as many levels in each of the two blocks as there are in the factor under study. Each level of the factor under study is combined once and only once with the intersections of the levels of the two blocking effects.

Interactions and Blocks: There can be no interactions between the block effect and the factor under study for valid conclusions to hold.

Segmented Design: An experiment that is easier to run since the random order is partially ignored. Unlike blocking, a segmented design requires knowledge and control of the unwanted sources of variation.

Split Plot Design: An experiment that is easier to run since the random order
                       is partially ignored. Unlike a segmented design, there is no
                       need to know the sources of the unwanted variation. The
                       statistical analysis compensates for the lack of
                       randomization. See Chapter 18.

10

Nested Designs

One of the uses of experimental design suggested in Chapter 1 was the "quantification of errors." All of the designs we have used so far have been concerned with the determination of a factor's functional effect on a response in a mechanistic or "fixed" sense. In addition to this fixed way of looking at the mechanism, we also set up the design to allow each of the levels of a factor to be present and in combination with the levels of all other factors. These factorial designs "cross" all the factors and levels. Table 10-1 shows a typical multi-level factorial experiment with replication. Observe how the levels of Factor #1 combine or cross with the levels of Factor #2.

TABLE 10-1

		Temperature (Factor # 1)			
		100°	200°	300°	400°
	10	R1	R2	R1	R3
	psi	R2	R1	R3	R2
		R3	R3	R2	R1
	15	R3	R1	R1	R1
Pressure	psi	R1	R2	R3	R3
(Factor #2)		R2	R3	R2	R2
	20	R3	R3	R2	R1
	psi	R1	R2	R1	R3
		R2	R1	R3	R2

In the above design, with the two factors of temperature and pressure, we find that for the 10 psi pressure setting, all four temperature levels are present. This allows us to study the effect of temperature at a specific level of pressure, and in fact do so at more than one level of pressure as we see in the rest of the experiment, for the pattern of temperature change repeats over all three levels of pressure. Similarly, we can see from Table 10-1 that the pressure is studied (or changed) over all levels of temperature.

The "crossed" factorial experiment has an important place in experimental techniques and has been the subject of most of this book, for it allows us to study and make inferences about the interactions between factors. However, as we look further into the design in Table 10-1, we find another "factor" with three "levels." This factor is not really a factor at all, but the replication used to determine the error or noise in the experiment. We have already discovered that the replicates must be randomly placed throughout the experiment. Therefore, each replicate has an equally likely chance of appearing anywhere in the experiment at any time. Because of this randomness, we ask the question, Does it make sense to compare the replicates' levels as we make comparisons across the temperature and pressure? Of course, the answer is no. The three levels of the replication cannot be combined to study the "effect" of Rl or any interaction between Rl and the other true factors of temperature and pressure.

The replicates are included in the experiment for the simple reason to quantify the random error involved in setting up the control factors and measuring the response variable. The labels Rl, R2, R3 are mere conveniences in the outline of our design. The set of three replicates contained in each treatment combination is complete in itself. The only remote relationship between each replicated cell we would like to see is that the amount of variation in each cell does not differ significantly from the variation in any of the remaining cells.

In the analysis of variance as shown in Chapter 13, we compute the pooled (or averaged) variance across the replicated cells. We can look at the error variance in each cell as contained or "nested" in that cell.

Figure 10-1 shows the way the crossed experiment involving temperature and pressure would look in diagrammatic form. In this diagram, a line is drawn from each level of Temperature to each level of Pressure and represents a single treatment combination or run in the experiment. Notice how the lines cross. This type of design is called the "crossed" factorial. If we were to take one of the treatment combinations from the factorial design and extend the diagram to include the replicates, then we would obtain the new diagram shown in Figure 10-2.

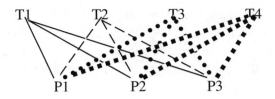

FIGURE 10-1

Notice that the three replicates do not cross over into another cell. This method of configuration is called "nesting" and can be extended to designs beyond the simple application of replication.

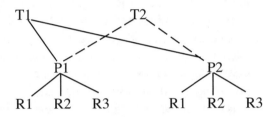

FIGURE 10-2

A Nested Design

In many manufacturing situations, we may have a process producing a product that is experiencing a greater degree of variation than can be tolerated by the functional specifications. The control chart may be in control, but the overall level of variance is too high. By merely taking the product as it emerges off the end of the line and measuring the response, we can compute the mean level of the quality parameter and its variance. This variance is probably the result of many influential sources. We would like to find the key source or sources of variance that contribute most to the overall observed variation. If we are able to do this detective work, we then have a good chance of controlling the overall variation and producing a product that is fit for use (that is, it meets the functional specifications).

Identifying and quantifying the sources of variation in an efficient manner is the job of the nested experimental design. In its most fundamental application, the concept of nesting is used in all replicated experiments to quantify the error variance. We have just shown in the previous example (Table 10-1) that even in a fully crossed, fixed factorial experiment the error is always nested within the cells created by the combinations of the factors under study. We can extend the nesting concept upward to all of the factors in the experiment and study the effect of the variation as well as the effect of the change in overall mean values.

Coals to Kilowatts

To illustrate the concept of a nested design, let's look at an example from the electric power industry. The problem is centered around the level of pollution producing components in solid fossil fuel (coal) used for producing steam to generate electricity.

The coal comes from strip mines and is loaded into hopper cars directly from the mine. The electric utility is able to sample the coal before it leaves the mine site. If the sulfur content is below 5%, the EPA will allow the coal to be burned. The electric company has observed a high degree of variation in past shipments of coal from this mine and wants to locate the sources of variation. The possible variance-inducing factors are hopper car to hopper car, samples within hopper cars, and analyses within samples. Figure 10-3 shows the diagram of this experiment, which is a completely nested design.

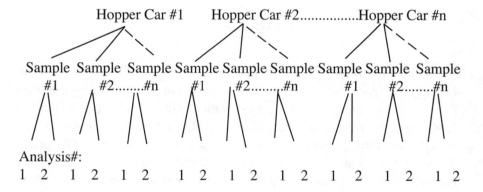

FIGURE 10-3

Each hopper car constitutes an experiment in itself. We can sample each car and find the variation induced in sulfur % from the chemical analysis. We can also find the variation from sample to sample taken within the car. If we look across many hopper cars, we can identify the amount of variation caused by car-to-car differences. Of course, all of the variation we observe is a function of the way the coal was formed in the ground, for the hopper cars only represent the loading of the coal from the mining operation. The nested design breaks the loading process down into a hierarchy of levels that can be isolated to pinpoint the sources of variation. The nested design is sometimes called a hierarchical experiment due to its ordered structure.

Getting back to the example, let's say that we randomly pick three hopper cars and take four samples per car. We will split the samples into two parts and replicate the measurements to quantify the levels of variation induced by the chemical analysis technique. Table 10-2 shows the data for this nested experiment.

TABLE 10-2

	HOPPER CAR #1								HOPPER CAR #2								HOPPER CAR #3							
SAMPLE	1		2		3		4		1		2		3		4		1		2		3		4	
ANALYSIS	1	2	1	2	1	2	1	2	1	2	1	2	1	2	1	2	1	2	1	2	1	2	1	2
SULFUR CONTENT	2	3	4	4	2	2	2	4	4	4	4	5	3	3	3	5	6	7	5	6	4	6	5	7

The average sulfur content across the 24 observations is 4.17, and the standard deviation is 1.52. If we can assume a normal distribution, then a portion of the coal will be expected to have a sulfur content above the 5% EPA limit. Based on the normal distribution model, we can expect that 30% of the coal from this mine to be in excess of the EPA standard for sulfur content. Since we have exceeded the specification, the real question is, where are the major sources of variation? Or is it the chemical analysis that is causing the variance to be so large? The samples we have taken in the above nested design can answer these questions, since the structure of this design allows us to split the overall variance into its component parts. In Chapter 18 we will complete the analysis of this experiment and pinpoint the likely sources of variation. We will also show what to do with this knowledge of the variance in controlling the differences between cars, samples, and chemical analyses.

Summary

The nested experimental design allows us to structure the treatment combinations into homogeneous groups that can be treated as separate sources of variation. In this way, we can dissect an overall source of variation into its component parts. With the knowledge of the major sources of variation, we can control the overall variance by controlling the individual components of variance.

Problems for Chapter 10

1. Show in a schematic outline an experimental design to study the components of variance for a chemical process which produces a granular product (1 to 5 mm beads). The process is run in batches, which are determined by the raw material. The product is packaged in large barrels with more than 100 barrels per batch. The property under study is the viscosity of the material when heated and extruded through a small orifice.

2. A product that is being produced on an "around the clock" production schedule is experiencing a greater-than-expected reject rate. How would you begin to attack this problem? What questions could be asked to identify the sources of variation?

3. A chemist has developed a new analytical method to determine the concentration of a harmful chemical in fish. Suggest a nested design for the following possible variance contributors: between sampling locations (lakes), between batches of reagents, between fish.

4. A market researcher wants to determine the best sampling scheme that will produce the lowest variance with the least cost She expects to find sample-to-sample variance due to cities, states, counties, and people. What type of experiment could help determine which source of variation is the biggest? Outline such an experiment.

APPENDIX 10

Fixed Effect: We study only the levels present in the design to determine their mechanistic influence on the response. We may interpolate between levels, but we do not make any general statement as we do about the random effect.

Random Effect: We are not interested in measuring a mechanistic effect and more emphasis is placed on the general influence of this type of factor on the overall variation imposed on the response. Instead of being picked at specific levels, the random factor's levels are chosen by chance from a large array of possibilities.

Nested Factor: A set of treatments that are complete in themselves and do not cross into another set of treatments. Random error found by replication is always nested within the treatment combinations.

Crossed Factor: The levels of this type of factor combine with all the levels of the other factors.

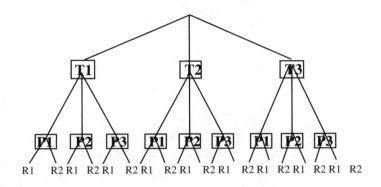

Nested Factors

T1			T2			T3		
P1	**P2**	**P3**	**P1**	**P2**	**P3**	**P1**	**P2**	**P3**
R1	R2	R1	R2	R1	R2	R1	R2	R1	R2	R1	R2	R1	R2	R1	R2	R1	R2

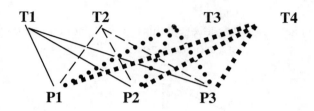

Crossed Factors

	T1	T2	T3	T4
P1				
P2				
P3				

11

Evolutionary Operation

All of the experimental designs we have studied so far have been "revolutionary" in their nature. In our study of process behavior or product characteristics, we have made rather large changes to the control factors to allow the observation of wide changes in the response variables. These experiments have revealed much information about the process or product, but the techniques have caused big changes to take place. Sometimes our experiments produce treatment combinations that actually make bad, or less than desirable, levels of the response variable's requirement. In such experiments, we welcome such "negative" information since it tells us how a process works. The contrast between the low level of a response and its higher level provides the information necessary to gain insight and understanding.

In two respects, then, these designs are <u>rev</u>olutionary. First, there is a great "<u>revolt</u>" or change in the levels of the factors under study. Second, because of this revolt we are able to "<u>reveal</u>" great quantities of information. Such revealing designs with "revolting" levels are useful in research projects, development endeavors, pilot plant trials, and just about any other experimental situation where we *can* and *may* make wide changes to the control factors under study.

The Prime Directive of Manufacturing

There is, however, a situation where wide ranging changes in the control variables are not allowed, for fear of producing "bad product." If a product is already being

225

manufactured and has not had the advantage of a good history of experimentation during its development cycle, then it may be advantageous to improve the process via some type of systematic investigation. The problem, however, in such a situation is that revolutionary changes will probably cause the process to go out of control and, worse yet, produce some bad product.

>**The *prime directive* of a production line is not to produce any more bad product than would be expected based on past history.**

Due to the nature of revolutionary types of experimentation, it is clear that such designs are incompatible with the basic goals of a manufacturing situation.

Evolutionary Operation

Instead of applying the relatively swift methods of a revolutionary experiment, we use the slow, systematic technique that allows us to evolve into improved process conditions. This methodology, called "Evolutionary Operation," or EVOP, adheres to the following concepts and allows us to explore the terrain of a response surface.

> 1. **PRIME DIRECTIVE**: No interruption of the process. No increase in defective production.
> 2. The investigation is based on small changes to the control factors under study.
> 3. The design is a 2-level factorial, plus a center point
> 4. Work is conducted in phases with a change to new conditions (a new phase) taking place only when we have statistical evidence of a better set of conditions.

The four EVOP concepts mesh together to form a closed loop. The small changes in the control factors assure us that the prime directive will not be violated. (If it were violated, we would soon be "out of the factory on our ear.") The systematic approach in our study of a response surface utilizes the tried-and-true 2^k factorial with the added benefit of a center point to allow for the investigation of curvilinear effects.

Figure 11-1 is a photograph of a response surface. This particular surface represents the yield of a process as a function of temperature and pressure. The higher the "mountain," the greater the yield. If we were to take a "slice" of the mountain parallel with the ground, we would obtain an iso-yield section. That is, all around the perimeter of the slice we would observe the same value of the response. We would continue to take slices of the mountain and much like a contour map transfer these sections to a two- dimensional plot (as shown in Figure 11- 2) that shows the relationship between yield and the two control variables of pressure and temperature. To use the information in the re- sponse surface diagram, we find the intersection of the pressure and the temperature that produces a given yield. We would probably be interested in obtaining the highest yield, so for this surface we find that where the temperature is between 79 and 81 and the pressure is between 22 and 23, we get a region that gives 100% yield. However, this surface is not parallel to the temperature axis and some of the combinations of temperature and pressure points will not give 100% yield. To be exact, the 81°, 22 psi combination gives only a 67% yield. The slope of the surface is very steep in this part of the drawing, as indicated by the closeness of the contours.

On the other end of the yield range, we find that there are many combinations of pressure and temperature that give us the lowest reading of 40% yield. Some of these are:

$$
\begin{array}{ll}
81° & 21.00 \text{ psi} \\
84° & 22.75 \text{ psi} \\
76° & 22.75 \text{ psi} \\
77° & 23.15 \text{ psi}
\end{array}
$$

One feature of this process is the large changes in yield as a function of the changes in temperature and pressure. The manufacturing process must be kept in tight control to hold the specific yield at all. Let's say that we are obtaining yields between 60% and 70% for a set point of 77°, 21 psi with tolerances of ±1° and ±0.5 psi. If we were to experiment and make large changes (say, 2 to 3 degrees and 1 to 2 psi) we would certainly get into the 100% yield region, but we would also fall into the 40% or less region. Remember, we are working in a manufacturing operation and must continue to deliver a good product!

Because we may not disrupt the process, we resort to the EVOP method to *guide* us up the slope of the surface to the higher yield region. Of course, we don't know what the shape of the surface looks like as we begin, so we must grope around and find the proper ascent vector from our current set point. To do so, we

FIGURE 11-1

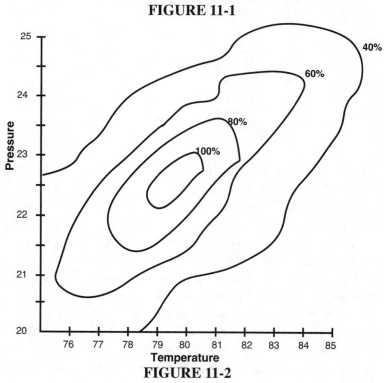

FIGURE 11-2

set up a small 2^2 factorial design with a center point. The center point is the current operating conditions of 77° and 21 psi. We stay within the tolerances of the process and use temperature excursions that range only over $\pm 0.5^{\circ}$. This means that our low level of temperature is 76.5°, while the high is 77.5°. The pressure is changed only $\pm .25$ psi. This produces levels of 20.75 psi and 21.25 psi.

While we attempt to set and hold the process conditions, we realize that such exact settings are probably impossible and combined with the natural measurement error in the response, we find that while we do our best, the yield is subject to variation and shows fluctuations even when we set the factors exactly the same. To compensate for this variation and to characterize its magnitude, we replicate the conditions. Replication is no problem, since we are in a production situation and we take advantage of the many times the process is run in a day.

A Slow Experience

To take advantage of the EVOP method, we must be very patient. The purposeful modifications in the manufacturing conditions will produce changes in the response that are very close to changes induced by random noise. However, by taking many observations, the variation between the averages we observe can be reduced by the "central limit theorem." Of course, by taking more and more samples, we build up enough data to create mean values that begin to stabilize. If there are differences between the means, then we will be able to detect these differences when the sample size gets big enough. EVOP has its own analysis technique that we shall watch as we complete this example involving yield.

Table 11-1 is the EVOP worksheet that allows us to gather data and complete the analysis at the same time. This form has been established for a two factor EVOP and the factors used to compute the variation apply only to this type of design. Most EVOPs are limited to two factors, so this form and the following example will handle most of the situations encountered. A blank form is found in the Appendix of this chapter.

Use of the EVOP Worksheet

Remember, EVOP takes a long time. For this reason, we first fill out the phase and cycle numbers. The phase refers to the set of conditions we will use in the factorial design. The phase will not change until we have evidence of a better set of operating conditions. Cycle is another name for a replicate. We will perform as many *cycles* as necessary to reduce the error limits to the point where the signal is big enough to be seen. Table 11-1 indicates that we are in Phase I, Cycle 1. The

TABLE 11-1

EVOP WORKSHEET

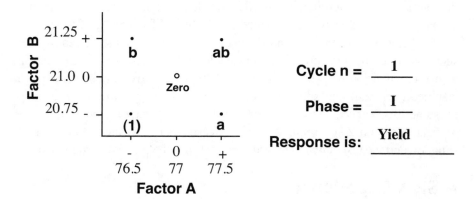

Cycle n = _____1_____

Phase = _____I_____

Response is: _____Yield_____

Calculation of Averages						Calculation of Std. Dev.
Treatment Combination	0	(1)	a	b	ab	
(i) Previous Sum						**Previous** $\sum s =$
(ii) Previous Average						**Previous Average s =**
(iii) New Observation	70	65	65	63	75	**New s = Range ($f_{k,n}$)**
(iv) Difference [ii-iii]						**Difference Range=**
(v) New Sum [i+iii]	70	65	65	63	75	**New** $\sum s =$
(vi) New Average $\dfrac{v}{n}$	70	65	65	63	75	**New Average s =** $\dfrac{\text{New } \sum s}{n-1} =$

DATE:	11/15	11/15	11/15	11/15	11/15
Time:	1PM	10AM	3PM	11AM	9AM

FOR 2^2 EVOP

Table of $f_{k,n}$

Effects:

$A = \frac{1}{2}((a+ab)-((1)+b)) =$

$B = \frac{1}{2}((b+ab)-((1)+a)) =$

$AB = \frac{1}{2}(((1)+ab)-(a+b)) =$

Center$^* = \frac{1}{5}((1)+a+b+ab)-(4(zero)) =$

*If negative, center is near max.

*If positive, center is near min.

Cycles	$f_{k,n}$
1	----
2	.30
3	.35
4	.37
5	.38
6	.39

ERROR LIMITS (n=# of Cycles)
For Averages and new Effects:
Error Limits= $2\bar{s}/\sqrt{n}$
For Change in Center Effect
Error Limits = $1.78\bar{s}/\sqrt{n}$

response is yield, and it is a good idea to fill in the date since we will be operating over a long period of time.

The pictorial diagram in the upper left corner helps us keep an account of the five treatment combinations in the EVOP. We fill in the levels of temperature and pressure, and now we are ready to begin the exploration of our process.

Each of the five treatment combinations are run in random order as in any experiment. The values are filled in on Line iii, the new observation. Since Cycle 1 has no replication, we move directly into Cycle 2 in Table 11-2. We transcribe the information from the previous worksheet before starting Cycle 2. The previous sum and the previous average are merely the first observations from Cycle 1. We then obtain new observations and now we can find the differences between the new observations and the previous average. This difference is placed in Line iv. We scan the range of differences and find the largest negative difference is -2. Note that we watch the algebraic signs in our finding of the differences. The largest positive difference is +2. The range of these differences (4) is recorded just to the right of Row iv. We will use this shortly to compute the standard deviation or *noise* in the experiment. We enter the sum of the observations in Row v and by dividing by n, the cycle number, we get the average in Row vi. The effects are computed by filling in the values of the new average at the appropriate points in the formulas below the data table. These formulas merely find the average *contrast* or change from the high level of the factor to the low level of the factor. So for Factor A (temperature) we find the sum of the responses for the high levels of A (i.e., a + ab) and subtract from it the low levels of A (i.e., (1) + b). Taking half of this value gives us the average effect of Factor A. Similar calculations are made for Factor B and the AB interaction effect.

The calculation of the effect for the center point *may* look complicated, but it is conceptually equivalent to finding the average of the four pieces of data in the factorial design and subtracting the zero point from this average.

$$\frac{1}{5} \, (\, (1) + a + b + ab - (4 \, (zero) \,) \,) \tag{11-1}$$

The value calculated in Expression 11-1 provides us with information on the "curviness" of the response surface in the region of experimental investigation. If the region is curved, then we will observe a large difference between the "conceptual average" obtained by taking the four corners of the factorial design and the actual zero data point in the center of the design. If there is no curvature, then the two values are the same (within experimental error). Because of the way

TABLE 11-2
EVOP WORKSHEET

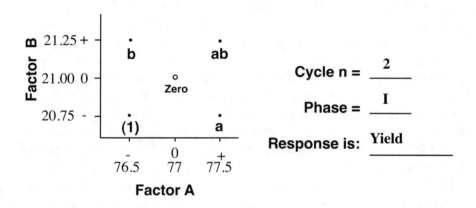

Cycle n = __2__

Phase = __I__

Response is: __Yield__

Calculation of Averages						Calculation of Std. Dev.
Treatment Combination	0	(1)	a	b	ab	
(i) Previous Sum	70	65	65	63	75	Previous $\sum s$ = None
(ii) Previous Average	70	65	65	63	75	Previous Average s = None
(iii) New Observation	68	67	65	64	74	New s = Range $(f_{k,n})$ 4 (0.3)= 1.2
(iv) Difference [ii-iii]	2	-2	0	-1	1	Difference Range= 4
(v) New Sum [i+iii]	138	132	130	127	149	New $\sum s$ = 1.2
(vi) New Average $\frac{v}{n}$	69	66	65	63.5	74.5	New Average s = $\frac{New \sum s}{n-1}$ =1.2

DATE:	11/15	11/15	11/15	11/15	11/15
Time:	1PM	10AM	3PM	11AM	9AM

FOR 2^2 EVOP

Table of $f_{k,n}$

Effects:
A = $\frac{1}{2}$((a+ab)-((1)+b)) = 5
B = $\frac{1}{2}$((b+ab)-((1)+a)) = 3.5
AB = $\frac{1}{2}$(((1)+ab)-(a+b)) = 6
Center*= $\frac{1}{5}$((1)+a+b+ab)-(4(zero)) =-1.4
*If negative, center is near max.
*If positive, center is near min.

Cycles	$f_{k,n}$
1	----
2	.30
3	.35
4	.37
5	.38
6	.39

ERROR LIMITS (n=# of Cycles)
For Averages and new Effects:
Error Limits= 2$\bar{s}$/ $\sqrt{n}$ =1.7
For Change in Center Effect
Error Limits = 1.78$\bar{s}$/ $\sqrt{n}$ =1.5

TABLE 11-3
EVOP WORKSHEET

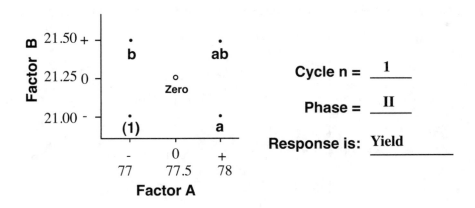

Cycle n = ____1____

Phase = ____II____

Response is: Yield

Calculation of Averages						Calculation of Std. Dev.
Treatment Combination	0	(1)	a	b	ab	
(i) Previous Sum						Previous $\sum s =$
(ii) Previous Average						Previous Average s =
(iii) New Observation	74	68	69	68	78	New s = Range $(f_{k,n})$
(iv) Difference [ii-iii]						Difference Range=
(v) New Sum [i+iii]	74	68	69	68	78	New $\sum s =$
(vi) New Average $\frac{v}{n}$	74	68	69	68	78	New Average s = $\frac{New \sum s}{n-1}$ =

DATE:	11/20	11/19	11/20	11/20	11/19
Time:	3PM	1PM	11AM	9AM	9AM

FOR 2^2 EVOP

Table of $f_{k,n}$

Effects:

$A = \frac{1}{2}((a+ab)-((1)+b)) =$

$B = \frac{1}{2}((b+ab)-((1)+a)) =$

$AB = \frac{1}{2}(((1)+ab)-(a+b)) =$

$Center^* = \frac{1}{5}((1)+a+b+ab)-(4(zero)) =$

*If negative, center is near max.
*If positive, center is near min.

Cycles	$f_{k,n}$
1	----
2	.30
3	.35
4	.37
5	.38
6	.39

ERROR LIMITS (n=# of Cycles)
For Averages and new Effects:
Error Limits= $2\bar{s}/\sqrt{n}$
For Change in Center Effect
Error Limits = $1.78\bar{s}/\sqrt{n}$

we subtract the zero from the average of the factorial points, a negative value indicates that the zero point is near a maximum, and a positive indicates that the zero is near a minimum. Figure 11-3 illustrates this concept by showing a side view of the surface with the average of the factorial and the zero point.

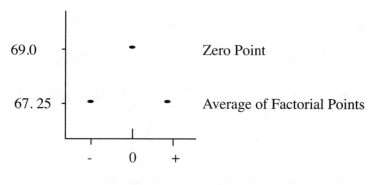

69.0   •   Zero Point

67.25   •   •   Average of Factorial Points

            -     0   +

FIGURE 11-3

If we subtract the zero point from the average of the factorial, we get a negative value for this example indicating that the zero point is higher than the rest of the experimental values as illustrated in Figure 11-3. Now, we don't really make such a simple subtraction of the average and the zero point but weight the 5 points according to Expression 11-1. We look at the average of the whole set of five values (that's why a divisor of 5) with the zero point multiplied by 4 to balance it against the four points of the factorial.

The reason for the measure of the position of the center point with respect to the total experiment is to tell us if we should stop the EVOP process because we have reached a maximum or a minimum value in the region of investigation. In our example we have an indication of being in a maximum region, but we need to now compute the variation or noise in the data to see if this is a *real* change taking place.

Signal to Noise in EVOP

To make the calculations of the variation encountered in an EVOP as easy as possible for the production people using the technique; the standard deviation is derived from a simple range calculation.

Again, Table 11-2 has the details. Since we have only reached Cycle 2, there are no previous sum of the standard deviations ($\sum s$) or a previous average standard deviation. We can compute a new standard deviation (new s) on the third line (just across from Row iii). The difference range of 4 for this example is multipled by the $f_{k,n}$ from the table at the lower right corner of the worksheet. This factor, $f_{k,n}$, is the multiplier that will convert a range statistic into a standard deviation. The k subscript refers to the number of treatment combinations in the EVOP (5 in our example), and the n is the cycle number (2 in this case). The appropriate $f_{k,n}$ value for this example is then 0.30. This gives us a new standard deviation of 1.2, which is also the new sum of the standard deviation (new$\sum s$) and new average standard deviation. Now, some "purists" will argue that you can't average a standard deviation, but for the type of inference that we are using in the EVOP system, the method works similarly to the use of the range in control charts. We will not go into the theory here, but further details are available in Box and Draper (1).

Time for Decision

With a measure of the effects and also armed with a standard deviation, we have all the elements for making a statistical inference via a signal-to-noise ratio. We will make this even easier by calculating "error" limits for the effects under study. If the effect we observe exceeds the error limit, then we can safely say (at a 95% level) that there is more than just chance influencing the response variable.

The error limits for the averages and the effects are calculated alike by taking twice the standard deviation and dividing by the square root of n. This produces an error limit of 1.7 for this example. The error limit for the change in the center effect is only slightly different. The standard deviation is multiplied by 1.78 since there are 5 values in the computation of the center point value. The error limit for the center point is 1.5.

If we compare the error limits to the effects calculations, we find that all the effects are greater than the error limit. This indicates that there is a significant change in the response due to the changes in temperature and pressure. However, the change in the center effect is less than (in absolute value) the error limit for the change in the center effect and we are at neither a maximum nor a minimum. We have a "GO" signal from these results and should move to a new phase of investigation. The direction that gives us the highest yield is toward the **ab** treatment combination. The difference between our current zero point and the best experimental point is (74.5 - 69) 5.5 yield units, which exceeds the error limit for an average. Therefore, we can make the move to the **ab** (77.5° temp., 21.25 pressure) treatment combination and use this as our new zero point in Phase II.

Phase II

In Phase II, our new conditions (Table 11- 3) are moved up in both temperature and pressure. In the second cycle (Table 11- 4) we observe large effects for all the factors and the interaction, but while these effects are greater than the error limits, the difference between the current set point (the zero point) and the next higher contender (the ab treatment combination) is only 1.5 (absolute) yield units. This is not enough to consider a change to the next phase since the 1.5 is not greater than the error limit for averages of 2.97. We therefore go to another cycle (Table 11- 5) and gather 5 more data points.

In Cycle 3 (Table 11- 5), the error limits are reduced by both better repeatability and because the divisor in the computation of this factor is increased. Now we can see how EVOP gets its sensitivity by many replicates. The error limits are of sufficient size to allow us to decide that the "**ab**" treatment is significantly larger than the "zero" treatment

There is one slight problem with a decision to move to another set of operating conditions. The effect of the change in the center indicates that the center is near a maximum. If we did not have a basic knowledge of the response variable, we might be inclined to halt the EVOP process here. However, we know that yield can reach 100%, so we move on to a third phase with the conditions of the old "**ab**" treatment becoming the conditions for the new center point in the third phase.

By now you should get the idea that EVOP is a slow process. We have spent about two weeks in only two phases and there is a lot more work to accomplish before we can reach our goal of 100% yield. Figure 11-4 is a summary of the completion of this EVOP. Notice how the technique follows a clear path up the side of the response surface and reaches the top without any detours.

Equations from EVOP

With the large quantity of data after an EVOP has been run, the experimenter might be tempted to load all the information into a regression computer program and create an equation of the surface. While this is a good thought, remember that the scope of the EVOP experiment is very narrow. The equation would be valid only in that portion of the response surface that was explored. Therefore, the equation would not be general enough to perform the robust predictions expected of it. Such an equation would be sufficient to help control the process in the region of interest. It could be used to help get a sick process back to health should it wander from the path worn in the mountainside by the EVOP investigation. However, if you look at the response surface, you will see that there are many possible paths to the top.

TABLE 11-4
EVOP WORKSHEET

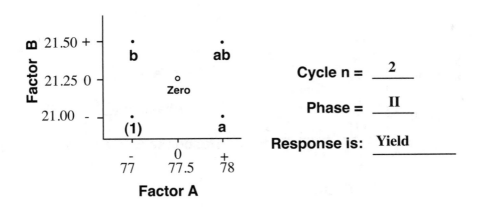

Cycle n = _____2_____

Phase = _____II_____

Response is: Yield

Calculation of Averages	0	(1)	a	b	ab	Calculation of Std. Dev.
Treatment Combination	0	(1)	a	b	ab	
(i) Previous Sum	74	68	69	68	78	Previous $\sum s$ = **None**
(ii) Previous Average	74	68	69	68	78	Previous Average s = None
(iii) New Observation	77	67	69	64	76	New s = Range $(f_{k,n})$ 7(0.3)=2.1
(iv) Difference [ii-iii]	-3	1	0	4	2	Difference Range= 7
(v) New Sum [i+iii]	151	135	138	132	154	New $\sum s$ = 2.1
(vi) New Average $\dfrac{v}{n}$	75.5	67.5	69	66	77	New Average s = $\dfrac{\text{New} \sum s}{\text{n-1}}$ = 2.1

DATE:	11/23	11/23	11/24	11/24	11/23
Time:	11AM	3PM	9AM	11AM	9AM

FOR 2^2 EVOP

Table of $f_{k,n}$

Effects:

$A = \frac{1}{2}((a+ab)-((1)+b)) = 6.25$

$B = \frac{1}{2}((b+ab)-((1)+a)) = 3.25$

$AB = \frac{1}{2}(((1)+ab)-(a+b)) = 4.75$

Center*= $\frac{1}{5}((1)+a+b+ab)-(4(zero))=-4.5$

*If negative, center is near max.

*If positive, center is near min.

Cycles	$f_{k,n}$
1	----
2	.30
3	.35
4	.37
5	.38
6	.39

ERROR LIMITS (n=# of Cycles)
For Averages and new Effects:
Error Limits= $2\bar{s}/\sqrt{n}$ =2.97
For Change in Center Effect
Error Limits = $1.78\bar{s}/\sqrt{n}$ = 2.64

TABLE 11-5
EVOP WORKSHEET

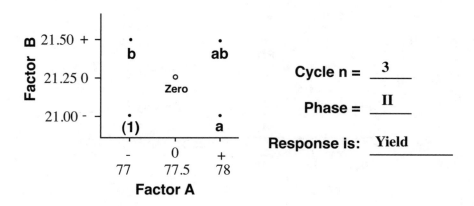

Cycle n = __3__

Phase = __II__

Response is: Yield

Calculation of Averages	0	(1)	a	b	ab	Calculation of Std. Dev.
Treatment Combination	0	(1)	a	b	ab	
(i) Previous Sum	151	135	138	132	154	Previous $\sum s = 2.1$
(ii) Previous Average	75.5	67.5	69	66	77	Previous Average s = 2.1
(iii) New Observation	75	68	67	66	79	New s = Range $(f_{k,n})4(0.35)=1.4$
(iv) Difference [ii-iii]	0.5	0.5	2	0	-2	Difference Range= 4
(v) New Sum [i+iii]	226	203	205	198	233	New $\sum s = 3.5$
(vi) New Average $\frac{v}{n}$	75.3	67.7	68.3	66	77.7	New Average s = $\frac{\text{New} \sum s}{n-1}$ =1.75

DATE:	11/30	12/1	11/30	12/1	11/30
Time:	9AM	1PM	1PM	9AM	11AM

FOR 2^2 EVOP

Table of $f_{k,n}$

Effects:
A = $\frac{1}{2}((a+ab)-((1)+b)) = 6.15$
B = $\frac{1}{2}((b+ab)-((1)+a)) = 3.85$
AB = $\frac{1}{2}(((1)+ab)-(a+b)) = 5.55$
Center*= $\frac{1}{5}((1)+a+b+ab)-(4(zero))=-4.3$
*If negative, center is near max.
*If positive, center is near min.

Cycles	$f_{k,n}$
1	----
2	.30
3	.35
4	.37
5	.38
6	.39

ERROR LIMITS (n=# of Cycles)
For Averages and new Effects:
Error Limits= $2\bar{s}/\sqrt{n}$ =2.02
For Change in Center Effect
Error Limits = $1.78\bar{s}/\sqrt{n}$ = 1.8

The path that is taken depends on where we start the process. We started in the South-West corner and moved in a North-East direction to get to the top. We could have just as easily started in the far North-East corner and moved in a South-West direction to reach the same finish point.

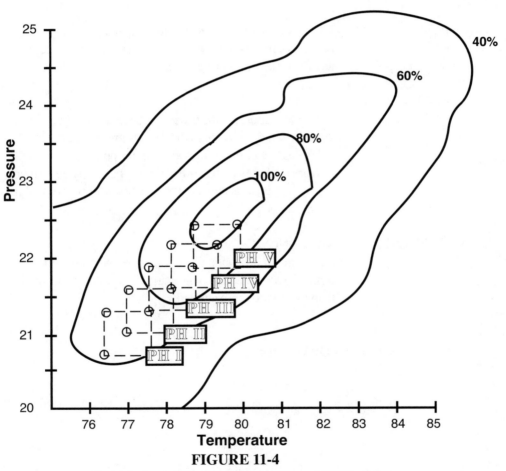

FIGURE 11-4

Completion of the EVOP. The Phases ($\mathbb{PH}$) of the process are shown as they climb the response surface.

EVOP REFERENCES

1. Box, G. E. P., and Draper, N. R. 1969. **Evolutionary Operation**, John Wiley & Sons, New York.
2. Bingham, R. S. 1963. Try EVOP for Systematic Process Improvement, *Industrial Quality Control*, Vol. XX, No. 3.
3. Box, G.E.P., and Hunter, J.S. 1959. Condensed Calculations for Evolutionary Operation Programs, *Technometrics*, Vol. 1, No. 1, 17-95.

APPENDIX 11

EVOP: EVolutionary OPeration: a method that allows improved product or process performance while maintaining the uninterrupted flow of a production line.

RULES OF EVOP:
                         1. No interruption of the process or increase in % defective.
                         2. Make small changes to the factors under study.
                         3. Use a 2-level factorial with a center point.
                         4. Work in phases with a shift to a new phase only when there is evidence of a statistically significant improvement in the response.

STATISTICAL SIGNIFICANCE:
                         Statistical significance is demonstrated when the error limits as calculated from the standard deviation are smaller than the effect being studied.

See next page for EVOP Worksheet.

EVOP WORKSHEET

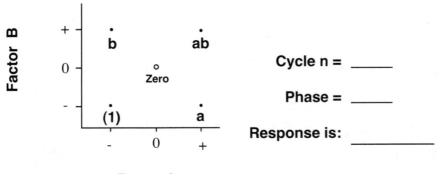

Cycle n = _____

Phase = _____

Response is: _____

Calculation of Averages						Calculation of Std. Dev.
Treatment Combination	0	(1)	a	b	ab	
(i) Previous Sum						Previous $\sum s =$
(ii) Previous Average						Previous Average s =
(iii) New Observation						New s = Range $(f_{k,n})$
(iv) Difference [ii-iii]						Difference Range=
(v) New Sum [i+iii]						New $\sum s =$
(vi) New Average $\dfrac{v}{n}$						New Average s = $\dfrac{\text{New} \sum s}{n-1} =$
DATE:						
Time:						

FOR 2^2 EVOP

Table of $f_{k,n}$

Effects:

$A = \frac{1}{2}((a+ab)-((1)+b)) =$

$B = \frac{1}{2}((b+ab)-((1)+a)) =$

$AB = \frac{1}{2}(((1)+ab)-(a+b)) =$

$Center^* = \frac{1}{5}((1)+a+b+ab)-(4(zero)) =$

*If negative, center is near max.

*If positive, center is near min.

Cycles	$f_{k,n}$
1	----
2	.30
3	.35
4	.37
5	.38
6	.39

ERROR LIMITS (n=# of Cycles)
For Averages and new Effects:
Error Limits= $2\bar{s}/\sqrt{n}$
For Change in Center Effect
Error Limits = $1.78\bar{s}/\sqrt{n}$

PART III

Sorting the Signal from the Noise

12

Simple Analysis

The simplest of all experimental analyses is just to take the data as we collect it and look to see if there is any difference in the responses between the sets of conditions we have investigated. However, in any and all experimental work, the results we observe are subject to variation. If there were no variation, then the discipline of statistical analysis would not need to exist and we could perform this simple analysis without any concern. But there is, in reality, variation that must be accounted for as we sift through the data to see if the differences we observe are bigger than the level of variation that surrounds them. This is the role of statistics in experimental design.

It is interesting to speculate on the origin of the word "statistics" and how the science evolved from early uses in feudal kingdoms. There are three roots to the word. The first, "stat," refers to the state or country where these methods were first utilized. The second, "ism," is a root that can take on the meaning of fact or condition. The third part, "ist," is a suffix meaning discipline or the practice of doing something. So, the word "statistics" first meant a discipline to understand the condition of the state. Indeed, in the early days of statistics, the prime duty of the royal statistician was to report population, wealth, and all the other important information that would help the kingdom gather the correct taxes. Today we still have our demographers and actuaries who "write about people" and gather facts for the government. Every 10 years, a national census of the population is taken at great cost and sometimes with controversial results.

While the counting of people and wealth is still practiced and is an important part of the statistical field, it is the discipline of statistical analysis that we shall treat in the third part of this book. Statistical analysis is a necessary ingredient in the understanding of the results of our experimental designs. The particular analysis technique we shall investigate is called "statistical inference." This branch of mathematics, which does something useful for people, consists of a body of methods for making wise decisions in the face of uncertainty. Such inference is based on the laws of probability.

We shall assume that the reader has a rudimentary idea of probability. You have undoubtedly gained such knowledge of probability from betting on the weekly football pool or from listening to the weatherman talk about "chances of rain." To understand the use of probability in statistical inference, we need go no further than to say that we would like to have a high probability (close to 1 or certainty) of making the "correct decision." (For those who are in need of a quick review of some basic probability axioms, Appendix 12 is appropriate reading.) The "correct decision" is a wise decision based on the application of the methods of statistical inference.

Another piece of background material that we must understand comes from basic statistical counting and distribution concepts. We have implied that *statistics* is a collection of methods of analysis. This is absolutely true when we use the word in this active mode. However, the word *statistics* can also be used with another meaning. In this sense, we mean estimates of the associated parameters of the whole population. Table 12-1 shows the names and symbols of the population parameters and the related sample statistics. We must realize that the values of the statistics based on the data we obtain are only estimates of the true population parameters. If we had the true values of the population parameters, we would not need statistical inference methods to help make decisions, and the probability of correct decisions would be 100%!

TABLE 12-1

	Measure of Centralness	Measure of Variation
Population *Parameter*	μ (mu)	σ (sigma)
Sample *Statistic*	$\overline{X}$ (X bar, average)	s (standard deviation)

Before we can look at the statistical methods that allow us to make wise decisions, we need to determine how much of a chance we can take in making a wrong decision. This is called our "risk" and the complement of the risk (1 - risk) is the probability of being correct. Ideally, we would like the risk to be zero. Practically, it is impossible to make the risk zero since we are working with variation-prone statistics, but the risk can be reduced to any level we require: .05 (5%), .01 (1%) or even less depending on the quality of the information obtained from the experiment and the number of runs we have made. This is why the design of the experiment is very important.

Hypothesis Testing

The cornerstone of statistical inference is the discipline known as "hypothesis testing." The hypothesis test stems directly from our prior knowledge of the process and the objective we write before we even gather one iota of information. In simple experiments, the hypotheses are stated about differences between mean values and also about differences in the variation. We shall treat the mean-value hypotheses in the following examples before getting into the methods for testing variation.

Example 1

Let's assume that a production line has been routinely making a piece-part with a specified dimension of 10 inches. If we were to take a sample from the production line over time and measure, say, 25 parts, we would hope to find a mean value of 10. More than that, we would like all of the parts to be exactly 10 inches! However, not all the parts will be exactly the same size because of process variation. Table 12-2 shows the size of 25 parts sampled from the process. Notice that the parts are all near the 10 inch mark, but none is actually 10 inches exactly. Since we do not know any better, we will form a hypothesis of optimism and say that there is no difference between the specification (which we shall call μ_0) and the mean of the production population (which we shall call μ). Note that the hypothesis is in terms of the population parameters rather than the statistics. The basis of statistical inference is to form a hypothesis about the population parameters and check its validity using the available statistics. The formal statement of the "null hypothesis" (hypothesis of no difference) is:

$$\mathbf{H_0}: \ \mu = \mu_0 = 10" \tag{12-1}$$

TABLE 12-2

10.06	9.76	10.07	10.15	9.85
10.02	9.85	10.03	10.16	10.12
10.08	9.98	10.04	10.15	9.95
9.98	9.82	9.92	9.95	9.81
9.84	10.08	9.76	9.93	9.81

The Alternative Hypothesis

Now, what can we say about the alternative to this null hypothesis? Without any extra knowledge about the process, we can only say that there is a difference between the all-time average and the value for the current production. This is stated as the "alternative hypothesis" or the hypothesis of difference, as follows:

$$\mathbf{H_1}: \mu \neq \mu_0 \tag{12-2}$$

Together, the null and alternative hypotheses provide the basis for what is called *hypothesis testing*. In such tests, we are comparing two populations of items and we have the simplest experiments with the simplest methods of analysis. While these are simple tests, the concept behind these tests is the same concept that applies to all statistical inference. We look at the change in the response (signal) due to the changes we have induced in the experiment and compare this signal to the random variation (noise) observed as we measure this signal.

In the case of the simple "t" test of $\mu = \mu_0$ that we are using, the signal is the difference between μ and μ_0. The noise is the variability in the measurement of the statistic $(\overline{X})$ that estimates μ and is expressed as the standard deviation (s).

When the signal is equal to or smaller than the noise, it is obvious that the signal is no different than the noise. We then say that there is no significant influence due to the conditions of our experiment. When the signal is vastly larger than the noise, we have no difficulty in deciding that the signal is significantly bigger than the noise and the change due to our experiment is real. It is in the region where the signal and noise are nearly the same, or when the signal is only two or three times as big as the noise, that there is uncertainty in our decision. This is where probability based decision rules help us make the wise decisions we wish to make consistently.

Let's look at the possible outcomes of our simple experiment, which compares the mean of our sample with the specification of the piece-part process. Since we are taking only a sample (a sort of a snapshot), we need to know how the information in this sample could compare to the truth of the situation.

TABLE 12-3

Sample says:

		Reject Null: (Parts Differ)	Do Not Reject Null: (Parts are Same)
T R U T H	Parts Differ	Want to be here if parts differ	Error if parts are really different **Type II error (β risk)**
	Parts are Same	Error if parts are really the same **Type I error (α risk)**	Want to be here if parts are the same

Alpha Risk

In Table 12-3 there are four alternatives before we make a decision based on the evidence before us. Two of the alternatives are good, and we would like to land in one of those boxes of the truth table most of the time. However, the lower left box, where the truth is that no difference exists between the parts (but our sample says that there is a difference), leads to an error of "the first kind." Before we make the decision leading to this error, we have a risk associated with this error. This risk is named the **alpha (α) risk**, after the first letter of the Greek alphabet.

Beta Risk

The box in the upper right corner of the truth table again presents us with an error if we fall into it. Here we sit back and say there is no difference, when in truth there is a difference. This is a "lazy" error and is an error of "the second kind." The risk we take in committing this error before we make any decision is called the **beta** (β) **risk**, after the second letter of the Greek alphabet. If you want to be able to remember the difference between these risks in a simple mnemonic manner, think of the somewhat lazy looking letter beta that is relaxing while it should be out finding differences. Since there are only two risks, the other must be alpha!

The "Lazy Beta"
FIGURE 12-1

No matter what we call the risks and the errors, the real lesson to be learned is that when faced with uncertainty, there is a risk that our decision will be in error due to the variation in the statistics we are working with. There are two ways to combat this error.

    1. We could attempt to improve the physical nature of the variation by working on the response variable to "clean it up " by making it less variable.
    2. Usually, we take a larger number of observations and make use of a consequence of the "central limit theorem."

Among other things, the central limit theorem assures us that no matter what the original distribution of our measurements, if we take sufficiently large numbers of observations from this population, then the distribution of averages determined from these sets of observations will approximate the normal distribution. This result is handy, since we will be using probability models based on the normal distribution to help separate our signal from the noise.

      Another important concept is that the estimate of the variance of the resulting distribution of averages will be equal to the variance of the original data divided by the number of observations (n) in the samples used to form the averages! In algebraic form this is:

$$s_{\bar{x}}^2 = \frac{s_x^2}{n}$$

(12-3)

And of course if we are dealing with the standard deviations, this equation becomes:

$$s_{\overline{x}} = \frac{s_x}{\sqrt{n}} \quad \text{(called the standard error of the mean)} \quad (12\text{-}4)$$

So if we compute the standard error of the mean for the data in Table 12-1, we get:

$$\overline{X} = 9.967 \quad s_x^2 = 0.01621 \quad s_x = 0.1273 \quad n = 25$$

$$s_{\overline{x}} = \frac{0.1273}{\sqrt{25}} = 0.0255$$

This means that by taking 25 observations, we have effectively reduced the uncertainty in the hypothesis test by a factor of 5. The resulting lower variation shows why an average value is a more stable estimate of the "truth" than a single observation.

Steps in Hypothesis Testing

Now how do all these concepts we have discussed come together? We want to make wise decisions in the face of uncertainty. We now have a mean value that is only one of many possible mean values. If we were to take another set of n observations, we would most likely obtain a different mean. If we were to keep taking n observations and determine many such means, we would have a distribution of means that as a consequence of the central limit theorem will have a normal shape and a tighter variance than the original distribution of individual observations. Since we have no other information before we begin the experiment and take observations, we hypothesize that the sample mean we will obtain from our experiment is a part of this distribution of means that is centered at 10 inches. The job of hypothesis testing is simply to determine if the mean value we have is a part of the distribution of means that would result if we kept taking observations and computing means.

Of course we do not need to establish the shape of the distribution of these mean values experimentally, since the central limit theorem has done this for us. We need only to take one sample mean from our population and see if it belongs to the hypothesized distribution. Without the central limit theorem, we would have to establish the shape of our sampling distribution every time we did an experiment. This would be a costly activity. The central limit theorem is a very important part of efficiency in statistical inference.

Let's follow the formality of 7 steps and put these concepts together.

Step 1. State the null and alternative hypotheses:
H_0: $\mu = \mu_0 = 10$ inches
H_I: $\mu \neq \mu_0$

Step 2. Understand the requirement that the errors are independent and normally distributed. (The errors we speak of here happen to be the "noise" we refer to and measure by the standard deviation.)

Step 3. State the level of alpha and beta risks:
alpha = 0.05 (this choice depends on the situation)
beta =? (as we shall see, beta is dependent upon other considerations)

Step 4. Determine the critical region based on the selected alpha risk and the sample size: To complete step 4, we need to look at the "Student t" table in the back of this book. With 25 observations, we have 24 degrees of freedom (df). Our risk for this test, which can reject the null hypothesis if the mean is significantly smaller or larger than the standard (a "two tail test"), is 0.05 which gives "t" critical values of ± 2.06. (± because this is a "two-tail" test since we could end up in the negative or the positive tail of the distribution.)

Step 5. Compute $\overline{X}$, s^2 and s.

Step 6. Calculate the "t" value for this problem based upon the statistics in hand:

$$t = \frac{\overline{X} - \mu_0}{s/\sqrt{n}} \qquad\qquad (12\text{-}5)$$

Step 7. Draw conclusions in *ordinary language*: For this case we cannot reject the null hypothesis. *There is no evidence that the process has changed.*

Let's look at our decision in the light of the risks involved. Is it possible to commit a type I error? If you look back at Table 12-2, you will note that the only time you can commit a type I error is when you reject the null hypothesis. So we can't make a type I error in the above decision and the probability of a type I error is zero. However, in not rejecting the null we have made a decision associated with the type II error. But we have not stated a beta risk, which is the risk of making a type II error! This is an unfortunate but all-too-common mistake in hypothesis testing of data that does not come from a designed experiment. The "design" in this experiment should have been to select a sample size sufficient to control both the type I error and the type II error.

A Designed Hypothesis Test

Let us revisit our problem once again and see how a simple design can be meshed with a "t test" to produce a wise decision.

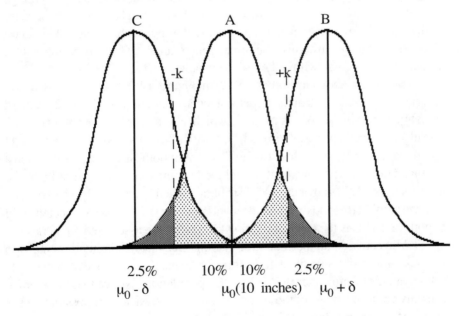

FIGURE 12- 2

Figure 12-2 shows the piece-part problem in terms of the possible distribution patterns we could encounter in our experiment. If there is no difference from the hypothesized true mean of 10 inches, then we will be in distribution "A" with the μ of our sampled population no different from the μ_0. If the population of parts has shifted upward from the 10 inch all-time average, then we would be in the "B" distribution. If the average has shifted downward, then we are in the "C" distribution.

The problem is to determine if we have made a "significant" move from the "A" distribution to somewhere else. Since we are using the sampling distribution of the mean and the consequences of the central limit theorem, we need to realize that the distributions we are observing are distributions of averages (which approximate normality). However, we only have one such average as our sample and this average could be anywhere in each of the three distributions. If the average we calculate as a result of our observed data is really in the "A" distribution, then it could be right on the 10 inch (μ_0) point. It could also be to either the right or left of the (μ_0). There is a small but measurable chance that the average we observe could be so far on either side of the "A" distribution that it is overlaps into the "B" or "C" distributions. This is a result of the fact that we have a *distribution* of averages and not a single value that keeps appearing over and over again. While we do not know exactly where the value of our sample average will fall, we know the shape of its distribution. It is a normal distribution (because of the central limit theorem), and we know that between -2 and +2 standard deviations of the mean this distribution will encompass 95% of all possible values of the sample average within the "A" distribution.

But what about the other 5%? In Figure 12-2 lines are drawn and designated by the letter k at the upper 2 standard deviation and lower 2 standard deviation points in the "A" distribution, which leaves a total of 5% cut off of "A." The placement of k puts 2.5% in each "tail" of the "A" distribution. When we set our alpha risk at 5% in such a "two-tail" test of hypothesis, we construct such a picture. This picture shows that should the average of our sample fall into the lower or upper 2.5% tail of the "A" distribution when it's still really a part of "A," then we would erroneously conclude it was not a part of "A" as we reject the null hypothesis! This is the error of the first kind (with the associated alpha risk).

You can see that the chance of making such an error is small ($\alpha=.05$), so we make a wise decision most of the time. But what if the average of our sample actually comes from either the "B" or "C" population? Notice that the average can only come from one of these populations. It is physically impossible for the average to be a part of both "B" and "C" at the same time. Let's say that the sample average really comes from the "B" distribution. This sampling distribution of the mean will be δ (delta) away from the "A" distribution point by point so the mean of "B" ($\mu_0 + \delta$) will be a distance δ from μ_0. How big is δ? That really depends upon the actual shift in the "B" distribution's average from the target μ_0. How much of a shift in "B" is important to detect? That depends upon our judgment of the importance of the shift. In other words, we must bring some prior knowledge to the situation in the form of an answer to the question, "how much of a difference makes a difference" to us. This difference is not based on statistics, but on our knowledge of the process. So, if the piece-part may vary ±0.015 inches without any adverse consequences, then we may say that we have an indifference zone of up to 0.015 inches. *The difference that makes a difference is 0.015 inches.*

Therefore, if we have a shift in the mean of the piece part production process as great or greater than 0.015 inches, we want to do something about it. How do we take this wise action? Recall that the standard error of the mean ($s_{\bar{x}}$) will shrink by the inverse of the square root of n, the number of observations in our sample. This is where our designed experiment comes into the picture. We will *select* a sample with enough observations to sufficiently control the risks of both types of errors at chosen, allowable levels. These levels of risk again depend on the prior knowledge we bring to the situation. For this example, we will pick an alpha of 5% and a beta of 10%.

With these risks, there is a 10% chance that we will erroneously think we are in the "A" distribution when we really are in the "B" distribution. This is the lightly shaded area of Figure 12-2 to the left of +k in the "B" distribution. Given the above picture of the situation, we are in a position to calculate the placement of k and at the same time compute the number of observations in our sample that will satisfy the risk levels we have set.

Looking at the "A" distribution, k will be 1.96 standard deviations away from the μ_0 point, so calculating the t statistic (t with infinite df) we find:

$$t = \frac{k - \mu_0}{s/\sqrt{n}} \qquad 1.96 = \frac{k - 10}{0.1273/\sqrt{n}} \qquad (12\text{-}6)$$

Now, in Equation 12-6 we do not insert the original sample size of 25 because we do not know if this sample size is sufficient to control the alpha and beta risks.

By rearranging Equation 12-6 we obtain:

$$1.96 \, (0.1273/\sqrt{n}) = k - 10 \qquad (12\text{-}7)$$

Now let's look at the "B" distribution and where +k falls within it. The point +k will be located where 10% of the values of "B" (in the lower tail) are less than k. From the normal distribution table, this is a point 1.28 standard normal units to the left of the mean of "B," so that t = −1.28. We substitute this information into the t formula as we did in Equation 12-6 and get:

$$-1.28 = \frac{k - 10.015}{0.1273/\sqrt{n}} \qquad (12\text{-}8)$$

Note that 10.015 is μ plus the delta of 0.015. This is the location of the mean of the "B" distribution.

When we rearrange Equation 12-8 we get:

$$-1.28 \, (0.1273/\sqrt{n} \,) = k - 10.015 \qquad (12\text{-}9)$$

By combining Equations 12-7 and 12-9, we have two equations in two unknowns that are easily solved by eliminating k as follows:

subtract Equation 12-9 from Equation 12-7:

$$1.96(0.1273/\sqrt{n}) = k - 10$$

(change signs) $\underline{+1.28(0.1273/\sqrt{n}) = -k + 10.015}$

and we get: $3.24 \, (0.1273/\sqrt{n}) = 0.015 \qquad (12\text{-}10)$

We continue with a little simple algebra:

$$\sqrt{n} = 3.24(0.1273)/0.015$$

$$\sqrt{n} = 21.4968$$

$$n = 756.07$$

Since we can't have a fraction of an observation, we round up to 757, which will detect a .015 difference with a standard deviation of 0.1273 and we can control the alpha and beta risks we have set. Now that's a hefty sample size. It might make you believe that statistics makes more work! Let's probe the problem once again and see why 757 observations are required. Notice that the sample size formula can be reduced to the more general form (12-11.)

$$n = \left[\frac{(t_{\alpha} + t_{\beta}) \, (s)}{\delta} \right]^2 \qquad (12\text{-}11)$$

Observe in Equation 12-11 that as δ (delta) becomes small, n gets larger. Also, if s is large in comparison with δ, then n also becomes larger. In our example, we have a rather large s (relative to δ) and a small δ (relative to s). We are trying to find differences smaller than the "natural variation" in the process, and this takes an enormous number of observations to accomplish.

Could we back off the risks to reduce the sample size? Let's say that we set the beta risk at 50% (or the same as the odds of a coin toss). Then the t value for beta would be zero and the sample size would be

$$n = [1.96 \, (0.1273) / 0.015]^2 = 288$$

This is still a big sample, and worse, when we make the decision not to reject the null hypothesis, there is still a 50% chance of sitting back and saying that there is no change in the population suggested by this sample when there really is a change! Notice that the original sample of 25 observations is only a drop in the bucket compared with the required number of observations to make a wise decision with these data that have so much variation.

If we were to back off on the difference (δ) that makes a difference to us, then the sample size will shrink. Let's say that we make δ as large as the standard deviation of the data we have. Effectively, Equation 12-11 then becomes:

$$\text{For } s = \delta: \qquad [(t_\alpha + t_\beta)1]^2 = (t_\alpha + t_\beta)^2 \qquad (12\text{-}12)$$

For our example with alpha= 0.05 and beta =0.10, the sample size is merely the t_a=(1.96) and t_b=(1.28) added together and squared! This gives us a sample size of 11. But in using the sample-size formula, there is a hitch. The t distribution is a function of the degrees of freedom in the sample you have taken. The original values we put into Expression 12-12 assumed a large sample size was needed (a large sample in a t distribution is usually considered above 30). See Table 2 in the back of the book, critical values of the **Student's t** distribution and notice that as the df increase, the **Student's t** values converge to the standard normal distribution values. But we have calculated that only 11 observations are needed when the standard deviation is equal to delta! So we must modify our sample-size-determination activity by recalculating the sample size based on the degrees of freedom for 11 observations. The values for t_a and t_b becomes 2.2281 and 1.37, respectively, for 10 degrees of freedom. This set of new t values produces a sample size of:

$$n = (2.2281 + 1.37)^2 = 12.94$$

We round up to 13 observations and again go through the calculation with 12 df. The t values are 2.18 and 1.36, which when added together and squared produce a value of 12.52, which rounds up to 13 observations.

This is the same as the previous computation, and thus the sample size has converged for the case when delta equals the standard deviation. Since the ratio of various deltas to standard deviations will always produce the same sample size, it is not necessary to go through this iterative procedure to find a sample size in designing an experiment. Table 3 in the back of this book gives the sample size for various levels of alpha and beta, given values of delta divided by the standard deviation.*

To illustrate the use of Table 3, let's find the sample size for a situation where the standard deviation as calculated from a preliminary sample of five observations is 22.68. The difference we want to detect (this is the minimum difference) is 14.17. To enter the table, we find D, which is the ratio of physical unit difference (delta) to the physical standard deviation.

$$D = \frac{\delta}{s}$$

$$D = 14.17/22.68 = 0.6247$$

The closest value of D in Table 3 is 0.60. If we set our alpha risk at 0.01 and our beta risk at 0.1, then the sample size will be found at the intersection of the D = 0.60, alpha = 0.01, and beta = 0.1. The number of observations is 45. Table 12-3 is an excerpt of Table 3 and shows the number of observations as a function of the alpha and beta risks. The sample size of 45 is the bracketed value in Table 12-3.

Let's investigate the effect of the risks on the sample size by looking at Table 12-3 where D is a constant 0.60:

TABLE 12 -3

		Alpha risk		
		0.01	0.05	0.10
	0.01	71	53	46
Beta	0.05	53	38	32
Risk	0.10	{45}	32	26
	0.20	36	24	19
	0.50	22	(13)	9

*The procedure to derive Table 3 is the same in principle as the conceptual method just shown, but it is based on the more exact non-central t distribution.

Popular opinion is that the beta risk controls the sample size. If we examine Table 12-3, we can see that both alpha and beta have the same effect on the sample size for equal risk levels. There must be a basis for this erroneous concept that beta controls the number of observations. The idea arises from the fact that we can always set a level of the t test at which we reject the null hypothesis (that is the alpha risk). If we reject the null hypothesis, then there is of course no beta risk! So, if we grabbed a small sample that was not based on any design at all and rejected the null hypothesis, we are unable to make an error of the second kind. However, if we did not reject the null hypothesis using such a small sample, then the beta risk would be very high and would force us to a larger number of observations to control the beta. To further illustrate this point, let's say that we set the alpha at 5% and take 13 observations (the value circled in Table 12-3).

If we reject the null hypothesis, then we are home free. But if we do not reject the null hypothesis, we are stuck with a 50% beta risk! To obtain a more reasonable level of beta, we need to increase the sample size to 32 or even 38 for a 0.1 or 0.05 probability of not sitting back and saying there is no difference when there is indeed a difference.

A few last thought on risks and errors before we look at other types of hypothesis tests. In all formulations of the risks, we have stated the risk value as if it were the actual risk involved in our hypothesis test. The numerical value of the risk we have stated is actually the worst (or highest) level of probability of making an error in the decision we make. Depending on the actual information we gather, this risk can be less than the value stated. A sort of "what if" diagram called an *operating characteristic curve* shows the actual risks for different levels of deviations from the mean in question. The appendix of this chapter shows the construction of this risk-defining tool.

Many computer programs will state the probability associated with the rejection risk (and not mention beta) as if this risk were an *exact* value. This is not the intent of the computer output, but is often misinterpreted as the exact risk. So, if you see p=.0467 on the output, do not think that the chance of rejecting the null hypothesis is exactly 4.67%. You must understand that the p value is calculated from the values of the statistics. You will recall that statistics are variable. Therefore p, itself a statistic, is also variable and would be different if we were to do the experiment all over again. Use the computer calculated p value as a guide and calculate the t to see if you have exceeded the "critical t" based on the worst risk you are willing to take.

One last thought related to the decision that is based on the calculated t value when compared with the tabulated "critical t" value. The calculated t is unlikely to be exactly the same value as the tabulated t value, but if this should happen we need to know the decision rule. Also, understanding this decision will help us understand the t test better. The tabulated "critical t" value is the largest we can expect and still be within the distribution of no difference. Therefore, if we obtain a t that is equal to or less than the "critical t," we have no reason to reject the null hypothesis. Only when we exceed the "critical t" value do we reject the null hypothesis. The situation is very similar to where the ball hits the court in tennis. The rule is "on the line is in" for tennis and also for statistical inference! Figure 12-3 shows this concept where the line on the tennis court is analogous to the "critical t" value.

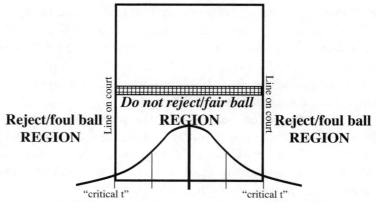

On the line is in for both tennis and statistical inference.
FIGURE 12-3

Tests Between Two Means

In our first example we looked at the difference between a standard value and a sample average. What happens if we want to compare two sample averages that could be different due to some process modifications? In this case, the null hypothesis would look like:

$$H_0: \mu_A = \mu_B$$

and the alternative would be:

$$H_I: \mu_A \neq \mu_B$$

To establish a proper design for this experiment, we need to look at the four items we freely select to establish the sample size. This will establish the maximum risks involved in making our decision.

The four items are:

- •alpha risk (α)
- •beta risk (β)
- •the amount of difference that makes a difference (d)
- •the standard deviation(s)

Given that we have established these four pieces of information, we may determine the sample size that will allow us to find the required difference within the variation of the process while we conform to the stated risks. However, this problem involves *two* populations. Each population's central tendency will be estimated by a statistic that is subject to variation. Further, the variation in the population will be estimated by a statistic. Because we have two sources of variation, rather than just the one source of variation when we tested against a standard value (which does not have variation), we will need to increase the number of observations to maintain the degree of belief (lack of risk) in our result. Table 4, Number of Observations for t-Test of Difference Between Two Means, is the appropriate guide for this activity.

But even before we look up a sample size in this table, let's answer a very realistic set of questions. How do we set up the risk levels, determine the important difference, and find the standard deviation? The levels of risk as well as the important difference are both functions of the problem in front of us. The amount of difference we need to detect depends on our prior knowledge of the process. This must be a finite value different from zero. If we want zero difference between means, then the number of observations would need to be infinite or equal to the entire population. Sometimes this unreasonable request is made in the field of health where any difference from a standard or from a control group would mean that a medical problem could arise. From a theoretical point of view, this is a proper way to treat the subject, but it causes sample sizes to grow to huge quantities, especially when the risks are set at very low levels. In such cases a compromise can be reached by establishing the difference that makes a difference as a reasonable fraction of the standard deviation involved in making the measurements.

This concept brings up the subject of the standard deviation. While we have a somewhat free choice of the risks and difference, the standard deviation is a function of the operating and measurement system we are working with.

Therefore, while it appears to be one of the numerical values we may select, it is really dependent on the way we conduct our experimental work. We may get an idea of the variation by looking at the history of the response. This history may appear in control chart statistics if we have been watching the response in this manner. We may also have other sources of historical information, such as previous experiments, that can provide the estimate of variation.

If no historical information is available, we will need to establish the level of variation by taking a preliminary sample consisting of between 3 and 5 observations. (This recommendation emerges from a "rule of thumb" based on the rapid change in t values between 1 and 2 df and the less rapid change in t values between 4 and 5 df in the Student's t distribution.) No matter how we get an estimate of the variation, if this variation that we observe from history or from our preliminary sample is excessive, then it might be worth the effort to examine the sources that contribute to this variation using the methods of nested design found in Chapter 10 and the analysis techniques of Chapter 18. Such efforts can pay off in the long run by allowing us to reduce the work in the final design. If we simply live with the existing variation (and it is excessive), the number of observations must grow greatly and we will either spend excessive resources doing the job correctly, or possibly do nothing at all, reverting to "gut feel" rather than the correct statistical analysis.

Buying a Copier

In this next example, we have narrowed our choices from a field of many brands of copiers to two particular devices that are priced within a few hundred dollars of each other. Brand A and brand B sales offices will agree to go along with our experiment and allow us to sample the output of a number of their machines so we may evaluate the quality of the output. Since we will be placing a large order for copiers from either of the vendors, we want to make sure that the device we choose will be of high quality.

• Response Variable

Using a national standard of measurement of quality and an appropriate test pattern, we will canvas a number of sites where the copiers are working on a day-to-day basis. If there is no significant difference between the copy quality of the two devices, we will purchase the less expensive one. If there is a big enough difference in the quality between the machines, then we will purchase the copier with the better quality. Having the correct response variable that is precise, quantitative, and meaningful is an essential part of our experimental design.

• Setting Risks

The next step in our design is to set the risks we are willing to take, considering that the observations we obtain are mere statistics that are subject to variation. Since this is not a life-or-death problem, we will set our maximum risks as follows: the chance of saying that there is a difference when there really is no difference (alpha risk) will be limited to no more than 5% (0.05). This means that one time in 20 we will not go for the less-expensive machine (assuming that it comes up with lower copy quality) when we should. This is a risk that will cost us some immediate resources, but will not have a long-term impact.

The chance of getting stuck with a poor performer over a number of years is associated with the beta risk. If we decide from the sample that there is no difference between the copiers, and there really is a difference, then we could buy the wrong device based on price (assuming that the low-price device actually has lower quality). So we want to protect ourselves from such a long-term risk and set the beta at 1% (0.01). This means that one time in a hundred we will say the devices are the same when they are really different.

• Setting the Important Difference

The amount of difference that makes a difference in the metric under study (copy quality value) depends on how fussy we are. This response is measured on a 0 to 100 scale with just perceptible differences equal to 2 units. That means that if shown a copy, a person could not distinguish a 78 from an 80. This gives us some insight into the difference we wish to detect. Obviously a change of 2 units or less is beyond our ability to see any difference. Being a fussy company, we decide that a difference of 5 units (or 2.5 times the perception limits) will be the amount of difference that is important to us.

• Determining Variation

What about the standard deviation? Since we have no prior knowledge in this situation, we must take a preliminary sample. We use the rule of thumb that suggests that the number of observations for such a preliminary sample should be at least 3 but no more than 5. For this example, where the cost of sampling is minimal, we will visit 5 installations at random to get our sample. Now all five of these observations must come from the same brand of copy machine, for if there were a difference between brands, this difference would inflate the variation. We will tentatively assume that the variation between brands is the same and use the estimate of the standard deviation from only one brand to set up our sampling scheme. We will choose the brand for the preliminary sample by tossing a coin. Let's say that brand A gets the toss and we obtain the five observations as follows:

82.3 72.4 67.9 86.2 72.9

which produces a standard deviation of 7.6.

• Finding the Number of Observations

Now we calculate the ratio of the important difference to the standard deviation:

$$D = 5/7.6 = 0.658$$

The closest entry in Table 4 is 0.65. The intersection of D, alpha (0.05) and beta (0.01), is at a sample size of 88 copies from each population of devices. Now this may appear to be a large sample, but look at the calculation of D. We are trying to find a difference that is 2/3 as small as the "natural" variation in the process. This is the price of being fussy.

Before we go out and gather this data in a random fashion, let's see where we are.

alpha $= 0.05$

beta $= 0.01$

5 units of quality are enough to make a difference

s $= 7.6$ (preliminary)

$H_0: \mu_A = \mu_B$

$H_1: \mu_A \neq \mu_B$

Now we actually collect the copy quality sample from randomly selected machines (Tables 12-4A & B) as follows.

TABLE 12-4A

MACHINE TYPE #1(A) PRICE: $1995

71.4	65.2	74.9	84.2	80.0	75.1	77.1	75.2	79.4	77.0	92.6
71.9	62.2	61.6	80.0	78.2	76.9	74.6	73.3	86.5	74.9	78.3
76.2	81.9	82.6	76.6	73.5	71.8	73.5	76.9	87.9	86.3	73.9
64.3	81.9	72.8	65.2	79.9	93.0	83.9	77.3	86.5	83.6	74.5
83.9	53.7	61.2	74.7	70.0	83.1	78.4	76.9	80.8	69.5	63.9
81.2	63.5	80.4	70.2	83.8	70.8	74.6	82.3	84.7	76.5	82.2
81.4	82.6	91.6	58.0	82.6	81.4	78.8	65.9	70.8	80.9	64.3
72.1	64.9	67.5	74.4	72.8	79.8	74.9	82.8	86.9	81.7	76.8

TABLE 12-4B

MACHINE TYPE # 2(B) PRICE: $2395

80.4	86.8	85.5	85.0	89.3	88.3	86.5	94.8	81.9	76.3	70.9
85.2	91.1	90.1	80.9	83.3	80.9	87.4	89.5	82.9	81.3	83.4
75.8	75.3	86.4	87.1	84.7	96.2	84.2	80.3	77.3	85.1	92.6
96.2	89.5	86.1	81.6	80.8	82.4	81.9	87.3	94.0	91.8	85.5
82.6	90.9	87.9	96.6	76.5	86.3	82.7	85.4	84.1	92.5	85.2
84.9	93.3	91.2	83.2	88.9	80.2	88.6	82.9	82.7	80.8	84.7
78.2	84.1	82.9	76.4	90.2	78.1	92.7	81.7	91.8	85.7	95.1
91.6	89.1	85.9	87.2	76.3	72.9	81.9	101.9	83.5	76.5	91.5

Using an appropriate computer program or our pocket calculator, we obtain the following statistical information for the data from the two machines:

$$\overline{X}_1 = 76.3 \quad \overline{X}_2 = 85.4$$

$$s_1 = 7.79 \quad s_2 = 5.85$$

$$s_1^2 = 60.7 \quad s_2^2 = 34.21$$

$$n_1 = 88 \quad n_2 = 88$$

While any computer program will calculate the value of the appropriate test statistic, let's see what is going on in such a program by looking at the "road map" of significance tests found in the Appendix of this chapter. We start at the branch of the decision tree marked "testing means." From there, we branch off to the type of hypothesis, which in this case is "$\mu_1 = \mu_2$." The next branch asks a question we are not yet prepared to answer with the knowledge we have obtained thus far. We need to know if the variances between the two populations are "significantly" different from each other. We can see that there is a numerical difference between s_1^2 and s_2^2, but is this a *real* difference or a difference just caused by the chance variation in the unstable statistics? Remember, statistics are *not* free from variation. If we were to take another set of observations, we would get a different set of values for the statistics.

To answer this important question about the variation, we will divert from our comparison of means momentarily and set up a test on variances using the "F" statistic. This involves a simple hypothesis test that looks at the ratio of the two variances. You may wonder why we are looking at the *ratio* of the variances.

Let's think about that idea for a moment. If we reached into a population and extracted a sample of n observations, computed the variance of that sample, and then repeated the operation all over again, would we expect that the numerical values of those two sample variances to be the same? Well they should be, since they came from the same population. But, of course, they are not the same numerically (unless we were very lucky *or* took every observation in the entire population). If the samples came from the same population, the variances are theoretically the same, and the ratio of these variances should be 1. But the observed sample variances are not the same (we got 60.7 and 34.2) and therefore their ratio will be different from 1. How much different from 1 can this ratio be and yet allow us to believe that the sample variances represent the same population? In other words, the variances are not different.

This must have been the type of question that Sir Ronald Fisher pondered as he developed the concept of the ratio of sample variances that was named in his honor as the "F" distribution. Figure 12-4 shows the shapes of various F distributions. The distribution is skewed (not symmetrical) and the length of the skew and the height of the peak depend upon the sample size. Fewer observations in estimating the population variance lead to a long tail with a flatter peak. As we increase the number of observations, we begin to approach the population itself, and sample variances from the same population will approach the true population variance and therefore their ratio will be 1!

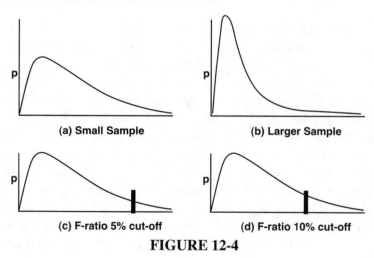

(a) Small Sample   (b) Larger Sample

(c) F-ratio 5% cut-off   (d) F-ratio 10% cut-off

FIGURE 12-4

Because of the way the F Table is constructed, we put the larger sample variance in the numerator of the ratio, which assures us F values equal to or greater than 1. We select a cut-off in the F-distribution table based on our degree of belief (probability) that if we *exceed* this tabulated ratio, we are not a part of this distribution. In Figure 12-4(c), the cut-off places 5% of the area to the right. In Figure 12-4(d), the F-ratio cut-off point leaves 10% of the area to its right and the F is smaller in magnitude than the F cut-off point in Figure 12-4(c). The exact values of the F cut-off points for various sample sizes and probability levels may be found in F tables such as Table 5 in the statistical tables section at the end of this book.

It is interesting to speculate on the basis of these F tables. While the fundamental F distribution includes fractional values, the smallest value in these tables is 1 (for infinite and infinite df). We shall see that the F distribution plays an important part of a more general analysis method called ANOVA which is the topic of the next chapter. In this method, the test statistic is always a single tail test. We just noted, "Because of the way the F Table is constructed, we put the larger sample variance in the numerator of the ratio, which assures us F values equal to or greater than 1." However, if we had set up the F with the smaller sample variance in the numerator, then the F value would be less than 1. In sampling two randomly selected populations for their variance, it is just as likely to have a fraction as a whole number. Therefore, we need to account for the probability of having an F less than 1. Figure 12-5 shows this in the F distribution that has both an upper tail as we originally saw in Figure 12-4 and now with a *lower* tail that accounts for the fractional F values. Since the F tables were developed for the majority application (ANOVA), they are set up for single-sided decisions with only an upper value.

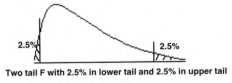

Two tail F with 2.5% in lower tail and 2.5% in upper tail

FIGURE 12-5

To adapt these single-sided tables to the need we currently have in testing variances that could have been found in a ratio less than one, we simply perform the F test as a upper tail test (i.e., put the larger sample variance in the numerator) and split the risk found in the F table in half. So, if we wanted to test at the overall level of 5% (0.05), we would look in the 2.5% (0.025) table. This is exactly what is illustrated in Table 12-5 where the overall total risk of 5% (0.05) is divided into the upper tail and lower tail.

Now with the idea of where to find the correct risk for our comparison of sample variances using the F statistic, we can go ahead and show how this statistical inference concerning variation is constructed and why this F test is important before we go ahead and form hypotheses about the mean values. It is important to note that the inference in the F statistic is about the variance, not the standard deviation. Do not make the often-made mistake of comparing standard deviations instead of the variances.

The "F" test is a formal hypothesis test with a null and alternative hypothesis as follows:

$$H_0: \sigma_1^2 = \sigma_2^2$$

$$H_1: \sigma_1^2 \neq \sigma_2^2$$

σ^2 is the population variance we are referencing. (Note that "σ" is the Greek letter [lower case] "sigma.") We compute the ratio of our sample variances (which are estimating the population variances) and compare this calculated value to the tabulated F value for 87 degrees of freedom (df) for the numerator and 87 df for the denominator, in our ratio at the 1% risk level. (It is necesary to interpolate to get this F). The test of this hypothesis shows that there is no significant difference between the two variances since the calculated F of 1.77 is to the left of the "critical cut-off F value." Hence, there is no evidence to conclude that these two populations of copiers have different levels of variation.

Calculated Value	**Tabulated Value**
F = 60.70/34.21= 1.77	$F_{0.01/2,\ 87,\ 87} = 1.83$

Since the variances are not significantly different, we move on in the flow diagram to the test of means. The test statistic for this situation looks somewhat complex, but it is merely the same signal-to-noise ratio we encountered in the simpler hypothesis test involving a single mean.

The signal part is the difference between the means. The noise part is the standard error of the means. Now it is the noise portion of the formula that seems to have grown more complex. Actually it is merely an expression that takes into account all the data we have gathered in this experiment and puts this good effort to use.

The denominator consists of two parts. The first part is the computation of the "pooled" standard deviation. The second merely divides this pooled standard deviation by the total number of observations we have obtained. Looking at the pooled standard deviation first, the computation simply takes the variance of each set of data and multiplies it by n-1. This turns the variance back into a "sum of squares."

Recall the formula for the sample variance is:

$$s^2 = \frac{\sum(x - \overline{X})^2}{n - 1}$$

The expression $\sum(x - \overline{X})^2$ in the numerator is the sum of the squares around the mean so if we multiply the sample variance (s^2) by $n - 1$, we can reconstruct this sums of squares.

Fundamental statistical concepts tell us that standard deviations are not additive, therefore we could not find an average standard deviation. So if we were to pick only one of the standard deviations, we would discard information about the other source variation (cost us resources to obtain) and we would not be very efficient. Fundamental statistical concepts further tell us that if the variances are independent, we may add them. However if the variances came from different sample sizes, we would want to give more weight to the larger sample and less weight to the smaller sample. Now, the sums of squares, which is essentially a weighted expression of the variance, is always additive whereas the sample variances are not additive when the sample sizes are unequal. Therefore the general pooling formula (12-13) will apply in any sample size situation.

$$s_p^2 = \frac{(n_1 - 1) s_1^2 + (n_2 - 1) s_2^2}{(n_1 + n_2) - 2} \tag{12-13}$$

Applying 12-13, we find that the pooled variance is 47.45 and the square root of this (the pooled standard deviation) is 6.89.

$$s_p^2 = \frac{(87) 7.79^2 + (87) 5.85^2}{(88 + 88) - 2} = 47.45$$

$$s_p = \sqrt{47.45} = 6.89$$

Now, s_p is a single estimate of the combined variation of the two populations.

Before we continue building the test statistic to make our decision, let's see what all of this work is leading to. We have hypothesized that there is no difference between the true means of the two populations. But all we have to test this hypothesis are variable statistics (sample averages). If the null hypothesis about the populations is correct, then there will be no difference between these population means. Figure 12-6 shows this difference as a point, with a value of zero. Again, that's great for a population, but we have *statistics* whose values are merely estimates and the difference between the sample means will not be zero all the time even if we have obtained data from the same population!

Now we can see why the drawing in Figure 12-6 is not just a spike, but rather a distribution that is centered around this zero difference point. If $\mu_1 = \mu_2$, the difference we observe between our sample averages will fall within the boundaries of this "zero-difference distribution." Now we have drawn the distribution as a normal. Is that correct? Recall the first consequence of the central limit theorem states that no matter what the parent population, the distribution of the sample averages will approximate a normal distribution. We are using averages, not individual observations to test our hypothesis, so we may make use of this powerful consequence that the distribution shape will be normal. This means that we may use the areas under the normal distribution curve to determine the probabilities of the errors in our decisions. If the observed difference between our sample means does not fall within reasonable bounds of the zero-difference distribution, we reject the idea that the two populations are the same.

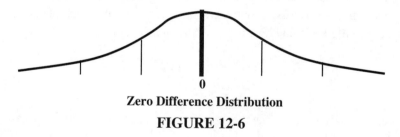

0

Zero Difference Distribution

FIGURE 12-6

All we need to do now is determine the spread of the zero-difference distribution. We have already found the best estimate of the overall (pooled) variance of the individual observations. But we are looking at a distribution of differences between sample averages. Another fundamental rule of basic statistics states that the variance of a linear combination of population means is the sum of the variances of these populations (providing that the variances are independent).

We are looking at the linear combination (the difference between μ_1 and μ_2) of two populations in Figure 12-6. Therefore, the variance of this difference (which is the variance of the zero-difference distribution) will be $s_p^2 + s_p^2$ (s_p^2 is, for our example, 47.45, which we just calculated from Expression 12-13).

What we are doing is analogous to what is called "additive tolerance stack-up." If I have an assembly made up of two parts that are mated together, then the overall height of the assembly will be the sum of the average size of each part. Let's say that part 1 has an average height of 1 cm, and part 2 has an average height of 2 cm. The average assembled height of the two parts will be 3 cm. But what will the variation in the height be? If there were no manufacturing variation, then the variation in the assembly would be zero. However, having no variation is an unrealistic situation. Let's say that the standard deviation of part 1 is 0.1 cm and part 2 has a standard deviation of 0.2 cm. If these two standard deviations are independent (which is likely in a manufacturing process), then the overall standard deviation (s_T) of the assembly is:

$$s_T = \sqrt{s_1{}^2 + s_2{}^2} \qquad\qquad (12\text{-}14)$$

And for our example:

$$s_T = \sqrt{0.1^2 + 0.2^2} = 0.2236$$

This is exactly the same thing that is happening with the zero-difference distribution. The only difference is that we have a common, pooled estimate of the variance *and* the central limit theorem working for us. So the variance of the difference between the two means is their sum, and the fact that we are finding the difference between two averages calculated from samples of 88 observations each reduces the spread of the zero-difference distribution as shown below:

$$s_{\bar{x}_p} = \sqrt{\frac{47.45}{88} + \frac{47.45}{88}} = 1.0385$$

What It All Means

Let's summarize the thoughts involved with the hypothesis test and then make our decision!

1. If the null hypothesis is correct, then the difference between the population means will be zero.
2. Since we do not have population means, but rather sample estimates of these means, the difference we observe will not be zero; rather, this difference will be part of a distribution around the zero point.

3. The shape of this distribution will be normal.
4. If the observed difference falls in a reasonable, predetermined part of this distribution, the null hypothesis is correct.
5. If the observed difference fails to fall in this "zero-difference distribution" the null hypothesis is rejected.

Now the moment we have been waiting for has arrived. Are the populations of copiers the same? We calculate the test statistic (t) as follows:

$$t = \frac{\overline{X}_1 - \overline{X}_2}{s_{\overline{x}_p}} = \frac{76.3 - 85.4}{1.0385} = -8.76$$

With a calculated t of -8.76, compared to the two-sided critical t for 174 degrees of freedom (87 df + 87 df), which is ±1.96, we reject the null hypothesis and conclude that there is a difference between the copy quality of the two machines.

We also note that Brand B gives better quality. While the prices are sufficiently different ($400) to warrant buying the cheaper one, we will choose the copier (Brand B) based on the quality difference demonstrated in this experiment.

Before we leave this chapter on "simple analysis," let's ask what the risk is in making the last decision. Remember that we rejected the null hypothesis. Therefore, while there was the potential of making either a type I error or a type II error before we made the decision, we can only make one error given that we have made only one decision. In this case, since we rejected the null hypothesis, we can only make a type I error.

The maximum risk involved with that error is the alpha risk that we designed into our experiment at 5%. With the amount of difference we observe between the two copiers, we can be quite sure that we have made the correct decision on the purchase of the "Brand B" copy machine. If we were to do this same experiment over and over again, our chances would be that in 20 repeats we would make the wrong decision, at most, only once. Those are pretty good odds!

Pooling Variances

One last note. If our variances had not been "equal," then the pooling of variances is not allowed. You can make the calculation, of course, but you get an incorrect representation of the variation in the pooled value. The larger variance would inflate the value of the smaller, and the smaller variance would deflate the larger variance in the pooled form. To skirt this problem and still allow an inference to be made, an approximate "t" statistic is calculated. This is the formula in the flow diagram called the Fisher-Behrens "t." To use this approximate "t" test, you simply find the square root of the two variances divided by their sample sizes and use this as the standard error of the mean. However, due to the lack of similarity between the variances, the degrees of freedom are adjusted downward in proportion to the difference between the variances. So the greater the difference in the variances, the fewer degrees of freedom. This test will give just as good validity as the "true t" test especially when the sample size is large (say, more than 30 observations from each population). As the sample does get smaller, however, the approximate "t" begins to lose sensitivity due to the reduction in the degrees of freedom.

Problems for Chapter 12

1. The following data are measurements of the force required to break a bond between two cemented metal parts. Two methods were used to join these parts together. Method A used one drop of adhesive that was applied after buffing the two mating surfaces thoroughly. Method B used two drops of adhesive after the surfaces were wiped clean with a dry rag. Method A requires 20 seconds of operator time per assembly; method B requires 1 second of operator time.

Method A	Method B
227	233
235	215
225	202
237	202
237	226
239	224
226	222
220	219
244	217
228	235

Use a t-test to test the hypothesis that the two methods need the same average force to break the bond.

2. A modification is planned to an existing process. The average, long-term output from this process has been 9.37. A pilot run with the change produces an average of 10.42 with a standard deviation of 1.22 from a sample of 25 observations. Set up the appropriate hypotheses and test them.

3. Two production lines produce the "same" part. The samples from these lines give the following values:

	Line A	Line B
$\overline{X}$	2.54 cm	2.83 cm
s	0.04 cm	0.03 cm
n	14	10

Set up the appropriate hypothesis test and answer the question with regard to significant differences between the two lines.

4. A development engineer wants to try out a new toner. He wants the particle size to be different by at least 2 microns. A preliminary sample shows that the standard deviation of the measurement is 0.5 microns. Given that he wants to limit his errors as stated, what size sample will be necessary to tell the difference between two batches of toner? (Limit the risk of saying toners are the same when they differ to 5%, and the risk of finding differences when there is no difference to 10%.)

5. In Appendix 12, "Steps in Hypothesis Testing" there is a note that says, "Do NOT use the data to decide on single sidedness. You must use outside (often economic) information to determine the direction." Explain the thought process behind this note. (Hint: statistics are variables, not constants.)

APPENDIX 12A
SOME BASIC RULES OF PROBABILITY

1. The sum of the probabilities in a single cause system must equal one. There may not be any negative probabilities.

2. If two events are mutually exclusive, then the probability that one or the other will occur equals the sum of their probabilities.

3. If two events are independent, the probability that both events will occur is the product of the probabilities of the individual events.

4. Odds may be related to probabilities by the following rule:
   For odds a to b and probability p,
   $a/b = p/(1-p)$
   and $p = a/(a+b)$

BASIC STATISTICAL FORMULAS
1. The mean or average

$$\overline{X} = \frac{\sum\limits_{i=1}^{n} x_i}{n}$$

2. The standard deviation

$$s = \sqrt{\frac{\sum\limits_{i=1}^{n} x_i^2 - (\sum\limits_{i=1}^{n} x_i)^2 / n}{n-1}}$$

STEPS IN HYPOTHESIS TESTING
1. State the null and alternative hypotheses, i.e.,

Population vs. Standard	Population vs. Population	
$H_0: \mu = \mu_0$	$H_0: \mu_A = \mu_B$	Double Sided
$H_1: \mu \neq \mu_0$	$H_1: \mu_A \neq \mu_B$	

- -

$H_0: \mu = \mu_0$	$H_0: \mu_A = \mu_B$	Single Sided*
$H_1: \mu > \mu_0$ OR	$H_1: \mu_A > \mu_B$	
$H_1: \mu < \mu_0$	$H_1: \mu_A < \mu_B$	

2. Decide upon the risks you are willing to live with.
   alpha risk (risk of rejecting H_0 when we should not) = _____
   beta risk (risk of not rejecting H_0 when we should) = _____

*Do NOT use the data to decide on single sidenness. You must use outside (often economic) information to determine the direction.

3. Determine how much of a difference is important in terms of physical variables involved.

$$\text{delta} = \underline{\hspace{3cm}}$$

4. Take a preliminary sample to determine the variation in your data or use a previously determined value of the standard deviation.

$$\text{Preliminary Std. Dev.} = \underline{\hspace{3cm}}$$

5. Determine the sample size based on the above criteria by using the appropriate table (Table 3 or 4, depending on your hypothesis). Choose "critical" t value based on this sample size and the alpha risk.

$$\text{Sample Size} = \underline{\hspace{2.5cm}}$$

$$\text{Set critical "t" value} = \underline{\hspace{2.5cm}}$$

6. Gather your data and compute values for the following statistics:

$$\overline{X}_A = \underline{\hspace{1.5cm}} \quad s^2 = \underline{\hspace{1.5cm}} \quad s = \underline{\hspace{1.5cm}}$$
(for $\mu_A = \mu_B$)
$$\overline{X}_B = \underline{\hspace{1.5cm}} \quad s^2 = \underline{\hspace{1.5cm}} \quad s = \underline{\hspace{1.5cm}}$$

7. Compute the value of the test statistic based on the "road map" and then draw conclusions. Always remember to use non-statistical words in your conclusion.

Construction of OC Curves

The operating characteristic (OC) curve is a type of "what if?" exercise. It shows the risk of accepting (or more properly, not rejecting) the null hypothesis for various possible locations of the true mean.

To construct the OC curve, we ask what if the true mean were at a certain point and then find the area or probability up to that point. This procedure is repeated for many points from which a smooth curve is plotted.

To illustrate the construction of an OC curve, we will use the following data.

$$\mu_0 = 8400 \qquad\qquad s = 760 \qquad\qquad n = 200$$

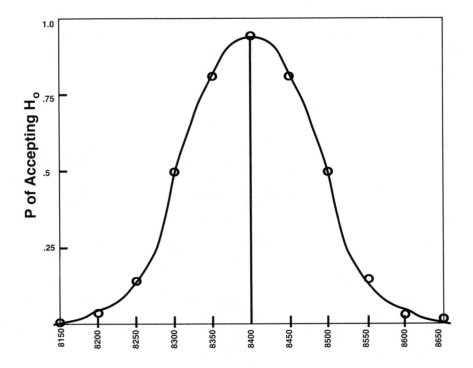

Construction of the OC Curve

To give the problem relevence, let's say that a shopping mall developer is gathering information to help make a decision concerning the placement of a new mall in a rural area. One of the criteria for a "go" decision is the annual spending habits of the customers. If the average annual spending of the potential customers is $8400, the developer will be encouraged to proceed. His economic model says that a difference of $100 is not enough to upset his plans. He can afford to engage a market research firm to survey 200 shoppers to obtain the estimate of the average annual spending. The results of the survey are shown below:

$$\overline{X} = \$8350 \quad s = \$760 \quad n = 200 \quad d = \$100$$

The t test for this situation would show no reason to reject the claim that the spending was $8400, and the mall developer would be encouraged to go ahead.

But what if the true value of the spending was really something different than the $8400 hypothesized?! An OC curve will tell us the probability.

We will start on the right side of the curve at the high dollar values. (Note that we will use the t values for ∞ df since the 200 observations converge to ∞ in the table. So in essence, we use the normal distribution table.) We will use the s_x based on the data and will consider the indifference value of $100.

$$\text{So,} \quad s_{\bar{x}} = s/\sqrt{n} = 760/\sqrt{200} = 53.7$$

What if true mean were?	t value	Probability of event (from normal table)
8650	$t = \dfrac{8500\text{-}8650}{53.7} = -2.793$	0.0026
8600	$t = \dfrac{8500\text{-}8600}{53.7} = -1.862$	0.0314
8550	$t = \dfrac{8500\text{-}8550}{53.7} = -0.930$	0.1762
8500	$t = \dfrac{8500\text{-}8500}{53.7} = 0$	0.50
8450	$t = \dfrac{8500\text{-}8450}{53.7} = 0.931$	0.8238
8400	$t = \dfrac{8500\text{-}8400}{53.7} = 1.862$	0.9686

Since the normal curve is symmetrical, we may find the probabilities on the left side by "reflecting" the values from the right side for the same difference from the hypothesized true mean of $8400.

8350 (same as 8450)	0.8238
8300 (same as 8500)	0.5000
8250 (same as 8550)	0.1762
8200 (same as 8600)	0.0314
8150 (same as 8650)	0.0026

Road Map of Significance Tests

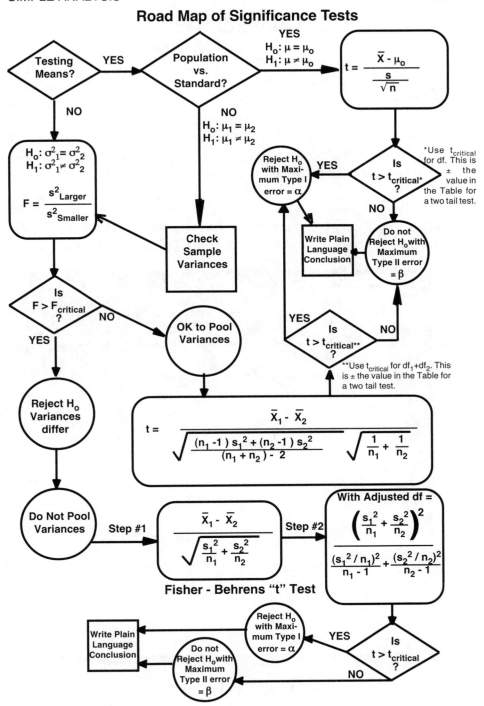

Fisher - Behrens "t" Test

APPENDIX 12B
USING MINITAB

Caution: *If you have jumped right to this appendix to be trained in the use of Minitab, be warned that without the educational perspective of the previous 35 pages, you will not be able to apply the principles of statistical inference in the general, universal way that is necessary to solve each different, individual problem that you encounter. Please read the chapter before using the computer software.*

Begin by launching Minitab. In the worksheet, enter the data (this example uses the copy machine comparison data from Tables 12-4A and 12-4B). A portion of the worksheet is shown in Figure APP 12B-1. Go to **Stat** dropdown and

	C1	C2	C3	C4
	Machine Type 1 (A)	Machine Type 2 (B)		
1	71.4	80.4		
2	71.9	85.2		
3	76.2	75.8		
4	64.3	96.2		
5	83.9	82.6		
6	81.2	84.9		
7	81.4	78.2		
8	72.1	91.6		
85	83.9	83.2		
86	82.2	84.7		
87	64.3	95.1		
88	76.8	91.5		
89				

MINITAB - 2sampletTest.MPJ - [Worksheet 1 ***]

File Edit Data Calc Stat Graph Editor Tools Window Help

FIGURE APP 12B-1

select **Basic Statistics**. Scroll down to **2-Sample t....** (Figure APP 12B-2) to open the dialog box. There are three ways to bring data into this test. The default is to have a single column with subscripts to indicate the two separate sets of samples. An easier way is to do as this example does it with two separate columns. Click on **Samples in different columns**,(Figure APP 12B-3) for this example. The final way is to enter an already calculated set of means and standard deviations. Note the box that says, **Assume equal variances**. If we do not check this box, Minitab will perform an approximate test assuming that the variances are not equal much like the Fisher-Behrens test found in the "Road Map" flow diagram. So, there is no automatic equality of variance testing built into the Minitab two-sample t-Test routine. To check the variances, we go to **Basic Statistics** again, but now scroll down to near the bottom of the list as shown in Figure APP 12B-4 and select **2 variances...** to obtain another dialog box (Figure APP 12B-5) that is much like the 2-sample t test. Since our data is in two separate columns, we click on that selection and enter the **First:** and **Second:** columns of our data.

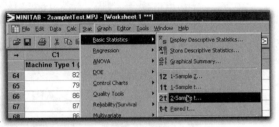

FIGURE APP 12B-2

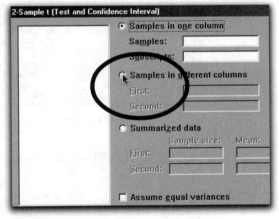

FIGURE APP 12B-3

     The options sub dialog box (Figure APP 12B-6) allows us to select the confidence level, which is 95%. We will be consistent with the 99% we had used before in the example and change the default confidence interval value to 99. The result in the session window (Figure APP 12B-7) shows the same result we found before.

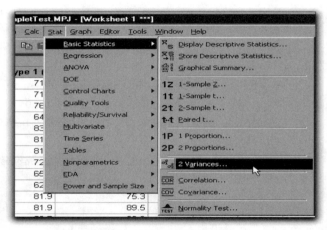

FIGURE APP 12B-4

FIGURE APP 12B-5

FIGURE APP 12B-6

Test for Equal Variances: Machine Type 1 (A), Machine Type 2 (B)
99% Bonferroni confidence intervals for standard deviations
                        N   Lower   StDev   Upper
Machine Type 1 (A) 88   6.40702 7.79032 9.84057
Machine Type 2 (B) 88   4.81003 5.84854 7.38775
F-Test (normal distribution)
Test statistic = 1.77, p-value = 0.008

FIGURE APP 12B-7

Now we can go back to the t-Test knowing that the variances are not different. In the main dialog box, we check **Assume equal variances**. Before leaving the main dialog box, click **Graphs** as shown in Figure APP 12B-8. This brings up a sub dialog box (Figure APP 12B-9) where we click on **Individual value plot**. This plot (Figure APP 12B-12) gives us an idea of the location and scatter of the individual data points. Click **OK** to return to the main dialog box where we

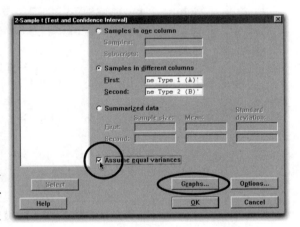

FIGURE APP 12B-8

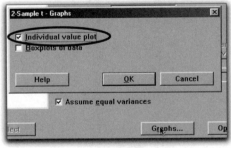

FIGURE APP 12B-9

choose **Options**, which allows us to select the confidence level for this test. This is similar to the options sub dialog in the variance test we just performed. You will notice a consistency in the Minitab dialog boxes that makes using this software intuitive and easy to use from one application to the next. Use 95% as we did before.

Now, simply click **OK** from the sub dialog box and **OK** from the main dialog box and Minitab does the rest! The results are shown in Figure APP 12B-11 and are exactly the same as we obtained when we did the t-Test "by hand."

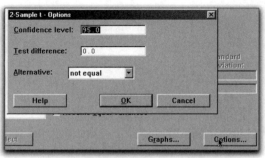

FIGURE APP 12B-10

Two-sample T for Machine Type 1 (A) vs Machine Type 2 (B)

	N	Mean	StDev	SE Mean
Machine Type 1	88	76.29	7.79	0.83
Machine Type 2	88	85.35	5.85	0.62

Diff. = mu (Machine Type 1 (A)) - mu (Machine Type 2 (B))
Estimate for difference: -9.06135

95% CI for difference: (-11.11092, -7.01180)

T-Test of difference = 0 (vs not =): T-Value = -8.73
 P-Value = 0.000 DF = 174
Both use Pooled StDev = 6.8882

FIGURE APP 12B-11

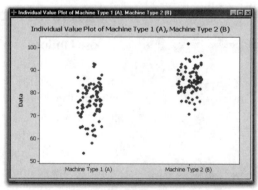

FIGURE APP 12B-12

Minitab will also compute the number of observations in a sample. To do so, scroll down to the bottom of the **Stat** dropdown (Figure APP 12B-13). Of the eight choices, we'll select **2-Sample t...** and see how it compares with the values found in the sample size tables. Figure APP 12B-14 is the main dialog box for this effort. We'll use a difference that is equal to the standard deviation (0.25 each) and opt for a power of .9 (power is 1 - β). Click **Options** to select the α risk and the number of and direction of the tails in the test. In Figure APP 12B-15, we stay with the default α of 0.05 and choose a two-sided test.

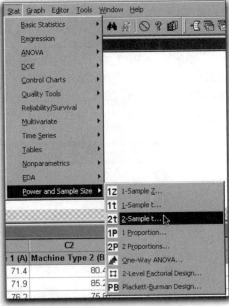

FIGURE APP 12B-13

FIGURE APP 12B-14

It is possible to enter multiple powers and differences in a "what if" type of exercise. It is also possible to enter a given sample size and see what the power is for a difference required.

Figure APP 12B-16 shows that the number of observations to achieve an alpha risk of 0.05, beta risk of 0.10 with a difference of 0.25, and a standard deviation of 0.25 is 23 from each sampled population. This is the same value as the sample-size table in the back of this book recommends.

FIGURE APP 12B-15

Power and Sample Size

2-Sample t Test
Testing mean 1 = mean 2 (versus not =)
Calculating power for mean 1 = mean 2 + difference
Alpha = 0.05 Assumed standard deviation = 0.25

| | Sample | Target | |
Difference	Size	Power	Actual Power
0.25	23	0.9	0.912498

The sample size is for each group.

FIGURE APP 12B-16

13

Analysis of Means by Using the Variance

While we were able to make wise decisions in the face of uncertainty using the simple analysis techniques of Chapter 12, the real problems we encounter are often of a more complex nature and require analysis techniques beyond the simple "t" test.

It is easy to see how the "t" test works, with its signal-to-noise ratio concept. The difference between the means is the signal in the numerator, and the variation is the noise in the denominator. But what do we do when there are more than two means to compare? We could set up a "t" test that looks at only the extreme values, but this would waste a lot of the information that costs us valuable resources. We could run "t" tests on all possible pairs of means in question, but as we shall see this leads to inflated risk problems, reducing our ability to make correct decisions.

Let's look at a specific example with three means being compared. If we were to use a "t" test on the following null hypothesis, we would not have one "t" test but a series of "t" tests depending on the number of ways the means can be paired together.

$$H_0: \mu_1 = \mu_2 = \mu_3$$

The alternative hypotheses would look like this:

$$H_1: \mu_1 \neq \mu_2 \qquad H_2: \mu_1 \neq \mu_3 \qquad H_3: \mu_2 \neq \mu_3$$

Now, if we set the risk at 0.05, then the probability of being correct (or the level of confidence) is one minus the risk, or in this case, $1 - 0.05 = 0.95$. The rules of probability state that for independent events, the probability of all events taking place is the product of the individual probabilities. If each "t" test were an independent test then we may apply the *probability product* rule to find the overall probability of being correct in our multiple decisions. This means that for all the decisions to be correct in our example, the first decision and the second decision and the third decision must all be correct. Applying the rules of probability, we continue our example as follows:

$$(0.95)\ (0.95)\ (0.95) = 0.857375 \tag{13-1}$$

Now, this resulting probability (0.857) is considerably less than the 0.95 level of confidence we had specified in any one single test. If we were to use the multiple "t" test method, we might not be making very wise decisions.

This Remedy Is Spelled A-N-O-V-A

So what do we do? Pack up and quit? No, what we need to develop is a hypothesis that tests multiple means, and does so with a single risk. The only way to assure a single risk is to have a single test. To accomplish this goal we set up the same null hypothesis as in the multiple "t" test situation, but the alternative hypothesis is different. Instead of setting up pairs of differences to make many alternatives as in the multiple "t" test, we make one blanket statement of the alternative hypothesis that simply states that the means differ. Now for the moment this may be a relatively unsatisfactory alternative hypothesis, since we really want to know which means differ; but as we progress, we shall see that all our burning questions will be answered, and will be answered with control over the probability of making the correct decision.

To see how we can control this probability, let's return to the original example and try to compare three populations. This is a single-factor, three-level experiment and is commonly called a "one-way ANOVA." ANOVA is an acronym for "ANalysis Of VAriance" which is the name of the technique we are about to develop. While the name says "analysis of variance," we are really performing an ANalysis Of MEans BY USing VAriances. Of course an acronym ANOMEBYUSVA would be correct, but unpronounceable. However, we need to keep in mind that this procedure is concerned with finding out if averages differ and only uses the variance as a tool to help make a wise decision.

Signal/Noise

This analysis of means via variances may sound like a roundabout method, but it accomplishes our goal of single risk inference in a multi-level experiment. The concept is to compute the variance within the levels under investigation (the noise) and compare this with the variance induced between the levels (the signal). If the resulting signal-to-noise ratio exceeds that which is expected by chance, we say that there is a difference between the means.

The test statistic used in ANOVA is the same simple "F" ratio that we used to test differences between variances in Chapter 12. To illustrate the concept of ANOVA, let's look at two numerical examples. The first will show a situation in which we *know* that the population means are not different, and the second will illustrate a situation where the means are different.

No Difference

Let's take 15 random, normally distributed numbers all from the same source population (factor), and randomly divide them into three equal portions (levels).

TABLE 13-1

Portion 1	Portion 2	Portion 3	
17.47	22.43	18.80	
21.25	19.12	16.72	
18.87	17.65	19.99	
23.76	23.72	23.08	
21.88	19.29	21.67	
$\overline{X}$: 20.646	20.442	20.052	$\overline{\overline{X}}= 20.38$
s^2: 6.206	6.392	6.115	
Σx: 103.23	102.21	100.26	Grand Total $=305.70$

If we just look at the averages, we see that there is very little difference among them. This is not unexpected, since the total of 15 numbers that came from the same cause system were just randomly divided into three parts. We would expect any one randomly selected portion to represent the whole.

We also notice that the variation (or noise) within each of the portions is fairly consistent. This is an important consideration, and, as we shall see later, is a requirement necessary to use ANOVA properly. The noise that we calculate is the pooled variance over the three portions. The pooling calculation is similar to that used previously in Chapter 12 in the simple "t" test.

While it is easy to see where the noise comes from in the ANOVA calculations, the source of the signal is a bit more obtuse. To find the signal variance, we need to go back to our null hypothesis and see what it is saying.

$$H_0: \mu_1 = \mu_2 = \mu_3 \tag{13-2}$$

Expression 13-2 states that all of the means are the same. If this is so, and we were to put all of the data together as if it came from one population, then the variance we would observe would be no greater than the variance observed within any portion of the data. These "any portion" variances are the variations within each of the individual populations we are studying. We should recognize these variations as the lowest order noise.

To compute the signal, we combine the individual populations of the null hypothesis together as if they came from only one larger population and compute the variance of this population. If there are no differences among the populations, then the variance we observe will be no different from the variance we observe in the individual "within" populations. If there is a difference among the populations under study, then the variance we compute from the one combined population will be larger than the variance based on the individual populations. Figure 13-1 illustrates the concept of the analysis of variance in the form of population diagrams.

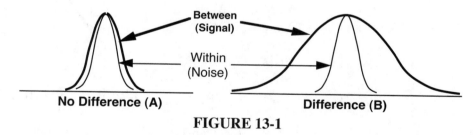

FIGURE 13-1

In Figure 13-1(A) the signal and noise populations are not different. Since they are the same, when we perform an "F" test we find that the signal-to-noise ratio, F, is only a chance happening and does not exceed the "critical" value in the table. In Figure 13-1(B), the spread of the "between" population is much greater than the spread of the "within" population. Here an "F" test shows that the between variance is indeed larger than the within variance.

In the second numerical example, we shall see how this (Figure 13-1 (B)) takes place. But for the moment let's concentrate on a numerical evaluation of the "A" portion of Figure 13-1 by going back to the data in Table 13-1.

We need to compute two variances for this example. Recall that the formula for variance may be expressed as follows:

$$s^2 = \frac{\sum x_i^2 - \frac{(\sum x_i)^2}{n}}{n - 1} \tag{13-3}$$

This formula says to take each x value, square it, and add it to the rest of the x squared values; then find the sum of all the x values and square this quantity, and then divide by n (the number of x's). Take the difference between the two quantities and divide by n–1 (the degrees of freedom). If we apply this formula to portion 1 of the example in Table 13-1, we get:

$$\sum x_i^2 = 17.47^2 + 21.25^2 + 18.87^2 + 23.76^2 + 21.88^2 = 2156.1123 \tag{13-4}$$

$$(\sum x_i)^2 = (17.47 + 21.25 + 18.87 + 23.76 + 21.88)^2 = (103.23)^2 = 10656.433$$

$$\text{and } (\sum x_i)^2/n = 10656.433/5 = 2131.2866 \tag{13-5}$$

When we find the difference between 13-4 and 13-5 we have the numerator of the expression for finding the variance. This is called the "sum of squares corrected for the mean." This "correction" is not to make up for any mistakes, but to center the variance around a mean of zero. The sum of the x's, quantity squared, divided by n is called the "correction factor." Without this correction factor, the magnitude of the sum of squares would be dependent upon the magnitude of the numbers we were using. By "correcting" for the mean, we are able to look at only the magnitude of the *variation* without the extra confusion of the mean level. In a more general sense, we are centering our calculation of variance around a mean of zero. Another, better name for the "correction" factor could be "centering" factor. Using the name "centering" factor eliminates the connotation of "mistake" attached to the word "correction" and communicates the concept a bit better. Continuing with the numerical example, we obtain the difference between Expressions 13-4 and 13-5, which is:

$$2156.1123 - 2131.2866 = 24.8257 \tag{13-6}$$

If we were to divide this result by the degrees of freedom ($5 - 1 = 4$), we would get 6.206, which is the variance reported in Table 13-1. However, we do not choose to divide at this point but to hold the sum of squares within the first portion and go on to compute the sum of squares within the two other portions. For portion 2 the sum of squares is 25.568, and for portion 3 the sum of squares is 24.460.

Now, why did we hold off in the computation of the variance? Since the "within" variation is really made up of the three portions of the data from the experiment, we should use all three in our calculation of the noise variation. To use all three, we must pool the variances together. This is the same operation we performed in Equation 12-13. In that operation of pooling, we had to multiply the variance by $n - 1$ and turn it back into a sum of squares. To eliminate this multiplication step, we simply leave all the calculations of variances in the ANOVA system in terms of sum of squares (Σsq) until the last possible moment and thus minimize the arithmetic operations. As mentioned in Chapter 12, given independence, the sums of squares will always be directly additive. Variances are additive, but the additivity is based on the weight of the n in the sample.

So now let's add up the "noise" or "within" sums of squares.

$$
\begin{array}{ll}
24.8257 & \text{within portion 1 noise} \\
25.5680 & \text{within portion 2 noise} \\
\underline{24.4600} & \text{within portion 3 noise} \\
74.8537 = & \text{"noise" } \Sigma sq \text{ within all 3 portions (pooled)}
\end{array}
$$

This is only one of the sources of variation in our data. We still need to compute the "signal" portion. To do this, we need to put together the three portions (levels) as if they came from one population. We could take the averages of each of the three portions and compute the variance among these averages, but this leads to a computational problem if the sample sizes are different for each portion. Instead, we square the *totals* of each portion. Then we divide each of these squares by the number of items that went into the total we squared. Then we sum these three terms. This is the sum of squares for the difference in means, but it still needs to be centered around the grand mean, so we take the grand total squared, divided by the total number of observations, and subtract this from the "raw" sum of squares between the means. Symbolically:

$$
\text{"Signal" } \Sigma sq = \frac{X^2_{p1}}{n_{p1}} + \frac{X^2_{p2}}{n_{p2}} + \frac{X^2_{p3}}{n_{p3}} - \frac{\text{Grand Total}^2}{n_{Total}} \tag{13-7}
$$

and with the numbers from Table 13-1:

$$\text{"Signal"} \ \Sigma sq = \frac{103.23^2}{5} + \frac{102.21^2}{5} + \frac{100.26^2}{5} - \frac{305.7^2}{15} = 0.9109$$

We can see that this is a very small sum of squares. It is in fact the extra sum of squares that would be introduced into the data if there truly were a difference between the means of the three portions. However, in this example, we obtained the data from one source of random normal numbers, and we would not expect to see any extra variation due to the "pieces" we broke out of the overall sample except by chance. The amount we got by chance in this case was a mere 0.9109.

As a final check, let's see if the two components we have just computed add up to the total sum of squares. The "noise" was equal to 74.8537 and the "signal" was 0.9109. Together they add up to a total of 75.7646.

To check this sum, we will simply take each individual observation, square it, and add all these squares together. (There will be 15 items). Then center around the mean by subtracting the grand total squared divided by the grand number of observations. You probably recognize this as the same set of calculations we would do if we were to compute the simple variance of the 15 pieces of data!

$$\text{So: } (17.47^2 + 21.25^2 + 18.87^2 + \ldots + 21.67^2) - \frac{305.7^2}{15}$$

Total $\Sigma sq = 6305.9304 - 6230.166 = 75.7644$

This is pretty close to the result we obtained (off by 0.0002 !) from the sum of the parts. This amount of rounding error can be expected when using even large-scale computers. All the calculations for this example were made using a pocket calculator with the facility to compute means and standard deviations and display the sum of the x's and the sum of the x^2's. This is a handy tool to use for such computations.

Now with all the necessary calculations finished, it is time to put the information in order and draw the conclusions about the data we have been investigating. This activity is called the construction of the "ANOVA table" which is shown in Table 13-2. This reporting format is used in computer programs such as Minitab, so it is important to understand its origin.

In Table 13-2 we identify ("source column") and record the sums of squares as computed on the previous pages. The third column contains the degrees of freedom (df) for each source of variation. Let's see where these df come from.

TABLE 13-2

ANOVA

	Source	Sum of Squares	df	Mean Square	$F_{calculated}$	$F_{critical}$
"SIGNAL"	Between portions	0.9109	2	0.4555	0.073	3.89
"NOISE"	Within portions (pooled)	74 .8 537	12	6.2378		
	TOTAL	75.7444	14			

Starting at the bottom of the table, we find the df for the total by subtracting one from the total number of observations in the data. Since there were 15 observations, we have 14 df. The concept behind these df is exactly the same as discussed in previous chapters. Whenever we compute a mean value (which is implicit in the computation of a variance) we "lose" a df. Or, more precisely, we *convert* one of the original pieces of information to the mean.

Going to the top of the table, we look at the number of levels in the "between portions" source of variation, and from our data table (Table 13-1) we observe that there were three levels. Since we put these levels together to compute an overall mean, we subtract a degree of freedom from the number of levels, so there are 2 degrees of freedom for the "between portions" source. In general, each factor under study will have degrees of freedom equal to one less than the number of levels for that factor. So if a factor had 6 levels there would be 5 df.

Now for the last source of variation. You could use an "easy" method and say that the within portion source must be the remaining 12 degrees of freedom. This is correct but it is worth the effort to see where the 12 df come from in this case and in general. Let's look into each of the portions in Table 13-2. There are 5 observations in each portion. When we compute the variance in each portion, we lose a degree of freedom and have 4 df left. We repeat this in each portion and therefore the degrees of freedom for the "within" (or "noise" source) is equal to the sum of the degrees of freedom from all portions, or 12 df for "noise."

Next we take the sum of squares and turn it into a variance by dividing by its df. In our ANOVA table this is called the "Mean Square" (i.e., average sum of squares). It is actually the variance we have been working to quantify, and it will be used to make our decision in the F test. The final computation is that of the F value which is done by comparing the signal (between portions) in the numerator with the noise (within portions [pooled]) in the denominator.

$$F = \frac{0.4555}{6.2378} = 0.073$$

This comparison is made using the ratio of mean squares, which are really the variances. We now compare this calculated F with the critical F in Table 5 (located among the tables in the back of this book) for 2 numerator degrees of freedom and 12 denominator degrees of freedom. At the 0.05 risk level, the critical F value is 3.89. Since the computed F is far less than the critical F, we do not reject the null hypothesis, and we conclude that the three portions all came from the same cause system. This confirms what we knew all along about this set of data, since we drew our sample from the same source.

Big Difference

While the last example showed how to compute all the components of the ANOVA table, there were no statistical differences among the samples. In this next example we will obtain samples from three *different* populations of random normal numbers. Let's see if the ANOVA technique can discover these differences. This example will produce a situation as pictured in Figure 13-1 (B) where the between-level variance is much greater than the within-level variance.

Again, we will look at three portions or levels. However, unlike the first example, the averages of the levels are "designed" to be different by the way we obtained the data. To begin our analysis, we will compute the noise sum of squares within the three different levels under study. This is accomplished by summing the squares of the five observations within each of the three levels. Next, subtract the sum of the five observations squared divided by 5 from each of these sums of squared observations.

TABLE 13-3

Level 1	Level 2	Level 3
20.23	30.85	26.27
20.71	28.96	25.76
20.05	29.71	24.83
21.47	29.55	24.33
20.67	31.68	25 .60

$\overline{X} =$ 20.625   30.150   25.358 $\overline{\overline{X}} = 25.378$

$s^2 =$ 0.302   1.20   0.591

Σx 103.13   150.75   126.79 Grand Total = 380.67

Calculation of the within (noise) sum of squares:

$$\Sigma x_i^2 \quad - \quad (\Sigma x_i)^2 / 5$$

for level 1: 2128.3693 – 2127.1594 = 1.2099
for level 2: 4549.9131 – 4545.1125 = 4.8006
for level 3: 3217.5283 – 3215.1408 = 2.3875

Now we sum (pool) the above results over the three levels:

$$\Sigma sq_{Noise} = \quad 1.2099 + 4.8006 + 2.3875 = 8.398$$

which is the sum of the squares for error or noise.

Next we find the sum of squares for the difference between the levels (signal) under study. This is done by squaring each total from the three levels, dividing by the number of observations in each of the squared totals, summing them up, and then subtracting the centering factor from this result.

$$\Sigma sq_{Signal} = \frac{103.13^2}{5} + \frac{150.75^2}{5} + \frac{126.79^2}{5} - \frac{380.67^2}{15} = 226.7695$$

Finally, to check our arithmetic, we compute the total sum of squares. To do this, take each observation, square it and sum all of these squares. Subtract the centering factor from this sum of squares, and the value we get is equal to the sum of the two parts, the noise sum of squares and the signal sum of squares.

So we take: $20.23^2 + 20.71^2 + \ldots + 24.33^2 + 25.60^2 - \dfrac{380.67^2}{15} = 235.7695$

The Analysis of Variance, Table 13-4, is now constructed from these calculations.

TABLE 13-4

ANOVA

	Source	Sum of Squares	df	Mean Square	F	$F_{critical}$
"SIGNAL"	Between Levels	226.7695	2	113.385	162.02	3.89
"NOISE"	Within levels (pooled)	8.398	12	0.6998		
	TOTAL	235.1675	14			

In the above example we can see that the calculated F value vastly exceeds the tabulated value; therefore, we conclude that the difference between the means of the levels is not a chance occurrence, but a real indication that the means differ. Now you could say, "Of course the means differ. I could see that without any ANOVA calculations!" Let's just look once more at what the ANOVA is saying. While there will probably always be differences observable (even in the first example there were differences), the real question is: "Are the differences we observe bigger than the natural variation within the data?" If the variation within the levels is greater than the differences among the means of these levels, then we are unable to believe that there are any differences among these means and must conclude that the means are all the same. However, if the differences among the means are greater than can be expected by chance variation within the levels, we are inclined to believe that the means differ. This is what we mean by a "significant" difference. The word "significant" does not mean big or important in a physical sense, but simply from a statistical/probability stance, that the difference we observe among the levels (signal) is greater than the variation within the levels (noise). Many researchers often attach the wrong meaning to the word *significance* and equate practical importance with this concept. The result may be "significant," meaning it has a greater signal-to-noise ratio (F) than expected by chance, but the differences among the means under study may not amount to anything practically important. Therefore, when the researcher draws the conclusions, both the statistical and the practical implications should be addressed. We can always state the *statistical significance* (based on the F test), but we must also think of the practical aspects of the problem and also state the *practical importance*. Let's look at another example, and expand the problem to a real-life situation to illustrate this important point concerning significance and importance.

An Apple a Day

Apple growing is an important upstate New York industry. The farmers harvest the crop in the late summer and early fall. However, people like to eat apples all year round. While there is an abundance of apples at low prices in the fall, the winter and spring bring higher and higher prices to the consumer. Besides being a supply and demand situation, one price-inflating factor is storage. There are different kinds of apples grown and there are different types of storage methods. In this problem, a large apple co-op is experimenting to determine optimum storage conditions. The response to be measured is the time to spoilage as measured in months. Table 13-5 shows the data from this experiment that shall be analyzed using ANOVA.

We look at the storage conditions and find the sum of squares among these conditions by squaring the sum of each column and adding these squares together. We then divide this sum by the number of items that went into the sum that we squared (6 in this case). From this result, we subtract the centering factor to obtain the sum of squares due to temperature conditions.

TABLE 13-5

Storage Condition

	36°	38°	40°	42°	44°	
	6.5	7.5	8.0	7.5	5.0	
	6.0	8.0	8.5	7.0	4.5	
	7.5	8.5	9.5	9.5	7.5	
	8.0	9.0	9.0	9.0	7.0	
	5.0	6.0	7.0	6.0	5.0	
	4.5	6.5	7.5	5.5	4.5	
$\Sigma=$	37.5	45.5	49.5	44.5	33.5	210.5

$$\Sigma sq_{cond.} = \frac{37.5^2 + 45.5^2 + 49.5^2 + 44.5^2 + 33.5^2}{6} - \frac{210.5^2}{30} = 27.875$$

The next computational task is to calculate the within (noise) sum of squares:

$$\Sigma x_i^2 \qquad - \qquad (\Sigma x_i)^2 / 6$$

for 36°:	243.75	−	234.375 = 9.375
for 38°:	351.75	−	345.042 = 6.710
for 40°:	412.75	−	408.375 = 4.375
for 42°:	342.75	−	330.042 =12.708
for 44°:	195.75	−	187.042 = 8.708

Now we sum (pool) the above results over the five temperatures:

$$9.375 + 6.710 + 4.375 + 12.708 + 8.708 = 41.876$$

which is the sum of the squares for error or noise. We will now introduce a new way to think of the noise or error. From a purely statistical approach, the error is the random variation within a set of observations. However, from a deterministic point of view we can consider the "error" as that portion of the variation to which we are presently unable to assign a cause. It is a sort of "left over" or **residual** source of variation that we may investigate to determine its cause(s.) From now on, we will think of the noise as the residual rather than as the error since "error" in statistical analysis refers to the purely **random** component of variation (called "pure error"). Further work with this example will expand on this concept and show why we have introduced the new idea of the *residual* variation.

Now we will compute the total sum of squares and see if it checks out. To do this, simply find the square of each individual data point and sum all 30 of these squares. Then subtract the centering factor to obtain 69.75.

$$\Sigma sq_{TOTAL} = 6.5^2 + 6.0^2 + \ldots + 5.0^2 + 4.5^2 - \frac{210.5^2}{30} = 69.75$$

If we sum (recall the sums of squares are additive) the components as follows, we get:

Storage conditions + Residual = Total
27.875   +   41.876 = 69.751

The total of the components is very close (within .001) to the independently calculated total sum of squares and well within the allowable round-off accuracy of the calculation.

We now build the ANOVA table for this storage condition experiment. First we name the sources of the variation as the storage conditions and residual. The sums of squares values that we just calculated are placed in the appropriate positions within Table 13-6. We now consider the degrees of freedom. Since there were five different storage conditions in the experiment, we will have 4 df for this factor. (Remember we calculated a "conceptual" average of the five levels to be able to compute the sum of squares and we lost a df in the process of computing this average.) The df for the residual is just a bit more complicated, but is based on the same idea. When we calculated the residual, we looked into each of the five conditions where there were 6 observations. We quantified the variation within these observations via the sum of squares and in doing so, reduced the df from the original 6 df (there were 6 observations) to 5 df due to the calculation of the "conceptual" average involved in the computation of this sum of squares.

The Mean Square is the next column and results from the division of the sum of squares by the df. It is the mean square that leads us to our decision. The mean square is, in reality, the variance that we use in the ANOVA technique.

TABLE 13-6

ANOVA

Source	Sum of Squares	df	Mean Square	$F_{calculated}$	$F_{critical}$
"SIGNAL" Storage Conditions	27.875	4	6.97	4.15	$4.177_{@.01}$
"NOISE" Residual	41.876	25	1.68		
TOTAL	69.75	29			

The final two columns contain the decision-making items. The $F_{calculated}$ is obtained by dividing the **signal** (Mean Square for Storage Conditions) by the **noise** (Mean Square for Residual.) We obtain a calculated F of 4.15 for this data. The $F_{critical}$ (or tabulated F value) that corresponds to a 1% risk of saying that there is a difference when there really is no difference is 4.177. Our calculated F value is very close to the value in the F table, but just a bit lower; therefore, we do not have evidence to reject the idea that the five storage conditions are the same. We therefore conclude that there is no evidence to believe there is a difference between the storage conditions. Is this conclusion correct?

Investigating the Residuals

Statistically speaking, the conclusion follows the "rules" and is correct. The amount of difference induced by the different storage conditions is no bigger than the residual variation within each condition. *The* residual (sum of squares) is the unknown variation. We can also define *a* residual as the difference between an actual observation and the predicted value of that observation. Since we have found no significant differences among the storage conditions, the grand average of all 30 observations is a reasonable predicted value for any individual observation. This grand average is 7.02. Just to see how good our decision is, let's find the differences for the 6 observations at the 36° storage condition. To do so, we will simply subtract the grand average from each observation. Table 13-7 shows the results.

TABLE 13-7

Observed for 36°:	6.5	6.0	7.5	8.0	5.0	4.5
Grand Average:	−7.02	−7.02	−7.02	−7.02	−7.02	−7.02
Residual:	−0.52	−1.02	0.48	0.98	−2.02	−2.52

When we look at these residual differences, we observe that there is an interesting pattern. The first two residuals are negative, the next two are positive, and the last two are negative. This might be a fluke. So let's look at another storage condition in the same way and see if any similarities occur.

TABLE 13-8

Observed for 38°:	7.5	8.0	8.5	9.0	6.0	6.5
Grand Average:	−7.02	−7.02	−7.02	−7.02	−7.02	−7.02
Residual:	0.48	0.98	1.48	1.98	−1.02	−0.52

While this pattern is not exactly the same as in Table 13-7, we do see positive, more positive and then negative values. Displaying the residuals on a chart is a better way to highlight such patterns. The procedure we are developing is called "residual analysis." Whenever we perform an analysis of data, we should look at the pattern of the residuals to see if any interesting patterns have emerged. If patterns are present, then the residuals are not merely random, chance fluctuations in the data, but some systematic sources of the variation that we should also consider. The residual is not independent!

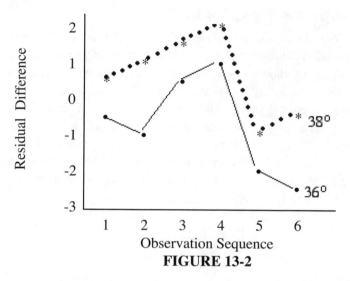

FIGURE 13-2

So what could cause such an up-and-down trend? The best way to answer such a question is to ask the people who set up the experiment. Of course, if *we* had set up the experiment, we would know the answer. Often, a set of data is brought to a statistical analyst and no clue is given about the design. This is not a good situation and should be avoided. It is important to involve the analyst from the start of the experimentation.

A Learning Example

In this apple storage experiment, we knew before we began the analysis that there was another source of variation. *However, to make the above point about the independence of the residuals and to show how the examination of residual values is a very revealing method for discovering the reasons behind our residual sum of squares, we have held back on the information about a second factor in this study!*

We are about to embark on a two-factor (2-way) analysis of variance. In such a situation, we do **not** complete the 1-way first and then go on to the 2-way. We do the 2-way as we show now right from the beginning. The approach we have taken is wrong when we know that there is more than one source of influence under study. We have done this wrong approach to show the consequences, not to set up a process to be followed.

The Correct Approach

Table 13-9 shows the three apple types that were in this study. The first group was Ida Red. The second group was McIntosh. The third group was Delicious. We might expect that different types of apples behave differently to storage conditions and might even have inherently different storage characteristics.

We now need to do the analysis correctly and include a second factor, the "apple type" in our study. This will help explain some of the large residual sum of squares we experienced in the one-way ANOVA and, of course, give us a much better insight into the physical causes of the differences in storage conditions. The residual is the unexplained part of the variation we are attempting to understand. Having more sources to help explain the residual is a valuable asset in our analysis.

TABLE 13-9
Storage Conditions

Apple Type		36°	38°	40°	42°	44°	
	Ida Red	6.5	7.5	8.0	7.5	5.0	68.5
		6.0	8.0	8.5	7.0	4.5	
	McIntosh	7.5	8.5	9.5	9.5	7.5	84.5
		8.0	9.0	9.0	9.0	7.0	
	Delicious	5.0	6.0	7.0	6.0	5.0	57.5
		4.5	6.5	7.5	5.5	4.5	
$\sum =$		37.5	45.5	49.5	44.5	33.5	210.5

To compute the contribution of the apple types to the total variation, we take the sums of the months to spoilage for each apple type, square these sums, and then divide by the number of observations in each sum that we are squaring. This operation produces a numerical value of 1513.875. From this "raw" sum of squares, we subtract the centering factor (the grand total squared divided by the grand number of observations). For this set of data, the value of the centering factor is 1477.0. This results in a sum of squares for the types of apples of 36.875. The formal calculation is shown below.

$$\sum sq_{type} \quad \frac{68.5^2 + 84.5^2 + 57.5^2}{10} - \frac{210.5^2}{30} = 1513.875 - 1477.0 = 36.875$$

Now, you might ask, how can we sum over the storage conditions as if they didn't exist in this calculation? You will notice that the experiment is balanced with the same number of storage conditions for each apple type. Because of this balance, we may ignore the effect of the storage as we independently study the types of apples. We will do the same "summing over" with the storage conditions. We will sum over the apple types as if they did not change since the same change in type of apple takes place over all temperatures.

The sum of squares for temperature conditions is obtained in the same manner as in obtaining the sum of squares of apple type.

$$\Sigma sq_{cond.} = \frac{37.5^2 + 45.5^2 + 49.5^2 + 44.5^2 + 33.5^2}{6} - \frac{210.5^2}{30} = 27.875$$

Next we compute the sum of squares due to the noise or lack of sameness in each treatment combination. Notice in Table 13-9 that in each cell or treatment combination the difference in storage life is exactly one-half month. We may simplify the arithmetic *in this example* by finding the sum of squares in one of the treatments and then since all treatment combinations are the same, extend this by multiplying by the number of treatment combinations (which is 15 for this example). In general, we would need to compute each cell Σsq and add all of them.

The sum of squares in all cells is: $\frac{6.5^2 + 6.0^2}{1} - \frac{12.5^2}{2} = 0.125$
(the same as in the first cell)

So the entire sum of squares for residual is: 15 (0.125) = 1.875

Now we will add the pieces and see if they equal the total sum of squares.

Conditions + Apple Types + Residual
27.875 + 36.875 + 1.875 = 66.625

Which is not the total of 69.75 found back on page 299. The combined sums of squares formed by the addition of the individual pieces above is short by a little more than 3 units. So where is this "lost" sum of squares?

There is one last source of variation that we have not accounted for. This is the source that accounts for the lack of similarity in the behavior of the different types of apples to temperatures. If all types of apples were affected in the same way by the storage conditions, we would say that the storage condition effect was a simple additive effect over the types of apples. If this additive effect does not exist, then we have what is called an "interaction" between conditions and apples. This interaction is the source of the "lost" sum of squares that is included in the total but not yet calculated as an individual contributor.

To compute the interaction sum of squares, we need to look into each cell of the data matrix. In Table 13-9, there are two observations per cell, and there are 15 cells which are made up of the intersections of the five columns of storage conditions with the three rows of apple types. So the first intersection involves the $36°$ temperature and Ida Red apple, which shows values of 6.5 and 6.0 months to

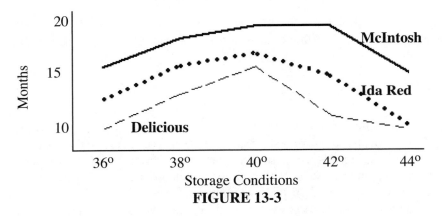

FIGURE 13-3

spoilage. We take these values and sum them to obtain the cell total. In the case of the first cell, the total is 12.5. In the second column and first row, the cell total is 15.5. The concept of the interaction is very much like looking at a family of curves. In such a family, we would like to see if the various apples behave the same for the different storage conditions. If all the apple types show the same behavior, then there is harmony in the family of curves. If, however, there is disagreement between the behavior of the apples to the various storage conditions, then there is disharmony or a "feud" going on in the family. Our ANOVA will tell us if the amount of disharmony is greater than we could expect by mere chance, given the residual variation among the observations. Figure 13-3 shows the interaction plot that illustrates the above concept. In it we have plotted the cell totals for each of the 15 combinations of apple type and storage condition. Now we take each cell total, square it, and add all these squares together. We now have the sum of squares for the interaction.

$$\Sigma sq_{Interaction} = \frac{12.5^2 + 15.5^2 + 16.5^2 + 14.5^2 + 9.5^2}{2} \quad + \quad \text{(first row)}$$

$$\frac{15.5^2 + 17.5^2 + 18.5^2 + 18.5^2 + 14.5^2}{2} \quad + \quad \text{(second row)}$$

$$\frac{9.5^2 + 12.5^2 + 14.5^2 - 11.5^2 + 9.5^2}{2} \quad \text{(third row)}$$

$$= 1544.875$$

Of course, we divide by the number of observations in the sum we are squaring, which is 2 in this case, and then subtract the centering factor to obtain the interaction sum of squares.

$$1544.875 \ - \frac{210.5^2}{30} = 67.867$$

But now this sum of squares is too large, since the total sum of squares is only 69.75, and we already have 66.625 in the two single effects and the residual.

The reason for this "extra" sum of squares comes from the way we computed the interaction sum of squares. We have actually included the sum of squares due to the storage conditions and the apple types in our computation of the interaction. Look at Figure 13-3. Notice that besides showing the different ways the apples react to the storage conditions, we also see how the apples react to storage in general! We also observe that the apples themselves show a difference. The McIntosh stores best and Delicious stores worst. We have already taken this into account with the single-effects sum of squares, but we do it all over again in our computation of the sum of squares due to the interaction. The plot of the data in Figure 13-3 shows the functional influence of storage temperature as well as apple types, in addition to the differential effect of these factors on storage.

To rectify this problem, we subtract the two single-effect sources of variation from the sum of squares that includes the interaction and the single effects, and we get the real interaction sum of squares, all by itself. This is possible of course, since sums of squares are always additive.

$$(\Sigma sq_{type} + \Sigma sq_{storage} + \Sigma sq_{interaction}) - (\Sigma sq_{type} + \Sigma sq_{storage}) = \Sigma sq_{interaction} \quad (13\text{-}8)$$

$$(67.87) \qquad\qquad - (27.875 + 36.875) = 3.117$$

We now have all of the components of the variation that add up to the total sum of squares. To complete the ANOVA table and draw the correct conclusions in the shadow of the uncertainty (the residual), we need to account for the information in the experiment via the degrees of freedom (df). The magnitude of the sum of squares is dependent on two considerations. First, if the factor has an influence, the sum of squares will be proportional to that influence. Second, the more opportunities we give the factor to influence the response, the larger the sum of squares. So, if we had taken more types of apples, and apple type has an influence, then the sum of squares for apple type would be greater. By dividing the sum of squares by the degrees of freedom, we scale the influence to reflect the number of opportunities afforded the factor being studied.

We determine the df as in the one-way example. Since we computed a sum of squares for each effect, and to do so requires an implicit calculation of an average, we convert one of the original df into information about the average. We therefore end up with one less df than the number of levels for the given factor.

	levels	for average	df for effect
With 5 storage conditions, the df are:	5 −	1 df =	4 df
With 3 types of apples, the df are:	3 −	1 df =	2 df

The df for the residual come from the same considerations as before. If we look at a segment of Table 13-9, we can see that there are two observations per combination of storage temperature and apple type.

SEGMENT OF TABLE 13-9
Storage Conditions _ _

Apple Type		36°	_ _ _ _ _ _
	Ida Red	6.5 6.0	_ _ _ _ _

When we computed the sum of squares for each of the cells, we again implicitly computed an average and converted one of the df to this type of information. Since there were 2 observations in each cell, we have one degree of freedom for each cell. We repeated this computation in 15 cells for this example, so the degrees of freedom for the residual is obtained by multiplying the degrees of freedom of the cell by the number of cells.

$df_{residual}$ = (Number of Observations/cell - 1) **x** Number of cells

15   = (2 − 1) **x** 15

The final effect for which we need to account for degrees of freedom is the interaction. Remember, this is the differential set of slopes associated with the different behavior of one single effect in conjunction with another single effect. Therefore, we are looking at information based on a change in the response as a function of *both* factors. To represent this information content, we find the product of the df of the interacting factors. For our case, we obtain:

$$df_{int} = df_{Factor\,A} \times df_{Factor\,B}$$

$$df_{int} = df_{Storage} \times df_{Apple\,Type}$$

$$8 = 4 \times 2$$

Now we will put it all together in the ANOVA in Table 13-10 and draw our conclusions.

TABLE 13-10

ANOVA

	Source	Sum of Squares	df	Mean Square	$F_{calculated}$	$F_{critical}$
"SIGNALS"	Storage Conditions	27.875	4	6.97	55.8	$3.0556_{@.05}$
	Apple Types	36.875	2	18.44	147.5	$3.6823_{@.05}$
	Interaction Cond.xType	3.117	8	0.39	3.1	$2.6408_{@.05}$
"NOISE"	Residual* (*estimate of error)	1.875	15	0.125		
	TOTAL	69.75	29			

From the table of critical F values, we obtain the F value that, when exceeded by the calculated F from the experiment, tells us that this is not a chance happening but the result of a systematic cause system. In this case, we see that all the critical F values have been exceeded, and we conclude that there is a difference between storage conditions and between types of apples, and that their interaction is also a significant effect. The significant interaction tells us that different apples store differently and do not behave the same under all storage conditions.

Now let's look at where the differences lie and if there is enough of a difference to warrant changes in the storage of the apples. To do this, we will compute the averages of each treatment combination in the experiment.

TABLE 13-11
Storage Condition

	36°	38°	40°	42°	44°
Ida Red	6.25	7.75	8.25	7.25	4.75
McIntosh	7.75	8.75	9.25	9.25	7.25
Delicious	4.75	6.25	7.25	5.75	4.75

When we plot the results of the computation of the averages vs. temperature in Table 13-11, we get a better picture of the way the storage conditions influence the life of each type of apple.

From the plot of the data (plot of averages), we see that there is a difference between each of the apple types, with the McIntosh exhibiting the longest storage life followed by Ida Red and Delicious. There also appears to be an optimum temperature for storage. However, some of the differences we see may not be significantly different given the amount of error or noise in this data. We cannot use a graphic display to draw our final conclusions, but must revert to another test of the data using methods that will go beyond ANOVA to tell us which levels are different.

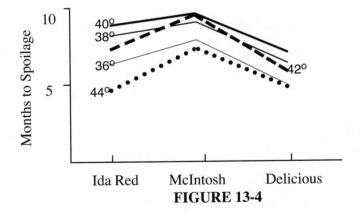

FIGURE 13-4

Comparisons of Individual Levels

There are a number of different methods to compare the individual means after an ANOVA has indicated that there is a difference among the means. In general, the technique used is to first rank order the means from lowest to highest and then compute a statistic that is related to the standard error of the mean. This statistic is usually called the shortest significant range (SSR) and represents the smallest difference that is statistically different when comparing the different mean values that have been ranked. Since we purposely rank the values, the probability increases of finding differences as we compare ranks that are farther apart. Therefore, the value of the SSR gets larger for each comparison we make across ranks that are further apart.

In the case of the apple experiment, we will look at the effect of each apple type at only one level of temperature at a time since there was a significant interaction in the results of our analysis. Had there been no statistically significant interaction, we would have compared apple types over all storage conditions averaged together.

In Table 13-12, we order the three types of apples in increasing average months to spoilage. We'll use abbreviations to simplify labeling the types of apples: McIntosh (M); Ida Red (I); and Delicious (D.)

TABLE 13-12
Months to Spoilage Ranked in Ascending Order

Temperature	Rank 1		Rank 2	Rank 3
36°	4.75(D)		6.25(I)	7.75(M)
38°	6.25(D)		7.75(I)	8.75(M)
40°	7.25(D)		8.25(I)	9.25(M)
42°	5.75(D)		7.25(I)	9.25(M)
44°	4.75 (D)	tie	4.75(I)	7.25 (M)

We next compute the shortest significant range (SSR) by finding the standard error of the mean from the mean square for the residual as follows:

$$s_{\bar{x}} = \sqrt{\frac{\text{Residual MS}}{\text{\# obs. in mean}}} = \sqrt{\frac{0.125}{2}} = .25$$

The divisor in the calculation of the standard error is simply the number of observations in the calculation of means we are comparing. In this case, there are only two observations per mean. If we had summed over the 5 different storage conditions, then the divisor would have been 10 since there are 2 replicates per cell. In fact, we did not sum over the temperatures because of the significant interaction.

Ultimately, we find the SSR by multiplying $s_{\bar{x}}$ by the upper points of the Studentized Range q found in Table 6 (among the tables in the back of this book). To enter Table 6, we need to know the df for the residual, which comes from our ANOVA table, and the number of means over which we are making the comparisons. In our case, the df for the residual is 15, so we move down to the row headed by "15" and then read across the maximum number of means in the rank. In our case, this is 3. The q values, then, are: 3.01 and 3.67 at the 5% level.

Therefore, the SSRs are:

TABLE 13-13

	Means in range (p)		
	2	3	
$s_{\bar{x}}$	0.25	0.25	
q	3.01	3.67	
SSR	0.7525	0.9175	SSR = q x-s_x

We now set up a "difference" table to be able to compare the actual differences in our data with the SSRs of Table 13-13.

We can see from Table 13-14 that all the apple types are different, except for the Ida Red and Delicious at the 44° storage condition. This inconsistency is probably the greatest contributor to the interaction. We also observe that the smallest differences occur at the 40° condition. This leads to the practical implication and conclusion that this is probably the best all-around condition for the storage of apples considering that we could not afford to set up different conditions for different apples. We also note that the 40° condition is also the condition that produces the longest storage life. This is a fortunate happening which makes the choice of storage conditions easier in this situation.

Let's look at some more practical implications of the data we have just analyzed. The difference table confirms that the results plotted in Figure 13-4 are not just numerically different but statistically different. That is, the signal we observe is significantly greater than the noise encountered. Since we have no control over the apples' inherent spoilage rates, and we see that the Ida Red and Delicious spoil sooner than the McIntosh, we can use this information to our advantage and attempt to "push" the quicker spoiling apples to market earlier or use them in other ways such as in cider production. The above insights are based on the practical importance of the data. Such practical conclusions must always be thought of and stated in any statistical analysis. All too often, the analyst will merely state the facts about the statistics and forget the reason for the experimentation in the first place. Always draw conclusions in words that relate to the problem at hand.

TABLE 13-14
Difference Table

	comparison of adjacent ranks	comparison over 3 ranks
36°	6.25(I) – 4.75(D) = 1.5* 7.75(M) – 6.25(I) = 1.5*	7.75 (M) – 4.75(D) = 3.0*
38°	7.75(I) – 6.25(D) = 1.5* 8.75(M) – 7.75(I) = 1.0*	8.75(M) – 6.25(D) = 2.5*
40°	8.25(I) – 7.25(D) = 1.0* 9.25(M) – 8.25(I) = 1.0*	9.25(M) – 7.25 (D) = 3.5*
42°	7.25(I) – 5.75(D) = 1.5* 9.25(M) – 7.25(I) = 2.0*	9.25(M) – 5.75(D) = 3.5*
44°	4.75(I) – 4.75(D) = 0.0 7.25(M) – 4.75(I) = 2.5*	7.25(M) – 4.75(D) = 2.5*

*indicates statistically significant differences

More Conclusions

While it appears that we have completed the analysis of the apple data, there are a few fine points yet to be covered. We drew some general conclusions about the storage temperatures, but did not back them up with any statistical tests. We will do so by applying the SSR concept to the temperatures to see which of them are different. From the table of means (Table 13-12) we will rank the average time to spoilage for different levels of temperature. Again, we will do so for isolated levels of apple type since there is an interaction between type and temperature. (That is, the apples respond differently to various storage conditions.) Since there are 5 levels of temperature in the storage-condition factor, there will be 5 ranks.

TABLE 13-15
Months-to-Spoilage Ranked in Ascending Order

Apple Type	Rank 1	Rank 2	Rank 3	Rank 4	Rank 5
Ida Red	4.75(44°)	6.25(36°)	7.25(42°)	7.75(38°)	8.25(40°)
McIntosh	7.25(44°)	7.75(36°)	8.75(38°)	9.25(40°)	9.25(42°)
Delicious	4.75(36°)	4.75(44°)	5.75(42°)	6.25(38°)	7.25(40°)

The $s_{\bar{x}}$ is the same value we computed for the apple differences since the averages in the above table (Table 13-15) are also based on the two observations per cell. We obtain the q values in the same way as before, but go out to a p of 5 since there are 5 levels of temperature. So, the SSRs are shown in Table 13-16.

TABLE 13-16
Means in range (p)

	2	3	4	5
$s_{\bar{x}}$	0.25	0.25	0.25	0.25
q	3.01	3.67	4.08	4.37
SSR	0.75	0.92	1.02	1.09

The critical question about the storage temperature information is which condition produces the longest storage. If two conditions are not significantly different, then we would pick the condition that is less costly to attain, which in this refrigeration example is a higher temperature. Looking back to Figure 13-3, which plots the spoilage data versus temperature, we would look at the peak storage life and ask if there is any difference between the peak and the next higher temperature. In this plot, the peak storage condition is 40°. The next condition (42°) shows a drop in life for two apple types, but no drop for the McIntosh type. From this, we may conclude that there is no difference between 40° and 42° for the McIntosh apples and use the lower cost alternative. For the Ida Red and Delicious, we go to the SSRs in Table 13-16 to decide if there is enough difference to warrant a significant difference between the temperatures. For both types of apples, the 40° condition is ranked highest and the 42° temperature is ranked third. This gives us a difference range over three items and a SSR of 0.92.

The difference between the 40° condition and the 42° condition for the Ida Red apple is 8.25 - 7.25 or just greater than the SSR, so we conclude that there is a significant difference and that the 40° condition should be used to store the Ida Red apples.

For the Delicious apples, the difference between the 42° and the 40° conditions is 7.25 – 5.75 = 1.5, which is greater than the SSR of 0.92. So, again, we conclude that there is a significant difference between conditions, and that we should use 40° for the storage of the Delicious apples.

We could have set up a difference table as was done with the apple types, but the extra work is not really necessary, since only certain questions needed to be answered in light of the practical implications of this study. The difference table for storage conditions would have been a complex set of numbers, and just might have confused the issue. A general rule in reporting results is to keep the analysis as simple as possible while still answering the questions that need to be answered.

Functional Relationships

The analysis of variance techniques is a very powerful method for answering questions involving comparisons of multilevels in an experiment. While we have shown how to compute the appropriate sums of squares, build the information into an ANOVA table from which we draw our conclusions, and then go beyond this table to actually compare individual means with a single risk, there is still more that can be extracted from the data in hand.

The storage condition factor in our example is a quantitative factor with equally spaced temperature levels. This was done in a fully planned manner when the experiment was designed. When we have a quantitative factor at equally spaced intervals, we may apply a method to investigate the polynomial functional nature of the data. That is, we may determine the shape of the curve represented by our data. There is a set of coefficients called "orthogonal polynomials" that when used to weight the computation of the sums of squares, will reveal from our ANOVA the curved (or straight-line) nature of the data. We shall now apply this concept to the example on storage of apples.

To do so, we will recompute the sum of squares due to storage conditions by restructuring the 4 degrees of freedom into four individual effects. There will be the linear (straight-line) effect; the quadratic effect (one-hump curve); the cubic (double-hump curve); and the quartic (triple-hump curve). A table of correct values for the orthogonal polynomials may be found in Table 7 at the back of this book. Before we use these orthogonal polynomial coefficients, we should look into their nature. We enter Table 7 through the number of levels (k) for the factor. Then we read the coefficients for each function across the rows of the table.

For our example, the values of the coefficients are as shown in Table 13-17.

TABLE 13-17

X:(read across the rows)	1	2	3	4	5
Function					
Linear	-2	-1	0	1	2
Quadratic	2	-1	-2	-1	2
Cubic	-1	2	0	-2	1

If we plot the coefficients from Table 13-17 on an equally spaced x-axis we obtain the plots of polynomial relationships shown in Figure 13-5.

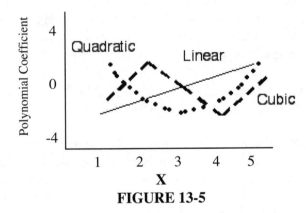

FIGURE 13-5

We can see the polynomial nature of these coefficients, but what about the "orthogonal" part? The word *orthogonal* is used to indicate the mathematical independence between any of the coefficients. Given independence, then all the covariances will be zero. (For a review of covariance, go back to Chapter 4, Expression 4-3.) Computing the covariance of the orthogonal polynomials is left as an exercise. The independence of these coefficients assures us that the sums of squares we obtain in the next step will be independent estimates of the polynomial effects. To obtain the sums of squares for the polynomial relationship, we will compute the sums of squares for storage conditions summed over types of apples, this time weighting the computation by the polynomial coefficients found in Table 13-17.

$$\Sigma sq_{linear} = \frac{(-2(37.5) \ -1(45.5) \ +0(49.5) \ +1(44.5) + 2(33.5))^2}{((-2)^2 + (-1)^2 + (0)^2 + (1)^2 + (2)^2)\,(6)} = 1.35 \quad (13\text{-}9)$$

Let's look at Expression 13-9 and see where the numbers come from. In the numerator, the values in parentheses are the totals of the temperature conditions from Table 13-9. These are multiplied by the linear orthogonal polynomial coefficients. We would normally divide this sum of squares by the number of observations in the sum being squared, and this is the "6" in the denominator. The other part of the denominator is merely the orthogonal polynomial coefficients squared and summed to compensate for their inclusion in the numerator. So, in general, we find the factor's influence weighted by the values of the orthogonal polynomials (the sum in the numerator), and then square this sum. Next find the sum of squares of the polynomial coefficients (in the denominator), and multiply this sum of squares by the number of observations in each total found in the numerator. In this manner, we are able to find a weighted sum of squares, and the weighting is based on the polynomial used. For the quadratic term:

$$\Sigma sq_{quadratic} = \frac{(\ 2(37.5) \ -1(45.5) \ -2(49.5) -1(44.5) + 2(33.5))^2}{((\ 2)^2 + (-1)^2 + (-2)^2 + (-1)^2 + (2)^2)\,(6)} = 26.3 \quad (13\text{-}10)$$

Similar computations are required to produce the sum of squares for the cubic component. On the basis of this new information we construct a new, revised ANOVA table. In each case, for the factor being broken down into its polynomial parts, there is only one degree of freedom for each polynomial sum of squares.

The revised ANOVA table now includes the sums of squares for the polynomial effects. With the calculated F's exceeding the critical F values in both the linear and quadratic effects, we conclude that a polynomial of second order is the appropriate functional form for this data. Notice that the sum of the four polynomial effects add up to the sum of squares for storage conditions (27.87) found in the previous ANOVA table (Table 13-10). In computing the polynomial sums of squares, we have simply partitioned the effects into their functional form.

We will leave ANOVA now but will return in Chapter 16 to further investigate this technique and show how it is a special case of a more general method called the general linear model (GLM) or the "regression" method.

TABLE 13-18

ANOVA

Source	Sum of Squares	df	Mean Square	$F_{calculated}$	$F_{critical}$
Storage Conditions					
Linear	1.35	1	1.35	10.8	$4.5431_{@.05}$
Quadratic	26.30	1	26.30	210.4	$4.5431_{@.05}$
Cubic	0.07	1	0.07	0.6	$4.5431_{@.05}$
Quartic	0.15	1	0.15	1.2	$4.5431_{@.05}$
Apple Types	36.875	2	18.44	147.5	$3.6823_{@.05}$
Interaction Cond.xType	3.117	8	0.39	3.1	$2.6408_{@.05}$
Residual	1.875	15	0.125		
TOTAL	69.75	29			

ANOVA Assumptions (Requirements)

In doing the Analysis of Variance, we must realize that there are three underlying assumptions that are really requirements that must be observed in order for the conclusions we reach to have validity.

The first assumption is more than an assumption; it is a *requirement* that influences the way we conduct our experiment. The assumption is that the errors are independent. That is, each test in the experiment should not have an effect on any other test. While this is an assumption, we can accomplish the desired effect by running the entire experiment in a random order to assure that any treatment combination has an equally likely chance of happening anywhere in the sequence of our testing. Therefore, any systematic variation outside the realm of the factors under study will not bias or be confounded with the functional results. This is verified by plotting the residual values in the time sequence of the observations.

Second, since we are pooling the error variances, we must make sure that these variances do not differ significantly from each other. If there is a large amount of difference between the variances we are pooling, then we could get a biased value of the pooled variance and our signal-to-noise ratio would become meaningless. So this second requirement for a valid conclusion to be drawn from an ANOVA is that the errors are homogeneous.

The third requirement for a valid ANOVA is based on the fact that the F statistic used to make decisions in ANOVA is actually a ratio of two independent χ^2 (chi-square) statistics divided by their respective df. In turn, these χ^2 statistics are sums of squared normal deviates. Therefore, the residual variances (errors) must be normally distributed.

While the first assumption is related to experimental conduct, the next two assumptions *can* and should be tested. The following statistical tests show how to check for homogeneity of variance and normality.

Test for Homogeneity of Variance

While there are many tests for checking homogeneity of variance, the Burr-Foster Q test has an advantage in as much as it is not sensitive to departures from normality. Therefore, it may be applied before a normality test without fear of drawing the incorrect conclusion.

To show this test in action, we will do a 4-level, one-way ANOVA with five replicates per treatment combination as shown in Figure 13-19.

TABLE 13-19

	100°	200°	Temperature 300°	400°
	90	150	250	400
	100	175	300	300
	85	125	325	500
	110	180	250	450
	90	150	225	375
$s^2 =$	100	492.5	1687.5	5750.0 $\sum s^2 = 8030$
$s^4 =$	10000	242556.25	284765625	33062500 $\sum(s^2)^2 = 36162712$
$\overline{X} =$	95	156	270	405
$s =$	10	22.19	41.1	75.8

The test statistic is q whose value is found by taking the ratio of the sum of the squares of the variances to the sum of the variances quantity squared.

$$q = \frac{\sum_{i=1}^{p} (s_i^2)^2}{(\sum_{i=1} s_i^2)^2}$$

Where s_i^2 is the variance of each population (i), and p is the number of populations we are studying.

In our example, p is equal to 4 (the 100°, 200°, 300°, 400° populations), so we sum the squared variances to obtain the numerator, then square the sum of the variances to get the denominator. The q value is the resulting ratio of these two quantities. Table 13-19 shows the values that make up the numerator and denominator, and the value of q is:

$$q = 36162712/8030^2 = 36162712/64480900 = 0.5608$$

The critical value of q that, if exceeded, signifies that there is non-homogeneity in the variances is found in Table 8 (at the back of this book). We enter Table 8 with the degrees of freedom for the populations under study. In this case there are 5 observations from each population, so there are 4 df. (The number of observations from all the populations must be the same in this test.) The other parameter necessary to get the critical q value is the number of populations we are investigating. In this example, there are p=4 populations. Therefore, the critical value of q for df=4 and p=4 is 0.549 with an alpha risk of 0.01.

It is usual to use very small alpha level for this test, since slight departures from homogeneity do not cause great consequences. ANOVAs with variances of a 9 to 1 ratio have been run with little change in the alpha risk. (The change that has been observed is from .05 to 0.06! [Box,1954].) In this example, we have exceeded the critical value of q and reject the hypothesis of homogeneous variances.

Here we see a typical problem that creeps into experimental data. If we look at the percentage error as calculated for each population by taking the standard deviation over the average, we see that the percentage error (averaging about 14%) is relatively constant over all populations. As the temperature increases, however, the size of the response increases, which leads to the non-homogeneity of the residuals. To combat this, we need to transform the data to bring all the values together. A simple "compressing" transformation is the logarithm (or log). In this example, we take the log of each value and then do the ANOVA. The only problem with such a transformation is that we "lose touch" with the original data. This is why we transform *only* when there is excessive heterogeneity in the data.

Test for Normality

The third requirement in any effort to draw valid conclusions from an ANOVA concerns the normality of the distribution of the errors. There are many statistical tests for normality, but the following Shapiro-Wilk test will be most useful for the kinds of small samples found in ANOVA problems. We will illustrate this procedure by testing for the normality of the replicated data at the $100°$-temperature level of the example in Table 13-20.

The set of hypotheses under study in this normality test is as follows:

H_0: The data have a non-normal distribution
H_1: The data have a normal distribution

Making the null hypothesis say "*non*-normal" is somewhat backward from the usual hypothesis set-up which would say that the distribution of the data is normal. Since many of the phenomena we study will follow a normal distribution, this hypothesis test has been constructed in such a way that when we find the data to have a normal distribution (the usual case), we must reject the null hypothesis. Since rejecting a null involves an alpha risk, such a risk may be stated. If the hypotheses were reversed, then not rejecting a normal distribution for the data (which would probably be most of the time) would involve a beta risk. The beta risk is not defined in the Shapiro-Wilk test, so by setting the hypotheses up as shown above, we can avoid having to deal with the beta risk when our data *is* normally distributed!

The first step in the procedure is to place the data being studied in ascending order. For the data in our example, this will result in the following set of values:

$$85 \quad 90 \quad 90 \quad 100 \quad 110$$

Next, we compute the sum of squares of this set of data:

$$(85^2 + 90^2 + 90^2 + 100^2 + 110^2) - 475^2/5 = 45525 - 45125 = 400$$

Now, we calculate the **b** value, which is found by taking the difference between the extremes in the data and multiplying this difference by a coefficient from

Table 9 (found at the back of this book). Continue by finding the difference between the next most extreme data points and, once again multiply by the appropriate coefficient. The process continues in this fashion until we have reached the center of the ordered data. Pictorially, the strategy looks like this:

$$\Delta = 25$$

$$85 \quad 90 \quad 90 \quad 100 \quad 110$$

$$\Delta = 10$$

$$X_1 \quad X_2 \quad X_3 \quad X_4 \quad X_5$$

To generalize, the above procedure can be reduced to and expressed by the following formula:

$$b = \sum_{i=1}^{k} a_{n-i+1} (X_{n-i+1} - X_i)$$

where k = (n − 1)/2 for odd n or k = n/2 for even n
       n is the number of items being tested
       X is the value of the item in the ascending order
       a_{n-i+1} is the factor from Table 9

For our numerical example:

$$b = 0.6646 \ (110 - 85) + 0.2413 \ (100 - 90) = 19.028$$

We now compute the test statistic W, which is the ratio of b^2 to the sum of squares of X.

$$W = b^2/\Sigma sq_X$$

$$W = 19.028^2/400 = 0.905$$

The critical value from Table 10 (at the back of this book) is 0.762 for an alpha of 0.05 for n equal to 5 (the number of items in the test). Since we have a calculated value greater than the critical value, we reject the null hypothesis and conclude that the data shows no departure from normality.

Testing for normality and homogeneity of variance may be done on the raw data as we have shown here or even more effectively on the residual values. "Hand calculation" of residual values is tedious. Minitab easily computes the residual values and performs the test for requirements on these residual values.

Problems for Chapter 13

1. A process engineer has set up the following experiment to determine if there is any difference among four methods for fastening two parts together. The methods are assigned in a random order to the design with as many observations per method as is affordable. The results are shown below, with the response being breaking strength in pounds of force. Complete the ANOVA and test for differences among means using the SSR. Draw the appropriate conclusions.

Cyanoacrylate Adhesive	"Contact" Cement	Hot Melt Glue	"Duco" Cement
20	16	24	12
28	14	26	14
17	15	25	12
20	17	26	13
21	16	25	12

2. A public relations person wants to test the effect of a series of letters explaining the company's stand on a recent "accident" at an atomic power station. The response is the reaction, measured in number of angry letters sent back to the company, protesting the event. Four different letters of explanation were published in local newspapers. The distance from the power plant was classified for each protest letter received. Complete the ANOVA on this data and draw conclusions about the letter factor and the distance factors. Why are there no replicates in this example?

Distance (Miles)

	0.5	1.0	2.0	4.0	8.0	16.0
A	20	10	5	2	1	0
B	50	24	10	5	2	1
C	15	6	3	1	1	0
D	100	49	20	10	2	1

LETTER

3. Perform an Analysis of Variance (ANOVA) on the following copy quality (0 is poor quality, 100 is excellent quality) data:

| | | **Vendor** | | |
Kodak	Ricoh	Xerox	Savin	Minolta
90	55	85	60	85
85	60	90	55	88
88	54	88	50	80
91	50	90	54	87
90	56	90	55	84

Is there a significant difference among the 5 levels of copy quality? If so, which vendors are different?

4. What can we conclude about the following skew responses? (Skew is the deviation from parallelism in a copy and is coded to produce whole numbers.)

Run-Out Level of Transport Rollers

		A	B	C	D	E
	5 ips	2	3	5	7	9
		1	5	6	6	8
Speed	10 ips	4	5	7	9	9
		5	7	6	9	10
	15 ips	1	5	7	6	10
		2	4	6	7	8

After running the ANOVA, test individual differences among levels of run-out and set up the orthogonal polynomials for the speed. Do not forget to state your conclusions.

5. Compute the cubic and quartic sums of squares for the apple storage problem. The data set is found in Table 13-9.

6. Computing the covariance between the orthogonal polynomial coefficients in Table 13-17.

7. Perform an Analysis of Variance (ANOVA) on the following data:

VENDOR

A	B	C	D	E
2	2	3	2	3
3	3	2	3	4
3	2	3	3	4
2	2	2	2	4
3	2	3	4	3

Are there significant differences among the five vendors for this measurement?

8. Interpret the interaction from the Apple Storage experiment by making a recommendation for best storage temperature for each apple type. See Figure App.13B-44 and Figure App.13B-45.

9. Perform an Analysis of Variance (ANOVA) on the following noise data from a experiment conducted with different sensor sensitivities of three digital sensor types used in digital cameras. The larger values are worse:

Sensor Sensitivity

	50	100	200	400	800
ccd	205 (15)	215 (34)	236 (16)	260 (39)	290 (10)
	215 (27)	220 (1)	234 (28)	265 (12)	285 (25)
	212 (7)	219 (21)	233 (4)	263 (42)	282 (13)
Super ccd	173 (32)	179 (31)	183 (17)	190 (38)	215 (30)
	167 (36)	180 (3)	186 (22)	196 (2)	224 (24)
	169 (6)	181 (19)	184 (14)	199 (35)	219 (29)
CMOS	122 (40)	121 (41)	126 (5)	129 (18)	140 (9)
	120 (33)	124 (20)	128 (44)	132 (23)	144 (26)
	121 (37)	122 (43)	127 (45)	131 (8)	143 (11)

Sensor Type

Write a complete report with Goal, Objective, Results, Conclusions, and Recommendations. Are the requirements for a valid ANOVA met? The run order is noted in parentheses.

10. After reading the analysis presented in Appendix 13B, consider if having more replicates "smooths out" the normality and homogeniety of variance tests based on an expanded set of data. Here is the Apple Storage experiment with 4 replicates instead of only 2 replicates with days to spoilage as the response.

Temperature

	36°	38°	40°	42°	44°
Ida Red	45.5	52.5	56.0	52.5	35.0
	42.0	56.0	59.5	49.0	31.5
	41.0	59.0	57.5	50.0	29.5
	40.0	60.0	60.5	51.0	33.5
McIntosh	52.5	59.5	66.5	66.5	52.5
	56.0	63.0	63.0	63.0	49.0
	55.0	60.0	64.0	65.0	50.0
	53.0	62.0	65.0	62.0	53.5
Delicious	35.0	42.0	49.0	42.0	35.0
	31.5	45.5	49.5	38.5	31.5
	33.5	43.5	52.5	39.5	33.5
	32.0	42.5	50.5	41.5	32.0

APPENDIX 13A

ANOVA ANalysis Of VAriance. (It really is an ANalysis Of MEans BY USing VAriance.)

Steps in computing an ANOVA:

Find the total sum of squares by taking each single piece of data, squaring it, and adding it to the rest of the squared data. From this sum, subtract the grand total squared, divided by the total number of data points, n.

$$\Sigma sq_{Total} = \Sigma\Sigma x_{ij}^2 - (\Sigma\Sigma x_{ij})^2/n \quad \text{(Total)}$$

Where x_i represents each data point.

Find the sum of squares for each effect in the experiment by first summing the values for the effect over all other effects ("summing over" means to sum as if the other factors did not exist). Now take each sum and square it. Divide this square by the number of items that went into the sum being squared. Sum all these squares and subtract the grand total squared divided by the total number of data points.

$$\Sigma sq_{Effect} = \Sigma x_{\cdot j}^2/n_j - (\Sigma\Sigma x_{ij})^2/n \qquad \text{(Effect)}$$

where $x_{\cdot j}$ represents the sum over all other factors, x_{ij} represents each data point, n_j is the number of observations in effect$_j$, and n is the total number of observations.

Find the residual sum of squares by extending the above concept into each treatment combination cell where there is replication to find the sum of squares in each cell. That is, find the Σsq of each cell element and subtract the sum of the cell elements squared, divided by the number of observations in this cell.

Construct an ANOVA table with the sums of squares you have computed. For each effect, there will be degrees of freedom equal to the number of levels in the factor less one. Any interaction will have degrees of freedom equal to the product of the number of degrees of freedom in the interacting terms. The degrees of freedom for residual will be the sum of the degrees of freedom from each cell, where the degrees of freedom for each cell will be one less than the number of items in that cell.

The table looks like this:

ANOVA

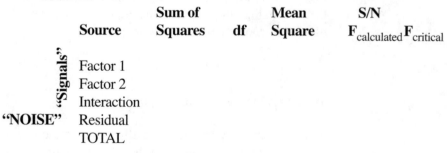

	Source	Sum of Squares	df	Mean Square	S/N $F_{calculated}$ $F_{critical}$
"Signals"	Factor 1				
	Factor 2				
	Interaction				
"NOISE"	Residual				
	TOTAL				

For "fixed-factor" designs, the $F_{calculated}$ is the ratio of the factor mean square to the residual mean square - the Signal to Noise (**S/N**).

Mean Square: Sum of squares divided by the degrees of freedom.

Fixed Factor: The levels of this factor are selected from only those that are available and include all the levels of interest.

Random Factor: As opposed to a fixed factor, the levels are picked at random from a population.(See Chapter 18 for more details and examples.)

SSR: Shortest significant range, a range statistic that helps determine which means differ after we have rejected the null hypothesis in an ANOVA. Any difference that is equal to or greater than the SSR is considered a statistically significant difference.

Orthogonal Polynomial Coefficients:

A set of values that are polynomial in nature and also independent. They are used after rejecting the null hypothesis in ANOVA to determine the functionality of quantitative factors. To be applied properly, the levels of the quantitative factor must be equally spaced in the design.

ANOVA Requirements:

Errors are independent; errors are homogeneous; and errors are normally distributed. These requirements are verified by plotting the residual values and/or performing tests of homogeneity and normality on the residual values. If any of these requirements are violated, the statistically based decisions could be in doubt.

APPENDIX 13B
Minitab & ANOVA

The tedious arithmetic calculations required for ANOVA can be accomplished with more assurance of being correct if a modern computing system is utilized. We will now show how Minitab is used to do the analysis of the 2-factor Apple Storage Condition Example. Table A13B-1 is a copy of the data we used earlier in this chapter.

TABLE A13B-1
Storage Conditions

	36°	38°	40°	42°	44°
Ida Red					
	6.5	7.5	8.0	7.5	5.0
	6.0	8.0	8.5	7.0	4.5
McIntosh					
	7.5	8.5	9.5	9.5	7.5
	8.0	9.0	9.0	9.0	7.0
Delecious					
	5.0	6.0	7.0	6.0	5.0
	4.5	6.5	7.5	5.5	4.5

Minitab supports a number of approaches and options for the ANOVA procedure. The more powerful and general method is the **General Linear Model (GLM)**. We begin by "building" the experimental design in the worksheet. Label the columns as shown in Figure App.13B-1.

FIGURE APP. 13B-1

The next step in building the design is to create the levels for each of the factors. While it is possible to manually enter the levels, using the **patterned data** feature of Minitab makes this

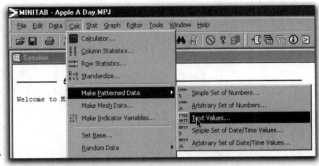

FIGURE APP. 13B-2

process much easier. As shown in Figure App.13B-2, drop down from the **Calc** heading to **Make Patterned Data** and over to **Text Values** for the Apple Type factor in column 1. In the dialog box (Figure App.13B-3) enter the names of the apples. Put them in quotes since the Ida Red type consists of two words. There are two observations for each apple type and there are 5 levels of temperature, so each word is listed 2 times and the entire set is entered 5 times. Click **OK** and the levels of the apple type factor are entered in the worksheet.

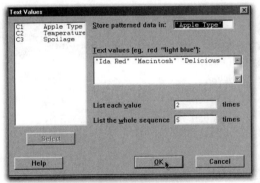

FIGURE APP.1 3B-3

Next, set up the levels for the temperature factor. Go to the dropdown **Calc** again, but this time move to **Single Set of Numbers** as shown in Figure App.

13B-4. In the dialog box shown in Figure App.13B-5, we enter the first value of temperature of 36 and the last value which is 44. The temperature is incremented in steps of 2 degrees, and we indicate that value in the **In Steps of:** entry box. The values of temperature are listed six times each and the sequence is only done once.

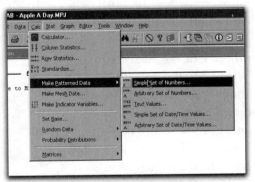

FIGURE APP. 13B-4

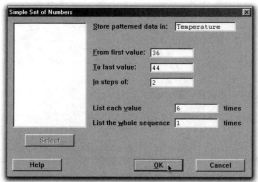

FIGURE APP. 13B-5

We click **OK** in this dialog box and view the completed experimental design in the worksheet as shown in Figure App.13B-6. The approach we have just taken is useful if we had not participated in the design of the experiment and merely needed to do the analysis. A better approach is to use the method shown in Chapter 5 for building a general factorial design, since the random run order is included in the design so generated.

We will continue with this example using the "hand made" design and realize that the order of running is an unknown. Next, we need to enter the response variable values in the column labeled "Spoilage." Care must be taken to be sure the proper response is matched with the apple type and temperature. This process is shown in Figure App. 13B-6 for about half of the data entries. The remaining entries are left as an exercise for the student. With the completed worksheet, we can now proceed to the analysis. While Minitab offers a simple two-way ANOVA option, the extent of this analysis is limited. Therefore we will use a more powerful technique called GLM.

↓	C1-T	C2	C3	C4
	Apple Type	Temperature	Spoilage	
1	Ida Red	36	6.5	
2	Ida Red	36	6.0	
3	Macintosh	36	7.5	
4	Macintosh	36	8.0	
5	Delicious	36	5.0	
6	Delicious	36	4.5	
7	Ida Red	38	7.5	
8	Ida Red	38	8.0	
9	Macintosh	38	8.5	
10	Macintosh	38	9.0	
11	Delicious	38	6.0	
12	Delicious	38	6.5	
13	Ida Red	40		
14	Ida Red	40		
15	Macintosh	40		
16	Macintosh	40		

FIGURE APP. 13B-6

To access GLM, we go to **Stat=> ANOVA => General Linear Model...** as shown in Figure App.13B-7. A dialog box opens, and we need to select the response (spoilage) and indicate the model we want to analyze. Many students miss this point and

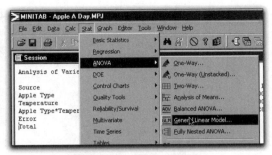

FIGURE APP. 13B-7

merely enter the factors without the interaction term. To enter the model, double click on the factors "apple type" and "temperature" and then to include the interaction, double click on apple type followed by an asterisk (*) and then double click the temperature. The asterisk is the symbol for multiplication.

We may skip the **Covariates** and **Options**

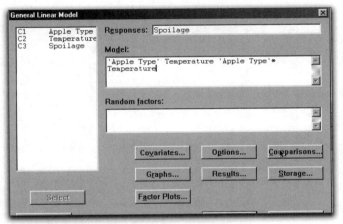

FIGURE APP. 13B-8

and go directly to the **Comparisons** as shown in Figure App.13B-9. Minitab does not support the type of multiple range tests described in this book. There is considerable debate in the statistical community over the "right" methods of multiple comparison. We will use the default Tukey test. We next go to graphs (Figure App.13B-10), which is a convenient option that allows us to plot the residual values. The **Four in one** option is not the best choice since it includes a time series plot (Residuals versus order) that is not meaningful in many instances, since we do not know the run order and the plots produced in this summary are too small for practical interpretation. A better approach is to store the fits and residuals (under **Storage** from the main dialog box) and use the normality test under **Basic Stat** and the test for equal variances under the **ANOVA** dropdown. We will do this later. The next sub-dialog box is the **Factorial Plots** (Figure App.13B-11) where we enter the factor names from the selection box

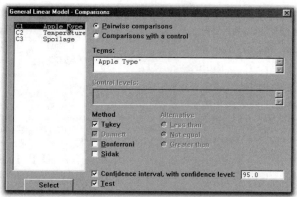

FIGURE APP. 13B-9

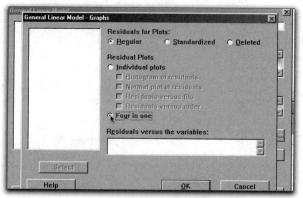

FIGURE APP. 13B-10

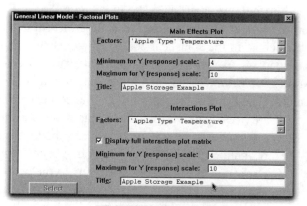

FIGURE APP. 13B-11

on the left and indicate the scaling we want. A title is an option. Much like the model specification, we need to indicate the factors in the interaction we wish to plot as well as the scaling and a title.

The last sub dialog box is **Storage** (Figure App.13B-12). We will store the **residuals** and the **fits** to do a more thorough analysis of the residuals than the four-in-one plot affords. After clicking **OK**, we return to the main dialog box and click **OK** to start this analysis.

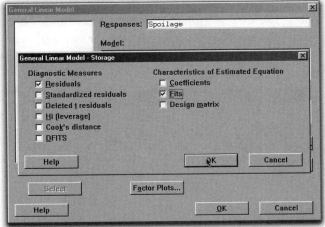

FIGURE APP. 13B-12

TABLE App. 13B-1

General Linear Model: Spoilage versus Apple Type, Temperature

Factor	Type	Levels	Values
Apple Type	fixed	3	Delicious, Ida Red, Macintosh
Temperature	fixed	5	36, 38, 40, 42, 44

Analysis of Variance for Spoilage, using Adjusted SS for Tests

Source	DF	Seq SS	Adj SS	Adj MS	F	P
Apple Type	2	36.8666	36.8666	18.4332	147.47	0.000
Temperature	4	27.8666	27.8666	6.9667	55.73	0.000
Apple Type*Temperature	8	3.1333	3.1333	0.3917	3.13	0.027
Error	15	1.8750	1.8750	0.1250		
Total	29	69.7416				

Tukey 95.0% Simultaneous Confidence Intervals Response Variable Spoilage
All Pairwise Comparisons among Levels of Apple Type

Apple Type = Delicious subtracted from:

Apple Type	Lower	Center	Upper	+---------+---------+---------+------
Ida Red	0.6897	1.100	1.510	(-----*-----)
Macintosh	2.2897	2.700	3.110	(-----*----)

```
                                    +---------+---------+---------+------
                                   0.70      1.40      2.10      2.80
```

Apple Type = Ida Red subtracted from:

Apple Type	Lower	Center	Upper	+---------+---------+---------+------
Macintosh	1.190	1.600	2.010	(-----*-----)

```
                                    +---------+---------+---------+------
                                   0.70      1.40      2.10      2.80
```

In the session window (Table App. 13B-1), the ANOVA table is displayed along with the multiple comparisons. The ANOVA is just like the one we completed earlier in Table 13-10. The four-in-one residual plots (Figure App.13B-13) show a non-random pattern. This is most likely due to the relatively crude measurement system that rounded to the nearest half month. If days to spoilage had been used, instead of months to spoilage, the systematic pattern in the residuals would have had more of an irregular pattern (no pattern!).

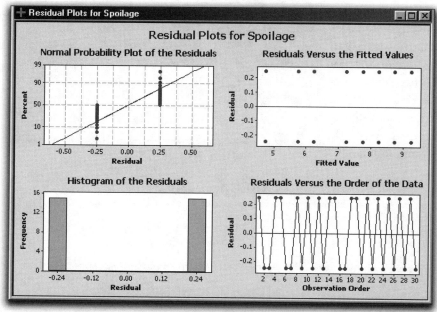

FIGURE APP. 13B-13

We will disregard the pattern in the residuals for this example since we have already indicated that the response variable is flawed in its rounding. Such a pattern in any experiment would suggest possible problems in the interpretation of the results. Further considerations and investigations (as we have done here) should be pursued to understand the pattern in the residuals before drawing conclusions. If it is possible to obtain "better" data, the experiment should be reanalyzed. **Again, the Apple Storage experiment is a teaching example to show some of the things that can go wrong and is not to be used as an example of a proper analysis procedure.**

Interaction and single effects plots are produced and shown in Figures App.13B-14 and 15. These show the same effects we saw previously in the analysis "by hand" approach (Figure 13-4).

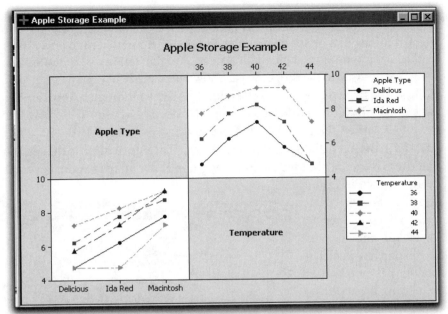

FIGURE APP. 13B-14

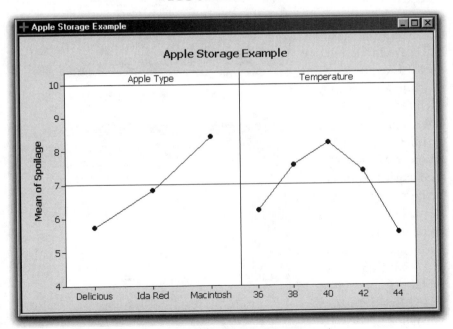

FIGURE APP. 13B-15

It is possible to go back in Minitab and refine a previous analysis using most of the basic settings from the previous analysis. Let's do that now and make use of routines that better display the residuals. Remember the analysis of residuals is important in being able to interpret the validity of the ANOVA.

We go back to the dropdown as shown previously in Figure App.13B-2 and from the main dialog box click **Graphs** as shown in Figure App.13B-16. Click on **Individual plots** and choose **Residuals versus fits**. We had previously stored the residuals and the fits (in Figure App.13B-12). Click **OK**, and in the main dialog box also click **OK**. The information in the session window is the same, but we have a new graph showing the residuals versus the fits shown in Figure App.13B-17.

This plot does not tell us too much, since the response (weeks to spoilage) is not as well defined as it should be. As we will see later when we use days to spoilage, the residual plots will have a bit more meaning.

We do not want any upward or downward trends in a residuals versus fitted values plot or any other regular pattern. We continue with our investigation of the residuals by checking the normality and the homogeneity. To do so we will utilize the stored residuals that have been placed in our worksheet.

We'll do the normality test first by going to **Stat=> Basic Statistics=> Normality**,

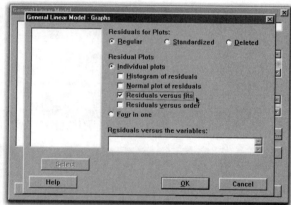

FIGURE APP. 13B-16

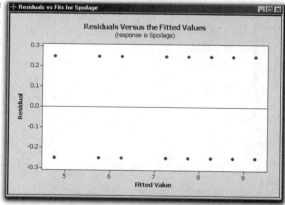

FIGURE APP. 13B-17

which is shown in Figure App. 13B-18. This brings up the dialog box (Figure App. 13B-19) where we select RESI1 as the variable. We may enter a title if we wish and then click on **OK**. The graph is produced (Figure App. 13B-22) along with the p-value, which is less than 0.01 for this apple storage example with weeks to storage as the response. Thus we reject the hypothesis and conclude

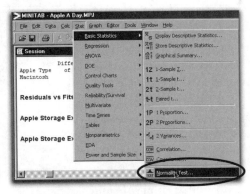

FIGURE APP. 13B-18

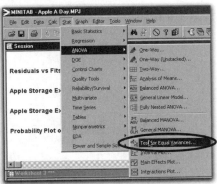

FIGURE APP. 13B-19

that the residuals are NOT normal. The addition of the hypothesis test in the Basic Statistics Normality test is a good reason to use it rather than to simply rely on the simple Four-in-one plot that does not have such hypothesis tests attached to it.

The next plot is to check for homogeneity of variances. This is done by using the test found under the **ANOVA** dropdown shown

FIGURE APP. 13B-20

in Figure App. 13B-20; this brings up the dialog box in Figure App. 13B-21. It is necessary to select a response that is the residuals (RESI1) as well as a factor (FITS1) to do this test for homogeneity of variances. Click **OK** and a plot (shown in Figure App. 13B-23) gives the confidence intervals and

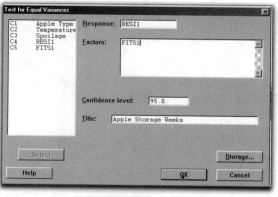

FIGURE APP. 13B-21

the p value for the Bartlett's Test. Unfortunately this is dependent on a normal distribution that we just found was not the case. An alternative Levine's Test is offered by Minitab, but was not able to be performed with this data. Thus we are unable to conclude if there is homogeneity of variances in this set of residuals even though the p value is 1.0. This is again a function of the data used in this example.

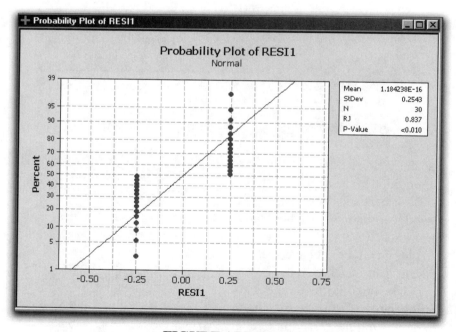

FIGURE APP. 13B-22

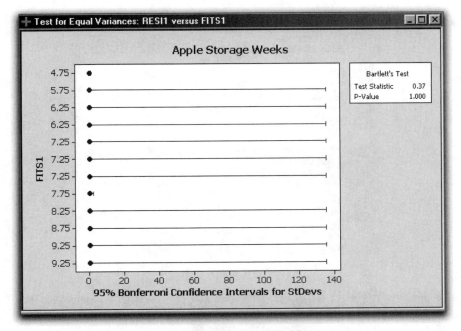

FIGURE APP. 13B-23

Using "Automated DOE" in Minitab

The same Apple Storage design was built in Chapter 4 using the automated routine in Minitab. Let's use that design with slightly better responses (days to spoilage). In Figure App.13B-24, we access the Minitab project file that we created earlier and open it from the folder where it had been stored (Figure App.13B-25). The response values are entered as shown in Figure App.13B-26. Note that these are new responses measured in days to spoilage instead of months, as used before.

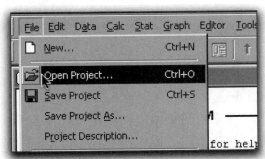

FIGURE APP. 13B-24

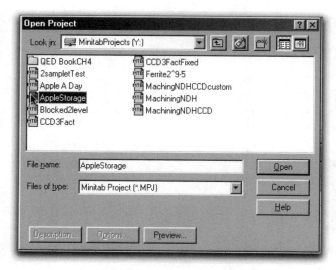

FIGURE APP. 13B-25

↓	C1	C2	C3	C4	C5	C6-T	C7
	StdOrder	RunOrder	PtType	Blocks	Temperature	Apple Type	Days to Spoilage
1	2	1	1	1	36	McIntosh	221
2	30	2	1	1	44	Delicious	152
3	28	3	1	1	44	Ida Red	150
4	19	4	1	1	38	Ida Red	221
5	17	5	1	1	36	McIntosh	242
6	1	6	1	1	36	Ida Red	194
7	26	7	1	1	42	McIntosh	285
8	29	8	1	1	44	McIntosh	224
9	25	9	1	1	42	Ida Red	223
10	15	10	1	1	44	Delicious	135

FIGURE APP. 13B-26

From the **Stat** dropdown (shown in Figure App. 13B-27), we go to the **Analyze factorial design** which brings up a dialog box (Figure App. 13B-28). We select the response, and click on **Terms,** which brings up a sub dialog box (Figure App.13B-29) and we shuttle the two single effects of Temperature and Apple Type as well as the AB interaction into the **Included Terms** side of the selection sub dialog box.

FIGURE APP. 13B-27

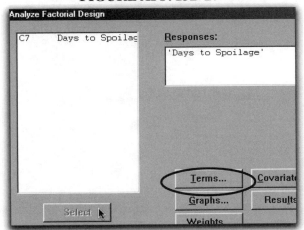

FIGURE APP. 13B-28

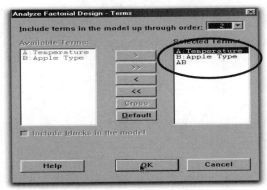

FIGURE APP. 13B-29

The next step is to select the residual graphs. This is exactly like the sub dialog box we saw in the GLM procedure. Select **Individual Plots** (as shown in Figure App.13B-30) along with the **Normal plot**, **Residuals versus fits**, and **Residuals versus order** (since we know the order in this experiment). Click **OK**, which brings up the main dialog box.

Click on **Storage** and store the Fits and Residuals as shown in Figure App.13B-31. Click **OK** and then click **OK** in the main dialog box. The analysis is completed as shown in Table App. 13B-2. This ANOVA is just slightly different from the previous analysis done via GLM since the response variable is the more refined days to spoilage rather than the cruder months to spoilage.

FIGURE APP. 13B-30

FIGURE APP. 13B-31

TABLE App. 13B-2

```
General Linear Model: Days to Spoilage versus Temperature, Apple Type

Factor         Type    Levels  Values
Temperature    fixed      5    36, 38, 40, 42, 44
Apple Type     fixed      3    Ida Red, McIntosh, Delicious

Analysis of Variance for Days to Spoilage, using Adjusted SS for Tests

Source                  DF    Seq SS    Adj SS    Adj MS       F       P
Temperature              4   25194.1   25194.1    6298.6   50.13   0.000
Apple Type               2   33949.2   33949.2   16974.5  135.11   0.000
Temperature*Apple Type   8    2737.4    2737.4     342.2    2.72   0.045
Error                   15    1884.5    1884.5     125.6
Total                   29   63765.3
```

Residual Plots

The time series residual plots in Figure App.13B-32 show a more random pattern than the previous plot (Figure App.13B-13) since the run order is known in this analysis rather than being confounded with the standard order. The residuals versus fitted values and the normal plot (Figures App.13B-33 and 34) show a better pattern than with the weeks to storage.

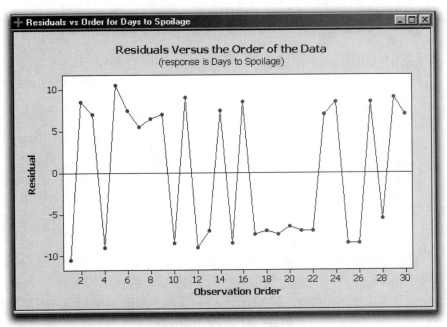

FIGURE APP. 13B-32

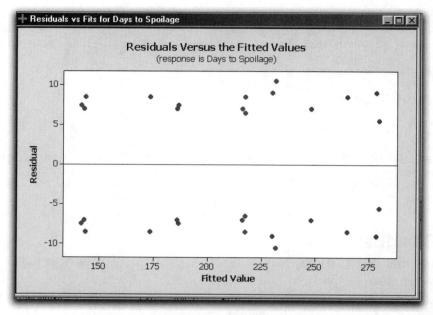

FIGURE APP. 13B-33

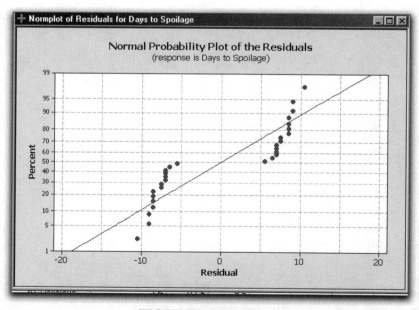

FIGURE APP. 13B-34

Since the normal plot from the simple analysis does not give any probabilities of rejection, we will use the normality test from the basic statistics dropdown shown in Figure App.13B-35. In the dialog box (Figure App.13B-36), we will choose the Anderson-Darling test (Ryan-Joiner is also a fine choice), which produces the plot (which looks like the plot we obtained before, but has a p-value of <0.01 which says that the residuals are not normal [Figure App. 13B-37]). This of course can make any conclusions from this experiment questionable with regard to the statistical significance.

FIGURE APP. 13B-35

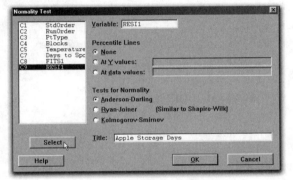

FIGURE APP. 13B-36

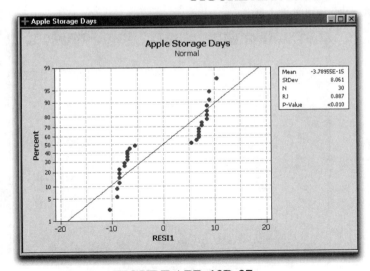

FIGURE APP. 13B-37

In addition, the final test of the residuals, which will use the Bartlett's test, is also in question since the Bartlett's test may be done only if the data is normal. We will go ahead with the test for this example data. The test for the homogeneity of variance is found under the **Stat => ANOVA** drop-down shown in Figure App.13B-38. In the dialog box (Figure App.13B-39), we enter the Residuals (RESI1) as the response.

FIGURE APP. 13B-38

The Factor is the fits (FITS1). Click **OK**, and the Bartlett's test is completed along with a set of simultaneous confidence intervals for the standard deviations for each treatment combination of the experiment as shown in Figure App.13B-40. Since the p-value is 1.0 (rounded from 0.99999), and the confidence intervals overlap, we have a reasonable chance of homogeneity.

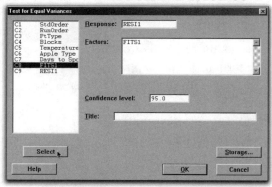

FIGURE APP. 13B-39

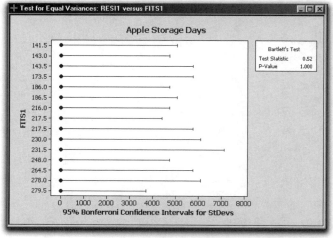

FIGURE APP. 13B-40

The last step is to obtain plots of the effects and the interaction. Figure App.13B-24 shows that this is accomplished by moving to **Factorial Plots** in the dropdown from **Stat**. In the dialog box (Figure App.13B-42), click on both the **Main** effects and **Interaction** options and use the **Fitted Means** instead of the default **Data Means** selection. It is necessary to set up the plots for both the main effects and the interaction. Click on **Setup...**, which takes you to the sub dialog box (Figure App.13B-43) where you select the response and the factors that you want plotted.

The same type of dialog box is also filled out for the interactions.

Figure App.13B-44 shows the interaction plots that should always be investigated before the single (*main effects* as Minitab insists) effects. Since the single effects are an average over the interaction plot, the added information of the interactions is much more important than just these average effects. Since we checked "Fitted Means" there is only one interaction plot shown that isolates the levels of the first factor (temperature) and plots versus the second factor (apple type). In this experiment, the interaction plot would have more insight if the apple type had been isolated as we have already seen in the plot shown in Figure App.13B-14. The interpretation of these plots (Figures App.13B-44 and 45) is left as an exercise for the reader in problem number 8.

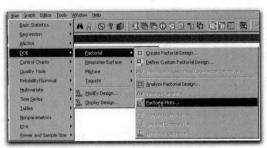

FIGURE APP. 13B-41

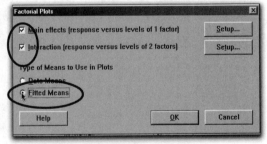

FIGURE APP. 13B-42

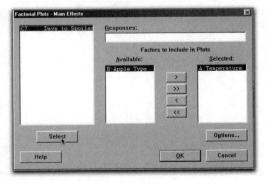

FIGURE APP. 13B-43

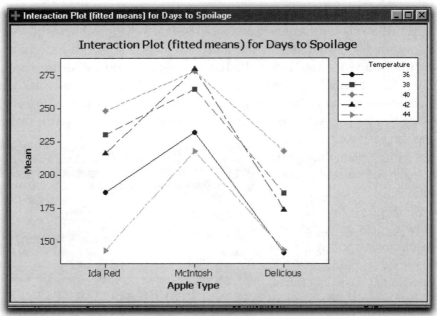

FIGURE APP. 13B-44

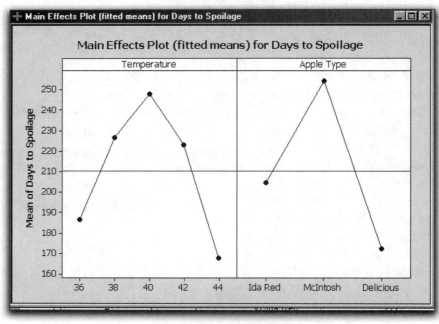

FIGURE APP. 13B-45

14

YATES Analysis
Analysis of 2^k and 2^{k-p} Designs

In our daily lives we use many algorithms to simplify the way we do business. While an algorithm is a step-by-step procedure that will simplify a rather involved activity, it has also been defined as a method that simplifies the complexity of a calculatio, but often mystifies understanding. We will attempt to remove some of the mystery by showing how this algorithm works.

The analysis of variance we studied in Chapter 13 can lead to some involved arithmetic. Fortunately, for the types of experiments that are used most of the time (the 2^{k-p}) and prove to be most efficient, there is a simplified algorithm for ANOVA. In Chapters 4 and 5 we learned how to design these 2-level factorial and fractional factorial experiments. Because of the 2-level characteristic of these experiments, we may use the YATES method to compute the sums of squares and other important decision-making statistics. While the YATES method may at first seem like a strange exercise, we will see that it produces the same results as a conventional ANOVA method and gives us a bonus piece of information. This bonus information will prove to be very valuable in the interpretation and presentation of the results from 2^k and 2^{k-p} designs.

In this chapter, we will learn how to make the computations for ANOVA using the YATES method which was developed by the pioneer experimental designer, Frank Yates. In Chapter 15 we will review the essential elements of matrix algebra to set the stage for Chapter 16, where we will demonstrate the common bond between ANOVA and the general

349

linear model and demonstrate how Yates developed his approach.

For now, let's see how the YATES algorithm works and try some examples. Here are the rules:

1. Place the response variable information from the 2^k or 2^{k-p} design in a column that is in Yates order. (For a review of YATES order, see Chapter 4.) If the experiment is fully replicated, sum the replicates first.

2. Set up n columns (where n=k-p) adjacent to the response column. Number the columns with Roman numerals (a convention).

3. a. Create entries for the first column (column I) by adding adjacent pair-wise sets of data from the response variable column.

    b. Complete the first column by going back to the start of the response variable column and subtracting adjacent pair-wise entries. The order of subtraction is to subtract the first response from the second response as they appear in YATES order.

4. Continue the pair-wise addition and subtraction using the newly created columns until you reach the last column (column n).

5. Compute the sum of squares by taking the individual entries in the last column (column n) and square each of them. Divide this square by r x 2^n (where n=k-p and r is the number of replicates). There is no "centering factor" needed in the YATES ANOVA since the algorithm is a "self-centering" method.

6. Compute the "half effects" for quantitative factors by taking the last column (column n) and dividing it by r x 2^n (r= # of replicates). If the factor is of a qualitative nature, then a "half effect" would be meaningless and the "full effect" should be calculated by dividing the last column by r x 2^{n-1} (r= # of replicates). The "half effect" is defined as the change in response over half the range of the factor under study. We shall treat the meaning of the half-effect further in this chapter and in Chapter 17.

These rules give the method for computing the ANOVA. We shall now look at a numerical example and show how to interpret the results of such an analysis.

For our example, we will use the data from Table 4-4.   In Chapter 4 we performed an "intuitive" analysis, and now by using the YATES ANOVA we will obtain similar results but with a bit more insight and a statement of our degree of belief in the answer.

The experiment in Chapter 4 investigated the effect of pressure, temperature, and time on the contamination of a polymer product. The response we are looking at is percent contamination.

The results from the experiment as run in random order are:

TABLE 14-1

Run	tc	% contamination
1	c	4.1
2	(1)	2.6
3	a	3.9
4	abc	3.2
5	ac	1.7
6	b	4.4
7	ab	8.0
8	bc	7.8

Step 1 puts the responses in YATES order:

TABLE 14- 2

Run	tc	% contamination
2	(1)	2.6
3	a	3.9
6	b	4.4
7	ab	8.0
1	c	4.1
5	ac	1.7
8	bc	7.8
4	abc	3.2

In this experiment there were three factors, so it was a 2^3 factorial with no fractionalization. Therefore, n = 3 – 0 or 3, and we set up 3 columns for the YATES approach to this analysis. We will do so in Table 14-3 and complete the pair-wise addition and subtraction for the first column in this table.

We continue and complete Table 14-3 by filling out columns II and III. In doing the arithmetic, be sure to watch the signs of the terms. A pocket calculator that has the change sign key makes this job easier.

TABLE 14-3

Run	tc	% contamination	I	II	III
2	(1)	2.6	6.5		
3	a	3.9	12.4		
6	b	4.4	5.8		
7	ab	8.0	11.0		
1	c	4.1	1.3		
5	ac	1.7	3.6		
8	bc	7.8	-2.4		
4	abc	3.2	-4.6		

Addition operation ——————————
Subtraction operation — — — — — — — —

TABLE 14-4

Run	tc	% contamination	I	II	III
2	(1)	2.6	6.5	18.9	35.7
3	a	3.9	12.4	16.8	-2.1
6	b	4.4	5.8	4.9	11.1
7	ab	8.0	11.0	-7.0	0.1
1	c	4.1	1.3	5.9	-2.1
5	ac	1.7	3.6	5.2	-11.9
8	bc	7.8	-2.4	2.3	-0.7
4	abc	3.2	-4.6	-2.2	-4.5

To produce the ANOVA table (Table 14-5) we will now take the last column (column III) of values from Table 14-4, square each entry and then divide each squared value by 2^n (since n=3, then $2^3= 8$ in this case) to produce the sum of squares. We also will take each value in this last column and divide it by 2^n to obtain the "half effects."

Now that we have done all the calculations, let's look into Table 14-5 to see what it all means. Each row of the table identifies an effect. We are able to identify the effects by looking at the treatment combination identifiers. The "a" identifier has the sum of squares and half effect for the "A" factor (pressure) next to it. The other effects are identified similarly. Note that the convention for measured effects uses an upper case letter while the treatment combinations use a lower case letter. While there is the link between these letters that identify the generic factor, remember that the tc identifier (lower case) merely has the value of the experimental data for that run of the experiment, while the effect (upper case) shows the average change in the response as a function of the change in the factor. Some people have made the error of thinking the response associated with the tc is the change due to that factor. Only by the analysis that looks at the change over the low and high levels of the factor do we gain an insight into the effect of this factor. Let's see what this experiment is telling us.

TABLE 14-5

tc ID	Observation % contamination	Sum of Squares	Half Effect	Measures (Name of effect)
(1)	2.6	159.3110	4.4625	Average
a	3.9	0.5513	-0.2625	**A** (pressure)
b	4.4	15.4013	1.3875	**B** (temperature)
ab	8.0	0.0013	0.0125	**AB** (press.-temp.)
c	4.1	0.5513	-0.2625	**C** (time)
ac	1.7	17.7013	-1.4875	**AC** (time-press.)
bc	7.8	0.0613	-0.0875	**BC** (time-temp.)
abc	3.2	2.5313	-0.5625	**ABC** (t-p-T)

In the above experiment there was no replication, so unless we can come up with a measure of residual, we cannot perform an F test to determine if there is a significant signal-to-noise ratio for the effects. It is sometimes possible to use an "outside" estimate of the residual. This residual estimate would come from previously run experiments or from long-term historical records such as control charts on the process. Let's assume that we have an outside estimate of this residual variance (σ^2) for this experiment of 0.9 with 20 degrees of freedom. Therefore, the critical F for 1 and 20 df at the 0.05 level is 4.35. Since in a 2-level experiment there is only one degree of freedom for each effect, the mean square is the same value as the sum of squares. The F's for each effect are computed by dividing the sum of squares for each entry (each effect) in Table 14-5 by the 0.9 residual mean square. If we had replicated the experiment, then we would do the same division operation, but use the internally generated residual mean square.

TABLE 14-6

Effect Measured	F	Conclusion
pressure	0.61	not significant
temperature	17.11	significant
press. x temp.	0.00	not significant
time	0.61	not significant
press. x time	19.67	significant
temp. x time	0.07	not significant
press. x temp. x time	2.81	not significant

Therefore we conclude that there is a significant temperature effect and a significant pressure-time interaction.

Using Minitab

To do an analysis in Minitab, it is necessary to have the experimental design. In Chapter 4 we built the contamination investigation experiment, and now we will retrieve that project as shown in Figure 14-1. To do so, go to **File** and **Open Project**. In the subsequent dialog box, we find the location of the design and highlight it. Click on **Open** (or double click on the file name) and the project is loaded into the worksheet and the session window. We need to add a column for the response (% Contamination), which is done in **C9**.

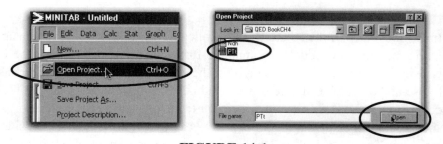

FIGURE 14-1

Figure 14-2(b) shows the entry of these responses that were taken friom Table 14-2. Now go to the dropdown **Stat => DoE => Factorial => Analyze Factorial Design** as shown in Figure 14-3.

C5	C6	C7	C8-T	C9
Pressure	Temperature	time	tc ID	% Contamination
100	70	10	(1)	
200	70	10	a	
100	90	10	b	
200	90	10	ab	
100	70	20	c	
200	70	20	ac	
100	90	20	bc	
200	90	20	abc	

C8-T	C9
tc ID	% Contamination
(1)	2.6
a	3.9
b	4.4
ab	8.0
c	4.1
ac	1.7
bc	7.8
abc	3.2

a   b

FIGURE 14-2

FIGURE 14-3

In the main dialog box (Figure 14-4) highlight the response and click **Select**. Then click on **Terms**, which brings up a sub dialog box where we essentially build the model we want to test for statistical significance. Figure 14-5 (a) shows that all the possible terms are included in the model by default, so we select Terms up to 2 factor interactions as shown in Figure 14-5 (b).

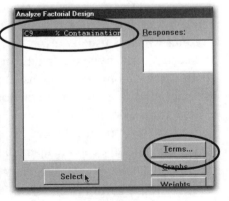

FIGURE 14-4

a   b

FIGURE 14-5

Click **OK** to return to the main dialog box. Since this is a preliminary model, we will not utilize the other options in Minitab at this point. Click **OK** in the main dialog box to produce the analysis shown in Figure 14-6.

```
Estimated Effects and Coefficients for % Contamination (coded units)
Term                   Effect     Coef  SE Coef      t      P
Constant                         4.463   0.5625   7.93  0.080
Pressure               -0.525  -0.262   0.5625  -0.47  0.722
Temperature             2.775   1.387   0.5625   2.47  0.245
time                   -0.525  -0.263   0.5625  -0.47  0.722
Pressure*Temperature    0.025   0.012   0.5625   0.02  0.986
Pressure*time          -2.975  -1.488   0.5625  -2.64  0.230
Temperature*time       -0.175  -0.087   0.5625  -0.16  0.902
```

FIGURE 14-6

Since none of the "p" values are less than or equal to 0.05 (a common 5% risk level), we might say there is nothing significant. However, there is only one df for the error in this analysis (it comes from the ABC interaction). If we look at the values of the t statistic, we see that Temperature and the Temperature * time interaction have the largest t values (outside of the constant which is usually always significant). We now should construct a "reduced model" by removing the insignificant interactions. There is some controversy among statisticians about removing the single effects, but Minitab will not allow an interaction to enter the model unless the single effect term is also included.

Now we return to the **Terms** sub dialog box and send the pressure*temperature and temperature*time interactions out of the model as shown in Figure 14-7. Click **OK** in the main dialog box and we see a much better model in Figure 14-8 with much lower "p" values. This is due to the fact that we have more df for error and therefore a more substantially based noise.

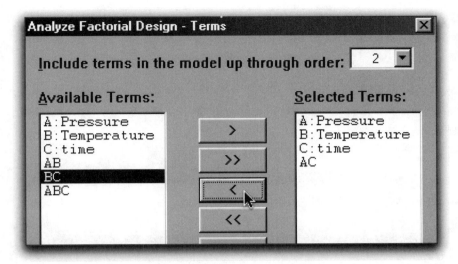

FIGURE 14-7

```
Estimated Effects and Coefficients for % Contamination (coded units)

Term            Effect    Coef   SE Coef      T      P
Constant                 4.463   0.3287    13.57  0.001
Pressure       -0.525   -0.262   0.3287    -0.80  0.483
Temperature     2.775    1.387   0.3287     4.22  0.024
time           -0.525   -0.263   0.3287    -0.80  0.483
Pressure*time  -2.975   -1.487   0.3287    -4.52  0.020
```

FIGURE 14-8

We still notice that there are two insignificant single effects. However, neither of these may be dropped since they are involved with the interaction. Minitab follows the rules of proper polynomials and will not allow an interaction to enter the model unless the single effects are also included. See problem #3 to confirm this.

With the analysis completed, we will have Minitab check for the error requirements and also plot the effects. Figure 14-9 is the sub-dialog box where we select the Pareto, Normal, Four-in-one, and set the alpha risk (the default is 0.05, and is fine for this problem).

Figure 14-10 shows that the error (as diagnosed via the residual values) adheres reasonably (but not perfectly) to the requirements of independence, normality, and homogeneity. We will proceed to interpret the results.

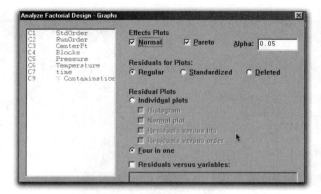

FIGURE 14-9

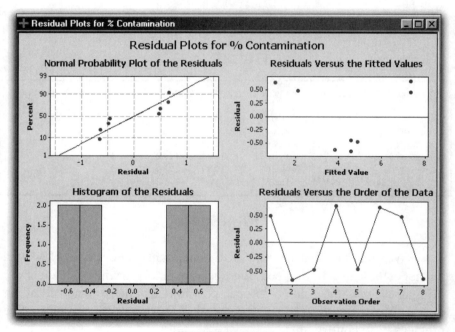

FIGURE 14-10

Following the same path that the t tests have shown, the Pareto and normal plots of the effects (Figure 14-11 a, b) indicate that there is one single effect and one interaction.

We now will proceed to plot these effects to help in our interpretation. Figure 14-12 follows the drop down to the **Factorial Plots**, which brings up the dialog box (Figure 14-13) where we click on **Main effects** and **Interactions**.

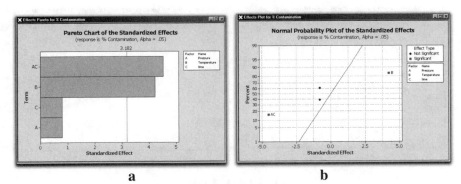

a   b

FIGURE 14-11

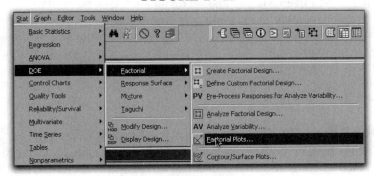

FIGURE 14-12

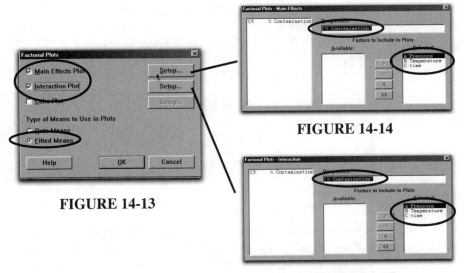

FIGURE 14-13

FIGURE 14-14

FIGURE 14-15

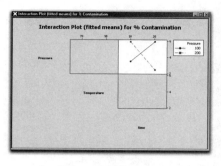 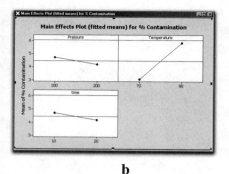

a   b

FIGURE 14-16

We will use the **Fitted Means** and set up the factors to be plotted. Figure 14-16a shows the interaction between pressure and time is a complete crossing within the experimental space. So, if the reaction is done at low pressure, a short time is required to reduce contamination. The opposite is true if the reaction is done at high pressure. Since time and pressure are both energy consuming, the low pressure and short time are best. In either case, temperature should be low (Figure 14-16b).

Application to 2^{k-p} Fractional Designs

The YATES method is completely applicable to fractional factorial designs. The method of computing the sums of squares is general enough to allow all the computations to be made as in the full factorial designs. The only difference comes in the interpretation.

In setting up the "measures" column at the end of the YATES ANOVA procedure, we start by listing the effects based upon the treatment combination identifiers from the *base design* used to generate the fractional factorial. For a review of this concept see Chapter 5. For the simplest 2^{k-p} design where k = 3, p=1, and the generator C≈AB; the ANOVA measures identifiers would look like this:

TABLE 14-7

tc	Sum of Squares	Measures	
l(c)		Average	
a		A, BC	for a 2^{3-1}
b		B, AC	
ab(c)		AB,C	1≈ABC

A more complex 1/4 fractional factorial as described in Chapter 5 has a much more complicated "measures" column. It is still built on the base design, which is a 2^5 in this case. Table 14-8 shows the outline of the YATES ANOVA for this design.

TABLE 14-8

tc	Response	I	II	III	IV	V	Sum of Squares	Half Effect	Measures
(g)									- - - - - - -
a(f)									**A**,BCDF,ABCDEG,FEG
b(f)									**B**,ACDF,CDEG,ABEFG
ab(g)									**AB**,CDF,ACDEG,BEFG
c(f)									**C**,ABDF,BDEG,ACEFG
ac(g)									**AC**,BDF,ABDEG,CEFG
bc(g)									**BC**,ADF,DEG,ABCEFG
abc(f)									ABC,**DF**,ADEG,BCEFG
d(f)									**D**,ABCF,BCEG,ADEFG
ad(g)									**AD**,BCF,ABCEG,DEFG
bd(g)									**BD**,ACF,CEG,ABDEFG
abd(f)									ABD,**CF**,ACEG,BDEFG
cd(g)									**CD**,ABF,BEG,ACDEFG
acd(f)									ACD,**BF**,ABEG,CDEFG
bcd(f)						Confounded 2-factor interactions•			BCD,**AF**,**EG**,ABCDEFG
abcd(g)									ABCD,**F**,AEG,BCDEFG
e(fg)									**E**,ABCDEF,BCDG,AFG
ae						Confounded 2-factor interactions•			**AE**,BCDEF,ABCDG,**FG**
be									**BE**,ACDEF,CDG,ABFG
abe(fg)	3-factor interactions may be used for residual error					⟶			ABE,CDEF,ACDG,BFG
ce									**CE**,ABDEF,BDG,ACFG
ace(fg)	3-factor interactions may be used for residual error					⟶			ACE,BDEF,ABDG,CFG
bce(fg)									BCE,ADEF,DG,ABCFG
abce	3-factor interactions may be used for residual error					⟶			ABCE,DEF,ADG,BCFG
de									**DE**,ABCEF,BCG,ADFG
ade(fg)	3-factor interactions may be used for residual error					⟶			ADE,BCEF,ABCG,DFG
bde(fg)									BDE,ACEF,**CG**,ABDFG
abde	3-factor interactions may be used for residual error					⟶			ABDE,CEF,ACG,BDFG
cde(fg)									CDE,ABEF,**BG**,ACDFG
acde	3-factor interactions may be used for residual error					⟶			ACDE,BEF,ABG,CDFG
bcde									BCDE,AEF,**G**,ABCDFG
abcde(fg)						Confounded 2-factor interactions •			ABCDE,**EF**,**AG**,BCDFG

The defining contrast for the above design is: $1 \approx$ ABCDF,BCDEG,AEFG

Looking at Table 14-8, of the 28 degrees of freedom needed to supply information on the 7 single effects and the 21 two-factor interactions, this design affords only 22 independent degrees of freedom. Three df are confounded among six of the two-factor interactions (indicated with dots in Table 14-8), and there are 6 df that measure three-factor interactions (indicated with arrows in Table 14-8). Since three-factor interactions are rare events and have expected values of zero, we may use these three-factor interactions to measure the residual error.

The usual practice in the analysis of fractional factorial designs is to do the analysis and determine the significant effects and then use the defining contrast to determine the confounding among the effects for only those effects that are significant. This saves a lot of extra effort in computing the confounding pattern from the defining contrast.

While an outline of an analysis as shown in Table 14-8 indicates the effects identified by code letters of the alphabet, in a real experiment it is difficult to follow the physical reality of the situation by using this alphabetical code. Therefore, it is recommended that the actual, physical names of the factors be substituted for the letters when communicating results.

Half Effects

The YATES ANOVA is able to help us make a decision about the statistical significance of the effects that are under study in our 2-level experiment. By looking at the half effects, we may also obtain a physical idea of the influence of these effects. This influence we are talking about is the change in the response over a change in the input factor. Such a change is measured numerically by a coefficient called a *slope*. The half effect is such a slope. In Figure 14-17, we show the change in the response over the distance from the low level of the input factor to the midpoint of the factor. In the design units (-1, +1) of the two-level factorial experiment this distance is half the range of the input factor. A slope is defined as the change in the response over a *unit* change in the input. So what we observe in Figure 14-7 is the slope of the line and since this slope takes place over half the range of the input factor it is called the *half effect*. We can use the half effect to construct an equation (in design units) describing the effects of all the significant factors from Table 14-5 on the response as shown in Expression 14-1.

$$Y = 4.4625 + 1.3875(\text{Temperature}) - 1.4875(\text{time x pressure}) \qquad (14\text{-}1)$$

We can solve this equation for different levels of temperature and time and pressure as long as we use the design units (-1, +1) as the levels of the factors.

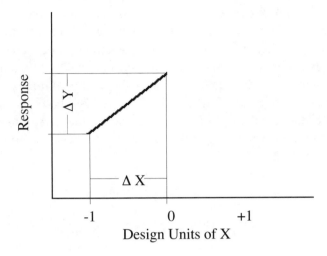

FIGURE 14-17

We will use this idea of the half effects in an equation in the next section where we will deconfound two-factor interactions.

Deconfounding Effects

In the example above and in many experiments, the resources do not match the required information and we end up with two-factor interactions confounded with each other. There are four methods of attack to resolve this problem.

>   1. Use prior knowledge to pick the likely interaction.
>   2. Run a subset experiment.
>   3. Run a minimum number of deconfounding runs.
>   4. Run the rest of the fraction of the experiment.

The next example will illustrate these methods. In this example we are investigating four factors in a 1/2 fractional factorial experiment with 8 treatment combinations. The factors are: temperature, pressure, time, and concentration. Using the ABC interaction from the base 2^3 design we obtain a defining contrast of:

$$1 \approx ABCD$$

from which we can see that all of the two-factor interactions are confounded together.

$$AB \approx CD \qquad AC \approx BD \qquad AD \approx BC$$

To understand the consequence of this confounding, let's take the first confounded pair of interactions and see exactly the meaning of AB ≈ CD. We can be very mathematical and state that the effect we observe for AB,CD is a linear combination of these two effects. But what does that mean? First of all, let's gain a little insight by looking at the pattern of minus and plus signs assigned to the AB and CD interactions. Table 14-9 shows the base design in YATES order with the signs for the four single effects. To obtain the signs of the interactions, we simply multiply the signs together following ordinary rules of algebra.

TABLE 14-9

A	B	C	D	AB	CD	Response	AB	CD
-	-	-	-	+	+	21	+21	+21
+	-	-	+	-	-	19	-19	-19
-	+	-	+	-	-	20	-20	-20
+	+	-	-	+	+	58	+58	+58
-	-	+	+	+	+	42	+42	+42
+	-	+	-	-	-	41	-41	-41
-	+	+	-	-	-	3 9	-39	-39
+	+	+	+	+	+	79	+79	+79

Sum for AB= 81

D≈ABC   Sum for CD= 81

Since the set of signs for AB are the same as the set of signs for CD, we can see that the result of summing the responses for these two interactions gives the same result and it is impossible to decide from this experiment which of these effects is producing the effect we observe.

Actually, it's worse than that. Let's say that there is a physical AB effect of +20 and a physical CD effect of -10. The resulting effect we would observe in this confounded experiment would be the linear combination of these two values which would be +10.

We would be misled by the result and unless we deconfound the effects we could end up with a process that fails to behave as we predicted. Let's continue the example and look at the experiment with actual physical values and actual responses rather than the design units as shown in Table 14-9.

The experiment was designed and run in random order. The results that we shall analyze are presented in Table 14-10 in YATES order.

TABLE 14-10

1 ≈ABCD

tc	A Temp.	B Press.	C Time	D Conc.	Yield
(1)	100	14	5	0.5	21
a(d)	200	14	5	1.0	19
b(d)	100	28	5	1.0	20
ab	200	28	5	0.5	58
c(d)	100	14	10	1.0	42
ac	200	14	10	0.5	41
bc	100	28	10	0.5	39
abc(d)	200	28	10	1.0	79

An analysis of this data set produces the YATES ANOVA in Table 14-11. Note that the physical names are attached to the effects rather than simply the letter identifiers. This is the recommended way all analyses should be presented.

TABLE 14-11

tc	Observation	Sum of Squares	Half Effect	Measures
(1)	21	12720.1	39.875	Average
a(d)	19	703.125	9.375*	**Temperature**,Press.×Time×Conc.
b(d)	20	666.125	9.125*	**Pressure**, Temp. × Time × Conc.
ab	58	820.125	10.125*	**Temp. × Press., Time × Conc.**
c(d)	42	861.125	10.375*	**Time**, Temp. × Press. × Conc.
ac	41	1.125	.375	**Temp. × Time, Press. × Conc.**
bc	39	1.125	-.375	**Press. × Time, Temp. × Conc.**
abc(d)	79	.125	.125	**Conc.**, Temp. × Press. × Time

*Important effects

We have put in boldface the likely effects and have ignored the unlikely three-factor interactions. The problem with this experiment is that all the two-factor interactions are confused with each other. By judging the size of the effects, it looks like there is a large temperature effect, a large pressure effect, and a large time effect. These effects have been singled out with an asterisk.

The concentration effect is very small. There is an interaction that exists either between temperature and pressure or between time and concentration. To deconfound this set of interactions, we could apply prior knowledge to the problem and choose the likely interaction based on our experience. Based on prior chemical considerations, we could expect a temperature-pressure interaction to take place. However, we do not have any *evidence* to support this claim.

Since we do not have a definite answer to the question of which interaction is real, we must gather more data. Gathering more data is reserved as a last resort since it requires the expenditure of resources. There are three options to get more data. We may do the following:

1. Run a minimum number of tc's
2. Run a subset experiment on the likely prior knowledge effect
3. Run the other half of the fractional factorial

Because we have some prior knowledge and a possible indication of the temperature-pressure interaction being the likely effect, we will use this knowledge and run a subset experiment involving only the pressure and temperature factors. This will be run in a 2^2 full factorial design and use 4 treatment combinations. The other option of running the other half of the 2 would have used 8 treatment combinations. If we can obtain the required information for 4 tc's we will be twice as efficient when contrasted with the more conservative approach of completing the rest of the experiment. As the number of runs in the experiment increases, the option to complete the rest of the experiment becomes less and less viable.

We will vary temperature and pressure over their original levels while holding time and concentration at their mid-point levels. The design is found in Table 14-12 and the analysis in 14-13.

TABLE 14-12

tc	Temperature	Pressure	Response
(l)	100	14	32
a	200	14	30
b	100	28	30
ab	200	28	68

TABLE 14-13

tc	Observation	Sum of Squares	Half Effect	Measures
(1)	32	6400	40	
a	30	324	9	Temperature
b	30	324	9	Pressure
ab	68	400	10	Temperature x Pressure

When we compare these results with the results in Table 14-11, we can see that the half effects for temperature, pressure, and their interaction (in the 10 region) are very close. We conclude that our contention that the interaction between temperature and pressure is valid since we observe the magnitude of the temperature-pressure interaction from the subset design is the same as the interaction of temperature and pressure from the fractional factorial design. If the 2^2 subset factorial had not shown similar results to the 2^{k-p} experiment, we would have had to run another 2^2 subset experiment involving the time and concentration factors.

Minimum Number of Deconfounding Runs

While running a subset experiment did a good job in this example to produce information that determines the true interaction, there is still the chance that such a deconfounding method will not work every time on the first try. If we need to iterate to obtain the answer, we may run out of funds before we have the required information. Let's look at another way to solve the confounding problem with one set of extra runs that gets the answer on the first try!

To understand this method, we need to think of the problem we have in resolving the difference between the two confounded interactions in Table 14-11. We need to contrast the Temperature-Pressure (T-P) interaction against the time-Concentration (t-C) interaction. This problem is reduced to a simple *one-factor-at-a-time* test to see if the interaction is the T-P or the t-C or possibly a combination of both! To make this contrast, we will run at most 2 more tc's that contrast these two effects by having different signs in the runs. Right now the set of signs for T-P are the same as for t-C (see Table 14-9 to confirm this).

Table 14-14 shows a set of runs that will contrast these two confounded interactions. Other runs could have been used as long as the resultant interaction contrasts are different for the two runs.

TABLE 14-14

Factor:	A	B	C	D	
tc	Temperature	Pressure	Time	Concentration	Yield
acd	+	-	+	+	40
(1)	-	-	-	-	21

TABLE 14-15

Factor:		A	B	C	D	
tc	Average[1]	Temperature[1]	Pressure[1]	Time[1]	Concentration[2]	Predicted
acd	39.875	+9.375	-9.125	+10.375	+0	50.5
(1)	39.875	-9.375	-9.125	-10.375	-0	11.0

[1]The average as well as the half-effects for temperature, pressure, and time come from Table 14-11.
[2] The half effect for concentration is not significant and is considered equal to zero. The half effects are slopes in design units and are multiplied by the design unit value of each factor to obtain the predicted response.

In Table 14-15, we calculate the predicted values of the responses based on the single factor effects found in the ANOVA. Notice how we may use the half effects as slopes in the model as long as we use design units as the factor levels. Table 14-16 shows the difference between the observed responses when the extra tc's were run, and the predicted responses based on the single effects only. This difference is due to the interaction.

TABLE 14-16

AB	CD	Observed	Predicted	Difference Due to Interaction
-	+	40	50.5	-10.5
+	+	21	11.0	10.0

$$\mathbf{AB} = \frac{-(-10.5) + 10.0}{2} = 10.25 \quad \mathbf{CD} = \frac{-10.5 + 10.0}{2} = 0.25$$

In Table 14-16 we find the difference between the observed and predicted, and then attribute this difference to the interaction effect by contrasting the difference according to the conditions in the interaction. So, for the AB interaction we find the algebraic sum of the differences divided by two (since there were two differences) which results in a value of 10.25. For the CD interaction we perform a similar algebraic summation and division, but here the result is only 0.25. Remember, we are treating the AB and CD interactions as effects, and we may separate their influence since these effects now have different levels, unlike the situation we originally encountered in Table 14-9 where AB and CD shared the same sets of levels. Also notice that it took only one extra run (acd) to separate the interactions since the other run (1) was a part of the original design. As we saw before, the interaction that is present is the AB or temperature-pressure effect since its half effect is very close (within experimental variation) to the half effect for the confounded pair of interactions in Table 14-11. Running only *one* extra experimental treatment combination has done an excellent job of separating the confounding in this design. It is possible-and recommended-that when an experiment is being designed, the team should anticipate the likely confounded interactions and include extra tc's along with the main experimental body to avoid the need to go back later to obtain data to resolve the confounding.

Fold-Over Designs and Analysis

There are some experimental situations that utilize "saturated designs." These design configurations allocate all the information (measured by df) to the single effects. If any interactions exist, they will be confounded with the single effects. A typical saturated design is an eight run, seven factor configuration as shown in Table 14-17.

TABLE 14-17

A	B	C	D	E	F	G	Response
-1	-1	-1	1	1	1	-1	-16
1	-1	-1	-1	-1	1	1	2
-1	1	-1	-1	1	-1	1	-8
1	1	-1	1	-1	-1	-1	14
-1	-1	1	1	-1	-1	1	-4
1	-1	1	-1	1	-1	-1	-2
-1	1	1	-1	-1	1	-1	-12
1	1	1	1	1	1	1	26

The YATES analysis of the above design shows that there are A, B, C, D, and G effects. However, these effects are confounded with 2-factor interractions as shown in Table 14-18.

TABLE 14-18

	ΣSq	Half Effect	Measures
	0.0	0.0	Average
E1	800.0	10.0	A + BD +CE + FG
E2	200.0	5.0	B + AD + CF+ EG
E3	200.0	5.0	AB + D + CG + EF
E4	32.0	2.0	C + AE + BF + DG
E5	0.0	0.0	AC + E + BG + DF
E6	0.0	0.0	BC + F + AG + DE
E7	128.0	4.0	ABC + G + AF + BE + CD

To resolve the confounding, we will run another experimental configuration (Table 14-19) with all of the signs reversed. The reason for reversing the signs will become clear as we take the results from each of the two segments of the experiment and combine them.

TABLE 14-19

A	B	C	D	E	F	G	Response
1	1	1	-1	-1	-1	1	12
-1	1	1	1	1	-1	-1	2
1	-1	1	1	-1	1	-1	12
-1	-1	1	-1	1	1	1	-18
1	1	-1	-1	1	1	-1	16
-1	1	-1	1	-1	1	1	-10
1	-1	-1	1	1	-1	1	0
-1	-1	-1	-1	-1	-1	-1	-14

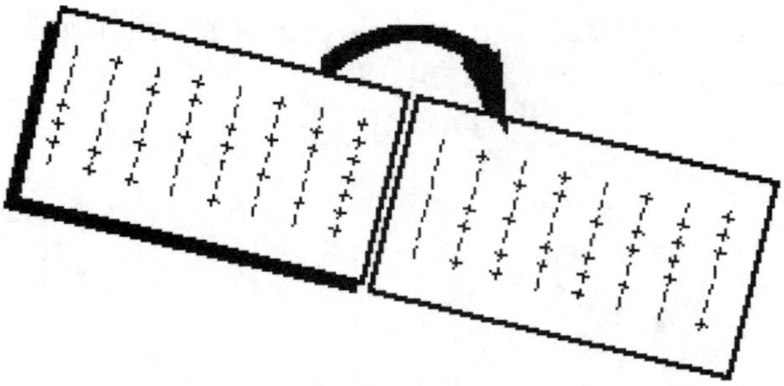

FIGURE 14-18

The reversal of signs leads to the name of this type of experimentation since when the two segments of the experiment are placed in line with each other it appears that we have "folded" the first experiment as shown in Figure 14-18.

TABLE 14-20

	ΣSq	Half Effect	Measures
	0.0	0.0	Average
E'1	800.0	-10.0	-A + BD +CE + FG
E'2	200.0	-5.0	-B + AD + CF+ EG
E'3	200.0	-1.0	-D + AB + CG + EF
E'4	32.0	2.0	-C + AE + BF + DG
E'5	0.0	0.0	-E + AC + BG + DF
E'6	0.0	0.0	-F + BC + AG + DE
E'7	128.0	4.0	-G + AF + BE + CD

Now we will see the power of the fold-over design in action. We will work with the effects (labeled E or E') of each combined factor/interaction. E1 contains the effects of A, BD, CE, and FG. E'1 has the -A, BD, CE, and FG effects. If we subtract E'1 from E1 and find the average of this difference, we obtain the effect of the single factor.

$$\frac{E1 - E'1}{2} = \frac{A + BD + CE + FG - (-A + BD + CE + FG)}{2} = \frac{2A}{2} = A$$

If we add E'1 to E1 and find the average of this sum, we obtain the effect of the interactions.

$$\frac{E1 + E'1}{2} = \frac{A + BD + CE + FG + (-A + BD + CE + FG)}{2} = \frac{2(BD+CE+FG)}{2}$$

Now we can see why we needed to reverse the signs in the second segment of the experiment. By doing so, we are able to take the linear combinations of the effects and since one is the negative of the other, we may obtain the single effects free of the confounded interactions.

Now, we could have done this by simply running an entire 16 run experiment, but if we had a problem with setting the levels of the factors in such an experiment (which can happen in exploratory research) we could have "blown" the entire 16 run experiment if our responses were not measurable. By running the fold-over in two segments, we are able to get some experience with the problem and either regroup or continue, depending on the results of the first segment.

Let's "deconfound" the numerical example we have been working with. We will start with the first effect. Remember that subtracting the second segment effect (E') from the first segment effect (E) produces the single effect, and adding the two effects produces the interaction effects.

$$\frac{E1 - E'1}{2} = \frac{10 - (-10)}{2} = 10 \quad \text{effect of A}$$

$$\frac{E1 + E'1}{2} = \frac{10 + (-10)}{2} = 0 \quad \text{effect of interactions}$$

$$\frac{E2 - E'2}{2} \quad = \quad \frac{5 - (-5)}{2} \quad = \quad 5 \qquad \text{effect of B}$$

$$\frac{E2 + E'2}{2} \quad = \quad \frac{5 + (-5)}{2} \quad = \quad 0 \qquad \text{effect of interactions}$$

$$\frac{E3 - E'3}{2} \quad = \quad \frac{5 - (-1)}{2} \quad = \quad 3 \qquad \text{effect of D}$$

$$\frac{E3 + E'3}{2} \quad = \quad \frac{5 + (-1)}{2} \quad = \quad 2 \qquad \text{effect of interactions}$$

$$\frac{E4 - E'4}{2} \quad = \quad \frac{2 - (-2)}{2} \quad = \quad 2 \qquad \begin{array}{l} AB + CG + EF \\ \text{effect of C} \end{array}$$

$$\frac{E4 + E'4}{2} \quad = \quad \frac{2 + (-2)}{2} \quad = \quad 0 \qquad \text{effect of interactions}$$

$$\frac{E5 - E'5}{2} \quad = \quad \frac{0 - (0)}{2} \quad = \quad 0 \qquad \text{effect of E}$$

$$\frac{E5 + E'5}{2} \quad = \quad \frac{0 + (0)}{2} \quad = \quad 0 \qquad \text{effect of interactions}$$

$$\frac{E6 - E'6}{2} \quad = \quad \frac{0 - (0)}{2} \quad = \quad 0 \qquad \text{effect of F}$$

$$\frac{E6 + E'6}{2} \quad = \quad \frac{0 + (0)}{2} \quad = \quad 0 \qquad \text{effect of interactions}$$

$$\frac{E7 - E'7}{2} \quad = \quad \frac{4 - (4)}{2} \quad = \quad 0 \qquad \text{effect of G}$$

$$\frac{E7 + E'7}{2} \quad = \quad \frac{4 + (4)}{2} \quad = \quad 4 \qquad \text{effect of interactions}$$

$$AF + BE + CD$$

To summarize our findings: There is an A effect, a B effect, a C effect, and a D effect. However, there are no E, F, or G effects. The indicated effect of the G factor was actually all in the interaction! Without the fold-over design, we would have done further work with G and ignored the real culprits in the interaction. Also note that what would have been interpreted as the D effect actually is apportioned to both D and the interaction effect. A similar mistake in interpreting the data would have taken place with these effects as we cited in the G factor.

The fold-over design is a powerful tool to help in our sequential process of building knowledge in an efficient manner. It is a hybrid of both design and analysis, and is highly dependent on our willingness to plan our experiments rather than just let the data happen.

Once again we have shown that it is the logical sequence of experiments built on prior information that provides us with an efficient approach to our understanding of a problem and allows us to solve it When we build on knowledge and use a structured plan of attack, the analyses will point the way to successful experimental interpretation.

Problems for Chapter 14

1.

Run No.	Treatment Combination	A Temp.	B Speed	C Conc.	Yield
1	a	170	90	12.5	296
2	c	120	90	21.5	421
3	b	120	110	12.5	351
4	(1)	120	90	12.5	421
5	abc	170	110	21.5	135
6	ab	170	110	12.5	127
7	ac	170	90	27.5	288
8	bc	120	110	27.5	352

a. Put in YATES order
b. Run YATES ANOVA
c. Plot any significant effects
d. Draw conclusions

2. Run a YATES ANOVA on the following data, and after obtaining the half effects decode back into physical units. Plot any of the significant effects.

Run No.	tc	Response (miles/ gallon)
1	b	98
2	ac	27
3	ab	46
4	(1)	68
5	bc	103
6	c	71
7	abc	51
8	a	22

	Low(-)	High(+)
A is speed	20 mph	40 mph
B is temperature	25°F	65°F
C is octane of gas	87	91.5

3. Attempt to do the analysis of the contamination example without including the pressure and time single effects in Minitab.

4. Write a report following the format shown in Chapter 2 for the machining process for wheel bearings analysis in the Appendix of this Chapter.

APPENDIX 14

Algorithm:	A method to simplify calculations that sometimes confuses understanding, but always gives the correct result.

YATES algorithm: A method used to construct an ANOVA from a 2^{k-p} design.

Half effect:	A change in the response over half the range of the factor under study.

Full effect:	A change in the response over the full range of the factor.

More on Using Minitab

Many of the ideas we have presented in this chapter regarding deconfounding can be accomplished without the tedious, manual arithmetic steps by using the advanced analysis ability of Minitab. We have demonstrated a simple analysis using this software earlier in this chapter. Using that example as a launching pad, we will now investigate the analysis of a fractional factorial experiment and the approach to deconfounding used in Minitab. In Chapter 5, we had constructed a 7-factor experiment with center points that investigated a machining process for wheel bearings. We will now open that Minitab project in Figure App. 14-1.

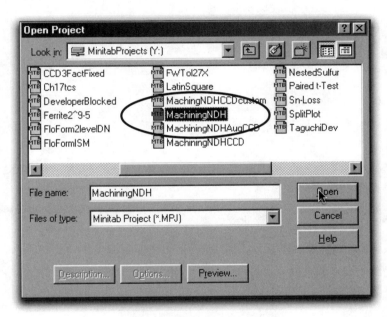

FIGURE APP. 14-1

In the worksheet (Figure App. 14-2) we add the response variable, chip rating. The chip rating is a psychometric (see Chapter 3) that grades the metal particles from the machining process on a scale from 1 to 5. A rating of 1 is considered a perfect chip and a rating of 5 is a long string of metal. String-like chips will clog the waste recovery system and should be avoided, so a chip rating of 1 is desired. With the responses entered, we follow the familiar path of **Stat=> Factorial=>Analyze Factorial Design** as shown in Figure App. 14-3 and select the response.

	C5-T	C6	C7	C8	C9	C10	C11	C12
	Material	Speed	Feed(Left)	Feed(Right)	HtCyTime	Punch/Die	HtPwr	Chp Rting
1	Soft	175	0.25	0.35	5.5	6	75	4.95
1	Hard	175	0.25	0.35	8.5	6	85	3.56
1	Soft	225	0.25	0.35	8.5	12	75	3.94
1	Hard	225	0.25	0.35	5.5	12	85	2.53
1	Soft	175	0.35	0.35	8.5	12	85	4.98
1	Hard	175	0.35	0.35	5.5	12	75	3.97
1	Soft	225	0.35	0.35	5.5	6	85	3.00
1	Hard	225	0.35	0.35	8.5	6	75	1.45
1	Soft	175	0.25	0.45	5.5	12	85	4.06
1	Hard	175	0.25	0.45	8.5	12	75	2.48
1	Soft	225	0.25	0.45	8.5	6	85	3.06
1	Hard	225	0.25	0.45	5.5	6	75	1.54
1	Soft	175	0.35	0.45	8.5	6	75	4.35
1	Hard	175	0.35	0.45	5.5	6	85	2.98
1	Soft	225	0.35	0.45	5.5	12	75	1.97
1	Hard	225	0.35	0.45	8.5	12	85	0.48
1	Soft	200	0.30	0.40	7.0	9	80	4.06
1	Hard	200	0.30	0.40	7.0	9	80	2.48
1	Soft	200	0.30	0.40	7.0	9	80	3.96
1	Hard	200	0.30	0.40	7.0	9	80	2.53

FIGURE APP. 14-2

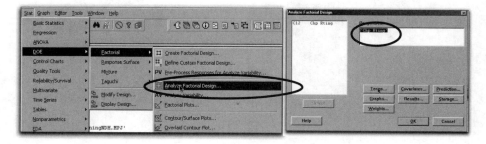

FIGURE APP. 14-3

In the **Terms** sub dialog box (Figure App. 14-4) change the default to 2nd order (2). Minitab will include only the two-factor interactions that are allowed by the available degrees of freedom *AND* will assign them in alphabetical order! This will make the analysis appear to select the interactions with the A factor (Material in this example) as the only ones that go into the model. This is often misleading to the analyst who should realize that there is a confounding pattern in such a fractional factorial design. In the preliminary analysis, it is sufficient to look at the magnitude of the T values since in this case, we have replicated centerpoints to provide an estimate of error. If there is no estimate of error, and all the df have been used to estimate effects, then a normal effects plot is needed to determine which effects are likely to be significant. We click **OK** in the main dialog box (Figure App. 14-5). Table App. 14-1 shows in bold the highest T values for the effects which include the material, speed, both feed rates and an indicated Material*Heat Cycle Time interaction. From an engineering view point, this interaction is ridiculous since the heat cycle time does not occur during the machining stage of this 3 stage process. From physical understanding, this interaction is eliminated.

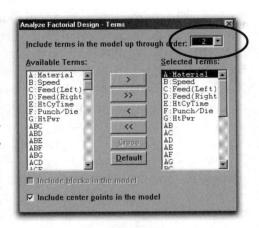

FIGURE APP. 14-4

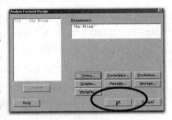

FIGURE APP. 14-5

TABLE APP. 14-1

Term	Effect	Coef	SE Coef	T	P
Constant		3.0812	0.02104	146.39	0.000
Material	**-1.4330**	**-0.7165**	**0.01882**	**-38.06**	**0.000**
Speed	**-1.6700**	**-0.8350**	**0.02104**	**-39.67**	**0.000**
Feed(Left)	**-0.3675**	**-0.1837**	**0.02104**	**-8.73**	**0.001**
Feed(Right)	**-0.9325**	**-0.4663**	**0.02104**	**-22.15**	**0.000**
HtCyTime	-0.0875	-0.0437	0.02104	-2.08	0.106
Punch/Die	-0.0600	-0.0300	0.02104	-1.43	0.227
HtPwr	0.0000	0.0000	0.02104	0.00	1.000
Material*Speed	-0.0775	-0.0388	0.02104	-1.84	0.139
Material*Feed(Left)	0.0600	0.0300	0.02104	1.43	0.227
Material*Feed(Right)	-0.0750	-0.0375	0.02104	-1.78	0.149
Material*HtCyTime	**-0.6750**	**-0.3375**	**0.02104**	**-16.03**	**0.000**
Material*Punch/Die	0.0425	0.0212	0.02104	1.01	0.370
Material*HtPwr	0.0275	0.0137	0.02104	0.65	0.549
Speed*Feed(Right)	-0.0350	-0.0175	0.02104	-0.83	0.452
Ct Pt		**0.1762**	**0.04706**	**3.74**	**0.020**

Table App.14-2 shows the generic confounding pattern for this fractional factorial. The interaction set (**AE + BC + DF**) is indicated in bold. AE has already been ruled out and DF which is the Feed (Right) - Punch/Die is also impossible from a physical view point since the Punch/Die factor is a part of another stage of this process. This leaves only the BC interaction as the possible (from physics) interaction. This then, is the interaction term we will select for our proposed reduced model (a reduced model only includes effects that are likely to be significant). Figure App. 14-6 shows the sub

TABLE APP.14-2

AB + CE + FG
AC + BE + DG
AD + CG + EF
AE + BC + DF
AF + BG + DE
AG + BF + CD
BD + CF + EG

dialog box with the selected terms in the reduced model. Now it is appropriate to go to the **Graphs** sub dialog box and choose the normal and Pareto as well as the four in one residual plots as shown in Figure App. 14-7. Click **OK** in the main dialog box to produce the analysis (part of which) is shown in Table App.14-3. This is an excellent model with very high T values and p values all far less than the usual cut-off point of 0.05. The Pareto (Figure App. 14-8 and normal (Figure App. 14-9) plots echo the T values in a visual manner. Pareto is an especially fine graphical way to show the influence of the effects.

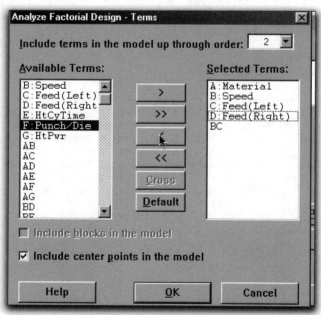

FIGURE APP. 14-6

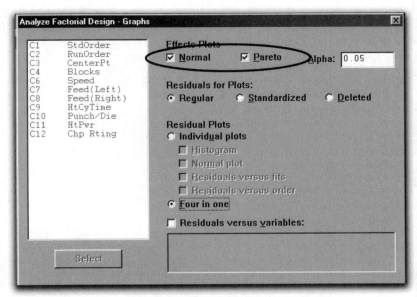

FIGURE APP. 14-7

TABLE APP.14-3

Factorial Fit: Chp Rting versus Material, Speed, Feed(Left), Feed(Right)

Estimated Effects and Coefficients for Chp Rting (coded units)

Term	Effect	Coef	SE Coef	T	P
Constant		3.0812	0.02680	114.95	0.000
Material	-1.4330	-0.7165	0.02397	-29.88	0.000
Speed	-1.6700	-0.8350	0.02680	-31.15	0.000
Feed(Left)	-0.3675	-0.1837	0.02680	-6.85	0.000
Feed(Right)	-0.9325	-0.4663	0.02680	-17.39	0.000
Speed*Feed(Left)	-0.6750	-0.3375	0.02680	-12.59	0.000
Ct Pt		0.1762	0.05993	2.94	0.011

S = 0.107222   R-Sq = 99.46%   R-Sq(adj) = 99.21%

Estimated Coefficients for Chp Rting using data in uncoded units

Term	Coef
Constant	-1.60624
Material	-0.716500
Speed	0.0476000
Feed(Left)	50.3249
Feed(Right)	-9.32499
Speed*Feed(Left)	-0.270000
Ct Pt	0.176250

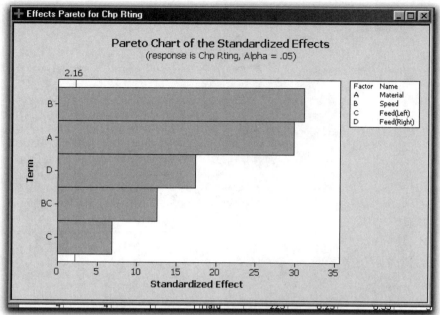

FIGURE APP. 14-8

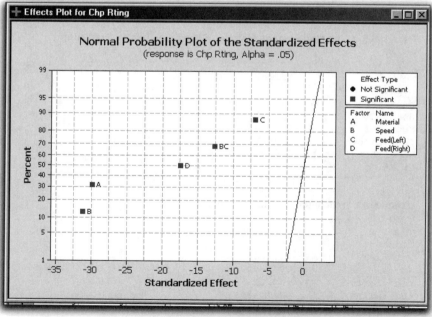

FIGURE APP. 14-9

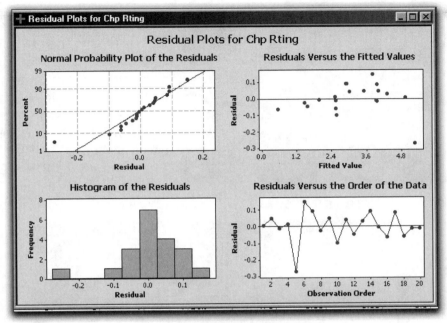

FIGURE APP. 14-10

The residual plots (Figure App. 14-10) are unremarkable and indicate an adherence to the requirements of a proper analysis.

The last touch we need to add to the analysis are the effects plots. To do so, follow the dropdown menu from **Stat=>DOE=>Factorial=>Factorial Plots** as shown in Figure App. 14-11.

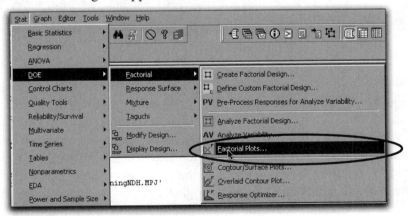

FIGURE APP. 14-11

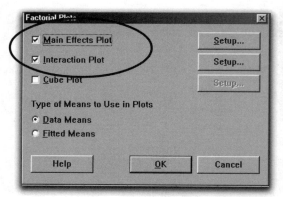

FIGURE APP. 14-12

Select **Main Effects Plots** and **Interaction Plots** in the dialog box (Figure App. 14-12) and then click **Setup** for the main effects and select only the factors that are significant using the shuttle as shown in Figure App. 14-13. Set up the interaction plot (Figure App. 14-14) for just the B and C factors involved in this interaction.

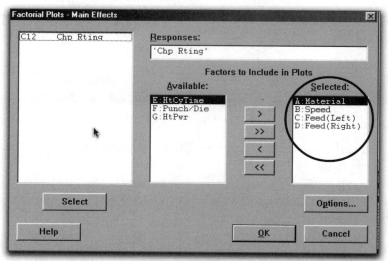

FIGURE APP. 14-13

Since a lower chip value is more desirable, the interaction plot (Figure App. 14-15) tells us that the higher speed, fastest left feed is the setting that delivers this result. The single effects plots (Figure App. 14-16) show that the hard material and fast right feed give the lower chip rating. Since we have already made our decision on the setting of the left feed and lathe speed, we do not need to interpret their single effects plots. However, notice how the left feed has a much shallower slope than the right feed. This is of course since the chip rating depends on both the feed and the speed as shown in the interaction. This is why it is always correct to make the decision from the interaction plot

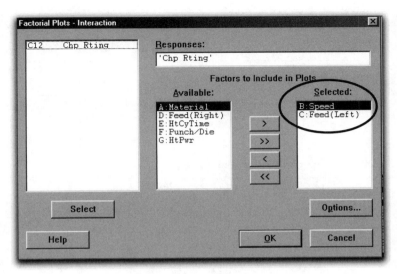

FIGURE APP. 14-14

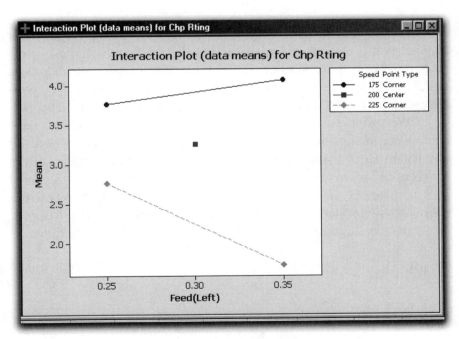

FIGURE APP. 14-15

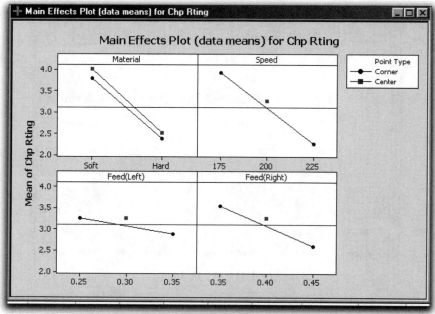

FIGURE APP. 14-16

and not the single effects plots if there is a significant interaction. So the rule of interpretation is to "look at the second order effects first" which may seem contrary to the usual interpretation of second order effects where the term "second order" has come to mean secondary or less important than first order. It is often the interaction (second order) that is the key to a successful, quality product.

We had physics on our side in deciding on which of the three confounded interactions was the real interaction. What if we did not have this physical insight? This is where the addition of a minimum number of extra runs is necessary to resolve the confounding. Table App.14-4 shows a set of such extra runs. Realize that a number of different solutions to the set of extra runs is possible. Here we used a simple 2^{3-1} fractional factorial design to develop the orthogonal contrasts for the three confounded interactions.

TABLE APP. 14-4

AE	BC	DF	A	B	C	D	E	F	G	Chip Rating
-	-	+	-	-	+	-	+	-	-	4.97
+	-	-	+	+	-	-	+	+	-	2.55
-	+	-	-	-	-	-	+	+	-	5.02
+	+	+	+	+	+	+	+	+	-	0.51

We now go the Minitab worksheet and add these new runs. As long as the Minitab coding for Standard Order, Run Order, Center Points, and Blocks is faithfully followed, the design does not need to be declared a "Custom Design." Figure App. 14-17 shows the physical level entries which were taken from the design unit values of Table App.14-4.

In Figure App. 14-18 we select a new model that can now include the three interactions that had been confounded before the new runs were added. If we had tried this model without the extra runs, Minitab would have rejected the BC and DF terms only including the AE term, solely because it is closest to the beginning of the alphabet. The analysis is completed and all three interactions are shown in Table App.14-5. Note that the Speed * Feed (BC) interaction is the only significant interaction term. This is echoed in Figure App. 14-19 which shows BC as the only significant interaction in the normal effects plot. A final reduced model (Table App.14-6) is constructed leaving out the insignificant interactions AE and DF as well as the single effects E and F that Minitab insisted be in the model if the interactions were to be tested. We will revisit this example since there is an indication of possible curvature with a T value of 3.0 for the center point (Ct pt) and a p value of 0.008, which is far less than the usual 0.05 minimum level of significance.

TABLE APP. 14-5

```
* NOTE * This design is not orthogonal.

Factorial Fit: Chp Rting versus Material, Speed, ...

Estimated Effects and Coefficients for Chp Rting (coded units)

Term                   Effect     Coef  SE Coef       T       P
Constant                        3.0731  0.02367  129.79   0.000
Material               -1.4361  -0.7180  0.02272  -31.59   0.000
Speed                  -1.6738  -0.8369  0.02537  -32.98   0.000
Feed(Left)             -0.3991  -0.1996  0.02367   -8.43   0.000
Feed(Right)            -0.9440  -0.4720  0.02380  -19.82   0.000
HtCyTime               -0.1038  -0.0519  0.02367   -2.19   0.047
Punch/Die              -0.0284  -0.0142  0.02367   -0.60   0.559
Material*HtCyTime       0.0564   0.0282  0.04726    0.60   0.561
Speed*Feed(Left)       -0.5835  -0.2918  0.04181   -6.98   0.000
Feed(Right)*Punch/Die  -0.1517  -0.0758  0.03832   -1.98   0.069
Ct Pt                            0.1844  0.05634    3.27   0.006


S = 0.102265   R-Sq = 99.67%   R-Sq(adj) = 99.42%
```

C1	C2	C3	C4	C5-T	C6	C7	C8	C9	C10	C11	C12
StdOrder	RunOrder	CenterPt	Blocks	Material	Speed	Feed(Left)	Feed(Right)	HtCyTime	Punch/Die	HtPwr	Chp Rting
17	17	0	1	Soft	200	0.30	0.40	7.0	9	80	4.06
18	18	0	1	Hard	200	0.30	0.40	7.0	9	80	2.48
19	19	0	1	Soft	200	0.30	0.40	7.0	9	80	3.96
20	20	0	1	Hard	200	0.30	0.40	7.0	9	80	2.53
21	21	1	1	Soft	175	0.35	0.35	8.5	6	75	4.97
22	22	1	1	Hard	225	0.25	0.35	8.5	12	75	2.55
23	23	1	1	Soft	175	0.25	0.35	8.5	12	75	5.02
24	24	1	1	Hard	225	0.35	0.45	8.5	6	75	0.51

FIGURE APP. 14-17

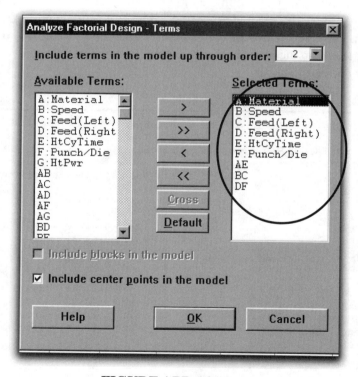

FIGURE APP. 14-18

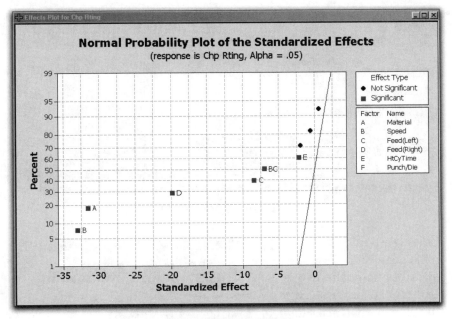

FIGURE APP. 14-19

Using the final model, we will utilize the Response Optimizer routine from Minitab. From the **Stat** dropdown menu, navigate to **Response Optimizer** as shown in Figure App. 14-20. In the first dialog box (Figure App. 14-21), we select the response. This optimization routine can find the trade-offs between multiple responses. We will look at this feature in an example later in the book.

TABLE APP. 14-6

Factorial Fit: Chp Rting versus Material, Speed, Feed(Left), Feed(Right)

Estimated Effects and Coefficients for Chp Rting (coded units)

Term	Effect	Coef	SE Coef	T	P
Constant		3.0714	0.02543	120.72	0.000
Material	-1.4194	-0.7097	0.02356	-30.11	0.000
Speed	-1.6530	-0.8265	0.02583	-31.98	0.000
Feed(Left)	-0.4108	-0.2054	0.02543	-8.07	0.000
Feed(Right)	-0.9217	-0.4609	0.02590	-17.79	0.000
Speed*Feed(Left)	-0.6468	-0.3234	0.02543	-12.71	0.000
Ct Pt		0.1861	0.06204	3.00	0.008

S = 0.113194   R-Sq = 99.48%   R-Sq(adj) = 99.29%

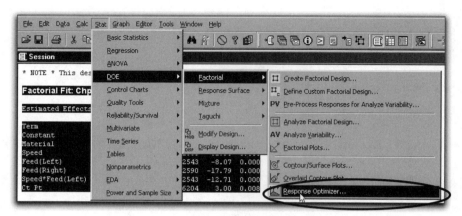

FIGURE APP. 14-20

In Figure App. 14-22 we set up the goal for the response which is a minimum in this case. Since the best chip rating is 1, we will make that the target with an upper bound of 1.5. We'll use the default weights. Actually the weights are not an issue when there is only one response variable. We click **OK** in the setup dialog box and the program thinks for a

FIGURE APP. 14-21

while before producing the output shown in Figure App. 14-23. The value of the response (y = 0.5456) is actually lower than the target (indicating an excellent chip). The beauty of this optimization process is the ability to try

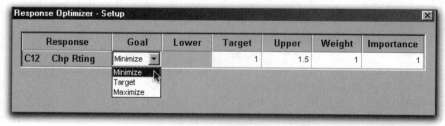

FIGURE APP. 14-22

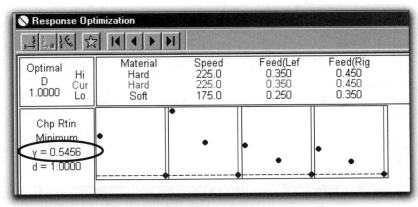

FIGURE APP. 14-23

Response Optimization				
Optimal Hi	Material	Speed	Feed(Lef	Feed(Rig
D Cur	Hard	225.0	0.350	0.450
1.0000 Lo	Hard	225.0	0.350	0.450
	Soft	175.0	0.250	0.350
Chp Rtin				
Minimum				
y = 0.5456				
d = 1.0000				

FIGURE APP. 14-24

different settings by merely sliding the indicator bar with the mouse as shown in Figure App. 14-24. Since material is a discrete factor, we may choose only one level or the other. So, if we wanted to see what the soft material would do, we just move the setting and in Figure App. 14-25 we see that the chip rating has jumped up to 1.9649 which is beyond the lower limit of 1.5 we had established. The conclusion is that the soft material would be making chips that were a bit on the "stringy" side and could have a bad effect on the material recovery process.

The optimizer is a powerful tool that goes beyond the derivation of the model and accomplishing the Objective by finding the functional relationship. The optimizer helps satisfy the Goal through the understanding of the model and is *applied* Statistical Experimental Design in action.

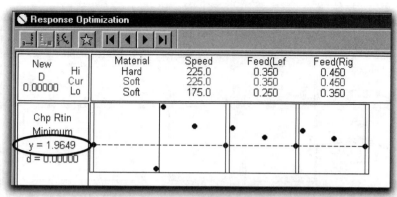

FIGURE APP. 14-25

## 15

Matrix Algebra

The YATES method and the general ANOVA techniques we have studied are both algorithms based on a more general system of analysis called regression analysis or the general linear model (GLM) approach to the solution of modeling problems.

To use the general linear model effectively, some of the rules of matrix algebra need to be understood. This brief chapter will simply show how to manipulate a matrix with no theoretical development of the concepts.

Matrix Defined

A matrix is simply a one-, two-, or n-dimensional rectangular array of numbers ordered in such a way that we can identify any element by a position number. The position number is called a subscript. In a very simple, one-dimensional matrix (commonly called a vector) we refer to the elements of the array as:

$$X_1, X_2, X_3, X_4, \ldots, X_n \qquad (15\text{-}1)$$

So we can call out the value of the third item of this array by finding X_3. This concept is especially useful in computer programming when we wish to build a general approach to the solution of a problem and there is an unknown number of values that will change each time the program is used.

The vector may be written as a row (across as shown above) or as a column (up and down).

A column vector looks like this:

$$X_1 \qquad\qquad\qquad\qquad (15\text{-}2)$$
$$X_2$$
$$X_3$$
$$X_4$$
$$.$$
$$.$$
$$.$$
$$X_n$$

Vectors are usually written in row fashion to conserve space.

The next step in looking at matrices is to look at a two-dimensional array which is usually called a "matrix." Again, the columns are the up-down elements and the rows are the across arrangement of the values. Now we have a double subscript. The first element of the subscript (usually the letter "i") refers to the row of the matrix. The second element of the subscript (usually "j") refers to the column of the matrix. So a matrix would look like this:

$$
\begin{array}{lllll}
X_{11} & X_{12} & X_{13} & X_{14} \cdots x_{1j} & \qquad (15\text{-}3)\\
X_{21} & X_{22} & X_{23} & X_{24} \cdots x_{2j} \\
X_{31} & X_{32} & X_{33} & X_{34} \cdots x_{3j} \\
X_{41} & X_{42} & X_{43} & X_{44} \cdots x_{4j} \\
. & . & . & . \\
. & . & . & . \\
. & . & . & . \\
X_{i1} & X_{i2} & X_{i3} & X_{i4} \cdots x_{ij}
\end{array}
$$

The essential parts of matrix manipulation we need to know to solve the equations involved in the general regression approach to experimental analysis are:

1. Matrix transposition
2. Matrix multiplication
3. Matrix inversion

Transposition

This is a very simple concept. To transpose a matrix, we just turn the matrix on its side. To be more exact, the rows are interchanged with the columns.

Given the matrix in Table 15-1, the transpose is shown in Table 15-2. The symbol for a transpose is $\mathbf{X}'$. This is usually pronounced "$\mathbf{X}$ prime." Notice that the matrix symbol is in boldface.

TABLE 15-1   TABLE 15-2

$$\mathbf{X} = \begin{bmatrix} a & b & c \\ d & e & f \\ g & h & i \\ j & k & l \end{bmatrix} \qquad \mathbf{X}' = \begin{bmatrix} a & d & g & j \\ b & e & h & k \\ c & f & i & l \end{bmatrix}$$

Multiplication

While any matrix may be transposed, there are restrictions on which matrices may be multiplied. Also, we must realize that the order of multiplication is important. The matrix that is written down first is called the premultiplier matrix and the matrix that follows is called the postmultiplier matrix. To perform matrix multiplication, the number of columns in the premultiplier matrix must equal the number of rows in the postmultiplier matrix.

If this does not occur, the matrices are not conformable for multiplication and the operation cannot be done. When specifying the matrix, we always look to its order, which is simply the maximum number of rows and columns. A matrix with 4 rows and 2 columns would be of order 4, 2. A matrix with 2 rows and 3 columns would be of order 2, 3. The two matrices could be multiplied together since they follow the above rule. Table 15-3 shows an example of these two matrices.

TABLE 15-3

$$\mathbf{A} = \begin{bmatrix} a & b \\ c & d \\ e & f \\ g & h \end{bmatrix} \qquad \mathbf{B} = \begin{bmatrix} a & b & c \\ d & e & f \end{bmatrix}$$

a 4 x 2 matrix   a 2 x 3 matrix

To multiply two matrices together, we start with the 1,1 element of the premultiplier matrix and multiply this value by the 1,1 element of the postmultiplier matrix. Go next to the 1,2 element of the premultiplying matrix and multiply this by the 2,1 element of the postmultiplying matrix. Add this result to the previous product, continue to find products across the first row and down the first column of the matrices and sum these to obtain a single value which becomes the 1,1 element of the resulting matrix.

Using the above **A** and **B** matrices with **A** as the premultiplier, we get the following results:

aa + bd, which becomes the 1,1 element of the resulting **C** matrix.

To obtain the 1,2 element of the **C** matrix, we stay in row 1 of the premultiplier but move to column 2 of the postmultiplier. We repeat the multiplication and addition across the row and down the column which produces the 1,2 element of the **C** matrix.

Continuing this example:

$$\mathbf{C}_{12} = ab + be$$

The procedure is repeated using row 1 of the premultiplier until there are no more columns left in the postmultiplier matrix.

For this example the first row of **C** is:

aa + bd   ab + be   ac +bf

$\mathbf{C}_{11}$ $\mathbf{C}_{12}$ $\mathbf{C}_{13}$

Having exhausted row one of the premultiplier, we move, as you might suspect, to row two of the premultiplier and repeat the whole thing over again. This is where computers come in handy since they don't tire of repetitive, boring operations. When all the rows of the premultiplier are "used up," the multiplication is finished. The resultant **C** matrix is found in Expression 15-4. While we have done the example with algebraic symbols, the method is less complicated when we have actual numerical values that can be combined rather than the strings of multiplications and additions that we see in Expression 15-4.

$$
C = \begin{bmatrix}
\begin{matrix} aa + bd \\ C_{11} \end{matrix} & \begin{matrix} ab + be \\ C_{12} \end{matrix} & \begin{matrix} ac + bf \\ C_{13} \end{matrix} \\[1em]
\begin{matrix} ca + dd \\ C_{21} \end{matrix} & \begin{matrix} cb + de \\ C_{22} \end{matrix} & \begin{matrix} cc + df \\ C_{23} \end{matrix} \\[1em]
\begin{matrix} ea + fd \\ C_{31} \end{matrix} & \begin{matrix} eb + fe \\ C_{32} \end{matrix} & \begin{matrix} ec + ff \\ C_{33} \end{matrix} \\[1em]
\begin{matrix} ga + hd \\ C_{41} \end{matrix} & \begin{matrix} gb + he \\ C_{42} \end{matrix} & \begin{matrix} gc + hf \\ C_{43} \end{matrix}
\end{bmatrix}
\qquad (15\text{-}4)
$$

The system used in this procedure is to go across the row and down the column. The subscript of the new row in the resulting matrix is supplied by the row number of the premultiplier matrix. The column number of the resulting matrix is supplied by the column number of the postmultiplier matrix. This is why there must be a match in the order of the matrices being multiplied.

Now let's look at the numerical example of the type we will encounter in the general regression situation. We will start out with a matrix and then transpose this matrix. The transpose will become the premultiplier matrix and the original matrix will be the postmultiplier matrix.

$$
X = \begin{bmatrix}
1 & 4 & 4 & 9 \\
1 & 2 & 4 & 9 \\
1 & 4 & 8 & 9 \\
1 & 2 & 8 & 5 \\
1 & 3 & 6 & 5
\end{bmatrix}
\qquad
X' = \begin{bmatrix}
1 & 1 & 1 & 1 & 1 \\
4 & 2 & 4 & 2 & 3 \\
4 & 4 & 8 & 8 & 6 \\
9 & 9 & 9 & 5 & 5
\end{bmatrix}
$$

$$
\underbrace{\begin{bmatrix}
1 & 1 & 1 & 1 & 1 \\
4 & 2 & 4 & 2 & 3 \\
4 & 4 & 8 & 8 & 6 \\
9 & 9 & 9 & 5 & 5
\end{bmatrix}}_{\substack{\mathbf{X'} \\ \text{Premultiplier} \\ 4 \times 5}}
\times
\underbrace{\begin{bmatrix}
1 & 4 & 4 & 9 \\
1 & 2 & 4 & 9 \\
1 & 4 & 8 & 9 \\
1 & 2 & 8 & 5 \\
1 & 3 & 6 & 5
\end{bmatrix}}_{\substack{\mathbf{X} \\ \text{Postmultiplier} \\ 5 \times 4}}
=
\underbrace{\begin{bmatrix}
5 & 15 & 30 & 37 \\
15 & 49 & 90 & 115 \\
30 & 90 & 196 & 214 \\
37 & 115 & 214 & 293
\end{bmatrix}}_{\substack{\mathbf{X'X} \\ \text{Resulting Matrix} \\ 4 \times 4}}
$$

In the above example we have taken a 4 x 5 matrix and multiplied it by a 5 x 4 to obtain a 4 x 4 resultant matrix.

Matrix "Division"

Matrix division is not done directly but by the inverse of multiplication. For this reason the operation is called *inverting* a matrix. This is one of the most numerically difficult operations in matrix algebra, and there are a number of methods used to accomplish this objective. We will simply indicate the concept behind the inversion of a matrix and supply a computer program in the Appendix to finish the job since no one does this operation by hand any more.

In ordinary arithmetic, we obtain the quotient by dividing the dividend by the divisor. So, if we wished to divide 8 by 2 we put 8 over 2 and obtain the answer 4. We could obtain the same result in a slightly different manner by multiplying the 8 by the reciprocal or *inverse* of 2, which is .5. We can reduce *all* division to multiplication by the inverse of the divisor. This is the idea in matrix algebra and the only way we can accomplish the division step in this system.

The basic rules to accomplish inversion are:
- Only square matrices may be inverted (i.e., rows = columns).
- The matrix may not be ill-conditioned or singular.
- The matrix times its inverse will produce the identity matrix.

When these conditions exist, then the matrix may be inverted. The symbol for an inverted matrix is "-1" in the superscript. So the inverse of $\mathbf{A}$ is symbolized by $\mathbf{A}^{-1}$.

The identity matrix (like any identity element in mathematics) is a matrix that will give back the matrix it premultiplies or postmultiplies. Or more generally, if you multiply or divide by the identity element, you do not change the value of the element being multiplied or divided. Let's look at an example of this concept in action. We will multiply a 4 x 4 matrix $\mathbf{A}$ by the 4 x 4 identity matrix.

The identity matrix as shown in Figure 15-1 is a diagonal matrix with ones on the diagonal and zeros elsewhere. When we premultiply $\mathbf{A}$ by the identity, the resultant is exactly $\mathbf{A}$. Post-multiplication of $\mathbf{A}$ by the identity returns the same result.

The criteria for checking an inverse of a matrix is if (1) the original matrix times its inverse gives the identity matrix, and (2) at the same time the inverse times the original matrix also gives the identity we have found in the inverse. The trick

Identity Matrix   A   Result

$$\begin{bmatrix} 1 & 0 & 0 & 0 \\ 0 & 1 & 0 & 0 \\ 0 & 0 & 1 & 0 \\ 0 & 0 & 0 & 1 \end{bmatrix} \times \begin{bmatrix} 2 & 4 & 6 & 8 \\ 1 & 3 & 5 & 6 \\ 4 & 8 & 6 & 2 \\ 3 & 5 & 1 & 6 \end{bmatrix} = \begin{bmatrix} 2 & 4 & 6 & 8 \\ 1 & 3 & 5 & 6 \\ 4 & 8 & 6 & 2 \\ 3 & 5 & 1 & 6 \end{bmatrix}$$

or

$$\begin{bmatrix} 2 & 4 & 6 & 8 \\ 1 & 3 & 5 & 6 \\ 4 & 8 & 6 & 2 \\ 3 & 5 & 1 & 6 \end{bmatrix} \times \begin{bmatrix} 1 & 0 & 0 & 0 \\ 0 & 1 & 0 & 0 \\ 0 & 0 & 1 & 0 \\ 0 & 0 & 0 & 1 \end{bmatrix} = \begin{bmatrix} 2 & 4 & 6 & 8 \\ 1 & 3 & 5 & 6 \\ 4 & 8 & 6 & 2 \\ 3 & 5 & 1 & 6 \end{bmatrix}$$

FIGURE 15-1

in inverting the matrix is to convert the original matrix into the identity matrix while changing an identity matrix to the inverse. This is a cumbersome task and not easily done by hand. The computer program in Appendix 15 does this quite well, and there are computer routines in other data processing systems that do so readily.

There is one inverse operation that is easily done by hand and is important in the realm of designed experiments. If we have a diagonal matrix to invert (i.e., a matrix that has elements only on its diagonal with the remaining elements zeros) the inversion is simply the inverse (or reciprocal) of the diagonal elements.

Original Matrix **A**   Inverse $\mathbf{A}^{-1}$

$$\begin{bmatrix} 4 & 0 & 0 & 0 \\ 0 & 4 & 0 & 0 \\ 0 & 0 & 4 & 0 \\ 0 & 0 & 0 & 4 \end{bmatrix} \qquad \begin{bmatrix} 1/4 & 0 & 0 & 0 \\ 0 & 1/4 & 0 & 0 \\ 0 & 0 & 1/4 & 0 \\ 0 & 0 & 0 & 1/4 \end{bmatrix}$$

FIGURE 15-2

This introduction to matrix algebra should be sufficient to support the subsequent chapters on the general linear regression approach to statistical analysis problems.

Problems for Chapter 15

1. From the following matrix **A**, identify the following elements:

   a. $a_{2,4}$

   b. $a_{4,2}$

   c. $a_{4,5}$

   d. $a_{1,3}$

   e. $a_{3,3}$

   f. The number 18 is the ——— row, ——— column

   g. The number 10 is the ——— row, ——— column

   h. The number 26 is the ——— row, ——— column

$$
\mathbf{A} = \begin{bmatrix}
2 & 4 & 6 & 8 & 10 \\
12 & 14 & 16 & 18 & 20 \\
22 & 24 & 26 & 28 & 30 \\
32 & 34 & 36 & 38 & 40
\end{bmatrix}
$$

2. Transpose the matrix found in Problem 1. This will be **A'**.

3. Multiply **A'** times **A** (**A** is from Problem 1, **A'** is from Problem 2).

4. Multiply the following matrix and vector.

$$
\begin{bmatrix}
1 & 2 & 3 & 6 & 3 & 7 & 1 \\
2 & 4 & 6 & 9 & 5 & 9 & 4 \\
4 & 6 & 8 & 2 & 7 & 9 & 9
\end{bmatrix}
\begin{bmatrix}
1 \\
4 \\
9 \\
36 \\
9 \\
49 \\
1
\end{bmatrix}
$$

5. Invert the following matrices:

$$
\mathbf{X} = \begin{bmatrix}
2 & -.6 \\
-1 & .4
\end{bmatrix}
\qquad
\mathbf{Y} = \begin{bmatrix}
14 & 3 \\
2 & 10
\end{bmatrix}
\qquad
\mathbf{Z} = \begin{bmatrix}
3 & 0 & 0 \\
0 & 3 & 0 \\
0 & 0 & 3
\end{bmatrix}
$$

APPENDIX 15

Matrix: An ordered rectangular array of numbers.

Subscript: The order number in a matrix, X_{35} is the number in the third row and fifth column.

Column: In a two-dimensional matrix, the values arranged vertically (up and down).

Row: In a two-dimensional matrix, the values arranged horizon tally (across).

Vector A one-dimensional matrix.

Element: The number in a matrix.

Postmultiplier: Second matrix in a left-to-right multiplication.

Premultiplier: First matrix in a left-to-right multiplication.

Identity Matrix: A matrix ($\mathbf{I}$) with diagonal elements of value l and all other elements of value zero. This matrix may premultiply or postmultiply another matrix ($\mathbf{A}$) and the result will always be the matrix $\mathbf{A}$.

Inverse: A matrix derived from another matrix ($\mathbf{A}$) such that the result of multiplying the inverse of $\mathbf{A}$ ($\mathbf{A^{-1}}$) by $\mathbf{A}$ produces the identity matrix $\mathbf{I}$.

$$\mathbf{I = A^{-1}\, A}$$
$$\text{and} \quad \mathbf{I = A\ A^{-1}}$$

Transpose: A matrix turned on its side. The rows are interchanged with the columns. The symbol is a "prime" sign. The transpose of $\mathbf{A}$ is symbolized by $\mathbf{A'}$.

Inversion: To invert a 2 x 2 matrix manually, one way is to find the determinant (D) of the matrix. This is accomplished by finding the cross product of the downward diagonals and subtracting the cross product of the upward diagonals. The diagonals are taken in the usual left-to-right direction. The determinant of $\mathbf{X}$ is then:

$$X = \begin{bmatrix} a & b \\ c & d \end{bmatrix} \qquad D = ad - cb$$

To complete the inversion, we rearrange the original matrix $\mathbf{X}$ by interchanging the positions of the downward diagonals and changing the signs of the upward diagonals. Then we divide the elements of this matrix by the determinant.

$$\mathbf{X}^{-1} = \begin{bmatrix} d/D & -b/D \\ -c/D & a/D \end{bmatrix}$$

A numerical example:

$$\mathbf{X} = \begin{bmatrix} 2 & 3 \\ 5 & 10 \end{bmatrix} \qquad D = 2 \times 10 - 3 \times 5 = 5$$

$$\mathbf{X}^{-1} = \begin{bmatrix} 10/5 & -3/5 \\ -5/5 & 2/5 \end{bmatrix} = \begin{bmatrix} 2 & -.6 \\ -1 & .4 \end{bmatrix}$$

Matrix Inversion Program in BASIC

```
4500 CLS:INPUT "What is the size of your Matrix";I1
4510 DIM A#(I1,I1),AB#(I1,I1),P(I1,I1)
4520 CLS:PRINT "Enter your matrix as prompted ROW WISE"
4540 FOR J=1 TO I1:FOR I=1 TO I1
4560 LOCATE 2,2:PRINT "Row ";J;" Column ";I;:LOCATE 2,20:PRINT "    ":LOCATE 2,
20:INPUT P(J,I)
4810 NEXT I:NEXT J
4980 FOR J=1 TO I1:FOR I=1 TO I1
5000 A#(J,I)=P(I,J)
5020 NEXT I:AB#(J,J)=1:NEXT J
5040 FOR J=1 TO I1:FOR I=J TO I1
5060 IF A#(I,J)<>0 THEN 5120
5080 NEXT I
5100 PRINT "SINGULAR MATRIX":END
5120 FOR K=1 TO I1
5140 S#=A#(J,K):A#(J,K)=A#(I,K):A#(I,K)=S#:S#=AB#(J,K):AB#(J,K)=AB#(I,K):AB#(I,K)=S#
5160 NEXT K
5180 TT#=1/A#(J,J)
5200 FOR K=1 TO I1
5220 A#(J,K)=TT#*A#(J,K)
5230 AB#(J,K)=TT#*AB#(J,K)
5240 NEXT K
5260 FOR L=1 TO I1
5280 IF L=J THEN 5400
5300 TT#=-A#(L,J)
5320 FOR K=1 TO I1
5340 A#(L,K)=A#(L,K)+TT#*A#(J,K)
5360 AB#(L,K)=AB#(L,K)+TT#*AB#(J,K)
5380 NEXT K
5400 NEXT L
5420 NEXT J
5440 CLS:PRINT "Your Inverted Matrix is:"
5500 FOR I=1 TO I1:FOR J=1 TO I1
5520 NN#(I,J)=AB#(I,J):P(I,J)=AB#(I,J)
5525 NEXT J:NEXT I
5526 CC=-11
5527 FOR I=1 TO I1:FOR J=1 TO I1
5530 CC=CC+12
5535 LOCATE 3+I,CC:PRINT USING "####.######"; P(J,I)
5540 NEXT J
5550 CC=-11
5560 NEXT I
```

16
Least Squares Analysis

In the realm of statistical inference used to sort the signal from the noise, there is a hierarchy of methods. We have studied these methods from the bottom up. We started with simple "t" tests and then moved up to the more general ANOVA. Now we just have one last method of analysis to discover. We have seen that "t" tests are restricted to investigations of dichotomies while ANOVA can go beyond the simple comparison of two populations at a time and look at "n" populations with a single risk. The general linear hypothesis based on least squares is at the top of the hierarchy and can do anything that ANOVA or "t" tests can plus a lot more.

Least Squares Developed

In investigating the relationships between variables, we would like to be able to create an equation that shows the dependence of a response on one or more control factors. Let's start with the simple case that involves only a single control factor (x) and a single response (y). The equation we create is a mathematical "short hand" that represents a state of nature we are investigating. If we were interested in learning the degree of dependence between these factors, we could use the concept of covariance introduced in Chapter 4. Recall the definition of the covariance was:

$$s_{xy} = \frac{\Sigma (x_i - \bar{X})(y_i - \bar{Y})}{n - 1} \tag{16-1}$$

The above defining formula can be converted into a formula that is easier for computing purposes. Notice in the defining formula (16-1) that the averages of x and y need to be computed and then subtracted from each value of x_i and y_i. This means that all the values of x_i and y_i must be stored in memory somewhere. Also, if the values of $\overline{X}$ and $\overline{Y}$ are not rounded properly or have repeating decimals that round poorly, the results of the calculations using Equation 16-1 could be in error. Therefore, the following formula, which is mathematically equivalent to 16-1, is a more accurate method to obtain the same results.

$$s_{xy} = \frac{n(\Sigma\, x_i\, y_i) - (\Sigma\, x_i)(\Sigma y_i)}{n(n-1)} \qquad (16\text{-}2)$$

Let's compute some covariances to get a feel for the kinds of numbers produced.

TABLE 16-1

| | **A** | | | | **B** | | |
x	y	x X y		x	y	x X y
1	4	4		10	40	400
2	6	12		20	60	1200
3	9	27		30	90	2700
4	10	40		40	100	4000
Total 10	29	83		100	290	8300

From Example **A** of Table 16-1, we obtain the following covariance:

$$s_{xy} = \frac{4(83) - (10)(29)}{4(4-1)} = 3.5$$

From Example **B** of Table 16-1, we obtain the following covariance:

$$s_{xy} = \frac{4(8300) - (100)(290)}{4(4-1)} = 350$$

Now we were led to believe that the magnitude of the covariance is an indication of the degree of relationship between the variables. We can see that the magnitude of **B** is 10 times that of **A**. However, if we were to plot the data as in Figure 16-1, we can see that there is no difference in the pattern of the relationship between either A or B. The only difference is in the magnitude of the numbers we are working with. **B** has a higher covariance because it starts with bigger numbers! The covariance is an *unanchored* measure of the relationship between the variables. To anchor this relationship, we need to scale the covariance by the product of the standard deviations of x and y. This new measure of the relationship will be called the correlation coefficient (r) and it is anchored between -1 and +1.

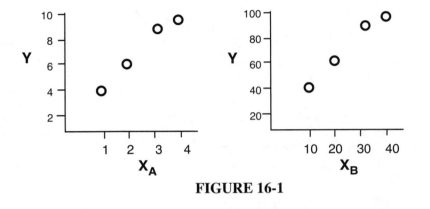

FIGURE 16-1

$$r = \frac{s_{xy}}{s_x \times s_y} \tag{16-3}$$

The correlation for **A** is:

$$r = \frac{3.5}{1.29 \times 2.75} = 0.987$$

The correlation for **B** is:

$$r = \frac{350}{12.9 \times 27.5} = 0.987$$

So we can see that the relationship between the two sets of data is exactly the same. The closer the absolute value of the correlation coefficient approaches 1, the better the linear relationship between the variables. The correlation coefficient is a first intuitive step in finding the relationships between variables. While it shows us the degree of relationship, it does not give us the equation or rate of change of y as a function of x. We could look back to Figure 16-1 and draw a "best eyeball fit" straight line through the data points. This line would follow the trend in the data putting as many points below the line as above as it pivots to follow the relationship which we can see, does not follow a perfectly straight line, but "bumps" along. We shall concentrate only on straight line relationships for the moment and sidestep the question of curving the line to better fit the data. Figure 16-2 shows such an eyeball fit.

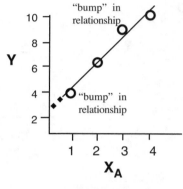

FIGURE 16-2

The change in y over a unit change in x is called the **slope of y on x**. The place where the line intersects the y axis at a value of zero for x is called the **y intercept**. We could obtain these values from the geometry in Figure 16-2.

$$Slope = 2.1$$
$$Intercept = 2.2$$

You probably remember this from the high school geometry course that presented the concept of the straight line relationship and gave a model for the line as:

$$y = mx + b \qquad (16\text{-}4)$$

where y is the value we are solving for
        m is the slope
        x is the known value

So, for our equation, we have: $y = 2.1x + 2.2$ (16-5)

We can "solve" our equation and see how well it predicts the actual values we had. The difference between the actual value (y) and the predicted value $(\hat{y})$ is a measure of how well we have drawn the line. To arrive at this set of values, we will solve the equation for each value of x. Table 16-2 shows these results and compares them with the observed y values to obtain the difference or "residual."

TABLE 16-2

x	y	$\hat{y}$	Residual $(y - \hat{y})$
1	4	4.3	-0.3
2	6	6.4	-0.4
3	9	8.5	0.5
4	10	10.6	-0.6

Notice that the sum of the residuals is equal to -0.8. It looks like we have biased the position of the "eyeball" line since, if we had indeed drawn it so that there were as many points below the line as above, then the algebraic sum of the differences would be zero. In drawing the line by sight we can use a trick that will enable us to only have to worry about one of the parameters of the line (the slope). By pivoting the line through the point where the average of x and y occur, we can adjust the slope to obtain the zero sum of the residuals criteria. The average of the x values is 2.5 and the average of the y values is 7.25. Let's try this out and see what happens.

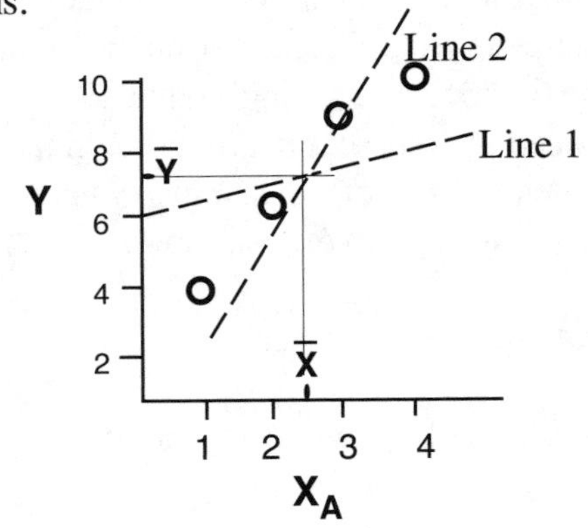

FIGURE 16-3

We have drawn two lines in Figure 16-3. Both pivot through the average of x and y. Both have a zero sum for the residuals! Both are different!

TABLE 16-3

LINE 1				LINE 2			
slope= 0.7	intercept = 5.5			slope= 3.0	intercept = -0.5		
x	y	$\hat{y}$	residual	x	y	$\hat{y}$	residual
1	4	6.2	-2.2	1	4	2.5	- 1.5
2	6	6.9	-0.9	2	4	5.5	- 0.5
3	9	7.6	1.4	3	9	8.5	0.5
4	10	8.3	1.7	4	10	11.5	1.5
			0.0				0.0

The problem we have encountered is that there are many lines that can be drawn using the criterion of pivoting through the average of x and y to produce a zero algebraic sum of the residuals. We do not have a **unique** line described by this process. There are other thoughts on the criteria for drawing a line through the data; however, they also lead to ambiguity and *many* possible lines rather than the unique solution we want.

The criterion generally used to fit a **unique** line to a set of correlated data (meaning that the x and y come in pairs) is called the method of least squares. In this method, we will fit a line that gives us the smallest sum of the squares of the residuals, while still pivoting through the average of x,y . Now, we could obtain the values for the intercept (called b_0) and the slope (called b_1) by trial and error application of least squares, but this would be a very tedious task.

Instead, we shall use a bit of calculus to find the minimum. By taking the first derivative of the function and setting it equal to zero we can solve for the minimum. All we need to do is set up a function in terms of the known values and complete this exercise. Let's investigate what we know about our function and apply this knowledge to our example.

The basic form of equation is: $y = mx + b$

Using the more general terms of b_0 (intercept) and b_1 (slope), this function becomes:

$$\hat{y} = b_0 + b_1 X_1 + \text{error}$$

We have added an error term since we do not expect a perfect straight line relationship between x and y. (Actually, we have not added an error, it always existed; we now simply recognize its existence!)

The error is the difference between the actual y value and the predicted $\hat{y}$ (symbolized by $\hat{y}$ and pronounced "y hat"). This difference is what we had called the "residual." We can build a table of what we know and what is unknown and then apply the calculus to compute the unknowns.

TABLE 16-4

x	y	$\hat{y} = b_0 + b_1 x_1$	Residual $(y - \hat{y})$
1	4	$b_0 + b_1 \times 1$	$4 - (b_0 + b_1 \times 1)$
2	6	$b_0 + b_1 \times 2$	$6 - (b_0 + b_1 \times 2)$
3	9	$b_0 + b_1 \times 3$	$9 - (b_0 + b_1 \times 3)$
4	10	$b_0 + b_1 \times 4$	$10 - (b_0 + b_1 \times 4)$

Since we want to find a unique line that will fit the data best we will square the residual error and then take the partial derivatives with respect to the b_0 term and then with respect to the b_1 term. Since we want the sum of these derivatives to be a minimum, we will sum the terms and then having two equations in two unknowns, we can solve for b_0 and b_1. Table 16-5 shows these steps.

TABLE 16-5

residual $(y - \hat{y})$ squared	partial derivative with respect to b_0	partial derivative with respect to b_1
$(4 - (b_0 + b_1 \times 1))^2$	$-2(4 - b_0 - b_1 \times 1)$	$-2(1)(4 - b_0 - b_1 \times 1)$
$(6 - (b_0 + b_1 \times 2))^2$	$-2(6 - b_0 - b_1 \times 2)$	$-2(2)(6 - b_0 - b_1 \times 2)$
$(9 - (b_0 + b_1 \times 3))^2$	$-2(9 - b_0 - b_1 \times 3)$	$-2(3)(9 - b_0 - b_1 \times 3)$
$(10 - (b_0 + b_1 \times 4))^2$	$-2(10 - b_0 - b_1 \times 4)$	$-2(4)(10 - b_0 - b_1 \times 4)$
SUM:	$-58 + 8b_0 + 20b_1$	$-166 + 20b_0 + 60b_1$

Now take these sums of the derivatives and set each equal to zero:

$$-58 + 8b_0 + 20b_1 = 0 \qquad (16\text{-}6)$$
$$-166 + 20b_0 + 60b_1 = 0 \qquad (16\text{-}7)$$

Rearrange these equations:
$$8b_0 + 20b_1 = 58$$
$$20b_0 + 60b_1 = 166$$

We have a set of simultaneous linear equations in two unknowns from which we may find the solution by elimination of one of the terms. We will eliminate b_0 by multiplying the second equation by 8/20,

$$8b_0 + 20b_1 = 58$$
$$\frac{8}{20}(20b_0 + 60b_1 = 166)$$

which produces:

$$8b_0 + 20b_1 = 58 \qquad\qquad (16\text{-}8)$$
$$8b_0 + 24b_1 = 66.4 \qquad\qquad (16\text{-}9)$$

subtract 16-8 from 16-9 to give:
$$4b_1 = 8.4$$

$$b_1 = \frac{8.4}{4} = 2.1$$

To solve for b_0 we simply substitute the value of b_1 into either of the original expressions created from the sum of the derivatives. Substituting into 16-8, we obtain:

$$8b_0 + 20\,(2.1) = 58$$
$$8b_0 = 58 - (20\,(2.1))$$

$$b_0 = \frac{16}{8} = 2.0$$

We may now write the least squares equation that described the four pieces of data as follows:

$$Y = 2 + 2.1X_1 \qquad\qquad (16\text{-}10)$$

The numerical procedure we have just completed obtains the correct least squares equation, but it is not the most convenient method of obtaining the results. Let's look into the numbers we have just calculated to see if there is a generalization that we can make and somehow set up a simpler formula to do the job on a routine basis.

If we look back to Table 16-5 and put the derivatives into an algebraic form, we obtain the following more general tabulation of the information.

TABLE 16-6

Residual Squared	Partial Derivative with Respect to b_0	Partial Derivative with Respect to b_1
$(y-(b_0+b_1x))^2$	$-2(y-b_0-b_1x)$	$-2(x)(y-b_0-b_1x)$
	.	.
	.	.
	.	.
SUM:	$2(-\Sigma y + \Sigma b_0 + \Sigma b_1 x)$	$2(-\Sigma xy + \Sigma x b_0 + \Sigma b_1 x^2)$

The multiplier of 2 may be discarded since it is common to both summations and we may now set these sums equal to zero and rearrange.

$$\Sigma b_0 + \Sigma b_1 x = \Sigma y \qquad (16\text{-}11)$$

$$\Sigma x b_0 + \Sigma b_1 x^2 = \Sigma xy \qquad (16\text{-}12)$$

We can use a simple mathematical concept to help solve these simultaneous equations. The sum of the b_0 is the same as all the b_0's added together, and since b_0 is a constant, we may just multiply by the number (n) of b_0's that we have. So $\Sigma b_0 = nb_0$. Using this concept we can rewrite 16-11 as follows:

$$nb_0 + \Sigma b_1 x = \Sigma y \qquad (16\text{-}13)$$

Now we will eliminate b_0 from the equation by multiplying 16-12 by $n/\Sigma x$ as shown in 16-14a. This multiplication results in 16-14b. (These operations are exactly the procedures we used on the numerical data.)

$$\frac{n}{\Sigma x}(\Sigma x b_0 + \Sigma b_1 x^2 = \Sigma xy) \qquad (16\text{-}14a)$$

which produces:

$$nb_0 + \frac{n}{\Sigma x}\Sigma b_1 x^2 = \frac{n}{\Sigma x}\Sigma xy \qquad (16\text{-}14b)$$

Now subtract Expression 16-13 from 16-14b to give 16-15:

$$\frac{n}{\Sigma x}\Sigma b_1 x^2 - \Sigma b_1 x = \frac{n}{\Sigma x}\Sigma xy - \Sigma y \qquad (16\text{-}15)$$

We can factor out a b_1 from the left-hand part of the expression. This produces:

$$b_1 \left(\frac{n}{\Sigma x} \Sigma x^2 - \Sigma x \right) = \frac{n}{\Sigma x} \Sigma xy - \Sigma y \qquad (16\text{-}16)$$

Expression 16-16 is pretty messy and does not resemble any familiar statistical formula. Let's see if we can clean it up and make it look like some old "friends." To do so, we will multiply through by $\Sigma x/n$ to obtain:

$$b_1 \left(\Sigma x^2 - \frac{(\Sigma x)^2}{n} \right) = \Sigma xy - \frac{\Sigma x \Sigma y}{n} \qquad (16\text{-}17)$$

and to be even more clever, we now multiply through by n to get:

$$b_1 \left(n \Sigma x^2 - (\Sigma x)^2 \right) = n \Sigma xy - \Sigma x \Sigma y \qquad (16\text{-}18)$$

Solving for b_1 we obtain the general formula for the slope of x on y:

$$b_1 = \frac{n \Sigma xy - \Sigma x \Sigma y}{n \Sigma x^2 - (\Sigma x)^2} \qquad (16\text{-}19)$$

Now the numerator in Expression 16-19 is the same as the numerator in Expression 16-2, which is the formula for the covariance! The denominator in Expression 16-19 is the same as the numerator in the formula for the variance of x. So, we can see that the solution to the general linear equation is simply the covariance divided by the variance of x. To find b_0, which is really a "centering" factor for average levels of x and y, we need only go to the basic math model and solve for b_0 at $\bar{X}$ and $\bar{Y}$.

$$\bar{Y} = b_0 + b_1 \bar{X}$$

and $\qquad\qquad b_0 = \bar{Y} - b_1 \bar{X} \qquad\qquad\qquad (16\text{-}20)$

Another way of looking at the b_0 point is to think of how the regression line pivots through the $\bar{X}$ and $\bar{Y}$ points. The degree of "pivoting" is the slope (b_1). If there is a slope of zero then there is no relationship between x and y and the b_0 term is the same as $\bar{Y}$. As the slope becomes larger and larger, the b term gets farther and farther away from the $\bar{Y}$. We should also recognize that because of the "center of gravity" effect of the averages, the b_0 term is also the point on the y axis where x has a value of zero.

Using the Formula

Let's go through our numerical example and use the formulas we have just derived. Table 16-7 shows the data and the required columns for the calculation.

TABLE 16-7

x	y	x^2	y^2	xy
1	4	1	16	4
2	6	4	36	12
3	9	9	81	27
4	10	16	100	40

$\bar{X} = 2.5$ $\bar{Y} = 7.25$ $\Sigma x = 10$ $\Sigma y = 29$ $\Sigma x^2 = 30$ $\Sigma xy = 83$ $n = 4$

Substituting the values into Expression 16-19:

$$b_1 = \frac{4(83) - (10)(29)}{4(30) + (10)^2} = \frac{42}{20} = 2.1$$

And by using 16-20 we obtain:

$$b_0 = 7.25 - 2.5(2.1) = 7.25 - 5.25 = 2.0$$

This is the same result that we obtained using the long, drawn-out procedure a few pages back.

The Role of the Matrix

Equations 16-19 and 16-20 form what are called the "normal equations." The term "normal" has nothing to do with the normal distribution function but means the "usual" equations for the solution by least squares.

Up to now we have worked with a single x variable, and the normal equations are not too complicated. However, what happens when there are more variables under study than just an x_1? What if there are two x's or even more? This is a very realistic question, since we need to study more than one factor at a time in our experimental designs. In these cases, the normal equations become larger and larger, and there are more pieces to them. So, if we had two x variables, there would be three equations in three unknowns. If there were 4 x's, there would be 5 equations in 5 unknowns. Note that there is always an extra equation to account for the b_0 or centering factor in our equations. While it is possible to solve such complex algebraic expressions, it is easier to have a more general solution technique for higher order problems involving multiple x factors. This branch of least squares analysis is called multiple regression and is handled systematically by matrix algebra.

Let's look into the normal equations and see what is there. Then we will generalize the quantities into a matrix form that will solve these problems by the application of a set of simple rules.

Since we know what is in the simple example (from Table 16-2) that we solved using algebra, we will continue to use these same numbers in our investigation of the matrix approach.

Let's rewrite the normal equations again and see if we can view them in terms of a matrix.

$$nb_0 + \Sigma x b_1 = \Sigma y$$
$$\Sigma x b_0 + \Sigma x^2 b_1 = \Sigma xy$$

In the left side of the equations we can see a common set of terms in the b's. Just as in algebra, we can isolate these terms by dividing through both sides of the equation. Since we will be working with matrix algebra we cannot really "divide," but can get the same effect by multiplying both sides by the inverse of all the parts that do not include the b's. This gives us the following:

$$\begin{bmatrix} n & \Sigma x \\ \Sigma x & \Sigma x^2 \end{bmatrix}^{-1} \begin{bmatrix} n & \Sigma x \\ \Sigma x & \Sigma x^2 \end{bmatrix} \begin{bmatrix} b_0 \\ b_1 \end{bmatrix} = \begin{bmatrix} n & \Sigma x \\ \Sigma x & \Sigma x^2 \end{bmatrix}^{-1} \begin{bmatrix} \Sigma y \\ \Sigma xy \end{bmatrix} \quad (16\text{-}21)$$

Since the inverse times itself gives the identity matrix, the left side of the equation reduces to the vector of b's, which of course is what we are looking for. Now the question is how to generalize the right-hand side of the equation. Here we have to be clever and a bit insightful into the form of the equation we are seeking. We need a set of b's that include a b_1 term, which is the rate of change in y with x; and we also need a b_0 term, which is a centering or averaging factor. The b_1 term comes from the x factors we have. But from where can we generate the b_0 term?

The Dummy Factor

It is possible to introduce a term into our equation by using what has been called a "dummy factor." In the case of the b_0 term, the dummy factor is simply a column of 1's. We can write the x data from our example as follows with the dummy 1's in the first position in a 4 x 2 matrix. We will also write this set of data in transposed form.

$$\mathbf{X} = \begin{bmatrix} 1 & 1 \\ 1 & 2 \\ 1 & 3 \\ 1 & 4 \end{bmatrix} \quad \mathbf{X'} = \begin{bmatrix} 1 & 1 & 1 & 1 \\ 1 & 2 & 3 & 4 \end{bmatrix}$$

If we multiply $\mathbf{X}$ by $\mathbf{X'}$ as follows, we get exactly what we are looking for.

$$\mathbf{X'X} = \begin{bmatrix} 4 & 10 \\ 10 & 30 \end{bmatrix} \quad \text{Which is:} \quad \begin{bmatrix} n & \Sigma x \\ \Sigma x & \Sigma x^2 \end{bmatrix} \quad (16\text{-}22)$$

And further, we can get the other vector of values by multiplying $\mathbf{X'}$ times the $\mathbf{Y}$ response vector as follows:

$$\mathbf{X'Y} = \begin{bmatrix} 1 & 1 & 1 & 1 \\ 1 & 2 & 3 & 4 \end{bmatrix} \begin{bmatrix} 4 \\ 6 \\ 9 \\ 10 \end{bmatrix} = \begin{bmatrix} 29 \\ 83 \end{bmatrix}$$

$$\text{Which is:} \quad \begin{bmatrix} \Sigma y \\ \Sigma xy \end{bmatrix} \quad (16\text{-}23)$$

Expressions 16-22 and 16-23 are exactly the parts of Expression 16-21 that we were seeking and lead us to the general formula in matrix notation terms for the vector of coefficients, which is the inverse of the X - transpose times the X, all times the X-transpose times the Y.

$$\boxed{B = (X'X)^{-1} (X'Y)} \qquad (16\text{-}24)$$

Expression 16-24 is one of the most useful formulas in statistical analysis since it expresses in the most general way the relationship between variables and forms the basis of the calculations needed for other approaches to statistical inference.

To find the inverse of the $X'X$ matrix (for this simple 2 x 2 case), we find the determinant of the matrix by taking the product of the downward diagonals and then subtract the product of the upward diagonals from this. The diagonals are in the usual left-to-right direction. For the general expression we have:

$$\begin{bmatrix} n & \Sigma x \\ \Sigma x & \Sigma x^2 \end{bmatrix} = n\Sigma x^2 - (\Sigma x)^2$$

Next we use the determinant to divide the original matrix after the downward diagonals are interchanged and the signs of the upward diagonals are changed. This gives:

$$\begin{bmatrix} \dfrac{\Sigma x^2}{n\Sigma x^2 - (\Sigma x)^2} & \dfrac{-\Sigma x}{n\Sigma x^2 - (\Sigma x)^2} \\ \dfrac{-\Sigma x}{n\Sigma x^2 - (\Sigma x)^2} & \dfrac{n}{n\Sigma x^2 - (\Sigma x)^2} \end{bmatrix} \qquad (16\text{-}25)$$

While the above expression is the $(X'X)^{-1}$, it is better numerical practice to save the division step for the last operation to avoid rounding error. Therefore, we will do the multiplication of the $X'Y$ by the $X'X$ before dividing the $X'X$ by the determinant.

$$\dfrac{1}{n\Sigma x^2 - (\Sigma x)^2} \underbrace{\begin{bmatrix} \Sigma x^2 & -\Sigma x \\ -\Sigma x & n \end{bmatrix}}_{X'X} \underbrace{\begin{bmatrix} \Sigma y \\ \Sigma xy \end{bmatrix}}_{X'Y} = \begin{bmatrix} \dfrac{\Sigma x^2 \Sigma y - \Sigma x \Sigma xy}{n\Sigma x^2 - (\Sigma x)^2} \\ \dfrac{n\Sigma xy - \Sigma x \Sigma y}{n\Sigma x^2 - (\Sigma x)^2} \end{bmatrix}$$

(ready to invert)

$$\begin{bmatrix} b_0 \\ b_1 \end{bmatrix} = \begin{bmatrix} \dfrac{\Sigma x^2 \Sigma y - \Sigma x \Sigma xy}{n\Sigma x^2 - (\Sigma x)^2} \\ \dfrac{n\Sigma xy - \Sigma x \Sigma y}{n\Sigma x^2 - (\Sigma x)^2} \end{bmatrix} \qquad (16\text{-}26)$$

The term for b_1 is exactly the same as we obtained from the algebraic approach to the solution of the least squares fit. If we substitute the values from our numerical example into the matrix form of the equation, we obtain the same values as before, 2.1 for the slope (b_1) and 2.0 for the intercept (b_0). The advantage of the least squares solution by matrix algebra is of course the expendability to more than one x variable.

Regression, ANOVA, and YATES

Since the least squares regression approach to statistical inference is the most general method, it is important to see how the ANOVA and particularly the YATES algorithm for ANOVA are related to this technique. We will take a simple 2^2 factorial design and analyze it using the general least squares approach and at the same time see how the YATES algorithm is indeed a simplification of the more general linear model approach.

Recall that the design matrix for a 2-level factorial can be written in its entirety (including interactions) as an ordered set of "-1's" and "+1's." We have just learned that we need to include a dummy variable (a column of +1's) in order to account for the intercept (or b_0) term in our equation. Therefore, the matrix that represents the X matrix for a 2^2 design is as follows:

$$\mathbf{X} = \begin{bmatrix} 1 & -1 & -1 & +1 \\ 1 & +1 & -1 & -1 \\ 1 & -1 & +1 & -1 \\ 1 & +1 & +1 & +1 \end{bmatrix} \qquad (16\text{-}27)$$

We will continue this numerical example by providing a set of responses which are arranged as a vector and are found in Expression 16-28. We will complete the ANOVA calculations by using the YATES method first. This analysis is found in Table 16-8.

$$Y = \begin{bmatrix} 2 \\ 4 \\ 2 \\ 2 \end{bmatrix} \qquad\qquad (16\text{-}28)$$

TABLE 16-8

Observation I	II	Half Effect	Sum of Sq.	Measures
2	6	10	2.5	–	Average
4	4	2	0.5	1.0	Factor A
2	2	-2	-.5	1.0	Factor B
2	0	-2	-.5	1.0	AB Interaction

Now we will use the general least squares approach to the solution of same numerical the example as found in Table 16-9.

TABLE 16-9

$$\mathbf{X'} \qquad\qquad \mathbf{X} \qquad\qquad \mathbf{X'X}$$

$$\begin{bmatrix} 1 & 1 & 1 & 1 \\ -1 & +1 & -1 & +1 \\ -1 & -1 & +1 & +1 \\ +1 & -1 & -1 & +1 \end{bmatrix} \begin{bmatrix} 1 & -1 & -1 & +1 \\ 1 & +1 & -1 & -1 \\ 1 & -1 & +1 & -1 \\ 1 & +1 & +1 & +1 \end{bmatrix} \begin{bmatrix} 4 & 0 & 0 & 0 \\ 0 & 4 & 0 & 0 \\ 0 & 0 & 4 & 0 \\ 0 & 0 & 0 & 4 \end{bmatrix}$$

$$\mathbf{X'} \qquad\qquad \mathbf{Y} \qquad\qquad \mathbf{X'Y}$$

$$\begin{bmatrix} 1 & 1 & 1 & 1 \\ -1 & +1 & -1 & +1 \\ -1 & -1 & +1 & +1 \\ +1 & -1 & -1 & +1 \end{bmatrix} \begin{bmatrix} 2 \\ 4 \\ 2 \\ 2 \end{bmatrix} \begin{bmatrix} 10 \\ 2 \\ -2 \\ -2 \end{bmatrix}$$

Notice that the final column (column II) in the YATES is exactly the same as the **X'Y** result. Also, notice that the **X'X** matrix is a diagonal matrix (only diagonal elements) with the diagonal elements equal to the number of

observations in the design. Now we begin to see the reasons why the
YATES method and the ANOVA method can work in the general linear
model. The fact that the design is **orthogonal** gives us the diagonal $\mathbf{X'X}$
matrix. The diagonal matrix is the easiest of all matrices to invert! To do
so, we simply take the inverse of each of the diagonal elements. The
inverse of the diagonal matrix in our example is then:

$$(\mathbf{X'X})^{-1} = \begin{bmatrix} 1/4 & 0 & 0 & 0 \\ 0 & 1/4 & 0 & 0 \\ 0 & 0 & 1/4 & 0 \\ 0 & 0 & 0 & 1/4 \end{bmatrix} \qquad (16\text{-}29)$$

We complete the solution of the least squares method by multiplying the $\mathbf{X'Y}$
matrix by the $(\mathbf{X'X})^{-1}$ matrix.

$$\begin{bmatrix} 1/4 & 0 & 0 & 0 \\ 0 & 1/4 & 0 & 0 \\ 0 & 0 & 1/4 & 0 \\ 0 & 0 & 0 & 1/4 \end{bmatrix} \begin{bmatrix} 10 \\ 2 \\ -2 \\ -2 \end{bmatrix} \begin{bmatrix} 2.5 \\ 0.5 \\ -.5 \\ -.5 \end{bmatrix} \qquad (16\text{-}30)$$

This is exactly the same as the half-effects from the YATES ANOVA! But
what about the sums of squares? It is possible to obtain the sum of squares
in matrix methods by the following formula:

$$\text{Sum of Squares} = \mathbf{B'} \, (\mathbf{X'Y}) \qquad (16\text{-}31)$$

For our example:

$$\begin{bmatrix} 2.5 & .5 & -.5 & -.5 \end{bmatrix} \begin{bmatrix} 10 \\ 2 \\ -2 \\ -2 \end{bmatrix} = \begin{bmatrix} 2.5 \\ 1.0 \\ 1.0 \\ 1.0 \end{bmatrix}$$

Note that we have not added the results of this vector multiplication to a scalar
value but have left the results as individual items that represent the sums of
squares of the three effects in this analysis.

Again, this is exactly the same as we obtained from the YATES ANOVA in Table 16-8. So, the YATES method takes advantage of the orthogonality of the design (which produces the diagonal $\mathbf{X'X}$ matrix) to simplify the computations. The general ANOVA method follows the same line of thinking to create the sums of squares used to draw inferences in the face of the variability of the data.

    As we shall see in the next chapter, we can make use of this common link between the least squares method and the ANOVA techniques to gain a better insight into the nature of the information in the experiments we perform.

Problems for Chapter 16

1. Using the algebraic formulas, find the slope and intercept of y on x for the following data:

x	y
5	6
8	3
3	8
5	7
6	5
9	2
10	1

2. Compute the correlation coefficient for Problem 1.

3. Use the matrix approach to find the slope and intercept for the data found in Problem 1.

4. Use the matrix approach to find the vector of coefficients for the following multiple regression situation.

X_1	X_2	X_3	y
−1	−1	−1	5
+1	−1	−1	6
−1	+1	−1	9
+1	+1	−1	6
−1	−1	+1	10
+1	−1	+1	12
−1	+1	+1	18
+1	+1	+1	12

APPENDIX 16

Covariance: $\quad s_{xy} = \dfrac{n(\Sigma\, x_i\, y_i) - (\Sigma\, x_i)(\Sigma y_i)}{n(n-1)}$

Correlation
coefficient, r: Measures the degree of linear dependence between two
variables. It is anchored between -1 and +1 with the highest
dependence as r approaches an absolute value of 1.
An r of zero indicates no dependence.

$$r = \frac{s_{xy}}{s_x \; s_y}$$

where: s_{xy} is the covariance.
s_x standard deviation of x.
s_y standard deviation of y

General Linear Model: $\hat{y} = b_0 + b_1\, x_1$

where: $\hat{y}$ (y hat) is the estimate of the value being
predicted.
b_0 is the value of y at x = 0 or the y intercept.
b_1 is the change in y due to x or the slope of the
line describing the relationship between x
and y.
x_1 is the particular value of x we are solving
for.

Slope b_1: $\quad b_1 = \dfrac{n\Sigma xy - \Sigma x\, \Sigma y}{n\Sigma x^2 - (\Sigma x)^2}$

Intercept: $\quad b_0 = \bar{Y} - b_1\, \bar{X}$
where $\bar{Y}$ and $\bar{X}$ are the averages of y and x.

Matrix Formula: $\mathbf{B} = (\mathbf{X'X})^{-1} (\mathbf{X' Y})$

                    Where: $\mathbf{X}$ is the matrix of control variables
                                 $\mathbf{Y}$ is the vector of responses
                                 $\mathbf{B}$ is the vector of coefficients

                    $\Sigma\text{sq} = \mathbf{B' X' Y}$

                    where: $\mathbf{X'}$ is the transposed matrix of control variables
                                 $\mathbf{Y}$ is the vector of responses
                                 $\mathbf{B'}$ is the transposed vector of coefficients

                    Note: To obtain the individual sum of squares for each effect,
                                 simply do not add the individual items in the above
                                 vector multiplication.

17
Putting ANOVA and
Least Squares to Work

Now that we have developed the ANOVA and least squares approach to statistical analysis and we can see the similarities in these two methods, let's find out how we can put these tools to work on an experimental design and analysis problem.

The following problem involving three factors and one response will serve as an example of the steps in the analysis of the central composite design. It will also serve as a general example of the approach to be taken in any statistical analysis of an experimental design. Table 17-1 shows the central composite design for the three factors of temperature, concentration, and speed.

The Goal is to maximize tensile strength. The Objective is to test the hypothesis that tensile strength is a function of temperature, concentration and speed.

The first step in the analysis will be to run a YATES ANOVA on the factorial portion of the design. This will identify the linear main effects and any of the interactions that are significant. Notice that the "zero" point has been replicated to produce an estimate of experimental error. We shall see later that this replication is important for a second reason. Table 17-2 is the computer generated YATES ANOVA for this problem.

The mean square of the replicates is found by computing the sum of squares for the four data points and dividing by the three degrees of freedom.

$$\Sigma sq = \Sigma x^2 - \frac{(\Sigma x)^2}{n}$$

$$\Sigma sq = (446^2 + 445^2 + 446^2 + 444^2) - \frac{(1781)^2}{4} = 2.75$$

$$\text{Mean Square (variance)} = \frac{2.75}{3} = 0.917$$

TABLE 17-1

| | | | | (Response) |
| | A | B | C | Tensile |
tc	Temperature	Concentration	Speed	Strength
(1)	120	30	6	230
a	180	30	6	384
b	120	45	6	520
ab	180	45	6	670
c	120	30	9	180
ac	180	30	9	330
bc	120	45	9	465
abc	180	45	9	612
$-\alpha_a$	100	37.5	7.5	260
$+\alpha_a$	200	37.5	7.5	507
$-\alpha_b$	150	25	7.5	206
$+\alpha_b$	150	50	7.5	689
$-\alpha_c$	150	37.5	5	495
$+\alpha_c$	150	37.5	10	400
zero	150	37.5	7.5	446,445,446,444

TABLE 17-2

No.	Observation	Sum of Squares	Half Effect	Measures	Index
1	230		423.875	Average	1
2	384	45150.1	75.125	A (Temp.)	2
3	520	163306	142.875	B (Conc.)	3
4	670	6.125	-.875	AB	4
5	180	5886.13	-27.125	C (Speed)	5
6	330	6.125	-.875	AC	6
7	465	10.125	-1.125	BC	7
8	612	.125	.125	ABC	8

The critical F value for 1 and 3 degrees of freedom is 10.128 at an alpha risk of .05. Using the replicates at the zero point to estimate the residual, we find statistically significant effects for the main effects of temperature, concentration and speed. There is also an interaction effect just significant (F = 11.04) for the BC interaction ,which is the concentration x speed factor. The remaining effects are not significant. If we did not have a residual estimate, we could have looked at the *magnitude* of the half effects to judge if there was enough of a change in the response to warrant further work or investigation. In the above case, we would probably have picked the three single effects, but would not have judged the interactions as making important contributions.

Using the Half Effects

We have shown that the half effects are the same quantity as the regression coefficients. We can use them to illustrate the trends in the experiment by plotting the change in the response as a function of the variables under study. The first half effect (Index 1 from Table 17-2) measures the average and is located at the "conceptual" zero point in the experiment. We call this "conceptual" since we don't really have a middle experimental point in a 2^{k-p} design. In these designs, we only have a -1 point and a +1 point. Since the half effect is the change in the response over a unit change in the control variable, the half effect measures the change from the center (or zero point) of the experimental space to either the -1 position or the +1 position. This is how the half effect gets its name since it is a measure of the change in the response over half the range of experimental space. Figure 17-1 illustrates this concept for the particular set of data we have been working with. The "x" axis has been laid out in the design unit space from -1 to +1 with zero indicated (but remember we don't have a real data point for the response from this part of the experiment). The "y" axis has the range of the response and the average is highlighted.

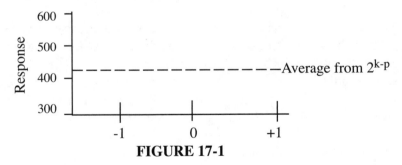

FIGURE 17-1

We can use Figure 17-1 as a model to plot the three main effects. The responses we plot will represent the average value for the factor under study at the low and high points. In our experiment, we could take the average of the four low values and the average of the four high values and plot them for each factor. We'll do this as an example for factor C, the speed. Afterwards, we'll show how this same task as was done here can be done using the half effects. To find the average low level of factor C, add the first four responses from Table 17-1 and then divide by 4. The average of the high level of factor C comes from adding the next four responses from Table 17-1 and dividing the result by 4.

$$\text{Low C} = \frac{230 + 384 + 520 + 670}{4} = 451$$

$$\text{High C} = \frac{180 + 330 + 465 + 612}{4} = 396.75$$

Now let's plot the average level of factor C at its low and high input points.

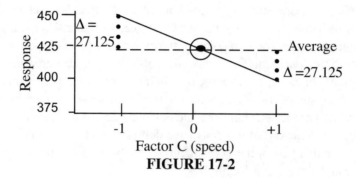

FIGURE 17-2

Notice that the difference from the average for both points is 27.125. This is exactly the half effect reported in the YATES ANOVA of Table 17-2. Therefore, we can obtain the average points for the single effect plots by simply adding the half effect and subtracting the half effect from the average value of the 2^{k-p} design. We will do this for Factors A and B, which we plot on one axis in Figure 17-3 along with Factor C.

Points for Factor A (Temperature)
low:	423.875 - 72.125	= 348.75
high:	423.875 + 72.125	= 499.00

Points for Factor B (Concentration)
low:	423.875 - 142.875	= 281.00
high:	423.875 + 142.875	= 566.75

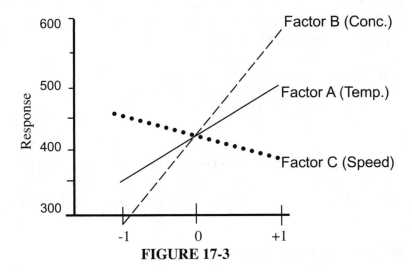

FIGURE 17-3

From Figure 17-3 we can see the relative effect of the change in the response as a function of the three single factors. In obtaining the points for this plot, we have used the half effects and actually solved a regression equation. The equation we have is a result of our YATES analysis and consists of the significant effects. The full equation is:

Tensile Strength = 423.875 + 72.125 x Temp. (A) + 142.875 x Conc. (B)
$\qquad$ -27.125 x Speed (C) - 1.125 x Conc. x Speed (BC)　(17-1)

To obtain the coefficients in Expression 17-1, we simply went to the YATES analysis and picked out the half effects that had significant F values. Now, the coefficients for temperature, concentration, and speed are in terms of the design units (-1, +1) and not in terms of the physical variables. Therefore, when we solve this or any other equation from a YATES analysis, we must put in the correct parameters. The correct parameters for solution are the coded levels of the design units which are -1 for the low level and +1 for the high level. When we obtained the points for Figure 17-2 we solved Equation (17-1) using the -1 when we subtracted and the +1 when we added the half effect to the average! We also ignored the rest of the equation in solving for just one main effect at a time. Actually by "ignoring" the rest of the equation, we have considered the other factors at their zero level, which essentially eliminates them from the computation.

To illustrate this point let's write the full equation down and solve for the low level of Factor A (Temp.). To do this we will put in the low level of A (which is a -1) and enter zero for all the other effects.

$$
\begin{aligned}
\text{Tensile Strength} &= 423.875 + 75.125 \times (-1) + 142.875 \times (0) \qquad (17\text{-}2)\\
&\quad - 27.125 \times (0) - 1.125 \times (0) \times (0)\\
&= 423.875 - 72.125\\
&= 348.75
\end{aligned}
$$

So, the rule to subtract to find the low level of a single effect and add to find the high level of the effect has its basis in the solution of the equation formed from the YATES analysis. This concept can now be extended into the solution for the points in the interaction.

Plotting Interactions

Factorial designs were specifically engineered to detect interactions. The analysis of variance will detect significant interactions, but the next step is to plot the functional form of the interaction to gain insight into the nature of the effect. We can obtain the coordinates of the interaction plot by solving the design unit equation from YATES. In our example, there was a significant BC interaction. To find the points to plot, we solve Expression 17-1 for all the terms involved with this interaction. Since this is a two-factor interaction with two levels per factor there will be 4 points to plot. We need to solve for the following conditions:

B	C
Low	Low
High	Low
Low	High
High	High

Since Factor A is not involved in the interaction, Expression 17-1 is reduced to:

$$
\begin{aligned}
\text{Tensile Strength} &= 423.875 + 142.875 \times \text{Conc.(B)} - 27.125 \times \text{Speed(C)}\\
&\quad -1.125 \times B \times C \qquad (17\text{-}3)
\end{aligned}
$$

Again, we have eliminated the factor not under study (in this case Factor A) by solving the equation using the center point of this factor, which is zero. Now we solve the above equation for the design unit levels for Factors B and C.

The interaction term (B x C) is found by finding the cross-product of the B and C single factor levels. Table 17-3 shows the work with plottable results.

Table 17-3

Levels			Substituted Into Equation	Result
B	C	B x C		
-1	-1	+1	423.875 + 142.875 x (-1) -27.125 x (-1) -1.125 x (+1) =	307.0
+1	-1	-1	423.875 + 142.875 x (+1) -27.125 x (-1) -1.125 x (-1) =	595.0
-1	+1	-1	423.875 + 142.875 x (-1) -27.125 x (+1) -1.125 x (-1) =	255.0
+1	+1	+1	423.875 + 142.875 x (+1) -27.125 x (+1) -1.125 x (+1) =	538.5

We can now construct the interaction table that will allow us to plot the interaction.

TABLE 17-4

S		Concentration(B)	
p		30 ppm (-)	45 ppm (+)
e	6 ips (-)	307	595
e			
d	9 ips (+)	255	538.5
(C)			

We now take the points from Table 17-4 and plot them. Since it is possible to form this plot with either of the factors on the "x" axis while the others are held at an isolated level (iso-level), we will plot the results both ways. Sometimes the interaction will "talk to us" in a more meaningful manner if plotted both ways.

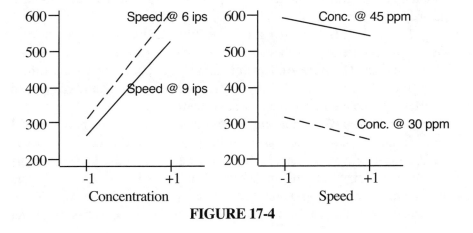

FIGURE 17-4

The amount of interaction is slight if we use the above plots to make this judgment. However, remember that the effect was significant in a statistical sense. That is, the lack of parallelism between the two iso-level lines is enough that when compared with the noise of the experiment we can say that there **is** non-parallelism. We may, however, judge the effect from a practical point and disregard the impact of the interaction based on our knowledge of the process. The lack of parallelism between these two iso-level lines is less than 10% (8.7%) with respect to the speed plot and only 1.6% with respect to the concentration plot. This percent lack of parallelism may be found by taking the ratios of the iso-slopes with the larger slope placed in the numerator. This ratio gives a relative, anchored feel for the magnitude of the impact of the interaction.

Plotting Curves

So far, the results of our analysis have indicated that there are significant effects for all three single effects and a single two-factor interaction. We still have the "bottom half" of the central composite design to draw upon for more information. The information from this part of the experiment is used to guide our thinking to the correct polynomial model. We will augment out linear plots with the alpha position values to see if a straight line continues or if there is evidence of a curve. If there is a curvature in the data, then there will be a difference between the zero value in the experiment (the one we actually ran a treatment combination on) and the "conceptual" zero that is calculated in the YATES analysis. Let's see if this is the case in our example and then determine why this takes place.

We will plot factors A, B, and C using the -1 and +1 points from the YATES analysis, which will give us the same points we plotted in Figure 17-3. However, in Figure 17-5 we will add the actual zero point and the two alpha points for each factor.

Figure 17-5 has some strange twists and turns and would lead us to believe that there are cubic effects for the factors under study. What has actually happened is the following. The average effect from the factorial portion of the experiment (found in the first line of the YATES output) is *lower* than the actual average value of the four zero points from the composite portion of the experiment. The average effect from the YATES is a mathematical result while the zero point is an actual observation. Why is the zero point bigger than the average? The answer lies in the functional form of the factors in **this** experiment. In another experiment, the zero point could be smaller than the average. In this experiment, two of the factors show a strong positive slope while the third shows a slight negative slope. There is a tendency with this data for the response to be higher as the factors' levels get bigger. Since the composite portion of the experiment is concentrated around the zero point, the positive effect of the two factors (A and B) is exerted to give a physical increase in the response for the composite treatments. Unfortunately, this does induce some confusion in the graphical analysis of the data in Figure 17-5. This confusion can be removed by adjusting the two parts of the experiment to a common zero point. While this adjustment could be achieved by either modifying the factorial portion or the composite portion, it is usually more convenient to change the composite values. While some people may argue that this does not reflect reality, remember we are only plotting the data to get a *qualitative* picture of the functional forms involved. The equation will be generated using the **actual** data with multiple regression.

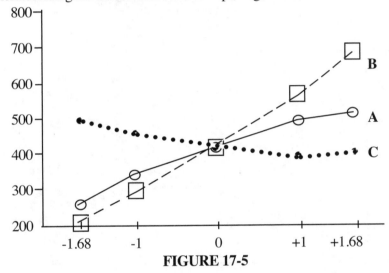

FIGURE 17-5

Adjusting the composite part of the experiment is convenient since we can still use the half effects to find the low and high plot points as described before.

Now let's re-plot Figure 17-5 and use the adjusted points from the composite portion. To find these adjusted points, we first find the difference between the YATES average and the zero point average of the composite.

$$\text{Delta} = \text{average (YATES)} - \text{zero (composite)} \qquad (17\text{-}4)$$
$$-21.375 = 423.875 - 445.25$$

To complete the adjustment, we add the delta (be careful of signs) to each point in the composite portion of the design as shown in Table 17-5.

TABLE 17-5

	Actual	Adjusted
tc	Response	Response
$-\alpha_a$	260	238.625
$+\alpha_a$	507	485.625
$-\alpha_b$	206	184.625
$+\alpha_b$	689	667.625
$-\alpha_c$	495	473.625
$+\alpha_c$	400	378.625
zero	445.25	423.875

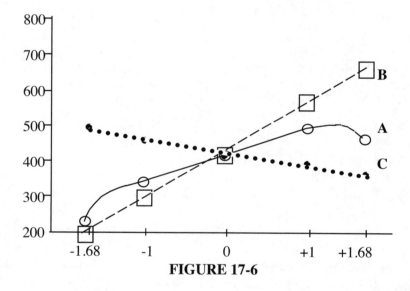

FIGURE 17-6

We have removed the "wrinkles" from our first plot and now see the functional forms of the polynomials for each of the factors. Factor A, the temperature, is quadratic in nature while B, the concentration, and C, the speed, are both linear. We now propose our model based on the analysis completed. There are three linear single effects, a quadratic (square) for one factor and a single two-factor interaction. The model is:

$$\text{Tensile Strength} = b_0 + b_1 \times \text{Temp.} + b_2 \times \text{Conc.} + b_3 \times \text{Speed} + \quad (17\text{-}5)$$
$$b_{11} \times \text{Temp.}^2 + b_{23} \times \text{Conc.} \times \text{Speed}$$

Using the Computer

The last step in our analysis is to process the data with a computer program for analysis. In Minitab, we need to set up the experiment before we can proceed to the analysis. Figure 17-7 shows the usual navigation from the dropdown to the **Create Response Surface Design**.

FIGURE 17-7

In the dialog box shown in Figure 17-8 we scroll down to select the number of factors in this Central Composite Design (CCD). Since there were three factors in the example found in Table 17-1, we select 3. Now go to **Designs** which brings up a sub dialog box found in Figure 17-9.

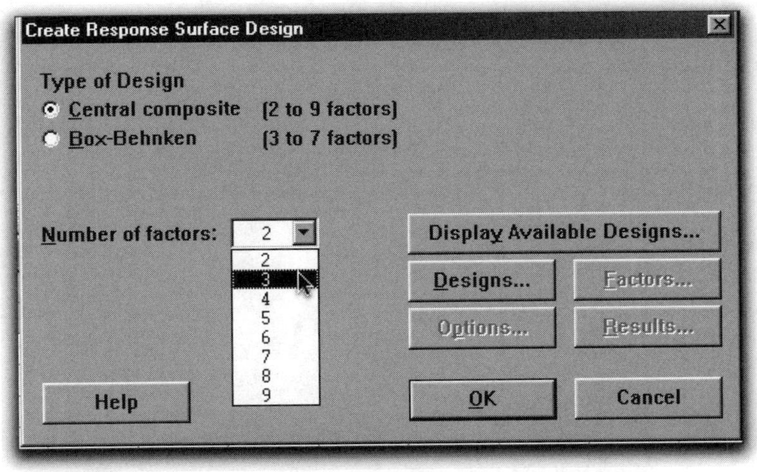

FIGURE 17-8

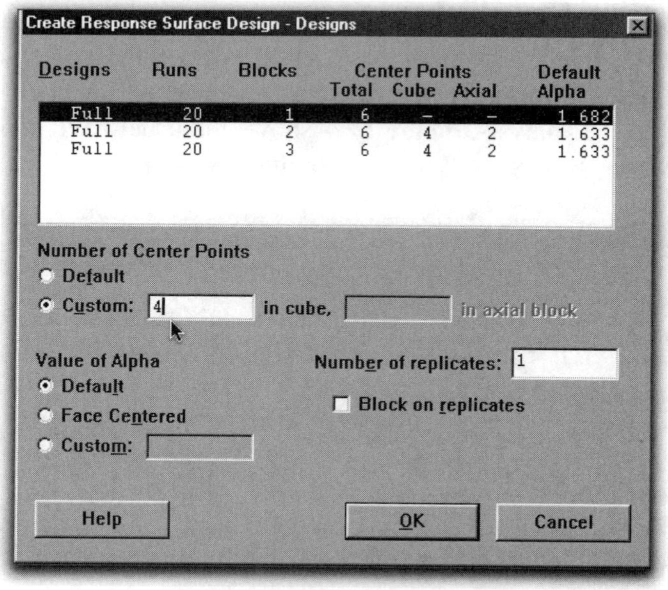

FIGURE 17-9

In the Designs sub dialog box select the **Full** with only one block. However, to match the design in Table 17-1, we will customize the number of replicated center points to 4 instead of the default of 6. We will use the default alpha of 1.682. Click **OK** and then in the main dialog box (Figure 17-8) click on **Factors**. If we enter the extremes of the working range, we need to click

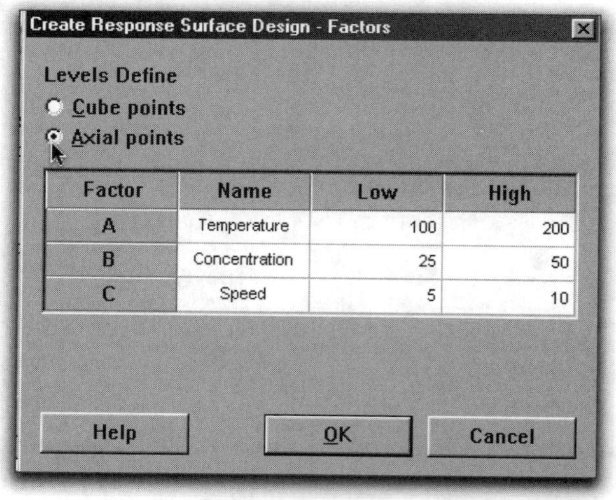

FIGURE 17-10

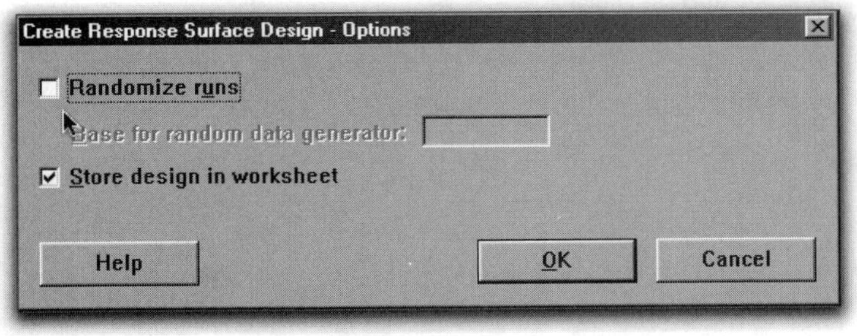

FIGURE 17-11

axial points in Figure 17-10. This is an easier option than computing the levels for the factorial portion of the design (Minitab calls these the Cube points"). After entering the factor names and the extreme levels, click **OK** and then from the main dialog box, click **Options**. To allow the viewing of the design, we will unclick the **Randomize Runs** option which after we return to the main dialog box and click **OK** we obtain the design shown in Figure 17-12. Although, while the levels in this design are exactly accurate for this CCD,

↓	C1	C2	C3	C4	C5	C6	C7
	StdOrder	RunOrder	PtType	Blocks	Temperature	Concentration	Speed
1	1	1	1	1	120.269	30.0674	6.0135
2	2	2	1	1	179.729	30.0674	6.0135
3	3	3	1	1	120.269	44.9324	6.0135
4	4	4	1	1	179.729	44.9324	6.0135
5	5	5	1	1	120.269	30.0674	8.9865
6	6	6	1	1	179.729	30.0674	8.9865
7	7	7	1	1	120.269	44.9324	8.9865
8	8	8	1	1	179.729	44.9324	8.9865
9	9	9	-1	1	099.999	37.4999	7.5000
10	10	10	-1	1	199.999	37.4999	7.5000
11	11	11	-1	1	149.999	24.9999	7.5000
12	12	12	-1	1	149.999	49.9999	7.5000
13	13	13	-1	1	149.999	37.4999	5.0000
14	14	14	-1	1	149.999	37.4999	09.9999
15	15	15	0	1	149.999	37.4999	7.5000
16	16	16	0	1	149.999	37.4999	7.5000
17	17	17	0	1	149.999	37.4999	7.5000
18	18	18	0	1	149.999	37.4999	7.5000

FIGURE 17-12

↓	C1	C2	C3	C4	C5	C6	C7	C8
	StdOrder	RunOrder	PtType	Blocks	Temperature	Concentration	Speed	Tensile Strength
1	1	1	1	1	120	30.0	6.0	230
2	2	2	1	1	180	30.0	6.0	384
3	3	3	1	1	120	45.0	6.0	520
4	4	4	1	1	180	45.0	6.0	670
5	5	5	1	1	120	30.0	9.0	180
6	6	6	1	1	180	30.0	9.0	330
7	7	7	1	1	120	45.0	9.0	465
8	8	8	1	1	180	45.0	9.0	612
9	9	9	-1	1	100	37.5	7.5	260
10	10	10	-1	1	200	37.5	7.5	507
11	11	11	-1	1	150	25.0	7.5	206
12	12	12	-1	1	150	50.0	7.5	689
13	13	13	-1	1	150	37.5	5.0	495
14	14	14	-1	1	150	37.5	10.0	400
15	15	15	0	1	150	37.5	7.5	446
16	16	16	0	1	150	37.5	7.5	445
17	17	17	0	1	150	37.5	7.5	446
18	18	18	0	1	150	37.5	7.5	444
19								

FIGURE 17-13

they are not practical. It is necessary to manually change the levels to reasonable values that would not be scoffed at by the operators who will execute this experiment. This is shown in Figure 17-13.

Obtaining the Data & Analysis

The data values are obtained by running the experiment. This is the same data found in Table 17-1 which is entered in C9, the Tensile Strength column. All of the analysis done by Minitab uses the method of least squares. It is possible (but tedious) to use the multiple regression routine under the **Regression** dropdown, but with a designed experiment, it is so much easier to simply follow the **DOE=> Response Surface=> Analyze Response Surface Design** dropdown.

If we were to utilize the Regression approach in Minitab, it would be necessary to construct the entire **X** Matrix with all the interaction and quadratic effects present, in the columns of the worksheet. When we use the DOE analysis, Minitab creates all the necessary columns of the **X** Matrix internally. This is a great time saver and also keeps the worksheet clutter free since most of the effects we would entertain as candidates are often not part of the final model. These effects would need columns in the worksheet and when not needed would be ignored or erased. The logistics of keeping track

FIGURE 17-14

of these effects can of-
ten lead to mistakes in
the analysis. So, we
follow the dropdown as
shown in Figure 17-14,
taking advantage of
Minitab's internal cre-
ation of the necessary
columns in the **X** Ma-
trix. The main dialog
box found in Figure 17-
15 asks for the response.

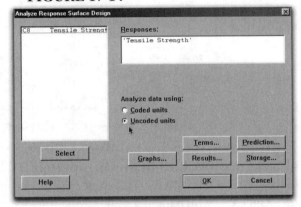

FIGURE 17-15

Minitab is able to analyze multiple responses. We will see this feature
in the *Case History* chapter in the *Derivation of Empirical Equations from
Statistically Designed Experiments* section of this book. For now, we will
select the single response, Tensile Strength. This may be done by highlighting the response and clicking on **Select** or simply by double clicking on the
response itself to move it into the response box.

Next click on the **Terms** button, which brings up the sub dialog box
shown in Figure 17-16. This is a shuttle system that allows the terms (effects)
to be moved from the left (not included in model side) or to the right (included
in model side). Selecting and clicking the arrow, or simply double clicking,
will move the term from one side to the other. For our initial model, we will allow
all the terms to enter. It is also possible to construct a "wholesale" model by

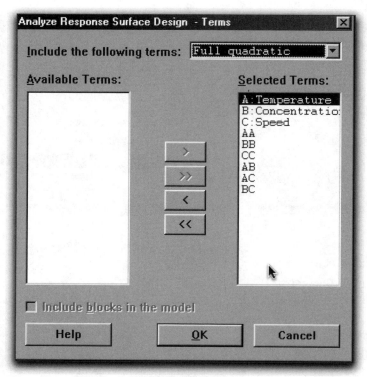

FIGURE 17-16

scrolling down the **Include the following terms** selector at the top of the sub dialog box. By default the **Full quadratic** is chosen by Minitab for this full factorial CCD which includes all the quadratic (squared) and all the interactions. Note how Minitab designates the quadratic terms as the factor's letter designation multiplied by itself (**AA** is the Temperature squared term).

To avoid missing any significant terms, it is advisable to include all the possible terms on the first iteration of the analysis as we are about to do. Click **OK** from the **Terms** sub dialog box and then click **OK** in the main dialog box. We will not look at residuals at this point, so the **Options** sub dialog box may be ignored for the moment.

The session window displays the coefficients and the corresponding t and p values. This is the best place to begin looking for the model. Since we have included all the terms in this model, there are no df for residual. However, since there were 4 replicates of the center points, there is some information on error that is used to test for the statistical significance of the effects. The three single effects are all significant with t values in excess of

TABLE 17-6

Estimated Regression Coefficients for Tensile Strength

Term	Coef	SE Coef	T	P
Constant	-1108.29	39.1032	-28.343	0.000
Temperature	10.25	0.2360	43.433	0.000
Concentration	19.64	0.9439	20.810	0.000
Speed	-16.05	4.7195	-3.400	0.009
Temperature*Temperature	-0.02	0.0006	-42.221	0.000
Concentration*Concentration	0.01	0.0094	1.203	0.263
Speed*Speed	0.28	0.2358	1.203	0.263
Temperature*Concentration	-0.00	0.0029	-1.328	0.221
Temperature*Speed	-0.02	0.0146	-1.328	0.221
Concentration*Speed	-0.10	0.0586	-1.707	0.126

S = 1.864   R-Sq = 100.0%   R-Sq(adj) = 100.0%

the critical t (0.05/2 for 1, 3 df). Only one interaction (the Concentration-Speed) has a possible influence, but its t value is only 1.707. We will not drop this interaction, but give it a chance to remain in the final model. Dropping too many terms too early can lead to a lack of fit. A rule of thumb is to allow effects with at a t value of least 1.5 remain for the next cut. We will bring the analysis dialog box up again and position it so that the session window is visible when we bring up the **Terms** sub dialog box. (Minitab allows the moving of the front-most window). This positioning allows the viewing of the model we are reducing along with the dialog box where this process takes place. This is shown in Figure 17-17 where the insignificant terms are shuttled to the left.

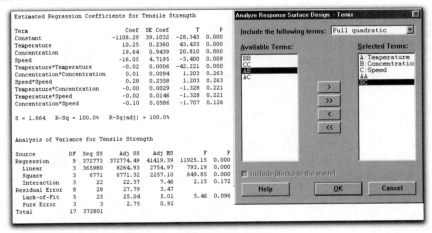

FIGURE 17-17

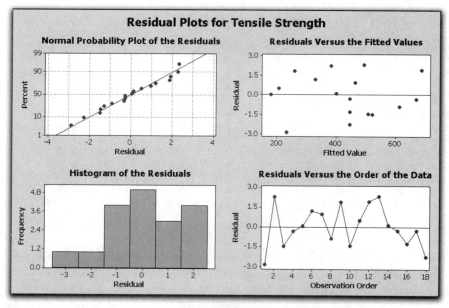

FIGURE 17-18

TABLE 17-7

Response Surface Regression: Tensile Stre versus Temperature, Concentratio, ...

The analysis was done using uncoded units.

Estimated Regression Coefficients for Tensile Strength

Term	Coef	SE Coef	T	P
Constant	-1100.67	22.7400	-48.403	0.000
Temperature	10.03	0.1850	54.204	0.000
Concentration	19.91	0.4790	41.568	0.000
Speed	-14.71	2.3949	-6.142	0.000
Temperature*Temperature	-0.03	0.0006	-40.943	0.000
Concentration*Speed	-0.10	0.0631	-1.584	0.139

S = 2.009   R-Sq = 100.0%   R-Sq(adj) = 100.0%

Analysis of Variance for Tensile Strength

Source	DF	Seq SS	Adj SS	Adj MS	F	P
Regression	5	372753	372753	74550.7	18478.42	0.000
Linear	3	365980	212235	70745.3	17535.19	0.000
Square	1	6763	6763	6763.0	1676.28	0.000
Interaction	1	10	10	10.1	2.51	0.139
Residual Error	12	48	48	4.0		
Lack-of-Fit	9	46	46	5.1	5.53	0.093
Pure Error	3	3	3	0.9		
Total	17	372801				

In the reduced model analysis where we leave out the quadratic terms for concentration and speed as well as the interactions between temperature and concentration and temperature and speed, we will look at the residuals plot as shown in Figure 17-18. These residuals have a reasonable adherence to the requirements and therefore we may interpret the analysis further. The single effects are all significant as well as the quadratic effect for the temperature. The interaction between concentration and speed is marginal, but we will keep it in the equation in deference to the "by hand" work we had done earlier in this chapter. The ANOVA table (in Table 17-7) is shown only to indicate that there is no significant lack of fit ($p = 0.093$). Remember the ANOVA table combines all the effects and is confusing because of this.

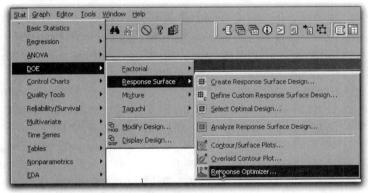

FIGURE 17-19

Now it is time to try the model out by using it to find the optimal settings of the three factors to produce a tensile strength that is maximized (recall the Goal, p. 421). The response optimization routine in Minitab takes much of the tedium out of this effort. To begin, follow the dropdown as shown in Figure 17-19 to **Response Optimizer** which brings up the main dialog box as shown

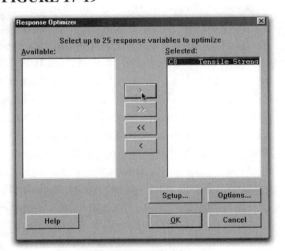

FIGURE 17-20

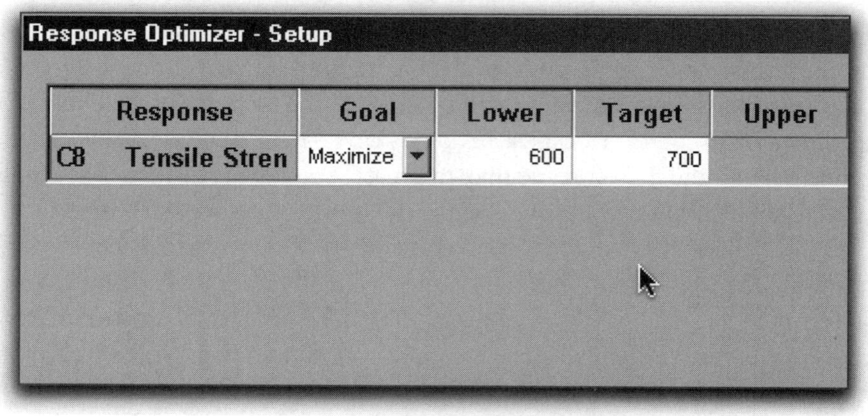

FIGURE 17-21

in Figure 17-20. All we need to do here is shuttle the response to the right. If we had multiple responses, we would select those we wanted to optimize simultaneously and shuttle them all to the right.

It is necessary to set up the optimization. Click **Setup** from the main dialog box. In the setup, it is possible to minimize, set a target, or maximize the response. Our goal is to maximize, so we choose this from the first selection box. We need to set a lower value and a target. If we maximize, there is no upper choice and this box is "grayed out." If there were multiple responses, we would choose the weights and importance for each response to help in the multiple response optimization. Since we have only one response in this problem, we can skip the weighting.

Once we click **OK**, the process begins and depending on the complexity of the optimization and the speed of the microprocessor in your computer, the computing effort may take a few seconds.

Figure 17-22 shows the optimization for the maximum, which turns out to be 796.9. This is higher than any of the observations in the data and nearly 100 units of tensile strength above the target we asked for in the Setup. Of course, this is a welcome outcome and reflects the fact that the design structure we have utilized while including a full factorial 2-level is really a fraction of the 125 runs that would have been necessary if we had done the full 5-level factorial experiment instead of the CCD.

The optimizer function in Minitab allows real time changes to be made as shown in Figure 17-23 where we ask, what if the speed were increased to 10? As we see, the tensile strength dropped by almost 100 units. This type of "What If?" questioning is an important part of the practical aspects of using

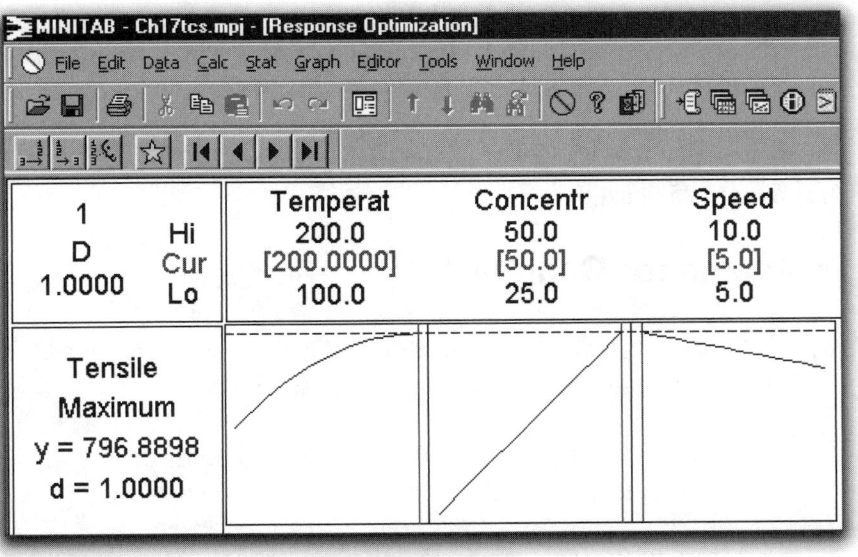

FIGURE 17-22

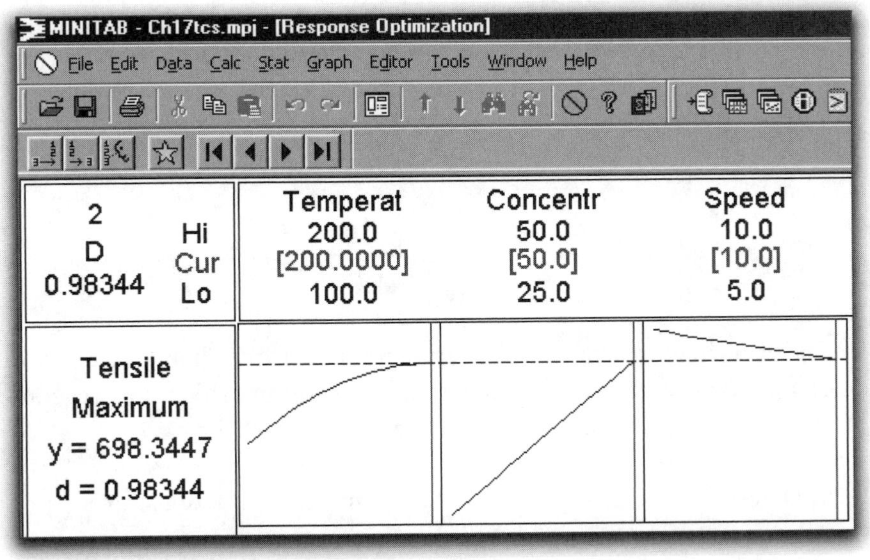

FIGURE 17-23

our experimental work, since it allows us to see if (as in this case a faster speed would mean more throughput) we can still accomplish the goal with other considerations such as cost in mind. The optimizer helps us visualize and make trade-offs in a very interactive manner.

Besides the Minitab computer program used to obtain the equation, there are many other commercially available software packages that will produce the same results. SAS, SPSS, JMP, and BMDP are all in common use on large mainframe computers and PC's. It is no longer necessary to labor over the calculations that were taught as an ordinary part of a design of experiments class a mere 25 years ago.

Problems for Chapter 17

1. A YATES analysis was performed on an experiment and produced the following half effects. Plot the effects as follows:

a) Single Effect A	c) Single Effect C
b) Interaction AC	d) Interaction AB

Half effect	Measure
200	Average
-10	A
20	B
5	AB
15	C
7	AC
0	BC
0	ABC

2. The experiment on the wheel-bearing problem from Chapters 5 and 6 was completed as expanded in Chapter 6. The resulting chip rating responses are found in the table on the next page.

a) Determine the factors that follow a curvilinear relationship.

b) Propose a complete model.

c) Find the equation using a computer program (such as Minitab).

d) Determine the optimum values to obtain a chip rating of 1 or less.

↓	C7	C8	C9	C10	C11	C12	C13	C14	C15	C16
	Feed(Left)	Feed(Right)	HtCyTime	Punch/Die	HtPwr	StdOrder_1	RunOrder_1	Blocks_1	PtType	ChipRating
1	0.25	0.35	5.5	6	75	1	1	1	1	3.816
2	0.25	0.35	8.5	6	85	2	2	1	1	2.303
3	0.25	0.35	8.5	12	75	3	3	1	1	3.829
4	0.25	0.35	5.5	12	85	4	4	1	1	2.268
5	0.35	0.35	8.5	12	85	5	5	1	1	6.375
6	0.35	0.35	5.5	12	75	6	6	1	1	4.872
7	0.35	0.35	5.5	6	85	7	7	1	1	4.815
8	0.35	0.35	8.5	6	75	8	8	1	1	3.290
9	0.25	0.45	5.5	12	85	9	9	1	1	3.782
10	0.25	0.45	8.5	12	75	10	10	1	1	2.313
11	0.25	0.45	8.5	6	85	11	11	1	1	3.792
12	0.25	0.45	5.5	6	75	12	12	1	1	2.306
13	0.35	0.45	8.5	6	75	13	13	1	1	6.418
14	0.35	0.45	5.5	6	85	14	14	1	1	4.894
15	0.35	0.45	5.5	12	75	15	15	1	1	4.797
16	0.35	0.45	8.5	12	85	16	16	1	1	3.314
17	0.30	0.40	7.0	9	80	17	17	1	1	4.371
18	0.30	0.40	7.0	9	80	18	18	1	1	2.885
19	0.30	0.40	7.0	9	80	19	19	1	1	4.403
20	0.30	0.40	7.0	9	80	20	20	1	1	2.914
21	0.30	0.40	7.0	9	80	21	21	1	1	3.683
22	0.30	0.40	7.0	9	80	22	22	1	1	2.156
23	0.20	0.40	7.0	9	80	23	23	1	1	1.101
24	0.40	0.40	7.0	9	80	24	24	1	1	4.571
25	0.30	0.30	7.0	9	80	25	25	1	1	4.898
26	0.30	0.50	7.0	9	80	26	26	1	1	4.882
27	0.30	0.40	4.0	9	80	27	27	1	1	2.879
28	0.30	0.40	10.0	9	80	28	28	1	1	2.897
29	0.30	0.40	7.0	5	80	29	29	1	1	2.843
30	0.30	0.40	7.0	15	80	30	30	1	1	2.892
31	0.30	0.40	7.0	9	50	31	31	1	1	2.859
32	0.30	0.40	7.0	9	100	32	32	1	1	2.876
33	0.30	0.40	7.0	9	80	33	33	1	1	2.863
34	0.30	0.40	7.0	9	80	34	34	1	1	2.868

3. Use the half effect results from the factorial design of Problem 1 and augment the following composite values. Plot each single effect and propose a model for regression analysis.

	Response
-αa	260
+αa	230
-αb	180
+αb	260
-αc	220
+αc	220
Zero	220

APPENDIX 17-A

Outline of the steps in the analysis of a central composite design:

1. Run YATES analysis on the factorial portion of the design.
2. Plot the results from the YATES analysis.
3. Modify the composite portion by adding the delta to the composite values of the response.

Delta = Average (YATES) - Zero (Composite)

4. Plot the results from the entire experiment. Do so by adding the alpha position values of the responses to the plot of the YATES results.
5. Propose a model for regression analysis.
6. Prepare the data for regression.
7. Run the regression.

Plotting results from half effects.
•For single effects, find the low level by subtracting the factor's half effect from the average. For the high level, add the factor's half effect to the average.

•For two-factor interactions, set up the equation in design units for the factors in the interaction. Solve the equation for the four points to be plotted.

APPENDIX 17-B
Calculating Regression Coefficients from Half Effects

In Table 17-2, the YATES ANOVA gives us the half effects. Now, remember that these values are regression coefficients, but they are in terms of design units. Back in Chapter 4 we showed how the design units were related to physical units. Recall that the design units of a 2-level factorial design are derived in general by the following expression:

$$\text{Design Unit} = \frac{X_{level} - \overline{X}_{of\ levels}}{\dfrac{X_{high} - X_{low}}{2}}$$

For the example in Chapter 17, then, the temperature factor would fit into this equation as follows:

$$\text{low design unit} \quad = \quad \frac{120 - 150}{30} \quad = -1$$

$$\text{high design unit} \quad = \quad \frac{180 - 150}{30} \quad = +1$$

The above relationships can be used to *decode* the design unit half effects into physical unit regression coefficients. You should notice that the regression coefficients in Table 17-6 do not match the half effects in Table 17-2. That's because the coefficients in 17-6 are in physical units and the half effects are in design units (i.e., -1 and +1). To make the transition from the design units to physical units, we first write the equation down in terms of the half effects from Table 17-2. We will use only the significant terms to keep the math to a minimum.

The equation (in design units) is:
$$Y = 423.875 + 72.125 \times \text{Temp.} + 142.875 \times \text{Conc.} - 27.125 \times \text{Speed}$$
$$-1.125 \times \text{Conc.} \times \text{Speed}$$

Now we substitute the design unit expressions into each term.

$$Y = 423.875 + 75.125 \times ((T - 150)/30) + 142.875 \times ((C - 37.5)/7.5)$$
$$- 27.125 \times ((S - 7.5)/1.5) - 1.125 \times ((C - 37.5)/7.5) \times ((S - 7.5)/1.5)$$

$$Y = 423.875 + \frac{75.125\ T - 11268.75}{30} + \frac{142.875\ C - 5357.8125}{7.5}$$
$$\frac{- 27.125\ S + 203.4375}{1.5} \quad - 1.125\left(\frac{(C - 37.5) \times (S - 7.5)}{11.25}\right)$$

$$Y = 423.875 + 2.504\ T - 375.625 + 19.05\ C - 714.375 - 18.08\ S + 135.62$$

$$\frac{- 1.125\ \ SC - 37.5\ S - 7.5\ C + 281.25}{11.25}$$

$$Y = 423.875 + 2.504T - 375.625 + 19.05C - 714.375 - 18.08S + 135.62$$
$$- .10SC + 3.75S + .75C - 28.125$$

Gather up terms:

$$Y = 558.63 + 2.504 \times T + 19.8 \times C - 14.33 \times S - 0.1 \times S \times C$$

This equation is still different from the regression equation found in Table 17-7. The main source of difference is the lack of the quadratic term for temperature afforded by the use of the data from the composite portion of the experiment. The b_0 term is also different, but this term is always fluctuating since it is a centering factor. This shows the value of the full CCD experiment as opposed to only a 2-level experiment. The technique we have just demonstrated shows that it is possible to do a complete regression without the use of a computer! The reason for this is, of course, that the design we began with is orthogonal. This orthogonality reduces the rather tough (matrix inversion) part of the regression computation to a simple algorithm.

18
ANOVA for Blocked and Nested Designs

In Chapter 8 we introduced the concept of blocking in factorial designs with the purpose of removing an unwanted source of variation. We will now show how such a source of variation will inflate the noise (or error) in an experiment unless it is treated properly. Table 18-1 is a reproduction of the random assignment of acres to the 16 treatment combinations as first found in Chapter 8. We have now added the response, plant growth in inches, and will do the analysis.

TABLE 18-1

RUN	TC	NITRO%	PHOSP%	POTSH%	TYPE PLANT	ACRE	HEIGHT
12	(1)	10	5	5	Corn	3	5.5
2	a	20	5	5	Corn	1	15.0
13	b	10	10	5	Corn	4	9.0
8	ab	20	10	5	Corn	2	12.0
14	c	10	5	10	Corn	4	7.5
4	ac	20	5	10	Corn	1	14.0
1	bc	10	10	10	Corn	1	9.5
9	abc	20	10	10	Corn	3	10.0
11	d	10	5	5	Tomato	3	3.5
3	ad	20	5	5	Tomato	1	13.0
10	bd	10	10	5	Tomato	3	4.0
15	abd	20	10	5	Tomato	4	12.0
16	cd	10	5	10	Tomato	4	5.5
6	acd	20	5	10	Tomato	2	8.5
5	bcd	10	10	10	Tomato	2	4.0
7	abcd	20	10	10	Tomato	2	9.0

The YATES analysis of variance is shown in Table 18-2. Since there were no replicates, we used the higher order interactions for estimate of the residual mean square. This was done by adding the sums of squares for ABCD, BCD, ACD, ABD, and ABC and then dividing this sum by their combined 5 degrees of freedom.

The results show that nitrogen (A) and the type of vegetable (D) have an effect on the height of the plant, while the other two fertilizer components do not significantly change the height. Remember, in this design the acres were randomly assigned to the treatments. If there is a systematic effect due to acres, then this effect has to show up in the residual. If the residual is big, then it will reduce the F ratio and our experiment is reduced in sensitivity.

TABLE 18-2

OBSERVATION	SUM OF SQUARES	"F"*	HALF EFFECT	MEASURE
5.5000	1260.2500	—.—	8.8750	Average
15.0000	126.5625	47.3	2.8125	A-Nitrogen
9.0000	0.5625	.2 NS	-0.1875	B-Phosp
12.0000	9.0000	3.4 NS	-0.7500	AB
7.5000	2.2500	.8 NS	-0.3750	C-Potash
14.0000	14.0625	5.3 NS	-0.9375	AC
9.5000	0.5625	—.—	-0.1875	BC
10.0000	1.0000	—.—	0.2500	ABC
3.5000	33.0625	12.4	-1.4375	D-Type Crop
13.0000	2.2500	—.—	0.3750	AD
4.0000	0.0000	—.—	0.0000	BD
12.0000	10.5625	—.—	0.8125	ABD
5.5000	1.5625	—.—	-0.3125	CD
8.5000	1.0000	—.—	-0.2500	ACD
4.0000	0.2500	—.—	0.1250	BCD
9.0000	0.5625	—.—	0.1875	ABCD

*Pool 3 and 4 factor interactions for error MS = 2.68 $F_{.05,1,5} = 6.6$

The design in Table 18-3 is a reproduction of the *blocked* experiment found in Chapter 8. We have added the response variable and use a YATES ANOVA for the analysis found in Table 18-4. In the blocked design, we equated the 4 blocks with three degrees of freedom representing the two primary blocks ABC, BCD, and the secondary block, AD. This means that the effect of blocks is confounded with these three interactions and we can see an accumulation of sums of squares for each block. If we compare Tables 18-4 and 18-2 we can see that the random assignment of the acres causes a "sprinkling" of the variation due to native fertility over the entire experiment, while in the blocked design the acre effects are neatly placed in only three positions. What is more important is that these three positions were determined by our choice in the original design and not by chance.

TABLE 18-3

RUN	tc	NITRO%	PHOSP%	POTSH%	TYPE PLANT	ABC	BCD	BLOCK	HEIGHT
12	(1)	10	5	5	Corn	-	-	1	10.0
2	a	20	5	5	Corn	+	-	2	11.5
13	b	10	10	5	Corn	+	+	3	6.0
8	ab	20	10	5	Corn	-	+	4	14.0
14	c	10	5	10	Corn	+	+	3	4.5
4	ac	20	5	10	Corn	-	+	4	12.5
1	bc	10	10	10	Corn	-	-	1	9.5
9	abc	20	10	10	Corn	+	-	2	11.0
11	d	10	5	5	Tomato	-	+	4	6.5
3	ad	20	5	5	Tomato	+	+	3	8.5
10	bd	10	10	5	Tomato	+	-	2	5.0
15	abd	20	10	5	Tomato	-	-	1	13.5
16	cd	10	5	10	Tomato	+	-	2	3.5
6	acd	20	5	10	Tomato	-	-	1	12.5
5	bcd	10	10	10	Tomato	-	+	4	6.0
7	abcd	20	10	10	Tomato	+	+	3	8.0

TABLE 18-4

OBSERVATION	SUM OF SQUARES	HALF EFFECT	MEASURE
10.0	1260.2500	8.8750	Average
11.5	100.0000	2.5000	A - Nitrogen
6.0	1.0000	0.2500	B - Phosp
14.0	0.0000	0.0000	AB
4.5	4.0000	-0.5000	C - Potash
12.5	0.0000	0.0000	AC
9.5	0.0000	0.0000	BC
11.0	42.2500	-1.6250	ABC - Block
6.5	16.0000	-1.0000	D - Crop Type
8.5	0.2500	0.1250	AD - Block
5.0	0.0000	0.0000	BD
13.5	0.0000	0.0000	ABD
3.5	0.0000	0.0000	CD
12.0	0.0000	0.0000	ACD
6.0	6.2500	-0.6250	BCD - Block
8.0	0.0000	0.0000	ABCD

The analysis of the 2^k blocked factorial is shown in Table 18-4. Just as in any YATES ANOVA, the sums of squares and half effects appear opposite the tc identifier for that factor or interaction. For instance, Potash (Factor C) has a sum of squares of 4 with a half effect of -0.5. This indicates that increasing potash concentration decreases the height of the crop. Notice that in this example, the effects are exactly zero for any interactions other than the interactions confounded with the blocks. The purpose of this example is to demonstrate the physical and mathematical meaning behind confounding and the effect of blocking. In our next example the data will contain random error as well as a systematic effect that will be "blocked" out.

Using Minitab

Now that we have seen the "theory" of blocking in a 2-level design, we need to see how to do the calculations using computer software. As in all Minitab calculations, we first need to set up the experiment before doing the analysis. Figure 18-1 shows the design in random order.

To set up the analysis, go to the **Stat** dropdown as shown in Figure 18-2 and navigate to **Analyze Factorial Design**. The main dialog box is shown in Figure 18-3 where we select the response variable, Height, as shown in the close-up in Figure 18-4.

↓	C1	C2	C3	C4	C5	C6	C7	C8-T	C9
	StdOrder	RunOrder	CenterPt	Blocks	Nitrogen	Phos	Potash	Crop	Height
1	4	1	1	1	20	5	10	Tomato	12.5
2	2	2	1	3	10	5	10	Corn	4.5
3	1	3	1	4	20	10	5	Corn	14.0
4	3	4	1	2	10	10	5	Tomato	5.0
5	13	5	1	2	20	5	5	Corn	11.5
6	16	6	1	3	20	10	10	Tomato	8.0
7	14	7	1	1	10	10	10	Corn	9.5
8	15	8	1	4	10	5	5	Tomato	6.5
9	8	9	1	4	10	10	10	Tomato	6.0
10	6	10	1	2	20	10	10	Corn	11.0
11	5	11	1	1	10	5	5	Corn	10.0
12	7	12	1	3	20	5	5	Tomato	8.5
13	10	13	1	4	20	5	10	Corn	12.5
14	9	14	1	3	10	10	5	Corn	6.0
15	11	15	1	1	20	10	5	Tomato	13.5
16	12	16	1	2	10	5	10	Tomato	3.5

FIGURE 18-1

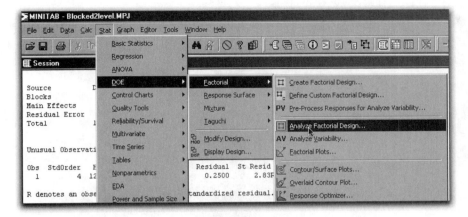

FIGURE 18-2

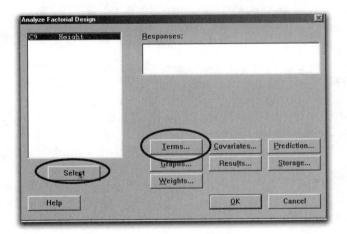

FIGURE 18-3

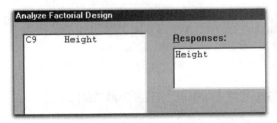

FIGURE 18-4

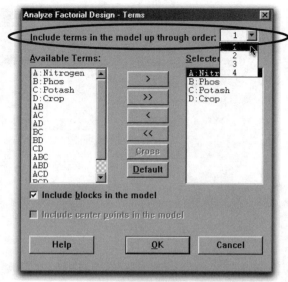

FIGURE 18-5

From the main dialog box (Figure 18-3), click **Terms**. The terms sub dialog box will show all the possible terms in the **Selected** right side. While it is possible to shuttle the unnecessary terms over to the "Available" left side by highlighting them and clicking on the arrow, it is also easier to simply **Include terms in the model up through order** (for this example order 1) as shown in Figure 18-5. Be sure to click **Include blocks in the model** for this blocked design.

Click **OK** in the **Terms** sub dialog box and return to the main dialog box. Click **OK** to produce the analysis as shown in Table 18-5 which compares with the previous "by hand" YATES analysis in Table 18-4 as follows:

- Minitab shows the statistical significance of each block.
- The "Effect" for each factor (which is twice the half effect shown in Table 18-4) are within round-off error of matching.
- Minitab creates an overall ANOVA lumping all single (main) effects.
- Minitab reports a residual value that is nearly 3 standard deviations (2.83) from the center. See boxed item in Table 18-5.

Comments on the Minitab Analysis

It is this author's opinion that blocks should not be tested for statistical significance. A block is a nuisance item and *not* a factor that is being considered for inclusion in a model or for making future decisions. Given that

philosophy, you should ignore the significance tests on the blocks shown in
the Minitab analysis.

The difference between the YATES half-effect and an effect is merely
in knowing if the influence occurs over the entire range of the experiment
(effect) or as a coefficient (defined as the change in the response over a single
unit change in the factor). Minitab also shows the coefficients (**Coef**), which
are exactly the same as the YATES half effects.

The ANOVA table is easily misleading since all the single effects are
combined and (if present) all the interactions are also combined. Therefore
it is impossible from the ANOVA to determine which effect is significant. It
is more informative to look at the significance of each effect separately to
determine if that particular effect should remain in the model.

TABLE 18-5

Factorial Fit: Height versus Block, Nitrogen, Phos, Potash, Crop

```
Estimated Effects and Coefficients for Height (coded units)

Term        Effect    Coef   SE Coef       T       P
Constant            8.906   0.03124  285.00   0.000
Block 1             2.469   0.05412   45.61   0.000
Block 2            -1.156   0.05412  -21.36   0.000
Block 3            -2.156   0.05412  -39.84   0.000
Nitrogen    5.000   2.500   0.03124   80.00   0.000
Phos        0.500   0.250   0.03124    8.00   0.000
Potash     -1.000  -0.500   0.03124  -16.00   0.000
Crop       -2.000  -1.000   0.03124  -32.00   0.000

S = 0.125   R-Sq = 99.93%   R-Sq(adj) = 99.86%

Analysis of Variance for Height (coded units)
Source          DF   Seq SS   Adj SS   Adj MS       F       P
Blocks           3   51.172   51.172  17.0572  1091.66  0.000
Main Effects     4  121.812  121.812  30.4530  1948.99  0.000
Residual Error   8    0.125    0.125   0.0156
Total           15  173.108
```

```
┌─────────────────────────────────────────────────────────────────┐
│Unusual Observations for Height                                    │
│Obs  StdOrder   Height      Fit  SE Fit  Residual  St Resid        │
│  1         4  12.4999  12.2499  0.0884    0.2500     2.83R         │
│R denotes an observation with a large standardized residual.       │
└─────────────────────────────────────────────────────────────────┘
```

```
Alias Structure
I
Blocks = Nitrogen*Crop + Nitrogen*Phos*Potash + Phos*Potash*Crop
Nitrogen
Phos
Potash
Crop
```

The residual value of 2.83 that is nearly 3 standard deviations from the center is an indication of a possible mistake in the data. This mistake could be an incorrect identification of the conditions used in that run of the experiment, or simply a clerical error in recording the data or entering it into the spreadsheet. In any case, it is important to check the records of the experiment in the lab notebook or other documentation to verify that there were no scientific or engineering errors or clerical mistakes. This type of "reality check" is always an important part of good experimental procedure. In this case, there were no apparent mistakes made and the analysis stands as shown.

We will continue the explanation of the basis for other blocked analysis and in Appendix 18 show how Minitab effectively accomplishes this work.

Complete Randomized Block

In this type of experiment, all of the possible treatments are present in each block. The analysis is accomplished via an ANOVA. We assume no possible physical interaction between the factor(s) under study and the items in the block. Therefore, there is a simplification of the ANOVA calculations since the interaction of block and factor does not need to be computed.

We should comment on the assumption of no interaction between the block and the factor or factors under investigation. The block is usually a nuisance factor such as day-to-day or batch-to-batch variation that we don't wish to study. Further, if there were an interaction, we would be hard pressed to attach a physical meaning to it since the block is usually a factor that varies at random rather than in a fixed or controlled manner. If we do not have control over a factor we can't control the interaction or predict what would happen at another randomly selected "level" of the block. As we shall see in the analysis of a randomized block experiment, the interaction (if any) will end up in the error term and if this interaction is of any substantial magnitude, the error will be inflated and thus will reduce the sensitivity of the experiment.

We will now analyze the developer solution design found in Chapter 9. Recall that we were investigating the effect of two different developer formulations on the density (how black) of the image. Also remember that the entire experiment could not be completed on a single day and we expected to find day-to-day variation creeping into the results. Table 18-6 shows the randomized block design.

We could test the analysis as a one-way ANOVA if we were unaware of the day-to-day source of variation. We shall try that approach and see what happens. We will do the usual ANOVA calculations.

TABLE 18-6

	Developer A	Developer B
Day 1	tc 3 (1.1)	tc 6 (1.2)
	tc 7 (1.0)	tc 2 (1.3) Total Day 1 = 9.3
	tc 1 (1.1)	tc 8 (1.2)
	tc 4 (1.3)	tc 5 (1.1)
Day 2	tc 4 (1.2)	tc 7 (1.4)
	tc 8 (1.4)	tc 1 (1.5) Total Day 2 = 10.9
	tc 5 (1.3)	tc 6 (1.4)
	tc 2 (1.2)	tc 3 (1.5)
	Total Dev. A = 9.6	Total Dev. B=10.6

We first will compute to total sum of squares.

$$\Sigma_{sq} \text{ Total} = 1.1^2 + 1.0^2 + \ldots + 1.4^2 + 1.5^2 - \frac{20.2^2}{16} \quad (18\text{-}1)$$

$$\Sigma_{sq} \text{ Total} = 25.84 - 25.5025 = .3375$$

Next we compute the sum of squares due to the developers:

$$\Sigma_{sq} \text{ Dev.} = \left(\frac{(9.6)^2}{8} + \frac{(10.6)^2}{8}\right) - \frac{20.2^2}{16} = .0625 \quad (18\text{-}2)$$

The residual sum of squares is the remaining part of this one-way ANOVA.

Total = Developer Effect + Residual
.3375 = 0.0625 + Residual $\quad (18\text{-}3)$
.275 = Residual Sum of Squares

The ANOVA table looks like that shown in Table 18-7.

TABLE 18-7

ANOVA	SOURCE	ΣSq	df	MS	F	$F_{.05,1,14}$
	Developers	0.0625	1	0.0625	3.18	4.60
	Residual	0.2750	14	0.0196		
	Total	0.3375				

Since the calculated F is less than the critical F, we fail to reject the null hypothesis and conclude that there is no significant difference between the two developers. However, is this conclusion correct in light of what we know about the day-to-day variation? If we look closely at Table 18-6 we can see that the variations induced by days is larger than the variation due to developers. In the one-way ANOVA we treated the day-to-day "factor" as a random source of variation and included its component of variance in the residual. Since day-to-day is systematic and accountable, we are able to extract it from the residual sum of squares. To compute the day-to-day variation, we simply treat this block effect as if it were another factor as follows:

$$\Sigma\text{sq Block} = \left(\frac{(9.3)^2}{8} + \frac{(10.9)^2}{8} \right) - \frac{20.2^2}{16} = .16 \quad (18\text{-}4)$$

The residual sum of squares is the remainder after we remove the developer sum of squares and block sum of squares from the total sum of squares.

$$\Sigma\text{sq Residual} \quad = .3375 - (.16 + .0625) = .115 \quad\quad (18\text{-}5)$$

The ANOVA table now considers blocks and produces a different conclusion as shown in Table 18-8.

TABLE 18-8

ANOVA	SOURCE	ΣSq	df	MS	F	$F_{.05,1,13}$
	Developers	0.0625	1	0.0625	7.07	4.67
	Blocks	0.1600	1			
	Residual	0.1150	13	0.0086		
	Total	0.3375				

We now reject the hypothesis that the developers are the same and conclude that developer B produces a higher density than developer A. Now none of the observations have changed, but we draw different conclusions depending on the ANOVA table we use! Of course we knew that the analysis in Table 18-7 is inadequate since we neglected to account for the systematic day-to-day source of change. The day-to-day makes up a large percentage of the residual sum of squares in Table 18-7. We can compute this percentage by taking the sum of squares for blocks from Table 18-8 and divide it by the residual sum of squares found in Table 18-7.

$$\frac{0.160}{0.275} \times 100 = 58.2\% \quad\quad (18\text{-}6)$$

Well over 50% of the noise is due to a systematic source that we can extract from the residual term and thereby increase the sensitivity of the analysis. Notice that we do **not** compute an F ratio for the blocks. To do so would be meaningless and only add confusion. The blocks are a source of variation that we simply wish to get rid of. We are not interested in making a study of the daily variation. If we were interested in treating the days as a factor, then we would have continued to compute the interaction between days and developers. However, from a physical perspective, such an interaction would be difficult to control because days is a qualitative factor over which we have no ability to select or control. The days just happen to us. Now, of course, if we could relate a physical parameter to the day-induced variation then we would be able to control the variation. In such a case we would no longer have a blocked experiment, but a two-factor factorial. Once again we can see that the analysis depends upon the original intent of the design. The more prior thought we put into an experiment, the more we can extract during the analysis. Goals, objectives and brainstorming are keys!

Paired Comparison

While ANOVA is the technique used most often to extract information from an experimental design, it is sometimes possible to use a simpler "t" test to do the analysis. The paired comparison experiment was introduced in Chapter 9 with an application aimed at bringing generality into an experimental situation. We could use ANOVA in the analysis of a paired comparison, but the traditional paired "t" test will do nicely.

TABLE 18-9

Car Type	Plain Unleaded	Gas-o-hol	Difference
Civic	31	34	3
Focus	33	37	4
Corolla	39	38	-1
Allero	28	30	2
PT Cruiser	26	27	1
VW "Bug"	39	43	4
Coup de Ville	18	21	3
Regal	23	26	3
Impalla	14	18	4
Taurus	18	20	2
	$\overline{X} = 26.9$	$\overline{X} = 29.4$	$\overline{D} = 2.5$
	$s = 8.75$	$s = 8.46$	$s_D = 1.58$

Table 18-9 shows the results of an experiment established to determine the difference between two types of fuel over many types of automobiles. If we were to run a simple "t" test between the averages of the two types of fuel we would find that the variation induced by the 10 different cars would make the residual so large, we would not be able to find a significant effect. This is exactly the same problem we have observed in other blocked designs. The only difference in this case is the fact that the "extra" source of variation (the cars) was *purposely* put into the experiment to give a broader inference space and make our conclusions about the fuels more general. But in doing so, we have inflated the noise in the experiment. However, the source of the noise is again systematic and **known**. We can remove the effect of the cars in the analysis. One way to do this is to concentrate on the response we are *really* interested in. It's not merely the miles per gallon (MPG), but the difference in MPG that is of interest. If we take the **difference** between the Gas-o-hol MPG and the plain unleaded MPG, we look at the response that is of interest to us, *and* remove the inherent variation caused by the differences in cars. In terms of a null and alternative hypothesis, this is:

$$H_0: \quad \delta = \delta_0 = 0 \tag{18-7}$$
$$H_1: \quad \delta > \delta_0 \qquad \text{where } \delta = \text{difference}$$

We use a single-sided alternative since we would believe that the Gas-o-hol will produce a better MPG than the plain unleaded gas. If the MPG is no different or lower, we will decide not to use the Gas-o-hol. The test statistic used in our decision process is similar to the expression used in the simple "t" test of the mean (i.e., $H_0: \mu = \mu_0$).

$$t = \frac{\overline{D} - \delta_0}{s_d/\sqrt{n}} \tag{18-8}$$

For our data we calculate:

$$t = \frac{2.5 - 0}{1.58/\sqrt{10}} = 5.0$$

A comparison with the single-sided t value at the .05 level of risk for 9 degrees of freedom (+1.8331) shows that we should reject the null hypothesis. We conclude that Gas-o-hol delivers a significantly better MPG rating.

The analysis we just completed via the "t" test can also be done using the more general ANOVA approach. We will find the sum of squares due to type of cars (the block) and the sum of squares due to fuels and subtract these two sources

TABLE 18-10

Car Type	Plain Unleaded	Gas-o-hol	Row Totals
Civic	31	34	65
Focus	33	37	70
Corolla	39	38	77
Allero	28	30	58
PT Cruiser	26	27	53
VW "Bug"	39	43	82
Coup deVille	18	21	39
Regal	23	26	49
Impalla	14	18	32
Taurus	18	20	38
Total	269	294	563 Total

of variation from the total sum of squares to obtain the "residual" or noise sum of squares. The results are shown in Table 18-11.

The ANOVA gives the same result as the "t" test although it requires more computation. An interesting observation of the calculated "t" and "F" values shows that the F is *exactly* the square of the t. This is always the case when we have 1 df for the numerator of the F distribution.

TABLE 18-11

Source	ΣSq	df	MS	F	$F_{.05,1,19}$
Fuel	31.25	1	31.25	25.0	5.1174
Cars (Block)	1322.05	9	—		
Residual	11.25	9	1.25		
Total	1364.55	19			

Latin Square

The same concept we have presented and developed for the blocked experiments holds for the Latin Square design. Recall that this design is able to block on two sources of variation at the same time while investigating **one** factor. In the example found in Table 18-12, we are interested in drawing inferences about types of photographic films but want to generalize the result by using different pictures. We also wish to remove the influence of variation between observers who will give a rating to the quality of the photographs.

TABLE 18-12

Block 1 - Picture Content

	Child	Model	Rural	Portrait	Old Man	Food	
	75(I)	85(Fj)	80(Ft)	50(Km)	65(A)	88(EK)	443
	97(EK)	43(Km)	73(A)	85(Fj)	60(Ft)	65(I)	423
	82(Ft)	67(I)	87(Fj)	97(EK)	37(Km)	75(A)	445
	82(A)	87(EK)	42(Km)	67(I)	72(Fj)	65(Ft)	415
	95(Fj)	70(Ft)	65(I)	80(A)	80(EK)	43(Km)	433
	57(Km)	77(A)	92(EK)	77(Ft)	57(I)	85(Fj)	445
	488	429	439	456	371	421	2604

(row label: Block 2 - Observers)

The ANOVA for the Latin Square follows the same path as other randomized block designs. We find the totals for each level of each block (row, column totals) and the totals for the effect we are investigating. We compute the usual total sum of squares and find the residual sum of squares by the difference between the 2 blocks plus the effect and the total.

For our example, the calculations are as follows:

$$\Sigma sq\ Picture\ =\ \frac{488^2+429^2+439^2+456^2+371^2+421^2}{6}\ -\ \frac{2604^2}{36}$$

$$\Sigma sq\ Picture\ =\ 189620.67 - 188356$$
$$=\ 1264.66$$

$$\Sigma sq\ Observers\ =\ \frac{443^2+423^2+445^2+415^2+433^2+445^2}{6}\ -\ \frac{2604^2}{36}$$

$$Ssq\ Observers\ =\ 189490.33 - 188356$$
$$=\ 134.33$$

Now we will need to find the sums for each of the 6 types of film used in our experiment. This is done by sweeping down each column and finding the response for each film (there will be only one film per column). So, for the first column we get a 75 for Ilford (I), in column 2 a 67 for Ilford, in column

3 we find a 65 for Ilford, etc. The sum for Ilford is 396. We continue to sum for each film and the results are as shown in Table 18-13.

TABLE 18-13

	Total	$\overline{X}$
Ilford (I)	396	66.0
Kodak (Ek)	541	90.2
Fotomat (Ft)	434	72.3
Agfa (A)	452	75.3
Fuji (Fj)	509	84.8
K-Mart (Km)	272	45.3

$$\Sigma \text{sq Films} = \frac{396^2 + 541^2 + 434^2 + 452^2 + 509^2 + 272^2}{6} - \frac{2604^2}{36}$$

$$\Sigma \text{sq Films} = 195870.33 - 188356$$
$$= 7514.33$$

$$\text{Total } \Sigma \text{sq} = 75^2 + 97^2 + ... + 43^2 + 85^2 - \frac{2604^2}{36}$$

$$\text{Total } \Sigma \text{sq} = 197340 - 188356 = 8984$$

So the residual is the total less the sums of squares of the blocks and the films.

$$\Sigma \text{sq Residual} = 8984 - (1264.66 + 134.33 + 7514.33) = 70.68$$

We now summarize this information in the ANOVA table (Table 18-14).

TABLE 18-14

Source	ΣSq	df	MS	F	$F_{.05,5,20}$
Films	7514.33	5	1502.87	426	2.599
Block 1 - Pictures	1264.66	5	—	—	
Block 2 - Observers	134.33	5	—	—	
Residual	70.68	20	3.53		
Total	8927.89	35			

From the above results, there is no question that the films produce different quality images. Notice that we do not test the two block factors since we are not interested in their effects and only compute their sums of squares to remove their influence from the residual term. If we had treated this analysis as a one-way ANOVA with the pictures and observers included in the random noise component, then the residual mean square would have been 48.9 and the F value for the films would have been 30.6 or more than an order of magnitude less than we observe in the doubly blocked Latin Square analysis.

While there is still a strong indication of difference between the films from the one-way ANOVA, we have a "clouded" conclusion due to the large contribution of the picture-to-picture factor.

We complete the analysis of the Latin Square by setting up the shortest significant range (SSR) which will tell us which individual films are different from each other. The standard error is computed and then multiplied times the studentized range statistic.

$$S_x = \sqrt{\frac{3.53}{6}} = 0.77$$

$$SSR_2 = 0.77 \times 2.95 = 2.26$$

Since all of the means in Table 18-12 are different by more than 2.26, we conclude that **all** the films are different in image quality.

Split-Plot Analysis

As we have noticed, some experimental designs are not the result of our planning but the result of circumstance. The split-plot is another of this type of design. Let's say that we are testing electrical insulating components in a "life" test as a function of temperature and time. In this test we will not wait until the insulators crack and are destroyed. Rather, we will measure an indicator variable, resistance, as our response variable.

Table 18-15 shows the resistance as a function of temperature and time. Notice that there are three replicates per combination of the two factors. Now, we all know how this experiment should have been conducted! Each of the 36 treatment combinations should have been set up at random. The temperature and the time are set and then the resistance value is measured. But this procedure would take an eternity! So, instead we set a temperature and put three components into the oven and pull them out after 30, 60, and 90 minutes. This is an easier method of gathering the data, but in doing so we have confounded the temperature with the set-up of the test. This is exactly the type of situation farmers encountered when they would plant more than one variety of crop in a specific section of land. The name of the method is inherited from early farming experimentation where the farmers would split a plot of land into sections for different crops, just like we have split the oven space into sections for our different bake times.

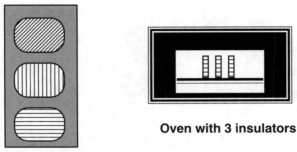

Plot of land split to grow 3 crops

Oven with 3 insulators

FIGURE 18-6

Figure 18-6 shows the similarity of the farmers' activity and the insulator investigation. Table 18-15 shows the data from the experiment in an organized fashion, while Table 18-16 shows the data collection order.

TABLE 18-15
Oven Temperatures (^{o}C)

		280	300	320	340
	30	12	9	9	6
		11	10	8	6
		10	8	6	4
Exposure Time (min.)	**60**	10	7	7	4
		9	8	6	3
		8	6	5	2
	90	9	6	5	2
		7	5	3	0
		5	4	4	1

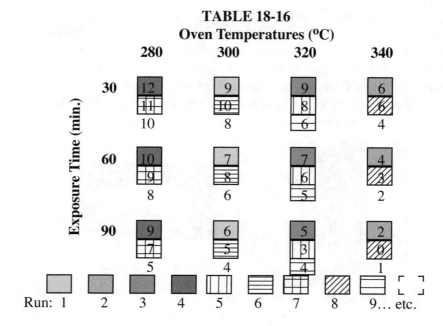

TABLE 18-16

The analysis of the split plot is conducted in a series of two-way ANOVAs. We begin by looking at the "whole plot," which in this case is the temperature factor. (Remember, like the land on which the crops were planted, the oven temperature is the whole plot.) We will "sum over" the exposure time, but since each replicate was not randomly set up, we will treat the replicate variation as if it were a factor. Our rearranged table becomes Table 18-17.

TABLE 18-17

Whole Plot Summation

	Oven Temperatures (°C)				
	280	**300**	**320**	**340**	**Row Total**
I	31	22	21	12	**86**
II	27	23	17	9	**76**
III	23	18	15	7	**63**
Column Total	**81**	**63**	**53**	**28**	**225**

The second arrangement of the data takes place in the split-plot summation found in Table 18-18. Notice that when we place the totals for the "replicates," we have summed over temperatures. However, unlike the exposure time factor, temperature did not change independently with the replicates. We will treat this problem in the final analysis.

TABLE 18-18
Split-Plot Summation

	Exposure Time (min.)			
	30	**60**	**90**	**Row Total**
I	36	28	22	**86**
II	35	26	15	**76**
III	28	21	14	**63**
Column Total	**99**	**75**	**51**	**225**

We need to do one more arrangement of the original data before we compute the sums of squares and make our decisions about the influence of the temperature and exposure time factors. The table we construct actually looks like the table we would build if we had not run a split-plot experiment.

TABLE 18-19

	Oven Temperatures ($^{\circ}$C)				
	280	**300**	**320**	**340**	**Row Total**
30	33	27	23	16	**99**
60	27	21	18	9	**75**
90	21	15	12	3	**51**
Column Total	**81**	**63**	**53**	**28**	**225**

We apply the ordinary formula for the sums of squares (Σsq) to each of the three tables and compute the Σsq for temperature, exposure time, the replication (which is confounded with temperature and is not a true indication of replication), and the interactions between temperature and time as well as the "interactions" between temperature and replicates and time and replicates.

$$\Sigma sq_{\text{Temperature}} = \frac{81^2 + 63^2 + 53^2 + 28^2}{9} - \frac{225^2}{36} = 162.9722 \qquad (18\text{-}9)$$

$$\Sigma sq_{\text{"Replicate"}} = \frac{86^2 + 76^2 + 63^2}{12} - \frac{225^2}{36} = 22.1667 \qquad (18\text{-}10)$$

$$\Sigma sq_{\text{Temp. x Repl.}} = \frac{31^2 + 22^2 + 21^2 + \ldots + 7^2}{3} - CF - (\Sigma sq_T + \Sigma sq_R) = 3.611 \qquad (18\text{-}11)$$

$$\Sigma sq_{\text{Exposure Time}} = \frac{99^2 + 75^2 + 51^2}{12} - CF = 96.0 \qquad (18\text{-}12)$$

$$\Sigma sq_{\text{ET x Repl.}} = \frac{36^2 + 28^2 + 22^2 + 35^2 + \ldots + 14^2}{4} - CF - (\Sigma sq_{ET} + \Sigma sq_R) = 3.333 \qquad (18\text{-}13)$$

$$\Sigma sq_{\text{Temp. x Time}} = \frac{33^2 + 27^2 + 233^2 + 16^2 + \ldots + 3^2}{3} - CF - (\Sigma sq_T + \Sigma sq_{ET}) = 0.4478 \qquad (18\text{-}14)$$

$$\Sigma sq_{\text{Temp. x Time x Repl.}} = \Sigma sq_{\text{Total}} - \text{All Other } \Sigma sq = 4.2193$$

TABLE 18-20

Source	Σsq	df	MS	$F_{\text{calculated}}$	$F_{\text{critical(.05)}}$
Replication	22.166	2	11.083	- - -	
Temperature	162.972	3	54.324	90.3	4.7571 Significant
Rep. x Temp.	3.611	6	0.602	- - -	
Time	96.000	2	48.000	57.6	6.9443 Significant
Rep. x Time	3.333	4	0.833	- - -	
Temp. x Time	0.448	6	0.075	0.2	2.9961
T x R x Tim	4.219	12	0.352		
Total	292.750	35			

In testing for statistical significance, we compare the effect to the interaction of that effect with the replicates. This is due to the fact that the replicates are not true replicates, but confounded with the temperature changes. While in some cases the analysis done in the ordinary manner-which assumes that the replicates are true, independent repeats of the runs with no confounding-will yield the same conclusions as the split-plot analysis, it is always safer to perform the correct analysis especially if we have violated the requirements of randomization due to restrictions on this important aspect of experimentation.

Missing Data

There is one final general area that is traditionally treated under the topic of restrictions on randomization (or blocking). Before the arrival of digital computers, it was necessary to create algorithms to perform the tedious arithmetic operations involved with experiments that were plagued with missing data. The criterion for these algorithms involved estimating the values of the missing data in such a way that the overall residual sum of squares was not changed due to the estimated values for the missing observations.

The technique used to accomplish this goal is the method of least squares. With modern computing equipment, we may simply use the general linear model (GLM) directly on the experiment with the missing data and dispense with the creation of dummy observations that balance the design so it may be subjected to the ordinary ANOVA which requires orthogonality.

Therefore, the recommendation for the analysis of experiments with missing data is to not use ANOVA, but to *carefully* use regression techniques as shown in Chapter 16 instead. Do so only as your experience grows.

This now completes the subject of restrictions on randomization. As we depart from randomization, the analysis becomes more complex as we have seen. It is much better experimental conduct and less of a hassle if we truly replicate in a random manner.

Nested Designs

Up to now, we have avoided any reference to a mathematical formality used in many books on ANOVA. We have not needed this formality of stating a mathematical model since all of our designs and analyses have had the same trivial underlying model. However, the nested design produces a different type of information than the crossed design and therefore needs a more formal guide to its analysis. Before we look at the model for the nested design, we should take a look at the models involved with other designs. We'll use the following notation in the construction of the models.

Y   Represents an observation and will have multiple subscripts to show its position in the design matrix.

μ Represents the grand mean of the experiment.

A,B,C,....   Represent the factors under study and will be singly subscripted with i,j,k,... etc.

ε Represents the residual experimental variation that is sprinkled over the entire experiment.

For a one-way (1 factor) experiment, the model is then:

$$Y_{ij} = \mu + A_i + \varepsilon_{j(i)}$$

We read the model by saying the observation Y is equal to the grand average (μ) plus an effect due to Factor A plus residual variation (ε) within the levels of Factor A (or nested.)

For a two-way experiment the math model is:

$$Y_{ijk} = \mu + A_i + B_j + AB_{ij} + \varepsilon_{k(ij)}$$

This reads the same way as the one-way but we have added another dimension (Factor B) to our sample space and the interaction between A and B.

We can expand these examples to as many factors as are in need of investigation, always adding the new factors, their subscripts, and the interactions that are appropriate.

For blocked experiments, we do not include terms in the model that depict an interaction between the factors under study and the blocks. So, for a simple randomized block with one factor, the math model is:

$$Y_{ijk} = \mu + A_i + Block_j + \varepsilon_{k(ij)}$$

And for a Latin Square:

$$Y_{ij} = \mu + A_k + Block1_i + Block2_j + \varepsilon_{ij}$$

In the Latin Square we play a little game with the subscript k. Since this design is only a 2-dimensional array, there really is no dimension k. We of course have superimposed the dimension k over the i rows and j columns. We usually do not "nest" any error in the treatment combinations of the Latin Square, but could do so, then the model would become:

$$Y_{ij1} = \mu + A_k + Block1_i + Block2_j + \varepsilon_{1(ij)}$$

While we may construct mathematical models for our experimental designs that will compactly state what we are testing (this is a form of an alternative hypothesis) we can also use the math model to determine where the sources of variation come from. This is their prime use with nested designs. The math model will help us in the determination of the expectation of the sources of variation, called the expected mean squares (EMS). Since a mean square can be reduced to a variance via the EMS, we can then quantify the sources of variation present in the process under investigation.

In Chapter 10 we had set up a hierarchical or nested type design that had as its objective the quantification of the variation of sulfur content in a coal analysis. Table 18-21 shows the sulfur content of the coal that comes from 3 randomly selected hopper cars, 4 random samples from each car and two analyses from aliquots of each sample. If we were to disregard the three possible sources of variation, we would compute the overall variation of the 24 pieces of data. This simple calculation is usually the trigger for further work if a specification is not met. In our case, the specification dictated by the Environmental Protection Agency of the U.S. government (EPA) states that the sulfur content may not exceed 5%. The average of the samples is below 5% (4.2%), but the standard deviation is 1.52% which (if we can assume a normal distribution) leads us to believe that 30% of our coal has sulfur above 5% as shown in Figure 18-7.

TABLE 18-21

	Hopper Car 1								Hopper Car 2								Hopper Car 3							
Sample:	1		2		3		4		1		2		3		4		1		2		3		4	
Analysis:	1	2	1	2	1	2	1	2	1	2	1	2	1	2	1	2	1	2	1	2	1	2	1	2
Sulfur Content:	2	3	4	4	2	2	2	4	4	4	4	5	3	3	3	5	6	7	5	6	4	6	5	7

$\overline{X} = 4.2\%, s = 1.52\%$

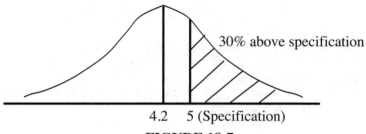

4.2   5 (Specification)
FIGURE 18-7

The question is what can we do about this problem? We first need to understand where the greatest sources of variation are present. We will do an analysis of the variance to pinpoint the major contributors. This is a real variance analysis, not an analysis of means like we have done before.

For this ANOVA we will "shake up" the hierarchy of our design and find the variance due to the hopper cars, the samples and the analyses. To do so we will compute "mini-ANOVAS" within each category. We'll start at the bottom of the hierarchy with the analysis-to-analysis variation. This calculation is nothing new to us since the method has been used in all other ANOVA calculations to compute the noise or residual. We find the sum of each of the two values and the sum of squares of each of the two values.

For Hopper 1, Sample 1, the two analyses were 2 and 3, so we get 5 for the sum and 13 for the sum of squared values. To get the sum of squares we square the sum and divide by the number of items that made it up (2) and subtract this from the sum of squared values.

$$\Sigma sq = 13 - \frac{5^2}{2} = 0.5$$

We repeat this type of calculation for all combinations of hopper cars and samples. There are 12 such combinations, which add up to a sum of squares of 10 and are shown in Table 18-22.

TABLE 18-22
Σ Squares due to Analyses

Sample		Hopper 1	Hopper 2	Hopper 3
	1	0.5	0.0	0.5
	2	0.0	0.5	0.5
	3	0.0	0.0	2.0
	4	2.0	2.0	2.0

We next move up to the sample-to-sample variation. Within each hopper car we will find the sum for each sample, the sum across samples, and the sum for squared samples.

TABLE 18-23

For Hopper 1:

Sample	Σ	Square
1	5	25
2	8	64
3	4	16
4	<u>6</u>	<u>36</u>
		141

$\Sigma sq = \dfrac{141}{2} - \dfrac{23^2}{8} = 4.375$

For Hopper 2:

Sample	Σ	Square
1	8	64
2	9	81
3	6	36
4	<u>8</u>	<u>64</u>
	31	245

$\Sigma sq = \dfrac{245}{2} - \dfrac{31^2}{8} = 2.375$

For Hopper 3:

Sample	Σ	Square
1	13	169
2	11	121
3	10	100
4	<u>12</u>	<u>144</u>
	46	534

$\Sigma sq = \dfrac{534}{2} - \dfrac{46^2}{8} = 2.500$

$$\overline{}$$
$$9.250$$

Remember that in all these calculations we are applying our general sum of squares formula which says to subtract the squared sum (divided by the number of items in the sum) from the sum of the squares (divided by the number of items in the sum that was squared).

$$\Sigma sq = \frac{\Sigma x_i^2}{n_{in\ x}} - \frac{(\Sigma x_i)^2}{n_{in\ x}} \qquad (18\text{-}15)$$

Observe that in Table 18-23 we have applied formula 18-15 individually to each hopper car since Sample 1 in Hopper 1 is a random sample and does not relate to the other Sample 1's. This is the main difference between the analysis of a crossed design and a nested design. After finding the sums of squares in each hopper due to the samples, we sum these sums of squares to obtain the overall (pooled) sum of squares for the sample-to-sample variation.

There is one last source of variation to compute that arises from different sulfur concentrations found in different hopper cars. We find the sum of each hopper car, square each sum and find the grand sum over all hopper cars. Then apply Expression 18-15.

$$\text{Hopper } \Sigma sq = \frac{23^2 + 31^2 + 46^2}{8} - \frac{100^2}{24} = 34.083 \qquad (18\text{-}16)$$

To check our arithmetic we will compute the total sum of squares in the usual way by squaring each individual observation and summing. Then find the grand sum and apply Expression 18-15.

$$\text{Total } \Sigma sq = \frac{2^2 + 3^2 + \ldots + 7^2}{1} - \frac{100^2}{24} = 53.333$$

We will put all the information into an ANOVA table, but before we do that we will go back to the math model for these components of variance experiment.

$$Y_{ijk} = \mu + HC_i + S_{j(i)} + A_{k(ij)} \qquad (18\text{-}17)$$

This model reads as follows: "The observation Y is equal to the grand mean plus an effect due to the hopper car (HC) sampling, plus an effect due to the samples (S) nested within hopper cars [this is how you read the parenthetical subscript] and an effect due to the analyses (A) nested within hopper cars and samples."

If we look back at our design, we see that within each hopper car we have 8 observations. Each observation has a variance associated with it, so we have 8 variance components for cars. Within each car there are 4 samples and within each sample there are 2 observations with a variance associated with each observation. Therefore each sample has 2 variance components. Each analysis is done once so there is only one component of variance for the analysis. If the above allocation of components of variance seems unclear, then it is best to use an algorithm to determine the sources of variation that make up each sum of squares from an experimental design.

The algorithm we shall use utilizes the math model as follows.

Set up column headings by using the subscripts and indicate the number of levels that the subscript represents. Set row headings by using the terms from the math model. Indicate if the factor is random (R) or fixed (F) over each column, as shown in Table 18-24.

TABLE 18-24

	R $i:3$	R $j:4$	R $k:2$
HC_i			
$S_{j(i)}$			
$A_{k(ij)}$			

To fill in the body of the table, start in the first column and place the number 1 where the subscript of a nested factor (in parentheses) *matches* the column heading subscript.

TABLE 18-25

	R* $i:3$	
HC_i		
$S_{j(i)}$	1	STEP 1
$A_{k(ij)}$	1	

Continue in column 1 and again look for a column-row subscript match. Where a match occurs place a 1 for a random factor or a 0 (zero) for a fixed factor. If there is no match, leave the space blank.

---

* A random factor's level is selected in a random fashion from a population. The inference from the levels in a random factor design may be generalized to that population. Most nested designs will incorporate random factors. The residual is always a random factor. If we pick a level with knowledge and purpose, it is a fixed factor. We may only make an inference about the levels chosen rather than the population statistics as in that of a random factor.

TABLE 18-26

	R	
	i:3	
HC_i	1	STEP 2
$S_{j(i)}$	1	
$A_{k(ij)}$	1	

Now repeat the same procedure for the remaining columns.

TABLE 18-27

	R	R	R
	i:3	j:4	k:2
HC_i	1		
$S_{j(i)}$	1	1	
$A_{k(ij)}$	1	1	1

In the remaining blank spaces bring down the column heading numerical values.

TABLE 18-28

	R	R	R
	i:3	j:4	k:2
HC_i	1	4	2
$S_{j(i)}$	1	1	2
$A_{k(ij)}$	1	1	1

We will work row-wise to build the EMS and determine the components of variance. We will first want to determine the first row or the Hopper Car (HC) contribution to the variation components. To do so, cover over (or ignore) the column with the subscript of the factor in question. In the first case, we ignore the first column, so Table 18-28 becomes 18-29.

TABLE 18-29

	R	R
	j:4	k:2
HC_i	4	2
$S_{j(i)}$	1	2
$A_{k(ij)}$	1	1

Look at each row heading. Wherever there is the subscript of the column we have ignored we may multiply the numerical values remaining in that row together to obtain the number of variance components for the row heading. If the factor of the row heading is random, then the component is random and is symbolized with a σ^2. If the factor of the row heading is fixed, then the component is symbolized by a Φ. We have all the random factors in our example, so they will have σ^2 components. Table 18-30 shows the EMS for the Hopper Car component. Note that the components are all a part of HC. The subscripts on the σ^2's come from the names of the factors.

TABLE 18-30

	R j:4	R k:2	EMS
HC_i	4	2	$8\sigma^2_{HC} + 2\sigma^2_S + \sigma^2_A$
$S_{j(i)}$	1	2	
$A_{k(ij)}$	1	1	

The EMS for Hopper Cars is 8 parts due to the cars plus 2 due to samples within the cars and 1 due to the analyses. This means that when we compute the mean square for HC, it is not just a variance due to cars but includes other sources of variation as well. This occurs because of the way the ANOVA of this hierarchical design sums from the bottom of the hierarchy up and in doing so includes all the lower components of variance in the upper factors. This is not a serious problem and, as we shall see, we can see the sum to determine the contribution of each single piece.

We next ignore the second as well as the first column and compute the EMS for all those factors with a j subscript (because both j and i are present for S).

TABLE 18-31

	R k:2	EMS
HC_i	2	$8\sigma^2_{HC} + 2\sigma^2_S + \sigma^2_A$
$S_{j(i)}$	2	$2\sigma^2_S + \sigma^2_A$
$A_{k(ij)}$	1	

Finally, we cover all three columns (since the last row has all three subscripts represented) and multiply by 1 for the component that has all three subscripts, namely the analysis. Now the EMS calculation is complete and is shown in Table 18-32.

TABLE 18-32

Source	EMS
Hopper Cars	$8\sigma^2_{HC} + 2\sigma^2_S + \sigma^2_A$
Samples	$2\sigma^2_S + \sigma^2_A$
Analyses	σ^2_A

Now let's go back and build an ANOVA table from the sums of squares we computed from the sulfur experiment. We will place the EMS along with the ANOVA table and then find the single value of the variance for each source.

TABLE 18-33

ANOVA

Source	Ssq	df	MS	EMS
Hopper Cars	34.083	2	17.044	$8\sigma^2_{HC} + 2\sigma^2_S + \sigma^2_A$
Samples	9.250	9	1.028	$2\sigma^2_S + \sigma^2_A$
Analyses	10.000	12	0.833	σ^2_A
Total	53.333	23		

To find the variance contribution of each source we will start at the bottom of Table 18-33. The analysis mean square is already reported as a single component and is 0.833, so there is no need to adjust it. However, the samples mean square has a component due to the analysis, so we must subtract 0.833 (σ^2_A) from the 1.028 ($2\sigma^2_S + \sigma^2_A$) to obtain 0.195 which is $2\sigma^2_S$. Therefore $1\sigma^2_S$ is 0.195/2 = 0.0975. The hopper car mean square has a component due to the samples and the analyses so we subtract 1.028 ($2\sigma^2_S + \sigma^2_A$) from 17.044 ($8\sigma^2_{HC} + 2\sigma^2_S + \sigma^2_A$) to get 16.016 which is $8\sigma^2_{HC}$. Therefore $1\sigma^2_{HC}$ is 16.016/8 = 2.002.

We will summarize the above information in Table 18-34 as an *analysis of the variation table*. Remember the average sulfur content was 4.2.

TABLE 18-34

Source	EMS (s^2)	% s^2	s	Coefficient of Variation $\left(\frac{s}{\overline{X}}\right) \times 100$
Hopper Cars	2.0020	68.3	1.415	33.7%
Samples	0.0975	3.3	0.312	7.4%
Analyses	0.8333	28.4	0.913	21.7%
Total	2.9328	100.0		

We can see from the $\%s^2$ column that the majority of the variation comes from the hopper cars followed by the analyses. Since the analysis has a better than 25% impact on the variation of the sulfur content, we would like to determine if something should be done to improve the repeatability of the method of analysis. The coefficient of variation is nearly 22%, which by most standards is far too great for such an analytical test method. We would conclude from this experiment that the analysis method will have to be improved by reducing the variability.

Using the Hierarchy

Now that we have the variances of the three "stages" in this sampling and measure system, it is possible to compute the standard error of our measurements and determine the number of samples from each stage.

The expression for computing the standards error is expanded to include a component for each stage of the system. For our problem, this is:

$$s_{\overline{x}} = \sqrt{\frac{s^2_{HC}}{i} + \frac{s^2_{S}}{ij} + \frac{s^2_{A}}{ijk}} \qquad (18\text{-}18)$$

$\quad$ i = Number of Cars
$\quad$ j = Number of Samples
$\quad$ k = Number of Analyses

If we were to sample 5 cars, with 2 samples in each car and run 2 analyses on each sample, the standard error would be:

$$s_{\overline{x}} = \sqrt{\frac{2.002}{5} + \frac{0.0975}{5x2} + \frac{0.8333}{5x2x2}} = 0.672 \quad (18\text{-}19)$$

If we sample 10 cars, 2 samples per car, but only 1 analysis per sample:

$$s_{\overline{x}} = \sqrt{\frac{2.002}{10} + \frac{0.0975}{10x2} + \frac{0.8333}{10x2x1}} = 0.497 \quad (18\text{-}18)$$

The $s_{\bar{x}}$ from the second (18-20) sampling requires the same number of analyses, but is 26% smaller (which is good) than the $s_{\bar{x}}$ of 0.672 from (18-19). We can build a table of standard errors for various combinations of cars, samples and analyses and pick the combination that gives the lowest standard error with least cost. In general, the standard error is reduced faster by increasing the sample size of high noise, early (higher) stage components in the hierarchy.

Problems for Chapter 18

1. Complete the analysis for the following blocked experiment. The primary blocks are BC and AC.

tc	Response
(1)	20.4
a	26.6
b	21.3
ab	21.7
c	15.8
ac	19.2
bc	16.2
abc	14.7

A: Type of Fuel (Regular, Premium)
B: Speed (30 mph, 60 mph)
C: Type of Automobile (Compact, Mid-Size)
Blocks: Driving conditions
Response is miles per dollar

a) How many block levels were there in this experiment?

b) If the blocks represented various parts of the nation (NE, SE, SW, NW), why do you think blocking was done?

c) Which combination of type of fuel, speed, and automobile gets the best economy?

2. Perform an ANOVA on the data in Table 18-15 (the split-plot design) as if the experiment were run as a completely randomized experiment.
   a) Why is the residual Σsq different from that shown in Table 18-20?
   b) Are the conclusions the same? Why?
   c) Was the extra effort of the split-plot analysis necessary?
   d) How does split-plot fit into the general "restriction on randomization" theme?

3. Analyze the following experiment. Response is judged degree of goodness of the hamburger served at 4 restaurants. (Scale 0=Poor, 10=Good.)

Observer	Bill Gray's	Vic & Irv's	Tom's	Don & Bob's
1	8.1	7.6	9.5	9.0
2	9.1	6.1	9.0	8.5
3	7.6	5.5	8.0	7.5
4	3.2	4.2	5.0	4.5
5	7.7	7.8	7.9	7.5
6	5.0	6.0	6.2	6.1
7	9.2	9.9	10.0	9.0
8	7.5	7.0	6.1	8.0
9	6.0	7.0	9.0	7.0
10	7.0	6.5	7.5	7.0

4. If the Latin Square analysis had not been used on the data in Table 18-12, then the mean square for residual would have been 46.7. In the application of the SSR using the 46.7, how would the conclusions change in contrast with the SSR conclusions made from the Latin Square analysis?

5. Compute the ANOVA and draw the appropriate conclusions for the following Latin Square design.

BLOCK 1

		1	2	3	4
	1	17.1 (E)	24.1 (M)	17.5 (C)	27.3 (H)
BLOCK 2	2	26.1 (H)	23.2 (E)	26.1 (M)	22.2 (C)
	3	18.3 (C)	25.3 (H)	18.2 (E)	24.1 (M)
	4	25.0 (M)	22.3 (C)	29.1 (H)	25.3 (E)

Block 1 represents regions of the country.
Block 2 represents different drivers.
The factor under study is brands of fuel.
The response is miles per dollar.
E=Exxon M=Mobil C=Citgo H=Hess

6. Apply the appropriate analysis technique to the following experiment

Lab A	Lab B
10.5	10.9
10.7	11.1
5.2	6.1
4.6	5.1
20.8	22.0
25.6	27.1
8.1	8.8
6.2	6.3

The response is particle size and each lab was given the same standard (there were 8 different standards).

7. Compute an ANOVA for the following data on nitrogen content of a fertilizer:

$$X_{ijk} = m + BAG_i + SAMPLE_{j(i)} + ERROR_{k(ij)}$$

BAG:		1				2				3		
SAMPLE:	1	2	3	4	1	2	3	4	1	2	3	4
Rep. 1	8	10	11	8	12	11	10	12	10	9	11	10
Rep. 2	7	9	9	7	11	12	12	11	10	9	9	11
Rep. 3	9	9	10	9	12	11	13	10	9	10	10	12

8. The ANOVA shown below, computed by the computer as if it were a cross design, came from a random-nested design experiment. Combine the appropriate terms, and from the EMS table, compute the variance components, the standard deviation, coefficient of variation and recommend a sampling and evaluation plan. (See Appendix 18 for the method of combining sums of squares and degrees of freedom to produce nested ANOVA.)

	Source	Sum of Squares	df	
A:	Sample to Sample	1,557,358	17	Grand
B:	Evaluators Within Samples	156,816	5	Average = 150
	A x B Interaction	420,816	85	
	Within Evaluators	431,925	108	

a) How many samples, evaluators, and replicates were there in the experiment?

b) Write the math model.
c) Find the EMS.
d) Compute the mean square for samples and between evaluators.
e) Set up components of variance table.
f) Build a sampling table and recommend the most efficient sampling/evaluation scheme.

APPENDIX 18

ANOVA

ANOVA for 2-level blocked factorial is done using the **Blocked** YATES algorithm. The block "effects" sums of squares are **Factorials** found in the line with the interactions confounded with blocks. These interactions are therefore not measurable.

Treatment of "Blocked" Factor

In a complete randomized block experiment, the block "factor" is treated as another factor in the ANOVA but the interaction between this "factor" and all others is not computed. Also the F test is not completed on any block "effect."

Latin Square

A Latin Square analysis is an extension of the blocked concept. No interaction is computed between the blocks nor between the blocks and the *single* factor under study.

Math Model

A mathematical model is a formality utilized to identify the sources of either fixed or random variation in the response variable resulting from changes in the control variables. Unlike a regression model, the math model does not seek coefficients, but merely cites the sources of variation. Insight gleaned from the math model helps establish the expected mean square (EMS). (Continue in this appendix for more details.)

Hierarchy   Hierarchy in sampling: To determine the standard error
                            which results from sampling a continuous variable at stages
                            or levels, we use the following relationship. We need to
                            know the variance of each stage. This may be determined by
                            a nested experiment.

$$S_{\bar{x}} = \sqrt{\frac{S^2_{TH}}{i} + \frac{S^2_{MH}}{ij} + \frac{S^2_{BH}}{ijk}}$$

Where $S_{\bar{x}}$ is the standard error.
S^2_{TH} is variance at top of hierarchy.
i is number of samples at top of hierarchy.
S^2_{MH} is variance at middle of hierarchy.
j is number of samples at middle of hierarchy.
S^2_{BH} is variance at bottom of hierarchy.
k is number of samples at bottom of hierarchy.

MORE ON EMS

Besides helping us allocate the components of variance from a nested experiment, the expected mean square is used in determining the proper terms in our
F test for significance. We mentioned that math models and EMS turn out to be
trivial formalities in most factorial experiments. We will now show why.

For a typical factorial fixed factor experiment, the math model is:

$$X_{ijk} = \mu + \tau_i + \beta_j + \tau\beta_{ij} + \varepsilon_{k(ij)}$$

and the EMS algorithm looks like this:

TABLE 18A-1

	Fixed i=a	Fixed j=b	Random k=n	EMS
A_i	0	b	n	$bn\Phi_A + \sigma^2_e$
B_j	a	0	n	$an\Phi_B + \sigma^2_e$
AB_{ij}	0	0	n	$n\Phi_{AB} + \sigma^2_e$
$e_{k(ij)}$	1	1	1	σ^2_e

To fill in EMS algorithm table:

1. Match rows and columns for subscripts:
    a. All nested (in parenthesis) subscript matches get a 1.
    b. All random column heading subscript matches get a 1.
    c. All fixed column heading subscript matches get a zero.

2. All non-match intersections bring down the number of levels in the factor under study found in the column heading (note that column i has a levels, j has b levels, and k has n levels).

3. To compute the EMS from the table, take each factor "row-wise."
    a. Ignore the column with the matching subscript.
    b. Look to each row that contains the subscript of the factor you are computing EMS for.
    c. Multiply the non-ignored terms of the row.
    d. For fixed factors, the EMS is Φ (phi) with a subscript of the factor's symbol (i.e., Φ_A).
    e. For random factors, the EMS is σ^2 with a subscript of the factor's symbol (i.e., σ^2_e).

4. Set up F test so the effect is compared to the next lower source of variation. If the effect is made up of $nb\Phi_a + \sigma^2_e$, the next lower source of variation is σ^2_e, and the F ratio is $(bn\Phi_a + \sigma^2_e)/\sigma^2_e$.

In the fixed model, we will always compare our effects to errors as the fixed effect EMS table (18A-1) shows. The random factor EMS is much different, and the F tests are made between main effects and the interaction, plus error as shown in Table 18A-2. The same rules are used to build Table 18A-2. Note that error is always random.

TABLE 18A-2

	Random $i=a$	Random $j=b$	Random $k=n$	EMS
A_i	1	b	n	$bn\sigma^2_A + n\sigma^2_{AB} + \sigma^2_e$
B_j	a	1	n	$an\sigma^2_B + n\sigma^2_{AB} + \sigma^2_e$
AB_{ij}	1	1	n	$n\sigma^2_{AB} + \sigma^2_e$
$e_{k(ij)}$	1	1	1	σ^2_e

A surprising result takes place in a mixed (some random, some fixed) model. Table 18A-3 shows that the random factor is tested against error alone, while the fixed factor is tested against the interaction and error. The symbol for an interaction that is part fixed and part random is expressed as σ^2 rather than Φ.

TABLE 18A-3

	Random	Fixed	Random	EMS
	i=a	j=b	k=n	
A_i	1	b	n	$bn\sigma^2_A + \sigma^2_e$
B_j	a	0	n	$an\Phi^2_B + n\sigma^2_{AB} + \sigma^2_e$
AB_{ij}	1	0	n	$n\sigma^2_{AB} + \sigma^2_e$
$e_{k(ij)}$	1	1	1	σ^2_e

Nested Sums of Squares from a Crossed ANOVA

Many computer programs for ANOVA simply compute the sums of squares assuming a crossed design since this is the most prevalent use of such analyses. However, it is possible to utilize the information from such an analysis in the case of a nested design by combining the appropriate sums of squares and degrees of freedom (df). The example in Table 18A-4 shows the procedure.

TABLE 18A-4

Crossed ANOVA

Source	Σsq	df	MS
Factor A	25	4	6.25
Factor B	30 ⟩ Combine=46	2 ⟩ Combine=10	15.00
A x B Interaction	16	8	2.00
Residual	15	15	1.00
Total	86	29	

Identify the factors in alphabetical order and in general, combine the interaction Σsq with the single factor ΣSq that matches the last letter in the interaction. Do the same for the df.

TABLE 18A-5

Crossed ANOVA ΣSq Rearranged as Nested ANOVA

Source	Σsq	df	MS
Factor A	25	4	6.25
Factor B$_{nested\ in\ A}$	46	10	4.60
Residual	15	15	1.00
Total	86	29	

APPENDIX 18 - MINITAB

General Blocked Design

Using the example (p. 455) of the blocked developer shown in Table 18-6, we will now complete both the design and analysis in Minitab.

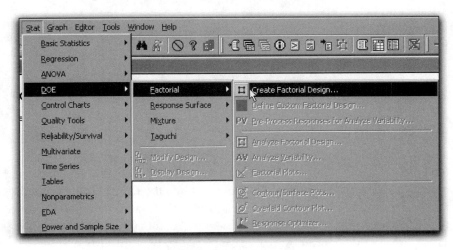

FIGURE APP-18 MTB -1

Since Minitab needs to have a design to accomplish the analysis, we begin by building the experiment from the usual dropdown as shown in Figure APP-18 MTB-1. Since this is a general factorial, we check that button in the main dialog box as shown in Figure APP-18 MTB -2. Click **Designs** in the main dialog box to bring up our design selection shown in Figure APP-18 MTB -3. We can name the factors and indicate the number of levels here. Notice that while the days are really a block, we will indicate at this

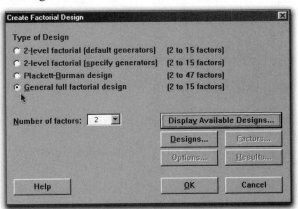

FIGURE APP-18 MTB -2

point that days are considered a factor. In the experiment from Table 18-6, there were 4 replicates per developer per day, so we indicate 4 as the number of replicates in the scroll box. It does not make any sense to block on replicates for this example, so we leave this selection button blank.

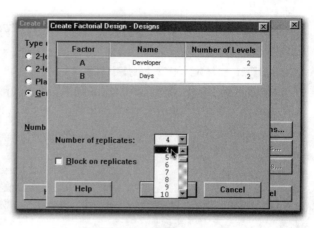

FIGURE APP-18 MTB -3

Click **OK** to return to the main dialog box and select **Factors** to bring up the sub-dialog box found in Figure APP-18 MTB -4. We need to change the type of factor (Minitab defaults to a numeric type) to **Text** and enter the names of the levels as shown. Notice that while the factor Days is actually numeric, the numbers are merely labels, and in the analysis we could not use the numbers to draw any inference about the functionality of the days. Therefore, the days are designated as text. As we will also see, we will eliminate the days as a factor since it is a block. We never draw inferences on blocks, so the days factor is correctly designated as a text factor.

Click **OK** in the Factors sub dialog box and then **OK** in the main dialog box. The design is entered into the worksheet as shown in Figure APP-18 MTB -5. We now need to manually modify the "Blocks" column (C4) to reflect that days are truly the block effect. There are a few ways to accomplish this change in Minitab. One would be to copy the column, Days, and paste it into the blocks column. Another would be to use the "Patterned data" command. The most direct method shown used here is to simply change the numerical values in C4 to correspond to the block levels.

FIGURE APP-18 MTB -4

↓	C1	C2	C3	C4	C5-T	C6-T	C7
	StdOrder	RunOrder	PtType	Blocks	Developer	Days	
1	1	1	1	1	A	1	
2	2	2	1	2	A	2	
3	3	3	1	1	B	1	
4	4	4	1	2	B	2	
5	5	5	1	1	A	1	
6	6	6	1	2	A	2	
7	7	7	1	1	B	1	
8	8	8	1	2	B	2	
9	9	9	1	1	A	1	
10	10	10	1	1	A	2	
11	11	11	1	1	B	1	

FIGURE APP-18 MTB -5

↓	C1	C2	C3	C4	C5-T	C6-T	C7
	StdOrder	RunOrder	PtType	Blocks	Developer	Days	OpticalDensity
1	1	3	1	1	A	1	1.1
2	2	12	1	2	A	2	1.2
3	3	6	1	1	B	1	1.2
4	4	15	1	2	B	2	1.4
5	5	7	1	1	A	1	1.0
6	6	16	1	2	A	2	1.4
7	7	2	1	1	B	1	1.3
8	8	9	1	2	B	2	1.5
9	9	1	1	1	A	1	1.1
10	10	13	1	2	A	2	1.3
11	11	8	1	1	B	1	1.2
12	12	14	1	2	B	2	1.4
13	13	4	1	1	A	1	1.3
14	14	10	1	2	A	2	1.2
15	15	5	1	1	B	1	1.1
16	16	11	1	2	B	2	1.5

FIGURE APP-18 MTB -6

Now we add the responses (from Table 18-6), and the final design is shown in Figure APP-18 MTB -6.

With the worksheet completely filled out, we can accomplish the analysis. It is possible to use the same **Analyze Factorial Design** approach we have utilized with the 2-level designs. Navigate there as shown in Figure APP-18 MTB -7 which brings up the analysis dialog box shown in Figure APP-18 MTB -8.

Select the response (Optical Density) and click on **Terms**. In Figure APP-18 MTB -9, we send all but the Developer (A) factor to the available (left) side since days represent the block and there cannot be any

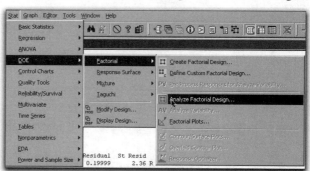

FIGURE APP-18 MTB -7

physical interaction with a block. Click **Include blocks in the model** and then **OK** in both the sub dialog box **Terms**, and in the main dialog box. The analysis of variance is produced in the session window and shown in Table APP-18 MTB -1. This is exactly (Minitab gets carried away with significant figures) within round-off the same as shown in the "by hand" analysis in Table 18-8 on p. 456.

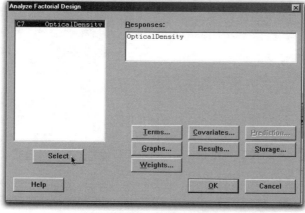

FIGURE APP-18 MTB -8

FIGURE APP-18 MTB -9

TABLE APP-18 MTB -1

General Linear Model: Optical Density versus Blocks, Developer

```
Factor      Type   Levels  Values
Blocks      fixed       2  1, 2
Developer   fixed       2  A, B


Analysis of Variance for Optical Density,
using Adjusted SS for Tests

Source       DF  Seq SS   Adj SS   Adj MS      F      P
Blocks        1  0.15999  0.15999  0.15999  18.09  0.001
Developer     1  0.06249  0.06249  0.06249   7.07  0.020
Error        13  0.11499  0.11499  0.00884
Total        15  0.33749
```

Paired Comparison in Minitab

The paired t-Test is built right into Minitab and does not need any clever manipulations as was necessary in the previous generalized block design.

It is necessary to enter the data in the worksheet as shown in Figure APP-18 MTB 10. There is no need to include the types of cars in this entry.

From the **Stat** dropdown go to **Basic Statistics** and select **Paired t** which brings up the main dialog box shown in Figure APP-18 MTB 12. Since we have the raw data, we will choose the first **Samples in column** option. We make the factor "Gas-o-hol" the first sample and the factor "Plain Unleaded" the second factor. This will give a positive result if Gas-o-hol does deliver higher mpg than Plain Unleaded since Minitab subtracts the second sample value from the first sample value. Next click **Options** to select the level of confidence and the type of alternative hypothesis.

↓	C1	C2
	Gas-o-hol	**Plain Unleaded**
2	37	33
3	38	39
4	30	28
5	27	26
6	43	39
7	21	18
8	26	23
9	18	14
10	20	18
11		
12		

FIGURE APP-18 MTB 10

We'll use the default 95% confidence level and set up an alternative hypothesis that is greater than. Click **OK** in both the sub dialog box and the main dialog box which produces the analysis found in the session window and in Table APP-18 MTB -2. The statistics are the same as we had produced

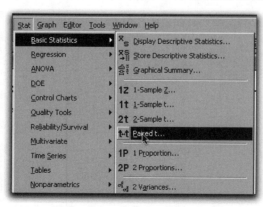

FIGURE APP-18 MTB 11

"by hand" in Table 18-9 (p.457) and the t-Test is exactly the same as the "by hand" t-Test we had done earlier. A calculated t value of 5.0 indicates a very low probability (0.000) that the fuels are the same. Since the average difference is positive, this says (because of the way we had the subtraction done) Gas-o-hol is better for more miles per gallon.

FIGURE APP-18 MTB 12

FIGURE APP-18 MTB 13

TABLE APP-18 MTB -2

Paired T-Test and CI: Gas-o-hol, Plain Unleaded

```
Paired T for Gas-o-hol - Plain Unleaded

                  N     Mean     StDev    SE Mean
Gas-o-hol        10   29.3999   8.4617    2.6758
Plain Unleaded   10   26.8999   8.7490    2.7667
Difference       10    2.49999  1.58113   0.49999

95% lower bound for mean difference: 1.58343
T-Test of mean difference = 0 (vs > 0): T-Value = 5.00
                                        P-Value = 0.000
```

Latin Square in Minitab

Unlike the paired t-Test that is built right into Minitab, a Latin Square design needs to be cleverly built. This can be accomplished by manually building the experiment – or as we show in this approach, by using the **General full factorial design** and entering just the responses that would have been part of the Latin Square.

We begin in the usual dropdown to reach **Create Factorial Design** as shown in Figure APP-18 MTB 14. Select **General full factorial design** button and indicate that there are 3 factors (all Latin Squares have 3 "factors"). Click **Designs** and fill in the names and number of levels as

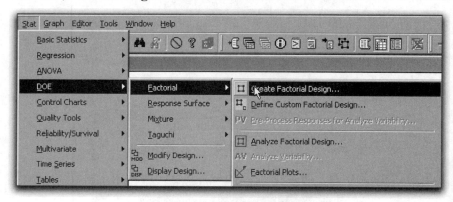

FIGURE APP-18 MTB 14

FIGURE APP-18 MTB 15

shown in Figure APP-18 MTB 16. We will use the example from Table 18-12 (p. 460) which is a 6 x 6 Latin Square, so all factors have 6 levels. Although it is completely arbitrary which factor is the factor being studied, we will make film brand the A factor and the two blocks (picture & observer) the B and C "factors."

Return to the main dialog box and click on

FIGURE APP-18 MTB 16

Factors, which brings up the sub dialog box shown in Figure APP-18 MTB 17. For this example, all of the factors are text types. Enter the names of the levels and then click **OK** to return to the main dialog box where you should click on **Options**. At this point we will not randomize the experiment, so unclick the **Randomize runs** button. Click **OK** and return to the main dialog

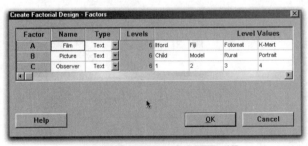

FIGURE APP-18 MTB 17

FIGURE APP-18 MTB 18

box. Once more click **OK** which puts the design into the worksheet which is shown in Figure APP-18 MTB 19. Because we set up a full factorial design, there are 216 runs. The Latin Square uses a subset (36) of these runs. By using Table 18-12, we can find the active runs and enter the response (Image Quality) in the subset from the Latin Square.

Figure APP-18 MTB 20 shows a portion of the entered data for this modified design. In order to make the response column the same length as the other columns (Minitab is insistent on this), make a dummy entry in the IQ column (C8) in row 217. This will allow fill *'s to complete column C8. Then erase the entry in row 217, C8. This tricks Minitab into thinking there are 216 entries in the response column.

Now we have the design and responses ready for analysis. Scroll through the dropdown from **Stat=>ANOVA=>GLM** as shown in

↓	C1	C2	C3	C4	C5-T	C6-T	C7-T	C8
	StdOrder	RunOrder	PtType	Blocks	Film	Picture	Observer	IQ
1	1	1	1	1	Ilford	Child	1	
2	2	2	1	1	Ilford	Child	2	
3	3	3	1	1	Ilford	Child	3	
4	4	4	1	1	Ilford	Child	4	
5	5	5	1	1	Ilford	Child	5	
6	6	6	1	1	Ilford	Child	6	
7	7	7	1	1	Ilford	Model	1	
8	8	8	1	1	Ilford	Model	2	
9	9	9	1	1	Ilford	Model	3	
10	10	10	1	1	Ilford	Model	4	
11	11	11	1	1	Ilford	Model	5	

204	204	204	1	1	Kodak	Portrait	6	
205	205	205	1	1	Kodak	Old Man	1	
206	206	206	1	1	Kodak	Old Man	2	
207	207	207	1	1	Kodak	Old Man	3	
208	208	208	1	1	Kodak	Old Man	4	
209	209	209	1	1	Kodak	Old Man	5	
210	210	210	1	1	Kodak	Old Man	6	
211	211	211	1	1	Kodak	Food	1	
212	212	212	1	1	Kodak	Food	2	
213	213	213	1	1	Kodak	Food	3	
214	214	214	1	1	Kodak	Food	4	
215	215	215	1	1	Kodak	Food	5	
216	216	216	1	1	Kodak	Food	6	

FIGURE APP-18 MTB 19

Figure APP-18 MTB 21, which brings up the GLM dialog box. We use GLM since it utilizes a general regression algorithm that does not require orthogonality. It is not even necessary to declare our modified full factorial as a "custom design."

In Figure APP-18 MTB 22 (the GLM dialog box), we select the response IQ (note that it is possible to have multiple responses). Then, we need to build the model, which is merely the three single factors: Film, Picture, and Observer. We realize that the Picture and Observer factors are not really

↓	C1	C2	C3	C4	C5-T	C6-T	C7-T	C8
	StdOrder	RunOrder	PtType	Blocks	Film	Picture	Observer	IQ
1	1	1	1	1	Ilford	Child	1	75
2	2	2	1	1	Ilford	Child	2	*
3	3	3	1	1	Ilford	Child	3	*
4	4	4	1	1	Ilford	Child	4	*
5	5	5	1	1	Ilford	Child	5	*
6	6	6	1	1	Ilford	Child	6	*
7	7	7	1	1	Ilford	Model	1	*
8	8	8	1	1	Ilford	Model	2	*
9	9	9	1	1	Ilford	Model	3	67
10	10	10	1	1	Ilford	Model	4	*
11	11	11	1	1	Ilford	Model	5	*
12	12	12	1	1	Ilford	Model	6	*
203	203	203	1	1	Kodak	Portrait	5	
204	204	204	1	1	Kodak	Portrait	6	*
205	205	205	1	1	Kodak	Old Man	1	*
206	206	206	1	1	Kodak	Old Man	2	*
207	207	207	1	1	Kodak	Old Man	3	*
208	208	208	1	1	Kodak	Old Man	4	*
209	209	209	1	1	Kodak	Old Man	5	80
210	210	210	1	1	Kodak	Old Man	6	*
211	211	211	1	1	Kodak	Food	1	88
212	212	212	1	1	Kodak	Food	2	*
213	213	213	1	1	Kodak	Food	3	*
214	214	214	1	1	Kodak	Food	4	*
215	215	215	1	1	Kodak	Food	5	*
216	216	216	1	1	Kodak	Food	6	*
217								

FIGURE APP-18 MTB 20

FIGURE APP-18 MTB 21

factors but the blocks in this experiment. To build the model, click in the model box and highlight the factors you want in the model and click **Select**. Now click on **Comparisons** in the main dialog box which brings up a sub dialog box shown in Figure APP-18 MTB 23. We can only make multiple

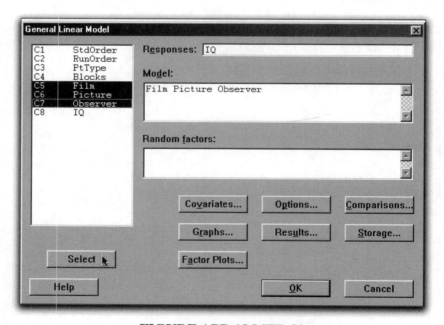

FIGURE APP-18 MTB 22

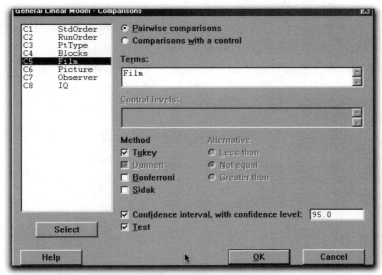

FIGURE APP-18 MTB 23

comparisons for the factor Film, which we send into the **Terms** box by highlighting and selecting. The default Tukey test is in the same set of tests as the SSR test we used in Chapter 13.

Click **OK** in both the comparisons and main dialog boxes and the ANOVA is produced in the session window and is shown in Table APP-18 MTB -3. If you compare this ANOVA with the one we did "by hand" in Table 18-14, you will see that they are exactly alike.

TABLE APP-18 MTB -3

General Linear Model: IQ versus Film, Picture, Observer

```
Factor     Type    Levels  Values
Film       fixed       6   Ilford, Fiji, Fotomat, K-Mart, Agfa, Kodak
Picture    fixed       6   Child, Model, Rural, Portrait, Old Man, Food
Observer   fixed       6   1, 2, 3, 4, 5, 6


Analysis of Variance for IQ, using Adjusted SS for Tests

Source    DF    Seq SS    Adj SS   Adj MS       F      P
Film       5   7514.32   7514.32  1502.86  425.34  0.000
Picture    5   1264.66   1264.66   252.93   71.58  0.000
Observer   5    134.33    134.33    26.87    7.60  0.000
Error     20     70.67     70.67     3.53
Total     35   8983.99
```

The multiple comparison has a long listing of comparisons between the six different film brands. Table APP-18 MTB-4 shows a sample with the Ilford compared with the other five brands. In this table, we see that all the other film brands differ from Ilford brand given the p-values of zero.

TABLE APP-18 MTB -4

```
Tukey Simultaneous Tests
Response Variable IQ
All Pairwise Comparisons among Levels of Film
Film = Ilford  subtracted from:

            Difference     SE of              Adjusted
Film        of Means    Difference   T-Value   P-Value
Fiji           18.83       1.085      17.35    0.0000
Fotomat         6.33       1.085       5.84    0.0001
K-Mart        -20.67       1.085     -19.04    0.0000
Agfa            9.33       1.085       8.60    0.0000
Kodak          24.17       1.085      22.27    0.0000
```

Nested Design and Analysis in Minitab

Minitab has a very nice analysis routine for a nested design. However, there is no direct way to create a nested design from a dropdown or dialog box. We will create the design manually. The coal/sulfur three-stage nested design Table 18-21 (p. 470) will serve as the example.

Begin by labeling the first three columns of the worksheet with the names of the factors in this investigation (Figure APP-18 MTB 24). Put the factor at the top of the hierarchy in the first column (the hopper car), followed in turn by the factor nested in the first factor (samples nested in cars), and then the next factor that is nested in the first two factors (analysis, nested in samples, nested in cars).

To build the design, we will use the **Make Patterned Data** option from the **Calc** dropdown menu as shown in Figure APP-18 MTB 25. It is possible to use either numbers to subscript the levels, or as we will in this example use **Text Values** to put actual names on the levels.

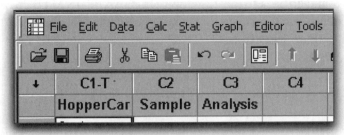

FIGURE APP-18 MTB 24

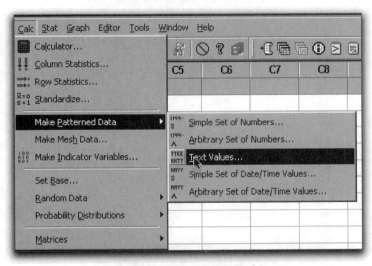

FIGURE APP-18 MTB 25

In the dialog box (Figure APP-18 MTB 26), indicate which column is to be filled and then enter the text values to be used. A space is the delineator between text values. Since there are 4 samples and 2 analyses per sample, each car must be listed 8 times (2 x 4) and the sequence is complete when done once.

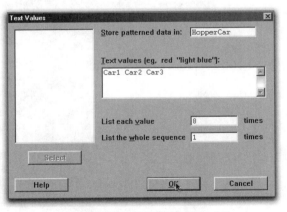

FIGURE APP-18 MTB 26

Click **OK** and go back to the dropdown (Figure APP-18 MTB 25) and repeat the procedure, but this time put the patterned data in the Samples column as shown in Figure APP-18 MTB 27. There are 4 samples and each is named "Sample #" with # going from 1 to 4. Since there are 2 analyses per sample, each sample label must be repeated twice. Since there are 3 Hopper Cars, the entire sequence must be repeated 3 times. Figure APP-18 MTB 28 shows the numeric entries for the analyses factor. Here is a way to determine if you have the right number of occurrences in setting the patterned data. The number of total observations (24 in this example) must be the product of the number of level labels, the number of times the label is listed, and the number of repeats

FIGURE APP-18 MTB 27

FIGURE APP-18 MTB 28

of the sequence. This is a quick check but will not always assure accuracy.
Figure APP-18 MTB 29 shows the design and the added response, % Sulfur.

↓	C1-T	C2-T	C3	C4
	HopperCar	Sample	Analysis	% Sulfur
1	Car1	Sample1	1	2
2	Car1	Sample1	2	3
3	Car1	Sample2	1	4
4	Car1	Sample2	2	4
5	Car1	Sample3	1	2
6	Car1	Sample3	2	2
7	Car1	Sample4	1	2
8	Car1	Sample4	2	4
9	Car2	Sample1	1	4
10	Car2	Sample1	2	4
11	Car2	Sample2	1	4
12	Car2	Sample2	2	5
13	Car2	Sample3	1	3
14	Car2	Sample3	2	3
15	Car2	Sample4	1	3
16	Car2	Sample4	2	5
17	Car3	Sample1	1	6
18	Car3	Sample1	2	7
19	Car3	Sample2	1	5
20	Car3	Sample2	2	6
21	Car3	Sample3	1	4
22	Car3	Sample3	2	6
23	Car3	Sample4	1	5
24	Car3	Sample4	2	7

FIGURE APP-18 MTB 29

With the design and the responses in the worksheet, we can commence to the analysis. This is really easy. From the **Stat** dropdown go to **ANOVA** and select **Fully Nested ANOVA** as shown in Figure APP-18 MTB 30.

In the dialog box (Figure APP-18 MTB 31), select the response by highlighting it and clicking on **Select**. Move the cursor to **Factors** and highlight the three factors (Hopper Cars, Samples, Analyses). Click **Select** to move them to the Factors area. There are no other options, so click **OK** and the ANOVA with the EMS and Analysis of the Variation table is produced and shown in Table APP-18 MTB -5. These results are similar to the "by hand" analyses shown in Tables 18-33 and 34.

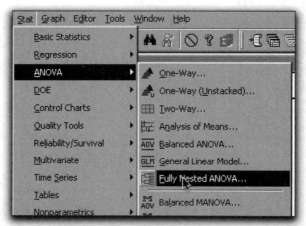

FIGURE APP-18 MTB 30

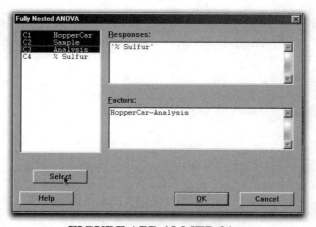

FIGURE APP-18 MTB 31

TABLE APP-18 MTB -5

```
Nested ANOVA: % Sulfur versus HopperCar, Sample, Analysis

Analysis of Variance for % Sulfur

Source      DF        SS        MS       F       P
HopperCar    2   34.0832   17.0416   16.581   0.001
Sample       9    9.2500    1.0278    1.233   0.360
Analysis    12   09.9999    0.8333
Total       23   53.3332


Variance Components

                       % of
Source      Var Comp.  Total   StDev
HopperCar      2.002   68.27   1.415
Sample         0.097    3.32   0.312
Analysis       0.833   28.42   0.913
Total          2.932           1.712


Expected Mean Squares

1  HopperCar   1.00(3) +  2.00(2) +  8.00(1)
2  Sample      1.00(3) +  2.00(2)
3  Analysis    1.00(3)
```

Split-Plot in Minitab

There is no designated analysis established in Minitab to accommodate a Split-Plot design. However, by clever assignment of "**random**," the random factor designation, we can accomplish a Split-Plot analysis. We will use the data as shown on p. 563 in this example.

It is necessary to build a design using the patterned data commands as we just did for the nested design. This process (not shown but similar to the nested design) produces the experimental design in the worksheet as shown in Figure APP-18 MTB 32.

By using the dropdown from **Stat=>ANOVA=>GLM** we obtain the dialog box for the General Linear Model (GLM). Select the response and build a model with the single effects terms and all the 2-way interactions. Do not put the 3-way interaction in the model. To indicate an interaction use the "*" (multiply) symbol. Designate the samples factor first and make it a random factor by selecting and clicking it in the "Random Factor" box. Making this factor the random factor is

↓	C1	C2-T	C3-T	C4-T	C5
	Run Order	Temperature	Time	Sample	Response
1	4	280	30	I	12
2	7	280	30	II	11
3	12	280	30	III	10
4	4	280	60	I	10
5	7	280	60	II	9
6	12	280	60	III	8
7	4	280	90	I	9
8	7	280	90	II	7
9	12	280	90	III	5
10	1	300	30	I	9
11	6	300	30	II	10
12	11	300	30	III	8
13	1	300	60	I	7
14	6	300	60	II	8
15	11	300	60	III	6
16	1	300	90	I	6
17	6	300	90	II	5
18	11	300	90	III	4
19	3	320	30	I	9
20	5	320	30	II	8
21	9	320	30	III	6
22	3	320	60	I	7
23	5	320	60	II	6
24	9	320	60	III	5
25	3	320	90	I	5
26	5	320	90	II	3
27	9	320	90	III	4
28	2	340	30	I	6
29	8	340	30	II	6
30	10	340	30	III	4
31	2	340	60	I	4
32	8	340	60	II	3
33	10	340	60	III	2
34	2	340	90	I	2
35	8	340	90	II	0
36	10	340	90	III	1

FIGURE APP-18 MTB 32

the key to setting up the Split Plot analysis in Minitab. To summarize, make the replicate (samples) the first in the hierarchy and designate it as a random factor.

Click **OK** and the analysis is completed as shown in Table APP-18 MTB-6. When we compare this ANOVA with the "by-hand" analysis of Table 18-20 (p. 466) we see that all of the sums of squares and df are correctly computed and the proper F tests have been done on the Temperature and Time factors.

FIGURE APP-18 MTB 33

FIGURE APP-18 MTB 34

TABLE APP-18 MTB -6

General Linear Model: Response versus Sample, Temperature, Time

```
Factor        Type      Levels   Values
Sample        random       3     I, II, III
Temperature   fixed        4     280, 300, 320, 340
Time          fixed        3     30, 60, 90


Analysis of Variance for Response, using Adjusted SS for Tests

Source               DF    Seq SS    Adj SS   Adj MS      F      P
Sample                2    22.167    22.167   11.083   10.23   0.019 x
Temperature           3   162.971   162.971   54.324   90.26   0.000
Time                  2    96.000    96.000   48.000   57.60   0.001
Temperature*Time      6     0.444     0.444    0.074    0.21   0.966
Sample*Temperature    6     3.611     3.611    0.602    1.71   0.202
Sample*Time           4     3.333     3.333    0.833    2.37   0.111
Error                12     4.222     4.222    0.352
Total                35   292.749

x Not an exact F-test.
```

Epilog

You have come to the end of this section of the book on analysis, and you are now equipped to embark on a wide variety of experimental designs and analyses that will serve you well in your quality and reliability improvement projects. In the next section, and the final section of the book, we will see examples of the use of designed experiments and applications that will inspire you to make great strides in science, engineering, or whatever is your discipline.

PART IV

The Derivation of Empirical Equations from Statistically Designed Experiments

A Case History of an Experimental Investigation

We have studied organization, experimental design techniques, and analysis methods. These are the three key elements of successful experimentation. Now is the time to see these concepts at work. In this chapter we will gain experience and learn by doing. We will look at actual experiments, observe how the planning, organization, and application of sound experimental methods produce the *required* information at the *least* expenditure of resources. This chapter will show us why we **design** experiments rather than just let information arrive at out doorstep in some haphazard manner.

We will look at three experiments that cover a diverse range of situations. The first, is right at home in the kitchen and shows how experimental design can be used in our daily lives. The second example is a chemical process, and the third is a mechanical design activity. In all cases, we will use the phase approach to orgainzation, experimental design, and analysis. In this manner, we are able to build on our prior knowledge in the most efficient way possible.

The Phase Approach

In Chapter 2 we introduced the concept of sequential or phased experimentation. The idea is to avoid putting all our experimental eggs into one basket. Rather than one big experiment, we work in phases. Each phase builds our knowledge in a systematic way. Figure 19-1 is an expanded outline of the phases of experimentation with the types of designs and the analysis techniques appropriate at each phase.

509

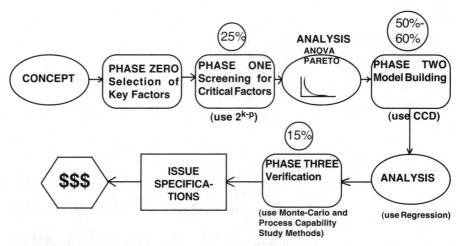

The building block approach: There are 8 steps in the building block approach to experimentation. These steps lead to a *toleranced* specification that produces a quality product that is on target with the least variation at the lowest production and development cost. During the concept phase, and throughout phase zero, we rely on our prior knowledge of the process. In phase one, we sharpen this knowledge with information derived from fractional factorial experiments (CH 5). The analysis of the factorial design (CH 14) leads us to phase two where we use multi-level designs (CH 6) which when analyzed (CH 16 and CH 17) produce a process model which we exercise (CH 20) to determine process set points and tolerances. We also verify our predictions in phase three and issue our final specifications. Notice the expenditure of resources circled near each phase. This is a guide for spending that assures good budgetary control and valid results.

FIGURE 19-1

Example one: A popcorn Formula

We will look at a comparatively simple experiment first and then two other more complex examples. Our first experiment involves the production of a tasty product, popcorn.

When first introduced,* the manufacturers of microwave ovens did *not* recommend making popcorn in such a device. Today we can buy ready-made packages of "Microwavable Popcorn," since it is a convenient method of quickly producing a single serving of this snack. We will show how to accomplish the same goal without the expense of the fancy packaging.

---

*The story goes that the cooking properties of microwave energy were discovered at a military installation where microwave communication beams were used. A box of unpopped corn got into the beam by accident and popped! Now an entire industry revolves around microwave cooking.

The concept phase of the experiment, although discovered by chance, needs to be developed to the point where we can obtain a uniform product. We would like our popcorn to pop completely and taste good. The first thing we must do is state a goal.

GOAL: To maximize the yield of popcorn and maintain taste.

Now it is time to bring our prior knowledge of the process together in such a way that we can write an objective for our experiment. This objective will be the basis for our experimental design.

To bring our knowledge together, we conduct a brainstorming session. To give the reader a better idea of the microwave popcorn process, here's a description of how it works.

A small dish is filled with popcorn and oil. To prevent a mess, a paper lunch bag is fitted over the dish. The bag is cut to fit into the oven and steam vents are cut in the bag. The dish is placed on a trivet to prevent the heat from being drawn off by the oven floor. The oven timer is set, and popcorn is usually produced.

During the brainstorming session, a large number of ideas are freely proposed. The team is made up of experts in corn popping and a statistical expert who "runs" the session.

The list of possible variables for the popcorn process from the brainstorming session is as follows:

TABLE 19-1

Factor	Name	Levels
A	Amount of Popcorn *	1 - 2 Tablespoon
B	Amount of Oil *	1 - 2 Tablespoon
C	Type of Oil *	Olive oil - Corn oil
D	Height of Bag *	6 - 9 inches
	Phase of the Moon	
E	Number of Holes in Bag *	2 - 4
F	Time of Cooking *	2 - 3.25 min.
	Trivet (Yes or No)	
	Placement In Oven	
G	Power Level *	Medium High
H	Type of Corn *	Jolly Orville

* selected factors

Notice that even for such a simple problem, we have a large number of factors to consider and control. After the brainstorming session, we use our prior knowledge to select only those factors we believe will have an influence on the popping.

Now we have enough information to write our objective. This objective will lead us to the experimental design that will get us the required information at the least expenditure of resources.

> **OBJECTIVE:** To test the hypothesis that yield and taste rating are functions of popcorn type and amount, oil type and amount, height of bag, number of holes in bag, cooking time and power.

Since there are eight factors, we will need at least 8 df (assuming a 2-level experiment) for the single effects and 28 df for the 2-factor interactions.

This means that for all these effects to be unconfounded, we would need a 2^{k-p} experiment with $28 + 8 = 36$ df. The closest integer power of 2 (without rounding down) is 64. With such a design, we would have 27 df in excess over the 36 we need. Therefore, we will discard such a large design and see what a 32 tc design will buy us.

With 32 treatments, we have an $n = 5$. Therefore, $p = 3$ since $k = 8$ (recall $n = k-p$). The 2^{8-3} design is a 1/8 fraction and we will select 3 higher order interactions to "generate" our design. We will pick three of the 4-factor interactions as follows:

$$\begin{array}{ll} F \equiv ABCD \\ G \equiv BCDE & \text{(19-1)} \\ H \equiv ABDE \end{array}$$

This gives the defining contrast of:

$$1 \equiv ABCDF, BCDEG, ABDEH, \underline{AEFG}, \underline{CEFH}, \underline{ACGH}, BDFGH \quad \text{(19-2)}$$

The design is a resolution IV, since the shortest "word" has four letters, which means that we have 2-factor interactions confounded together. We can get some of the 28, 2-factor interactions free of other 2-factor interactions, but not all of them. Since this is a screening experiment, we probably won't find all the factors making major contributions and we can "live" with this.

We could, and will, take this premise one step further to reduce the size of our experiment to only 16 tc's, which is more in line with the resources we have. Such a design would be a 2^{8-4} (1/16 fraction) and is a resolution IV with all 2-factor interactions confounded with each other. Since we have limited resources for the screening phase, this 2^{8-4} should strike a balance between information and resources.

The generators and defining contrast for this design are as follows:

$$E \equiv ABC$$
$$F \equiv ABD \tag{19-3}$$
$$G \equiv ACD$$
$$H \equiv BCD$$

$$1 \equiv ABCD, ABDF, ACDG, BCDH, CDEF, BDEG, ADEH,$$
$$BCFG, ACFH, ABGH, AEFG, BEFH, DFGH,$$
$$CEGH, ABCDEFGH \tag{19-4}$$

This design is also a resolution IV, but *all* of the 2-factor interactions are confounded with each other. Again, we suspect that the 8 factors will "collapse" to a lesser number and, in turn, justify the high degree of potential confounding. The justification is, of course, based on the hypothesis that a number of effects will not be important and neither will their interactions. Remember, our objective for this first phase of experimentation is to find the factors that contribute most to the functional relationship. We do not expect to find the best set of conditions among the 16 runs that we make. We will find the direction to the best set of conditions.

Before we begin to gather data, let's see if we have all the elements of a good experiment. Our *prior knowledge* has helped gather the factors. We have a clearly stated *Goal and Objective*. The *response* variable, yield, is quantitative and precise since it is based on an exact count of the number of popped and unpopped kernels. We will also measure the taste of the popcorn by having an expert give a rating on a 1 to 10 scale. With these elements in place, we are now ready to gather data.

Various statistical computing software may be used to conveniently set up the table of treatment combinations as shown in Table 19-2. The fractional factorial design template, T16-8 from Chapter 5 may also be used to construct this design structure. However, note that we have used a different set of generators for this problem than template T16-8 uses. Both design structures are equally effective in obtaining the answers we seek. In fractional factorial designs we have many options in the structure. Since in this case we are taking 1/8 of the runs, we have at least eight possible choices (or fractions) to choose from.

TABLE 19-2

Run	tc	AmtPC	AmtOil	TyOil	HBag	Holes	Time	Power	TyCorn
1	(1)	1tbsp	1tbsp	Olive	6 in	2	2 min	Medium	Jolly
2	bd	1	2	Olive	9	4	2	High	Jolly
3	ac	2	1	Corn	6	2	3.25	Medium	Orville
4	bc	1	2	Corn	6	2	3.25	High	Jolly
5	abcd	2	2	Corn	9	4	3.25	High	Orville
6	c	1	1	Corn	6	4	2	High	Orville
7	ab	2	2	Olive	6	2	2	High	Orville
8	abc	2	2	Corn	6	4	2	Medium	Jolly
9	b	1	2	Olive	6	4	3.25	Medium	Orville
10	ad	2	1	Olive	9	4	2	Medium	Olive
11	cd	1	1	Corn	9	4	3.25	Medium	Jolly
12	d	1	1	Olive	9	2	3.25	High	Orville
13	bcd	1	2	Corn	9	2	2	Medium	Orville
14	abd	2	2	Olive	9	2	3.25	Medium	Jolly
15	acd	2	1	Corn	9	2	2	High	Jolly
16	a	2	1	Olive	6	4	3.25	High	Jolly

Our experimental plan is shown in Table 19-2. The first run is: 1 tablespoon of Jolly popcorn with 1 tablespoon olive oil in a 6 inch bag with 2 holes at medium power for 2 minutes. The results are shown in Table 19-3 and are somewhat disappointing for this first run. There were only 13 out of 167 (7.8%) popped kernels and they had a poor, oily taste. Run 2 was a bit better at 60.5% and a medium taste. But if we were to look at each run to see which is the best, we miss the whole point of the *experiment*.

We must look at the whole result by applying the proper coherent analysis technique. In this case we use the YATES ANOVA and report the results in Table 19-4, which shows the effect on percent popped and Table 19-5, for taste result. There are two very large effects, power and time. Both act positively on the yield, but time has a negative effect on taste. This result is not unexpected since the longer time treatments had a tendency to <u>burn</u>, especially at the higher power levels. The interaction graph shown in Figure 19-2 displays this effect. The power increase does improve the taste as long as the time does not get excessive and burn the already popped corn. These two big effects account for most of the change in yield and taste. We also see that there

TABLE 19-3

Experiment with responses in YATES order

Run	tc	AmtPC	AmtOil	TyOil	HBag	Holes	Time	Power	TyCorn	% Popped	Taste	Comments
1	(1)	1tbsp	1tbsp	Olive	6 in	2	2 min	Medium	Jolly	7.8	1	smelly
16	a	2	1	Olive	6	4	3.25	High	Jolly	86.1	1	burned
9	b	1	2	Olive	6	4	3.25	Medium	Orville	25.5	2	smelly
7	ab	2	2	Olive	6	2	2	High	Orville	58.6	8	smelly
6	c	1	1	Corn	6	4	2	High	Orville	62.7	8	fluffy
3	ac	2	1	Corn	6	2	3.25	Medium	Orville	25.8	5	fluffy
4	bc	1	2	Corn	6	2	3.25	High	Jolly	89.3	6	burned
8	abc	2	2	Corn	6	4	2	Medium	Jolly	4.3	4	soggy
12	d	1	1	Olive	9	2	3.25	High	Orville	83.5	5	burned
10	ad	2	1	Olive	9	4	2	Medium	Olive	2.6	6	smelly
2	bd	1	2	Olive	9	4	2	High	Jolly	60.5	3	smelly
14	abd	2	2	Olive	9	2	3.25	Medium	Jolly	46.7	4	smelly
11	cd	1	1	Corn	9	4	3.25	Medium	Jolly	34.7	3	hard
15	acd	2	1	Corn	9	2	2	High	Jolly	66.4	9	small
13	bcd	1	2	Corn	9	2	2	Medium	Orville	0.8	1	small
5	abcd	2	2	Corn	9	4	3.25	High	Orville	83.6	8	burned

TABLE 19-4

Percent Popped OBSERVATION	SUM OF SQUARE	HALF EFFECT	MEASURES	Percent Contribution
7.8	–	46.181	Average	–
86.1	5.4056	0.581	A: Amount of Corn	0.03
25.5	0.0056	-0.019	B: Amount of Oil	0.00
58.6	38.7506	1.556	AB	0.25
62.7	0.8556	-0.231	C: Type of Oil	0.01
25.8	36.3006	-1.506	AC	0.23
89.3	32.7756	-1.431	BC	0.21
4.3	22.3257	-1.181	E: Holes	0.14
83.5	21.8556	1.169	D: Bag Height	0.14
2.6	57.3807	1.894	AD	0.37
60.5	5.1756	0.569	BD	0.03
46.7	2795.7660	13.219	F: Time	18.10
34.7	8.8506	-0.744	CD	0.06
66.4	12237.8900	27.656	G: Power	79.22
0.8	173.5807	-3.294	H: Type of Corn	1.12
83.6	10.7256	-0.819	ABCD	0.07

TABLE 19-5

Taste (1:poor, 10:good) OBSERVATION	SUM OF SQUARES	HALF EFFECT	MEASURES	Percent Contribution
1	–	4.625	Average	–
1	16.0000	1.000	A: Amount of Corn	14.58
2	0.2500	-0.125	B: Amount of Oil	0.22
8	4.0000	0.500	AB	3.64
8	12.2500	0.875	C: Type of Oil	11.16
5	0.0000	0.000	AC	0.00
6	6.2500	-0.625	BC	5.69
4	1.0000	-0.250	E: Holes	0.91
5	1.0000	0.250	D: Bag Height	0.91
6	12.2500	0.875	AD	11.16
3	9.0000	-0.750	BD	8.20
4	2.2500	-0.375	F:Time	2.05
3	4.0000	-0.500	CD	3.64
9	30.2500	1.375	G: Power	27.56
1	9.0000	0.750	H: Type of Corn	8.20
8	2.2500	0.375	ABCD	2.05

is a trade-off between yield and taste. If the yield comes from a combination of high power and longer time, the taste is reduced due to burning. The next biggest effect is caused by type of corn. Unfortunately, in this case, there is conflict between the two responses. The yield is negatively changed by the type of corn, while the taste is improved with the plus level. It was observed during the experiment that the tastier corn (Orville) was producing bigger morsels than the Jolly. Now that we have looked at the big effects we need to judge the contribution of the other factors to the results. Both bag height and number of holes had a minuscule effect on the two responses. The amount of corn had a positive effect on taste, but did not change yield. From a practical point, the olive oil had such

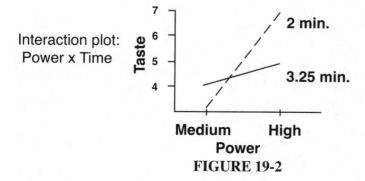

Interaction plot:
Power x Time

FIGURE 19-2

a smell that it was decided to use only corn oil in future phases of experimentation.

The screening experiment has pointed the direction for our next phase of work. We shall include power level, time, type of corn, amount of corn, and amount of oil. Now the amount of oil did not show large effects, but *prior experience* had suggested that this could interact with amount of corn to change yield. Due to the extensive confounding in this experiment, this effect could have been combined with another factor.

Phase II

In this phase of the experiment, we will narrow the range and move our power up to include medium-high and high but cut the time back a bit to avoid burning at high temperatures. We will still use the same levels of oil and corn and will also continue to compare the two types of popping corn. This results in 5 factors of which most are restricted to only 2 levels. A reasonable design for this number of factors could involve either 8 or 16 runs. The design with 8 treatment combinations has the problem that the single effects are confounded with the 2-factor interactions while the 16 tc design provides information on both single effects and 2-factor interactions. The two strongest interactions we might expect would be between Power and Time and another between Amount of Oil and Amount of Corn. An information analysis tells us that we need 5 df for the 5 single effects and 2 df for the two, 2-factor interactions. A design with 8 runs has exactly 7 df but there is a mismatch of the application of the degrees of freedom and at least one of the required interactions will be confounded with a single effect. (Do Problem 1 to show the reasoning to the above result.)

The consideration at hand is to either run the larger design (16 tc's) or live with confounding of the 2^{5-2}, or drop a factor to eliminate the confounding. Since we already have strong indications of the power factor's positive influence on both taste and yield, we will drop this factor and run a 1/2 fraction on 4 factors which will only require the 8 tc's that we can afford. This is a practical example of reducing the size of the experiment by eliminating the need for marginal information.

We use an appropriate computer program to set up the design structure. The defining contrast for this 2^{4-1} is 1≈ABCD.

Table 19-6 shows the factors and levels and Table 19-7 the YATES ordered runs with the yield, taste, and a new response "volume" as measured by the number of cups of popped corn per tablespoon of raw corn.

TABLE 19-6

FACTOR		LEVELS	
A	Time	2 min 15 sec	2 min 45 sec
B	Amount of Corn	1 Tablespoon	2 Tablespoon
C	Amount of Oil	1 Tablespoon	2 Tablespoon
D	Type of Corn	Jolly	Orville

The inclusion of new factors (especially a new response) is one of the innovative benefits of sequential experimentation. In the first phase of this experiment, we observed that the Orville corn was bigger. This is an important aspect of popcorn quality and can be measured by the volume metric. While not a statistical consideration, it is important that experimenters use their keen sense of observation throughout all phases of experimentation. Don't let the SED structure put "blinders" on your good engineering judgment.

The results of the second phase of popping (in Table 19-9) show a tighter range of yields than experienced in the first phase. There are some excellent results, and we might be inclined to just "pick the winners" at this point.

TABLE 19-7

Run	tc	Time	AmtPC	AmtOil	TyCorn	Volume	% Popped	Taste
6	(1)	2:15	1tbsp	1tbsp	Jolly	1.33	60.9	5
5	a	2:45	1	1	Orville	2.00	84.0	6
1	b	2:15	2	1	Orville	2.00	76.3	7
4	ab	2:45	2	1	Jolly	1.50	77.7	6
8	c	2:15	1	2	Orville	1.75	66.7	4
2	ac	2:45	1	2	Jolly	1.50	73.9	3
3	bc	2:15	2	2	Jolly	1.33	66.3	5
7	abc	2:45	2	2	Orville	2.00	81.3	7

However, we must look at the experiment as a whole via the ANOVA to detect the trends. We run the YATES analysis on the 3 responses to reveal the following results on volume, yield, and taste.

•Volume

The largest effect on volume was the type of corn. Orville produced a half-cup more popped corn per tablespoon of raw corn than Jolly. The other effects were much smaller than this major driver and did not conflict with the results for the other two responses. Our recommendation from this volume response is to use Orville's popping corn.

•Yield

The yield response showed time to be the biggest single effect followed by type of corn, and then amount of corn, and an interaction between time and amount of corn, or (due to confounding) another possible interaction between the amount of oil and the type of corn was next in influence and last, the amount of oil.

From physical considerations, it was decided that the time and amount of corn interaction was more likely to be the cause of the difference in yield rather than the amount of oil and type of corn. Again, there were no inconsistencies between the recommendations based on yield and the other two responses. These recommendations are to use 2 minutes, 45 seconds, with 2 tablespoons corn in 1 tablespoon oil, and use Orville's corn.

•Taste

The taste response confirmed the choice of Orville's corn, and showed an interesting interaction between amount of corn and amount of oil.

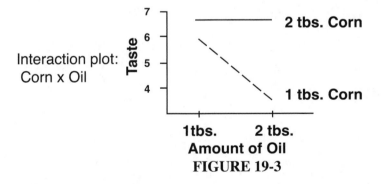

Interaction plot:
Corn x Oil

FIGURE 19-3

Figure 19-3 shows the plot of this interaction, which illustrates a subadditivity. The taste quality drops more than expected when 1 tablespoon of corn is immersed in 2 tablespoons of oil. The problem is that the popped corn comes up soggy and oily when too much oil is used. We may also look at this result in another way. If we are able to control the amount of popcorn to 2 tablespoons, then there will be less need to control the amount of oil. The 2 tablespoons of corn is considered the "robust" level of this factor, since it gives us a greater tolerance to changes in the oil concentration. Here the interaction has not just told us what is best, but what is best *all of the time*! This idea of robustness is a Taguchi concept.

The final recommendation based on the complete analysis is as follows:

FACTOR	LEVEL
Type Corn	Orville
Amount of Corn	2 Tablespoons
Amount of Oil	1 Tablespoon
Time	2 min, 45 sec (2:45)

Now if we look back to Table 19-7, we notice that this treatment combination was *not* in the design. This is not an unusual result. Because we have used only a *fraction* of the possible design configurations, it would be unexpected to find the *exact* optimum among the treatment combinations. Instead, we have been given the *direction* to obtain the optimum.

A series of repeated confirmation runs with the optimum formula for microwave popcorn produced the results in Table 19-8:

TABLE 19-8

RUN	YIELD	VOLUME	TASTE
1	93.7	2.3	10
2	94.0	2.1	10
3	93.0	2.3	10
4	93.5	2.1	10
5	94.0	2.3	10
6	94.1	2.3	10

The results of the confirmation runs give a consistent and better product than any of the trials in the experiment. This fact again illustrates why the total analysis must be completed on the *whole* design rather than just picking the best run. We have *interpolated* to a result that is better than a simple pick of the "best" tc's in the original design.

The principle of statistical experimental design (SED) is to be able to investigate a large population of possible outcomes (all the ways to make popcorn in this example) with a smaller sample (the experimental design structure) and from the analysis, find the best conditions in the population without having to try all the possible conditions in the population. That's the power of SED!

Example Two:
A Photographic Process

In our first example, many of the factors were of a qualitative (nonnumeric) nature and it was not necessary to go through the formalities of generating a mathematical model. The simple graphical analysis provided all the insight into the optimization of the process that produced good popcorn. In many other situations, like the one we are about to embark upon, the system is more complex and requires a more refined expression of the functional relationships. A mathematical model is a very concise way to accomplish this task. The math model is *not* the purpose of the experimental effort, but the by-product that will help us accomplish the purpose. It is unfortunate that many statistically oriented experimenters labor to simply create the model and stop short of using it. In this part of the chapter, we will encounter the various phases of a more complex experiment, create a model, and then in Chapter 21 we will put this model to use to engineer quality into the product described by this model. In this way, we become more than statistical engineers, we become quality engineers.

The Concept Phase

No experiment can begin without some prior knowledge of the process under investigation. In photographic processing, the exposed film is developed with a specially formulated solution of chemicals. The idea is to convert silver halide crystals that have been exposed to light into black metallic silver. This process is a well-known one that has existed for over 100 years. The real trick is to make the silver image black in proportion to the amount of light striking the film. Also, we want the image to be high in its information content and free from unwanted image defects. All of these characteristics may be measured separately by automated equipment or by visual methods. It is further possible to "roll up" the elements of image quality into a single value metric called the IQ (image quality) of the picture. We will use the IQ metric for this example.

Based on prior knowledge of the photographic process, the team of experimenters assembled the following factors at a brainstorming session.

Some factors could have an effect on the quality of the developed film and some (like the phase of the moon) are the types of factors that come up at a freewheeling brainstorming exercise.

TABLE 19-9

Temperature	Humidity
pH	Color of Solution
Weeks Before Change	Odor of Solution
Time of Development	Speed of Film Travel
Concentration of Metol	Phase of the Moon
Concentration of Hydroquinone (HQ)	KI Concentration
Day of the Week	Number of Rolls of Film
Replenishment Cycle	Agitation
Type of Film Developed	

There are 17 possible factors listed in Table 19-9. After the brainstorming session, the team rationalized the contribution of each proposed factor and decided what should be kept in the experimental efforts, what should be discarded completely, and what will be held constant. Phase of the moon was discarded as well as the subjective color and smell of the solution. Those factors related to developer solution wear-out were set aside for this part of the effort. It was decided to use new, fresh batches for each trial. Only one type of film was to be developed and a constant type of image would be exposed on it. These decisions reduced the number of factors to a manageable number.

Still, further decisions had to be made regarding some factors that are redundant. The agitation and time of development factors became redundant if an automatic processor is used with constantly moving film. It was decided that an automatic processor conforms more to the practice involved with the development of the film in question and therefore, the speed of film travel can replace the agitation and time factors. The above decisions leave us with seven factors. This is typical of the reduction possible after a brainstorming session. The seven factors are:

TABLE 19-10

Temperature	Humidity
pH	Speed of Travel
Concentration of Metol	Concentration of HQ
KI Concentration	

Up to this point in our experimental design efforts, we have not used a bit of statistics. We have relied upon our prior knowledge of the process to identify the factors influencing the response. We are now ready to state the goal and objective of this experiment.

GOAL: To obtain a maximum IQ (image quality) of 100 with the least variation.

OBJECTIVE: To test the hypothesis that the IQ is a function of Temperature, pH, Humidity, Speed, and Concentration of: Metol, HQ, and KI.

Defining the Required Information

The objective concisely states the factors about which we need to gather information. If we take the prescribed path in this investigation, an experiment that will identify the quantitative relative importance of the factors is in order. A 2-level fractional factorial design is the appropriate screening experiment for this purpose. We now need to determine the most efficient approach to this part in the sequence of experiments. Since we have seven factors we must allocate the information associated with these seven factors. There will be one degree of freedom required for each single factor effect in a 2-level design. We should also make provision for the possible 2-factor interactions. There will be 21 of these combined effects. Therefore, we will need 28 df in the screening design. The closest 2^{k-p} design would contain 32 treatment combinations and would be based on a 2^5 full factorial.

While we have just defined the required information, we have not balanced our definition of efficiency by looking at the resources necessary to get this information. With 32 treatments (or runs), it will take more than eight months to get the "required" information. From a psychological view, this is just too long. In eight months, the project could be cancelled or we could just lose interest. Management would probably lose interest sooner and give up completely on experimental design methods. We have two possible means of cutting the extent of this experiment. We could use a smaller fraction (with more confounding among the factors). We could also attempt to cut down on the actual physical time involved in the running of the tests by reducing the number of rolls of film or possibly using overtime to prepare for the next run in the sequence. It turns out that the original estimate of a week per run was very conservative and in actuality, it really only takes two days per run: one day to set up the conditions and another to develop and test the film. Now it looks like we can do two runs a week with a little "breathing space" to spare. The 2^{7-2} would now take 4 months, however a 2^{7-3} could take only 2 months, which is more to management's liking.

If we decide to run the 2^{7-3} there will be less information available in the form of the 2-factor interactions. In fact, all of the 2-factor interactions will be confounded with each other. Here is where some prior knowledge of the process is very important. Of the 21 possible interactions, we would expect only a few interactions to have a large influence on the quality of the development. From basic chemistry we can discount all of the interactions between the humidity factor and the other factors. This eliminates 6 of the 21, bringing the count to only 15 interactions. We could *expect* a Metol-HQ interaction since these two developers tend to "super-add" in their combined effect. It is now time to look at the statistics of the experimental designs and see exactly the amount of information available in different fractional factorial designs.

TABLE 19-11
2^{7-3}

Defining Contrast: 1~ABCE,BCDF,ACDG,ADEF,BDEG,ABFG,CEFG

Run	tc	A:Temp.	B: pH	C:Metol	D:KI	E:Humidity	F:Speed	G: HQ
6	(1)	Lo	Lo	Lo	Lo	Lo	Lo	Lo
5	a	Hi	Lo	Lo	Lo	Hi	Lo	Hi
7	b	Lo	Hi	Lo	Lo	Hi	Hi	Lo
14	ab	Hi	Hi	Lo	Lo	Lo	Hi	Hi
1	c	Lo	Lo	Hi	Lo	Hi	Hi	Hi
3	ac	Hi	Lo	Hi	Lo	Lo	Hi	Lo
9	bc	Lo	Hi	Hi	Lo	Lo	Lo	Hi
15	abc	Hi	Hi	Hi	Lo	Hi	Lo	Lo
12	d	Lo	Lo	Lo	Hi	Lo	Hi	Hi
4	ad	Hi	Lo	Lo	Hi	Hi	Hi	Lo
8	bd	Lo	Hi	Lo	Hi	Hi	Lo	Hi
10	abd	Hi	Hi	Lo	Hi	Lo	Lo	Lo
11	cd	Lo	Lo	Hi	Hi	Hi	Lo	Lo
2	acd	Hi	Lo	Hi	Hi	Lo	Lo	Hi
13	bcd	Lo	Hi	Hi	Hi	Lo	Hi	Lo
16	abcd	Hi	Hi	Hi	Hi	Hi	Hi	Hi

The information available in the two designs may be compared by looking at the defining contrasts for each design. In the 2^{7-3} design (Table 19-11), the defining contrast is made up of all 4-letter words. Due to management considerations we will probably have to use this design, so we should understand the level of confounding. All of the single effects will be confounded with 3-factor interactions and should not cause a problem. The 2-factor interactions will be confounded in sets of 3 as shown in Table 19-12.

Table 19-12

1≈ ABCE, BCDF, ACDG, ADEF, BDEG, ABFG, CEFG

$$AB \approx CE \approx FG$$
$$AC \approx BE \approx DG$$
$$BC \approx AE \approx DF$$
$$AD \approx CG \approx EF$$
$$BD \approx CF \approx EG$$
$$CD \approx BF \approx AG$$

$$ABCD \approx DE \approx AF \approx BG$$

Now it is time to see if the assignment of the factors to their exact columns works, given the pattern of confounding observed in Table 19-12. If we look at the 2^{7-3} design structure in Table 19-11, the most likely interaction (Metol-HQ) is CG. CG is confounded with AD and EF. Both of these effects are unlikely to take place, so the design looks pretty good, so far. Since we already stated that none of the effects would interact with the humidity, we can reduce Table 19-12 as follows and make wise, information-based judgments about the possibility of the effects taking place.

Table 19-13

AB ≈ FG	neither is likely
AC ≈ DG	most likely is AC (temp.-Metol)
BC ≈ DF	most likely is BC (pH-Metol)
AD ≈ CG	most likely is CG (Metol-HQ)
BD ≈ CF	neither likely
CD ≈ BF ≈ AG	none likely to occur
AF ≈ BG	most likely AF (temp.-speed)

Table 19-13 shows the summary of our prior expectations on the possibilities of 2-factor interactions taking place. We do not go so far as to state the quantitative effects, but only indicate if we would expect or not expect such an effect to take place. We will need the experiment to *quantify* our educated guesses. Based on this up-front analysis, we can safely say that the 2^{7-3}, 1/8 fractional factorial design with only 16 tc's will get us the required information in the short time of 2 months.

Adding Levels

Our design has only identified the factors with low and high levels so far. We now need to add the final pieces of information so the actual tests can be run. Again, prior knowledge of the process helps us set the levels from the "working ranges" over which we know a film will develop. Some of these working levels have come from one-factor-at-a-time experiments and some from information gathered over the years. Others have come from chemical considerations. In selecting the levels, we do not use the entire range for fear of finding combinations that will not work at all. In this example, we have set levels half way between the midpoint of the range and the upper and lower limits. We will see that the choice will be helpful in the second phase of the experiment.

We now use a computer program or the design template in Chapter 5 to generate the exact design for the factors and levels described in Table 19-14. We will run the 16 tc design plus center points which will allow us to check for any curved effects in the data and also by replicating the center point we will have a measure of the residual error in our experiment. Table 19-15 is the design with its random order with five repeats of the center point.

TABLE 19-14

FACTOR	NAME	WORKING RANGE	LEVELS
A	Temperature	70 - 90 degrees F	75, 85
B	pH (by alkali)	40 - 80 g/l Alkali	50, 70
C	Metol	8 - 16 g/l	10, 14
D	KI	1 - 5 g/l	2, 4
E	Humidity	20 - 80%	30, 70
F	Speed	10 - 20 ips	13, 17
G	Hydroquinone (HQ)	10 - 20 g/l	13, 17

Table 19-15 also shows the responses. These response values did not appear out of the "blue" but were the result of many hours of carefully controlled experimentation over a 2 month period. We have invested a lot of time in the design phase of this effort and now it begins to pay off. The analysis is easily accomplished using the YATES method and appears in Table 19-16.

TABLE 19-15

Run	tc	A:Temp.	B: pH	C:Metol	D:KI	E:Humidity	F:Speed	G: HQ	Response Image Quality
6	(1)	75	50	10	2	30	13	13	72.0
5	a	85	50	10	2	70	13	17	79.6
7	b	75	70	10	2	70	17	13	68.2
14	ab	85	70	10	2	30	17	17	75.9
1	c	75	50	14	2	70	17	17	77.8
3	ac	85	50	14	2	30	17	13	75.6
1	bc	75	70	14	2	30	13	17	81.9
15	abc	85	70	14	2	70	13	13	81.0
12	d	75	50	10	4	30	17	17	71.0
4	ad	85	50	10	4	70	17	13	71.8
8	bd	75	70	10	4	70	13	17	74.6
10	abd	85	70	10	4	30	13	13	75.2
11	cd	75	50	14	4	70	13	13	77.0
2	acd	85	50	14	4	30	13	17	86.4
13	bcd	75	70	14	4	30	17	13	72.2
16	abcd	85	70	14	4	70	17	17	82.0
4a	zero	80	60	12	3	50	15	15	82.2
7a	zero	80	60	12	3	50	15	15	83.4
9a	zero	80	60	12	3	50	15	15	84.0
12a	zero	80	60	12	3	50	15	15	83.0
15a	zero	80	60	12	3	50	15	15	81.8

Let us examine the ANOVA. First, because of the replication of the center point, we have an estimate of error which is 0.8 with 4 degrees of freedom. The critical F value for 1 and 4 degrees of freedom at the 10% risk level is 4.5448. Five of the effects exceed this critical level and have been marked and identified with their physical names. We can display the relative importance of the factors with a "Pareto chart" as shown in Figure 19-4. While a number of metrics could be used for the y-axis of this Pareto analysis, we will use the half effects. By using the half effect, it is possible to see the relative ranking of the influence of the factors as well as their practical importance since the half effect shows the physical change in the response as a function of the factor under study.

Note that there is a steady decrease in the half effect as we add more factors. If we had plotted all of the factors instead of just the significant ones, we would have observed an exponential decay function between half effect and the factor. This is the sort of thing Pareto (an Italian economist) observed when he plotted the wealth of individuals in his nation vs the number of people holding that wealth.

There were the few rich and the many poor. We have found the *vital* few factors among the *trivial* many in our experiment.

TABLE 19-16

ANOVA

Sums of Squares	Half Effect	Effect Measured		df	% Contribution	F
–	76.388	Average		–	–	–
67.2400	2.050	A	Temp.	1	18.92	84.050
0.0025	-0.013	B	pH	1	0.00	0.003
0.1600	0.100	AB		1	0.05	0.200
129.9599	2.850	C	Metol	1	36.56	162.450
0.0225	-0.038	AC		1	0.01	0.028
0.0400	0.050	BC		1	0.01	0.050
0.2025	0.113	ABC~E RH		1	0.06	0.253
0.2025	-0.113	D	KI	1	0.06	0.253
4.4100	0.525	AD		1	1.24	5.513
1.1025	-0.263	BD		1	0.31	1.378
0.0900	-0.075	ABD		1	0.03	0.113
1.2100	0.275	CD		1	0.34	1.513
81.9026	2.263	ACD~G HQ		1	23.04	102.378
68.8899	-2.075	BCD~F Speed		1	19.38	86.112
0.0225	-0.037	ABCD		1	0.01	0.028
355.4531		Total		15		

Outside estimate of error = 0.8

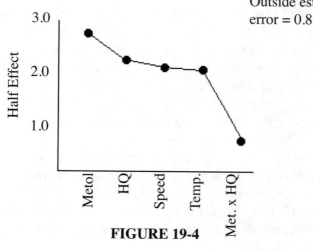

FIGURE 19-4

The Second Phase

The importance of the analysis we have just completed lies in the reduction of the number of factors for the next phase of experimentation. We know that there will be a next phase since we suspect a curved relationship with at least one of the variables. This is indicated by the fact that since the average of the factorial (76.388) differs significantly from the average of the 5 replicates (82.9). We can test for significance by a simple "t" test of the mean.

$$t = \frac{82.9 - 76.4}{.894/\sqrt{5}} = \frac{6.5}{0.4} = 16.25$$

Now we have a choice to make concerning the remaining experimentation. We may use a central composite design if we can get five levels for each of the factors. If we can't, then a Box-Behnkin three level would be an alternative. If we use the CCD, we could simply just *add* the alpha star points to the factorial portion that we already completed. In many situations this is a very economical way of running a sequence of experiments. Caution must be exercised to be sure that nothing has changed between the two segments of experimentation to bias the results. By running the center point replicates in the first phase, we have a way of checking for a change, since additional replicates of this same center point will be run during the alpha star point tests.

We will take the more efficient route and just add the alpha star points to the original set of data. We will also add 3 more replicates of the center point to check for commonality between the two segments of the experiment. We will of course run the new tests in random order. Table 19-17 shows these additional runs and the image quality results. Note that we use the extremes of the working ranges for the levels of the alpha points. Also notice that when we picked the plus and minus levels in the fractional factorial, we did so to allow the alpha distance to be ±2 in design units. This may seem to be fortuitous, but is a demonstration of good statistical engineering based on our prior knowledge of the process.

After running the 11 additional tests that form the alpha start points of the CCD, we first check the average of the zero points from segment two against the average of the zero points form the factorial portion of the first segment. We find that there is no significant difference. Had there been a significant difference, then we could have "adjusted" one of the sets of data to the other to avoid bias in the complete analysis.

TABLE 19-17

Run	tc	A:Temp.	B: pH	C:Metol	D:KI	E:Humidity	F:Speed	G: HQ	IQ
11	$-\alpha_a$	70	60	12	3	50	15	15	50.7
10	$+\alpha_a$	90	60	12	3	50	15	15	61.0
16	$-\alpha_b$	80	40	12	3	50	15	15	82.2
13	$+\alpha_b$	80	80	12	3	50	15	15	83.0
7	$-\alpha_c$	80	60	8	3	50	15	15	78.2
4	$+\alpha_c$	80	60	16	3	50	15	15	87.0
14	$-\alpha_d$	80	60	12	1	50	15	15	82.2
18	$+\alpha_d$	80	60	12	5	50	15	15	82.0
12	$-\alpha_e$	80	60	12	3	20	15	15	82.2
17	$+\alpha_e$	80	60	12	3	80	15	15	82.0
15	$-\alpha_f$	80	60	12	3	50	10	15	81.3
6	$+\alpha_f$	80	60	12	3	50	20	15	76.9
2	$-\alpha_g$	80	60	12	3	50	15	10	77.6
9	$+\alpha_g$	80	60	12	3	50	15	20	88.4
1	zero	80	60	12	3	50	15	15	83.9
5	zero	80	60	12	3	50	15	15	82.1
8	zero	80	60	12	3	50	15	15	82.8

Plotting the Results

By following the procedures of Chapter 17, we are able to construct the function plots in Figure 19-5(a-g). These plots show that there is a quadratic effect for the temperature and linear effects for speed, HQ, and Metol. The other three factors do not have any effect. Since we have an indication of an interaction between the Metol and Hydroquinone, we also plot its function in Figure 19-6.

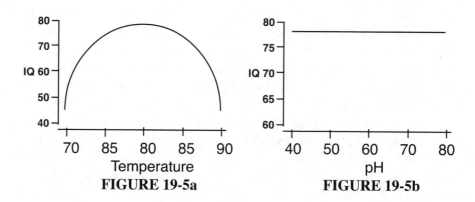

FIGURE 19-5a   FIGURE 19-5b

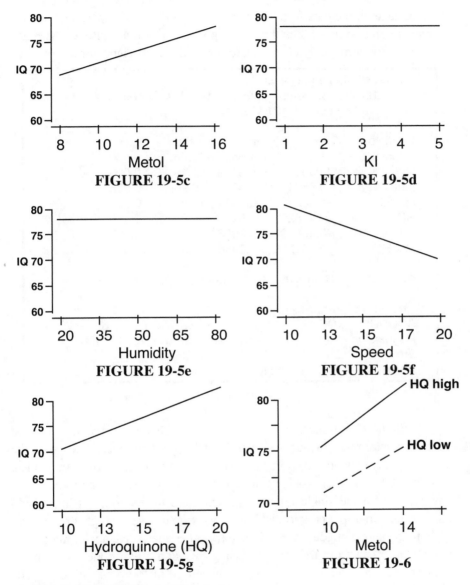

FIGURE 19-5c

FIGURE 19-5d

FIGURE 19-5e

FIGURE 19-5f

FIGURE 19-5g

FIGURE 19-6

Constructing the Model

We are now in a position to propose a model for consideration. Our proposal will include 4 linear terms, an interaction, and a quadratic (second order) term.

$$IQ = b_0 + b_1 \times \text{Temp.} + b_2 \times \text{Metol} + b_3 \times \text{Speed} + b_4 \times \text{HQ}$$
$$+ b_{11} \times \text{Temp.}^2 + b_{24} \times \text{Metol} \times \text{HQ} \qquad (19\text{-}5)$$

We now take all the data from the factorial portion of the experiment and the composite portion as well as the center points and use a multiple regression computer program (such as Minitab) to obtain the coefficients for our model.

The regression equation is:

IQ = - 1661 + 43.4 temp. - 0.652 metol - 1.08 speed - 0.466 HQ
        - 0.268 t sq + 0.131 M x HQ

Predictor	Coef	Stdev	t-ratio	p
Constant	-1661.49	34.64	-47.96	0.000
temp.	43.3517	0.8419	51.49	0.000
metol	-0.6521	0.6791	-0.96	0.346
speed	-1.08246	0.0674	-16.05	0.000
HQ	-0.4662	0.5443	-0.86	0.400
t sq	-0.268167	0.0053	-50.99	0.000
M x HQ	0.13125	0.0450	2.92	0.007

s = 0.7201   R-sq = 99.3%   R-sq(adj) = 99.2%

Analysis of Variance

SOURCE	DF	ΣSq	MS	F	p
Regression	6	1911.75	318.63	614.46	0.000
Error	25	12.96	0.52		
Total	31	1924.71			

TABLE 19-18

When we examine the model and the significance of the coefficients that have been determined in the physical units of the problem, we see that two of the single effects involved in the interaction have "t" test values less than the critical value of 2.05. Now there could be two approaches to resolving this problem. We could eliminate the single effects and retain the interaction, but this would produce an "improper polynomial," since we would have the higher order term without the single effects. We will take the other alternative and eliminate the interaction effect from the model. Using this simplified model, we run the regression analysis. These results show that all the factors are significant. The difficulty we have observed in this regression is caused by the small interaction effect and the fact that ,when working with physical units, there is a slight loss of orthogonality between interaction effects and the single effects making up the interaction. Another way to obtain the regression equation would be to derive the coefficients in design units and then decode the design unit coefficients into physical coefficients. In the above example, there really will be no difference in

The regression equation is
IQ = -1685 + 43.4 temp. + 1.32 metol - 1.08 speed + 1.11 HQ - 0.268 t sq

Predictor	Coef	Stdev	t-ratio	p
Constant	-1685.11	38.24	-44.07	0.000
temp.	43.3517	0.9557	45.36	0.000
metol	1.31667	0.0834	15.78	0.000
speed	-1.08246	0.0766	-14.14	0.000
HQ	1.10877	0.0766	14.48	0.000
t sq	-0.26817	0.0060	-44.92	0.000

s = 0.8174   R-sq = 99.1%   R-sq(adj) = 98.9%

Analysis of Variance

SOURCE	DF	ΣSq	MS	F	p
Regression	5	1907.34	381.47	570.87	0.000
Error	26	17.37	0.67		
Total	31	1924.71			

TABLE 19-19

the result of either analysis method since the interaction effect is so slight. So while prior knowledge gives us reason to believe that there is an interaction between the Metol and the HQ, for this response and this particular set of levels, it would be best to leave the interaction out of the equation.

Our final equation is then:

$$IQ = -1685.11 + 43.3517 \times \text{Temp.} - 0.268167 \times \text{Temp.}^2$$
$$+1.31667 \times \text{Metol} + 1.10877 \times \text{HQ} - 1.0825 \times \text{Speed}$$

At this point we probably would congratulate ourselves for the fine job of finding the equation that describes our process. **But** we have only accomplished our objective. There is more work to do. Now we must complete the task and take care of the **goal**. Our goal is to get an image quality of 100. None of the experimental runs produced an IQ of 100. However, we have the power in the form of our equation to find the combination of factors to achieve the IQ of 100. All we have to do is solve the equation to get this value.

By inspection, we can see that to get the highest IQ, we will have to put in the higher levels of Metol and HQ. The speed will have to be reduced because of its negative coefficient. The temperature has a quadratic term, so other than plotting the function, we can use some very simple calculus to find the maximum of this function. We take the partial derivative with respect to the temperature (all the other terms drop out) and set this equal to zero

$$\frac{\delta IQ}{\delta t} = \frac{\delta}{\delta t} \{43.3517t - 0.268167t^2\} = 43.3517 - 2(0.26167t)$$

...and setting this equal to zero $43.3517 - .53633t = 0$

which becomes: $0.53633t = 43.3517$

$$t = \frac{43.3517}{0.53633} = 80.82$$

We will now solve the equation for the optimum levels of each of the four factors. We will use Temperature at 80.82, Metol at 16 g/l, HQ at 20 g/l, and the speed will be held at 10 ips. Putting these values into the equation we get:

IQ=-1685.11+43.3517x80.82-.268167x80.82²+1.31667x16+1.10877x20-1.0825x10
IQ=-1685.11+3503.6844-1751.6326+21.067+22.1754-10.825
IQ = 99.36

The 99.36 is very close (actually close enough) to the goal of 100. We have now completed our work and have produced a formula for a product that meets the targeted quality goals and we have a process equation to allow troubleshooting when problems arise in the future. However, we are still not finished in our quest for a quality product. We have not yet asked the question concerning the variability of the product. We know that it is almost impossible to make a product without variation, but is there a way of finding out how much tolerance may be allowed in the factors that influence the product quality? The next chapter will show the way to answer that important question. We will revisit this photographic process in the last chapter and make our equation work even harder for us, with a greater impact on quality.

A Final Note on the Developer Experiment

While we have found 4 of the 7 factors that have an influence on the IQ response, we must realize that the other factors must be included in the formulation of the developer. Since KI, Alkaki (pH control), and humidity did not influence the IQ, we may use any level of these that fit other constraints, such as cost. Therefore, given a cost consideration, we would put the chemical factors in at their lowest levels and let the humidity change as it will with the weather. Information on factors that do not have an influence on the process is sometimes as important as information on factors that have a functional relationship with the response. We should never consider factors that "drop out" of the equation as negative information. On the contrary, this kind of information is very valuable.

Example Three: The KlatterMax

The report on the next page is an executive summary of the results of the effort that is described in the remainder of this chapter. This summary is placed in this position to illustrate the proper order of assembly of a technical report. Many reports are poorly organized with all of the details of the technical and statistical work too close to the front of the document and the answers too far at the end. Most readers are anxious to know what was done, why it was done, and what were the results. By placing the summary in the first part of the report, the reader can see the effects of the experimentation and then, if interested, will continue with the details of the operation and read the remainder of the report.

Background

A KlatterMax is a very old wooden toy with European origins. The play value is in the motion and the noise the little wooden figures make as they descend the ladder. Figure 19-7 is a photograph of a KlatterMax with one "Max" poised to descend while the other Max is in motion. A fully decorated Max might be depicted as a fireman or a sailor or any other profession that uses a ladder.

In designing and manufacturing the Maxes, it is important that the customer (typically a child from 3 to 5 years old) is able to place the Max on the ladder and watch the head-over-heels action without interruptions from the Max getting hung up or stuck on a rung or flying off the ladder before completing all the rungs. Time of descent is also important.

A contest might be held between Maxes with the fastest Max being declared the winner.

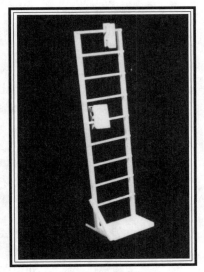

FIGURE 19-7

KlatterMax Experiment Report

To: Folk Art Guild
Subject: KlatterMax Experiment

Goal: To design and construct a KlatterMax toy that descends the ladder
            in the shortest time without any hang-ups or fly-offs.

Objective: To test the hypothesis that time of descent, hang-ups, and fly-offs are
            functions of length, thickness, width, slot length, inside length, slot
            width, and hole diameter.

Results: All factors influenced the performance of the KlatterMax.
            Short length increased time, hang-ups, and fly-offs. Thicker
            Maxes showed less tendency to fly-off but there was a
            thickness-slot length interaction that produced less variation
            in time with a thinner Max. Wider Maxes flew-off with less
            frequency. Longer slot lengths decreased time, hang-ups, and
            fly-offs. A shorter inside length improved time, hang-ups, and
            fly-offs. If the slot width is opened up, time is shortened, but fly-
            offs are increased. A larger hole will decrease fly-offs, but will
            also increase time when a shorter inside length is used.

Conclusions: While most of the factors may be set without conflicting results on
            the three responses, slot width and hole diameter are in need of a
            trade-off setting. Given the negative effect of the hole diameter on
            time, and the impact of slot width on fly-offs, a trade-off position
            would be to use the smaller hole with the smaller slot width.

Recommendations:
            The best settings for the "UltraMax" are (all dimensions in inches):

Length	2.75
Thickness	0.75
Width	1.50
Slot Length	0.375
Inside Length	0.75
Slot Width	0.157
Hole Diameter	0.375

The report on the previous page spells out all the information necessary to understand the effects of the seven factors on the three responses. Notice how the executive summary answers the following questions:

What am I doing? **Goal**
What am I investigating? **Objective**
What did I find? **Results**
What does it mean? **Conclusions**
What can I do with it? **Recommendations**

However, users of such information are often very interested in how such information was developed. The appendix to such an executive summary is a necessary part of every report. The following is such an appendix to the executive summary report.

KlatterMax: The Details (Appendix to Report)

In many situations, the time required to conduct an experiment is short and we may not know enough about the physics of a situation to select the influential factors as we have done in the last two examples. This example fits this situation.

A team of designers brainstormed the problem using the following goal as a guideline.

Goal: To design and construct a KlatterMax toy that descends the ladder in the shortest time without any hang-ups or fly-offs.

Table 19-20 shows the list of factors, the average ranking of these factors, as well as the standard deviation(s) of these ranks. The importance of including the variation measure(s) is to cite the controversial factors which have a higher standard deviation. The team addresses these high variation factors first and works to a consensus. So the standard deviation is an indicator of agreement (a small s) or disagreement (large s). The higher average rank places that factor closer to the top of the "cut list" with a greater likelihood of being one of the factors in the experiment. From the highest rank, least variable factors, the team is able to select the factors considered most important and then construct an experimental design structure that will be efficient. The brainstorming and reasoning process is the most important part of any experimental situation. If we pick the wrong factors, we can never obtain a rational, functional relationship.

Elevation of Factors Voting Form

FACTOR NAME	Type of Factor (C) Controlling (N) Noise (R) Response	Ranking of importance of the factors. 1: low rank; 5: high rank — Team Member's Ranking M1	M2	M3	M4	M5	M6	Average	Std. Dev.
Width (C)								4.6	0.84
Length (A) {tallness}								4.9	0.27
Height (B) {thickness}								4.1	0.86
Depth of slot (D)								4.4	0.76
Slot width (F)								4.9	0.36
Inside length (E)								4.8	0.43
Hole diameter (G)								4.4	0.84
Angle of wedge								3.3	0.91
Surface roughness of hole								2.4	1.22
Straight or curved at slot length								2.4	0.84
Tipping direction								1.2	0.43
Straightness of ladder	CONSTANT								
2 piece vs. 1 piece								1.9	1.07
Height of rungs	CONSTANT								
Width of ladder	CONSTANT								
Type of paint								1.3	0.68
Slot length	DUPLICATE							3.1	1.23
Material density	DUPLICATE							3.1	0.95
Type of wood	DUPLICATE							2.1	1.46
Symmetry across length	CONSTANT							3.0	1.24
Color								1.1	0.27
Mass of Max	DUPLICATE							3.4	1.45
Groove of Max								1.8	1.19
Shape of hole								3.4	1.09
Placement on ladder								1.8	1.05
Thickness of rungs	CONSTANT								
# of Maxes going down								1.6	1.01
Glue type								1.3	1.07

THESE FACTORS HAD THE HIGHEST AVERAGE RANK WITH THE LOWEST VARIATION AND WERE SELECTED FOR THE EXPERIMENT

TABLE 19-20

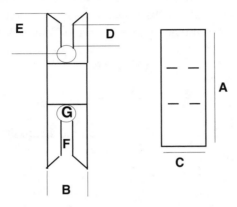

FIGURE 19-8

Since time is short, the team decides to construct a "saturated" design. Such a design configuration uses all its degrees of freedom to study the single effects. However, in this experiment, there may be interactions among the seven factors. To resolve these interactions, we will use the "fold-over" design, which is a concept introduced in Chapter 14.

The idea of a fold-over is to run a "mirror image" (signs are changed) follow-up experiment to resolve the confounding in the first saturated design. This will separate the single effects from the 2-factor interactions. A fold-over design is done in two segments with a completion of the analysis of the first design before the second segment is begun. Table 19-21 shows the first segment of the KlatterMax experiment based on the factors selected from the brainstorming and rationalization meeting.

The ladder for this experiment will be held constant with rung thicknesses of 0.125 inches and rung heights of 0.375 inches. The distance between rungs is 2.75 inches. This leads to the specific restrictions on the dimensions of the factors in the experiment.

The responses are measured as follows. The Max under test is placed on the top rung of the ladder and allowed to pivot down. At the same time the stopwatch is activated. When the Max reaches the bottom, the stopwatch is stopped. The time is recorded as the descent time. If the Max gets hung-up, it is given a nudge and allowed to continue. The stopwatch is not stopped during this nudging activity. If the Max flys off the ladder, the stopwatch is stopped, and the Max is repositioned on the ladder at the point of fly-off. The watch is started again as the Max begins to descend from that point. The responses for the first segment of the design are found in Table 19-22.

TABLE 19-21

Design Generators: D ≈ AB E ≈ AC F ≈ BC G ≈ ABC

tc	Length A	Thick B	Width c	SlotLg D	In Lgth E	SlotWi F	HoleD G
def	2.50	0.75	0.75	0.375	1.00	0.188	0.375
afg	2.75	0.75	0.75	0.125	0.75	0.188	0.500
beg	2.50	1.00	0.75	0.125	1.00	0.157	0.500
abd	2.75	1.00	0.75	0.375	0.75	0.157	0.375
cdg	2.50	0.75	1.50	0.375	0.75	0.157	0.500
ace	2.75	0.75	1.50	0.125	1.00	0.157	0.375
bcf	2.50	1.00	1.50	0.125	0.75	0.188	0.375
abcdefg	2.75	1.00	1.50	0.375	1.00	0.188	0.500

TABLE 19-22

	Descent Time				No. of Hang-ups				No. of Fly-offs			
Observation:1	2	3	X	1	2	3	X	1	2	3	X
def	13.00	7.42	12.61	11.0	0	2	0	0.7	9	4	9	7.7
afg	8.09	7.28	8.31	7.9	1	0	1	0.7	0	0	0	0.0
beg	17.41	12.89	11.79	14.0	6	4	3	4.3	2	4	5	3.7
abd	7.96	7.44	7.81	7.7	1	0	1	0.7	0	0	0	0.0
cdg	12.56	12.21	12.40	12.4	3	2	4	3.0	2	4	0	2.0
ace	10.38	10.62	8.84	9.9	4	4	4	4.0	2	2	1	1.7
bcf	8.41	7.47	8.10	8.0	2	2	1	1.7	1	1	4	2.0
abcdefg	7.13	6.90	6.44	6.8	0	0	0	0.0	0	0	0	0.0

ANOVAs are done for each of the three responses, and the results are shown in Table 19-23. While the results show that all the factors influence at least one of the responses, it is impossible to determine if the effects we are observing are due to the single effect or the interactions that are confounded with each of these single effects. The advantage of the fold-over design is the fact that with only half of the experiment completed, we are able to observe meaningful results and are encouraged to continue with the remainder of the experiment.

If we had observed that many (2 to 4 in this case) of the treatment combinations failed to perform (say they all flew off because they were too short), then we could redesign our experiment and not lose as many runs as we would if we had built and run the entire experiment, which would have been 16 tc's in this case.

TABLE 19-23

	Descent Time			Hang-ups			Fly-offs		
Factor	Half Effect	Percent Contrbtn.	F Ratio	Half Effect	Percent Contrbtn.	F Ratio	Half Effect	Percent Contrbtn.	F Ratio
Average	9.73	–	–	1.88	–	–	2.08	–	–
A: Length	-1.63	35.6	25.1*	-0.54	10.0	9.9*	-1.67	40.2	29.6*
B: Thick	-0.58	4.6	3.2	-0.21	1.5	1.5	-0.67	6.4	4.7*
AB~D: Slot Lg	-0.24	0.8	0.5	-0.79	21.3	21.2*	0.25	0.9	0.7
C: Wdth	-0.44	2.6	1.8	0.29	2.9	2.9	-0.67	6.4	4.7*
AC~E: In Lgth	0.73	7.1	5.0*	0.38	4.8	4.8*	1.08	17.0	12.5*
BC~F: Slot Wi	-1.30	22.6	16.0*	-1.13	43.0	42.9*	0.25	0.9	0.7
ABC~G:Hole	0.56	4.2	2.9	0.13	0.5	0.5	-0.67	6.4	4.7*

The residual variance for time is: 2.53 with 16 df, %Contribution: 22.7
The residual variance for hang-ups is: 0.71with 16 df, %Contribution: 16.1
The residual variance for fly-offs is: 2.25 with 16 df, %Contribution: 21.7
*The "critical F" for 1 and 16 df is 4.5 and effects that exceed this value are marked.

The advantage of the sequential approach is the prevention of an "all is lost" situation. We will now continue the sequence and construct and execute the fold-over part of this experiment. Table 19-24 shows the 8 runs that will resolve the confounding between the single effects and the 2-factor interactions. Notice that each run has exactly the opposite conditions of the original run.

TABLE 19-24

tc	Length A	Thick B	Width C	SlotLg D	In Lgth E	SlotWi F	HoleD G
abcg	2.75	1.00	1.50	0.125	0.75	0.157	0.500
bcde	2.50	1.00	1.50	0.375	1.00	0.157	0.375
acdf	2.75	0.75	1.50	0.375	0.75	0.188	0.375
cefg	2.50	0.75	1.50	0.125	1.00	0.188	0.500
abef	2.75	1.00	0.75	0.125	1.00	0.188	0.375
bdfg	2.50	1.00	0.75	0.375	0.75	0.188	0.500
adeg	2.75	0.75	0.75	0.375	1.00	0.157	0.500
(1)	2.50	0.75	0.75	0.125	0.75	0.157	0.375

The final eight KlatterMaxes are constructed using the same tools that were utilized in the construction of the original pieces. The three response variables were obtained in the same manner described before. These details are very important, since if any other factors change between the two segments of the fold-over, the results we obtain in the final analysis could be erroneous. The responses are found in Table 19-25 and the ANOVA summary is in Table 19-26.

TABLE 19-25

Responses from Segment 2

Observation:	Descent Time 1	2	3	$\bar{X}$	No. of Hang-ups 1	2	3	$\bar{X}$	No. of Fly-offs 1	2	3	$\bar{X}$
abcg	7.40	8.27	6.74	7.5	0	1	0	0.3	0	0	0	0.0
bcde	8.18	10.65	9.79	9.5	0	2	4	2.0	0	1	1	0.7
acdf	3.72	3.74	3.62	3.7	0	0	0	0.0	0	0	0	0.0
cefg	7.86	10.67	7.82	8.8	2	3	3	2.7	1	2	0	1.0
abef	9.76	7.67	9.59	9.0	3	2	4	3.0	0	3	3	2.0
bdfg	5.63	4.95	6.35	5.6	6	3	3	4.0	3	2	1	2.0
adeg	8.09	7.13	6.78	7.3	0	0	1	0.3	0	2	0	0.7
(1)	8.57	7.92	5.59	7.4	3	3	0	2.0	5	4	7	5.3

TABLE 19-26

Analysis from Segment 2

Factor	Descent Time Half Effect	Percent Contrbtn.	F Ratio	Hang-ups Half Effect	Percent Contrbtn.	F Ratio	Fly-offs Half Effect	Percent Contrbtn.	F Ratio
Average	7.35	–	–	1.80	–	–	1.46	–	–
A: Length	0.48	5.7	4.6*	0.88	27.0	12.3*	0.72	18.4	13.4*
B: Thick	-0.56	7.8	6.3*	-0.54	10.4	4.7*	0.29	2.5	1.9
AB~D: Slot Lg	0.80	15.9	12.9*	0.21	1.5	0.7	0.63	11.4	8.3*
C: Wdth	-0.02	0.0	0.0	0.54	10.4	4.7*	1.04	31.8	23.1*
AC~E: In Lgth	-1.31	42.7	34.5*	-0.21	1.5	0.7	0.38	4.1	3.0
BC~F: Slot Wi	0.57	8.1	6.6*	-0.63	13.8	6.3*	0.21	1.2	0.9
ABC~G:Hole	0.05	0.1	0.0	-0.04	0.1	0.0	0.54	8.6	6.3*

The residual variance for time is: 1.20 with 16 df, %Contribution: 19.8
The residual variance for hang-ups is: 1.5 with 16 df, %Contribution: 35.3
The residual variance for fly-offs is: 1.13 with 16 df, %Contribution: 22.0
*The "critical F" for 1 and 16 df is 4.5 and effects that exceed this value are marked.

Now we will separate the single factor effects from the interactions. Recall from Chapter 14 that to obtain the single factor effect, we subtract the half effect from the second segment of the fold-over from the half effect of the first segment of the fold-over and divide by two. To obtain the effect of the interactions we add the half effects from the two segments and divide by two. We will for simplicity call the effects from the first segment "**E** (1-7)" and the effects from the second segment "**E'** (1-7)." Table 19-27 shows the deconfounding for the time of descent response. Table 19-28 deconfounds the hang-up response. Table 19-29 is the deconfounding for the fly-offs response.

TABLE 19-27
Time of Descent Response

1st Segment	2nd Segment	Difference/2	Physical Effect
E1 = -1.63	**E1'** = 0.48	**E1-E1'** = -1.05*	Length (A)
E2 = -0.58	**E2'** = -0.56	**E2-E2'** = -0.01	Thickness (B)
E3 = -0.24	**E3'** = 0.80	**E3-E3'** = -0.52•	Slot Length (D)
E4 = -0.44	**E4'** = -0.02	**E4-E4'** = -0.21	Width (C)
E5 = 0.73	**E5'** = -1.31	**E5-E5'** = 1.02*	Inside Length (E)
E6 = -1.30	**E6'** = 0.57	**E6-E6'** = -0.94*	Slot Width (F)
E7 = 0.56	**E7'** = 0.05	**E7-E7'** = 0.26	Hole Diam. (G)

Sets of Interactions

1st Segment	2nd Segment	Sum/2	
E1 = -1.63	**E1'** = 0.48	**E1+E1'** = -0.58*	BD + CE + FG
E2 = -0.58	**E2'** = -0.56	**E2+E2'** = -0.57*	AD + CF + EG
E3 = -0.24	**E3'** = 0.80	**E3+E3'** = 0.28	AB + CG + EF
E4 = -0.44	**E4'** = -0.02	**E4+E4'** = -0.23	AE + BF + DG
E5 = 0.73	**E5'** = -1.31	**E5+E5'** = -0.29	AC + BG + DF
E6 = -1.30	**E6'** = 0.57	**E6+E6'** = -0.36•	AG + BC + DE
E7 = 0.56	**E7'** = 0.05	**E7+E7'** = 0.31•	AF + BE + CD

*indicates statistical significance (at the .05 risk level): $s^2 = 1.865$
•indicates interest, but not statistical significance

To determine the statistical significance, take the half effect and convert it back to the sum of squares, and then divide by the residual mean square. To convert from half effect to sum of squares, multiply the half effect by $r(2^{k-p})$ (where r is the number of replicates.) Next, square this result and then divide by $r(2^{k-p})$. For example, the half effect of length (A) is -1.05. Therefore, with r=3 and 2k-p =8, the sum of squares is:

$$\Sigma sq = (3(8) \times -1.05)^2/3(8) = 26.46$$

The F value for length is then 26.46/1.865 = 14.2.

TABLE 19-28
Hang-up Response

1st Segment	2nd Segment	Difference/2	Physical Effect
E1 = -0.54	E1' = 0.88	E1-E1' = -0.71*	Length (A)
E2 = -0.21	E2' = -0.54	E2-E2' = 0.17	Thickness (B)
E3 = -0.79	E3' = 0.21	E3-E3' = -0.50*	Slot Length (D)
E4 = 0.29	E4' = 0.54	E4-E4' = -0.13	Width (C)
E5 = 0.38	E5' = -0.21	E5-E5' = 0.30	Inside Length (E)
E6 = -1.13	E6' = -0.63	E6-E6' = -0.25	Slot Width (F)
E7 = 0.13	E7' = -0.04	E7-E7' = 0.08	Hole Diam. (G)

1st Segment	2nd Segment	Sum/2	Set of Interactions
E1 = -0.54	E1' = 0.88	E1+E1' = 0.17	BD + CE + FG
E2 = -0.21	E2' = -0.54	E2+E2' = -0.38	AD + CF + EG
E3 = -0.79	E3' = 0.21	E3+E3' = -0.29	AB + CG + EF
E4 = 0.29	E4' = 0.54	E4+E4' = 0.42	AE + BF + DG
E5 = 0.38	E5' = -0.21	E5+E5' = 0.08	AC + BG + DF
E6 = -1.13	E6' = -0.63	E6+E6' = -0.88*	AG + BC + DE
E7 = 0.13	E7' = -0.04	E7+E7' = 0.04	AF + BE + CD

*indicates statistical significance (at the .05 risk level): $s^2 = 1.105$

TABLE 19-29
Fly-off Response

1st Segment	2nd Segment	Difference/2	Physical Effect
E1 = -1.67	E1' = 0.79	E1-E1' = -1.23*	Length (A)
E2 = -0.67	E2' = 0.29	E2-E2' = -0.48•	Thickness (B)
E3 = 0.25	E3' = 0.63	E3-E3' = -0.19	Slot Length (D)
E4 = -0.67	E4' = 1.04	E4-E4' = -0.85*	Width (C)
E5 = 1.08	E5' = 0.38	E5-E5' = 0.35	Inside Length (E)
E6 = 0.25	E6' = 0.21	E6-E6' = 0.02	Slot Width (F)
E7 = -0.67	E6' = 0.54	E7-E7' = -0.60*	Hole Diam. (G)

1st Segment	2nd Segment	Sum/2	Set of Interactions
E1 = -1.67	E1' = 0.79	E1+E1' = -0.44	BD + CE + FG
E2 = -0.67	E2' = 0.29	E2+E2' = -0.19	AD + CF + EG
E3 = 0.25	E3' = 0.63	E3+E3' = 0.44	AB + CG + EF
E4 = -0.67	E4' = 1.04	E4+E4' = 0.19	AE + BF + DG
E5 = 1.08	E5' = 0.38	E5+E5' = 0.73*	AC + BG + DF
E6 = 0.25	E6' = 0.21	E6+E6' = 0.23	AG + BC + DE
E7 = -0.67	E6' = 0.54	E7+E7' = -0.06	AF + BE + CD

*indicates statistical significance (at the .05 risk level): $s^2 = 1.69$
•indicates interest, but not statistical significance

Table 19-30 is a summary of the results from the three responses. We have included only the single effects in this summary. Specification of the extra runs to resolve the interactions is the next exercise. Note: A negative change is good.

TABLE 19-30

Single Factor		Response		
		Time of Descent	Hang-ups	Fly-offs
Length (A)	2.50			
	2.75	-1.05	-0.71	-1.23
Thickness (B)	0.75		0.17	
	1.00	no effect		-0.48
Width (C)	0.75		-0.13	
	1.50	-0.21		-0.85
Slot Length (D)	0.125			
	0.375	-0.52	-0.50	-0.19
Inside Length (E)	0.75	1.02	0.30	0.35
	1.00			
Slot Width (F)	0.157			no effect
	0.188	-0.94	-0.25	
Hole Diameter (G)	0.375	0.26	no effect	
	0.500			-0.60

From the summary in Table 19-30, we obtain the following settings for the construction of the best KlatterMax based on single effects only. There were no conflicts in the selection of the factors' levels. This is fortunate, since we do not need to make any trade-offs among the settings for the three responses. In some situations, it is necessary to select a slightly suboptimal level of a factor for one response to improve another response. This is the reason for placing the half-effect values in the summary table.

The optimal length is at the 2.75 inch setting. This makes physical sense, since the distance from the top of a rung to the bottom of the next rung is 2.75 inches. With the shorter Max, it is more likely that it will swing out of the space between the rungs and fly off or get stuck as it glances the top of the next rung.

Thickness has no effect on time or hang-ups, but has a slight effect on fly-off, with the best setting at the thicker Max.

A wider Max goes down a bit faster and and prevents fly-offs. Physically, this may be due to the greater mass of the wider object and the fact that there is more of the Max to catch the next rung with the wider body.

The slot length of 0.375 inches has a positive influence on time and hang-ups. Having the longer slot allows the Max to stabilize before falling to the next rung. If there were no slot at all, the Max would swing off the ladder.

The inside length needs to be short for a faster Max. With a longer distance to travel inside itself, the Max wastes time in stabilizing, thus reducing the effective time of descent.

Slot width needs to be open enough to prevent too much friction from slowing the descent. Slot width has no effect on hang-ups or fly-offs.

The hole diameter had no effect of descent time nor hang-ups. However, a smaller hole did contribute to more fly-offs.

So the best MAX (based on single effects only) is:

Length:	2.75 inches
Thickness	1.00 inches
Width	1.50 inches
Slot Length	0.375 inches
Inside Length	0.75 inches
Slot Width	0.188 inches
Hole Diameter	0.50 inches

One of the runs from the second segment of the experiment came closest to this specification. It is the acdf treatment and had the best time (3.69 seconds average) with no hang-ups or fly-offs. While this result is encouraging, we have not yet looked at the information associated with the interactions. Remember, the reason for factorial designs is to be able to investigate the nonadditivities (or interactions) involved with the factors we study.

To investigate the effect of the interactions, we need to resolve the confounding. In Chapter 14 we showed a method to treat the interaction effect as a single factor and by using essentially a one-factor-at-a-time approach, we can determine and separate the confounded influences. We will apply that concept to this experiment. Since we would like to minimize the number of runs added to this experiment, we will build a table of resolving and observe if there can be any common runs between the three responses. Table 19-31 helps with this task.

TABLE 19-31

Responses			
Time of Descent		Hang-ups	Fly-offs
I BD + CE + FG	-0.58		
II AD + CF + EG	-0.57		
III			AC + BG + DF 0.73
IV AG + BC + DE	-0.36	-0.88	
V AF + BE + CD	0.31		

Table 19-32 shows the sets of contrasting signs for the interactions that will separate these confounded effects as well as the average responses for three replicate runs. While there are a number of different ways to obtain these sets of signs, we want to build a minimum of new Maxes. To do so, we have made an effort to be sure some Maxes are common between the sets to save effort. We will also construct the Max that the single factors predicted would be the best. This is the "abcdfg" treatment combination. This again "kills two birds with one stone" and saves effort, since we were probably curious about that particular prediction.

Confounded interaction set IV is common to both time and hang-ups. When we construct the extra Maxes to resolve this interaction, we will do so for both of these responses. This is also a saving of effort. We are able to use two of the Maxes from the second segment of the fold-over design (the (1) and the acdf) to save even more effort.

Note that it is necessary to set up orthogonal sets of contrasts for interaction resolution. While in this case we have only three items per set to separate, which would require a minimum of three contrasting runs, the orthogonality requirement forces us to set up four runs per confounded set. With all of these considerations, it will take only eight new Maxes to separate the interactions and gain the information we need to create the "UltraMax" we strive to build. In Table 19-32, the boldfaced tc's are the new builds and the italicized tc's are the already built Maxes. Note that many of the "new builds" are used more than once.

TABLE 19-32

I	BD + CE + FG	A	B	C	D	E	F	G	tc	Time	Hang	Fly
	+	-	-		+	-	-	-	+	+	-	**aef**	6.87	2.67	2.00
	-	+	-		+	-	-	+	-	-	+	**adg**	11.33	2.67	0.33
	+	+	+		-	-	-	-	-	-	-	*(1)*	7.36	2.00	5.33
	-	-	+		-	-	+	+	-	+	+	**cdfg**	7.58	0.66	0.33

II	AD + CF + EG											
	-	+	-		-	-	+	+	-	+	+	**cdfg**	7.58	0.66	0.33
	+	+	-		+	+	+	+	-	+	+	**abcdfg**	6.21	0.66	0.66
	+	+	+		-	-	-	-	-	-	-	*(1)*	7.36	2.00	5.33
	-	+	+		+	+	+	-	-	+	-	**abcf**	6.53	2.33	1.00

III	AC + BG + DF											
	-	-	+		+	-	-	-	+	-	+	**aeg**	11.55	4.00	1.33
	+	+	+		+	-	+	+	-	+	-	*acdf*	3.69	0.00	0.00
	+	-	+		+	-	+	+	-	+	+	**acdfg**	7.60	1.13	0.38
	-	+	+		-	+	+	-	-	-	+	**bcg**	11.04	1.66	1.00

IV	AG + BC + DE											
	-	+	-		+	-	-	-	+	+	-	**aef**	6.87	2.67	2.00
	+	+	-		+	+	+	+	-	+	+	**abcdfg**	6.21	0.66	0.66
	+	+	+		-	-	-	-	-	-	-	*(1)*	7.36	2.00	5.33
	-	+	+		-	+	+	-	-	-	+	**bcg**	11.04	1.66	1.00

V	AF + BE + CD											
	-	+	+		-	-	+	+	-	+	+	**cdfg**	7.58	0.66	0.33
	-	-	+		+	-	-	-	+	-	+	**aeg**	11.55	4.00	1.33
	+	-	+		+	-	-	-	+	+	-	**aef**	6.87	2.67	2.00
	+	+	+		-	-	-	-	-	-	-	*(1)*	7.36	2.00	5.33

Table 19-33 shows the predicted times for the four newly constructed Maxes as well as the predicted times for the previous built Maxes based on only the single effects. To obtain the predicted value in Table 19-33, we simply solve the equation below for the design unit (+1 and -1) values of the treatment in question.

$$Y_{time} = 8.55 - 1.05 \times A + 0 \times B - .21 \times C - .52 \times D + 1.02 \times E - .94 \times F + .26 \times G$$

TABLE 19-33
Time of Descent

tc Int.	A	B	C	D	E	F	G	Obsrvd.	- Predicted=	Due to
aef	+	-	-	-	+	+	-	6.87	8.05	-1.18
adg	+	-	-	+	-	-	+	11.33	7.37	+3.96
aeg	+	-	-	-	+	-	+	11.55	10.45	+1.10
abcdfg	+	+	+	+	-	+	+	6.21	5.07	+1.14
abcf	+	+	+	-	-	+	-	6.53	5.59	+0.94
cdfg	-	-	+	+	-	+	+	7.58	7.17	+0.41
bcg	-	+	+	-	-	-	+	11.04	10.09	+0.95
acdf	+	-	+	+	-	+	-	3.69	4.55	-0.86
(1)	-	-	-	-	-	-	-	7.36	9.99	-2.63

The differences we observe are due to the interactions. We will apply these differences to each confounded set of interactions to find the influential interaction. This is done by contrasting the difference due to the interaction. For the proper signs, see Table 19-32. We will consider an effect that exceeds 0.6 (the minimum half effect to be statistically significant) as the likely interaction influence after the deconfounding is completed.

Set I: BD= [+(-1.18) -(3.96) + (-2.63) - (0.41)]/4 = -2.05 likely
CE = [-(-1.18) +(3.96) + (-2.63) - (0.41)]/4 = 0.53
FG = [-(-1.18) -(3.96) + (-2.63) + (0.41)]/4 = -1.25 likely

The thickness (B) and slot length (D) interaction is most likely with the possibility of a slot width (F) and hole diameter (G) influence. We will look at both of these interactions in Figure 19-9.

Set II: AD= [-(0.41) + (1.14) + (-2.63) - (0.94)]/4 = -0.71 likely
CF = [+(0.41) + (1.14) + (-2.63) + (0.94)]/4 = 0.31
EG = [-(0.41) - (1.14) + (-2.63) + (0.94)]/4 = -0.81 likely

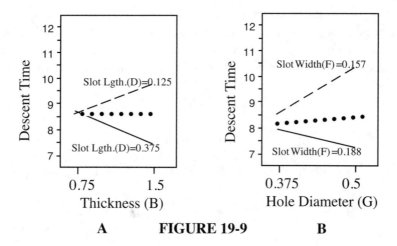

A   **FIGURE 19-9**   B

The interaction revealed in Figure 19-9 shows why in the analysis of the single effects, we found no influence due to thickness and hole diameter. On the average, there is a slope near zero for both of these factors (the dotted lines in Figure 19-9A and B.) But this is not the real functional influence of thickness nor hole diameter. Time increases with shorter slots as a function of thickness, but decreases with longer slots. What is even more important is the insensitivity to slot length variation with thinner Maxes that have smaller holes. Since slot length and slot width are difficult to control, we will use the more robust thickness (0.75) and more robust hole diameter (0.375) settings.

The inside length (E) and hole diameter (G) is an interesting interaction from set II. We will plot it and the length (A) slot length (D) interaction in Figure 19-10A and B.

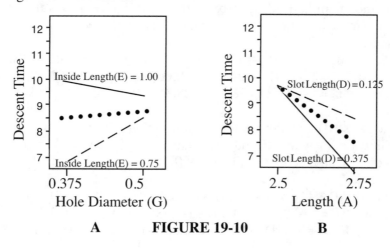

A   **FIGURE 19-10**   B

The interaction plot in Figure 19-10A shows if we use an inside length of 0.75, then we should use the 0.375 hole. Figure 19-10B indicates that a shorter time is the result of using a longer Max with a longer slot length. Both of these results will lead to the shortest descent times and are consistent with other findings thus far.

The final deconfounding for the time response is done for sets IV and V. There are no significant interactions in set IV. The length (A) and slot width (F) effect is the only significant interaction. This is plotted in Figure 19-11.

Set IV:AG = [-(-1.18) +(1.14) + (-2.63) - (0.95)]/4 = -0.32
       BC = [+(-1.18) +(1.14) + (-2.63) + (0.95)]/4 = -0.43
       DE = [-(-1.18) - (1.14) + (-2.63) + (0.95)]/4 = -0.41

Set V: AF= [- (0.41) -(1.10) + (-1.18) + (-2.63)]/4 = -1.33 likely
       BE = [+ (0.41) -(1.10) - (-1.18) + (-2.63)]/4 = -0.54
       CD = [+ (0.41) +(1.10) + (-1.18) + (-2.63)]/4 = -0.58

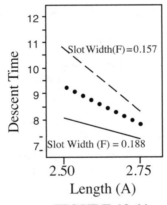

FIGURE 19-11

The length-slot width interaction indicates that the longer (2.75) Max is best because of shorter overall time as well as the lower sensitivity to slot width variation. This concept of using the setting that produces the flatter slope as a function of the second factor is an important Taguchi concept and is a part of his *parameter design* methodology.

We have completed the analysis of the time factor, and if we were to base our choice of levels on just this factor we would specify a Max with the dimensions found in Table 19-34.

TABLE 19-34

Specifications for the *acdf* Max, based on Time Response

Length (A)	2.75
Thickness (B)	0.75 (this produces less variability)
Width (C)	1.50
Slot Length (D)	0.375
Inside Length (E)	0.75
Slot Width (F)	0.188
Hole Diameter (G)	0.375 (this gives a faster time)

The only difference between these specifications and the ones based on the single factors is the choice of dimensions for thickness and hole diameter. However, we still need to look at the other two responses to decide on the final configuration. Deconfounding information is found in Table 19-35 for the hang-up response and Table 19-36 for the fly-off response.

$$Y_{hang-up} = 1.84 - .71 \times A + .17 \times B - .13 \times C - .50 \times D + .30 \times E - .25 \times F + .08 \times G$$

TABLE 19-35
Hang-ups

tc Int.	A	B	C	D	E	F	G	Obsrvd.	-	Predicted=	Due to
aef	+	-	-	-	+	+	-	2.67		1.56	+1.11
abcdfg	+	+	+	+	-	+	+	0.66		0.20	+0.46
(1)	-	-	-	-	-	-	-	2.00		2.88	-0.88
bcg	-	+	+	-	-	-	+	1.66		3.12	-1.46

$$Y_{fly-off} = 1.77 - 1.23 \times A - .48 \times B - .85 \times C - .19 \times D + .35 \times E + .02 \times F - .60 \times G$$

TABLE 19-36
Fly-offs

tc Int.	A	B	C	D	E	F	G	Obsrvd.	-	Predicted=	Due to
aeg	+	-	-	-	+	-	+	1.33		1.79	-0.46
acdf	+	-	+	+	-	+	-	0.0		0.25	-0.25
acdfg	+	-	+	+	-	+	+	0.38		-0.95	1.33
bcg	-	+	+	-	-	-	+	1.00		0.89	+0.11

The deconfounding for the hang-up response is done for set IV. These results are plotted in Figure 19-12A. The only interaction that exceeds the critical F value is DE, the slot length-inside length. This shows that if we use the shorter inside length (which the time response says is best), we should use the shorter slot length. This is contrary to the time recommendation.

$$Set\ IV:AG= [-(1.11) +(0.46) + (-0.88) - (-1.46)]/4 = -0.20$$
$$BC = [+(1.11) +(0.46) + (-0.88) + (-1.46)]/4 = -0.19$$
$$DE = [-(1.11) -(0.46) + (-0.88) + (-1.46)]/4 = 0.98\ likely$$

Deconfounding for the fly-off response is done for set III. The result is plotted in Figure 19-12B. None of these interactions are statistically significant, but we will plot the largest numerical effect since we have had problems with fly-offs and we are not intending to use the larger hole as suggested by the single factor analysis. The length–width interaction confirms that a wider Max has less variation as a function of length and also gives an overall lower level of fly-off. As Figure 19-12C shows, the slot length–slot width interaction, while not significant, is important. It shows why when the wide slots are used, more fly-offs take place with the longer slot length. For fewer fly-offs, we must use the narrow slot if the slot is long.

$$Set\ III: AC= [-(-0.46) + (-0.25) + (1.33) - (0.11)]/4 = 0.36\ largest$$
$$BG = [-(-0.46) + (-0.25) - (1.33) + (0.11)]/4 = -0.25$$
$$DF=[+(-0.46)+(-0.25)+(1.33)+(0.11)]/4=0.18\ interesting$$

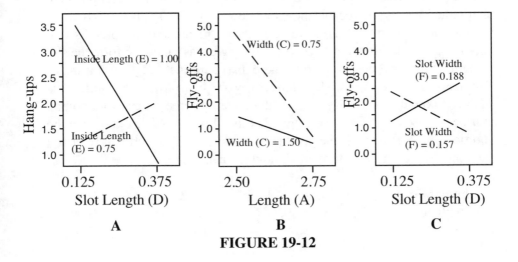

A   **B**   **C**

FIGURE 19-12

Now we may make our final recommendation for the "UltraMax" based on the all responses. Length (A) should be longest (2.75); the Max should be thin (B) (0.75), and wide (C) (1.5); the slot length (D) should be long (0.375) to avoid fly-offs; inside length (E) should be shortest (0.75); slot width (F) must be narrow (0.157) to prevent fly-offs; and the hole diameter (G) should be smaller (0.375) to go along with the choice of the inside length. This Max is the *acd* treatment. Five *acd* Maxes were built and tested to see if it is possible to manufacture such a configuration consistently. The results are shown in Table 19-37 for 10 runs of each Max.

TABLE 19-37

Serial Number		Time	Hang-ups	Fly-offs
MM001	Average	4.65	0.40	0.70
	Std. Dev.	1.09	0.70	0.95
MM002	Average	4.21	0.0	1.30
	Std. Dev.	0.69	0.0	1.06
MM003	Average	5.22	0.40	0.0
	Std. Dev.	0.54	0.52	0.0
MM004	Average	4.40	0.0	0.0
	Std. Dev.	0.40	0.0	0.0
MM005	Average	4.90	0.20	0.0
	Std. Dev.	0.37	0.42	0.0

Some of the builds performed better than others. An inspection of the individual Maxes showed that those that flew-off were slightly (by 1/16 inch) shorter at the corners. This was due to the manufacturing process that involved hand holding the Max while the angle was sanded to form the slot length. Since we observed from the experiment that length has a major influence on fly-offs, a slightly longer (by 1/8 inch) Max was built. This showed no tendency to fly-off, but did hang up three times. Later, upon further inspection it was discovered that the "Long Max" had a burr in the portion of the hole that is the pivot point for swing. This burr was removed and the "Long Max" performed without mishap.

Long Max	Average	4.91	0.30	0.0
(with burr)	Std. Dev.	0.71	0.48	0.0

We have completed the analysis of the KlatterMax experiment. The total number of Maxes built amounted to 24 for the experiment and another 6 for confirmation. For such a complex physical problem, this small expenditure of resources is a testimony to the power of Statistical Experimental design. Had an engineer used the "poke-and-hope" approach, which is so typical, the number of Maxes would have been far greater, and worse, the information obtained would be far less.

For average effects we have found a good design specification. However, there is still variation in the responses. We will investigate the KlatterMax problem in Chapter 21 and delve further into the manufacturing operation and see how different considerations can possibly lead to other design specifications that will make the manufacturing operation more robust or tolerant to manufacturing variation.

Problems for Chapter 19

1. Show why all 7 effects (5 single and 2 two-factor interactions) cannot be obtained with a 2^{5-2} fractional factorial design as discussed in the popcorn experiment, Phase II.

2. Show how dropping the power factor solves the confounding problem in the second phase of the popcorn experiment.

3. Run the YATES ANOVA on the 3 responses in Table 19-8 to confirm the findings reported in the conclusions.

4. If temperature and humidity were likely to interact as well as pH and Metol, would the design shown in Table 19-13 be adequate? State your reasons and suggest a way to get around this difficulty.

5. Compute the error mean square from the replicates in Table 19-19.

6. Solve the modeling problem described on the next page. Use the simulation computer program supplied to generate the responses.

Practice Problem: Photographic Emulsion

Response: Image Quality (0-100)

Controlling Variables:

	Working Range	Variation[*]
AgNO$_3$	5 to 15 g/l	0.2
Gelatine	20 to 40 g/l	0.5
KB	5 to 15 g/l	0.2
KI	1 to 5 g/l	0.2
Temperature	90° to 120°	0.5
Relative Humidity	50% to 80%	2.0
Speed of Dumping	Fast (1) or Slow (0)	–
Length of Mixing	5 to 20 min.	0.25
Size of Batch	1 to 10 liters (l)	0.25

[*] The variation is expressed as 1 sigma. So, 0.2 means that in setting the KBr it is possible to set 5 g/l from a low of 4.4 (3 sigma lower limit) and as high as 5.6g/l (3 sigma upper limit).

Costs: $500 per treatment combination

Gains: $2000 for each unit of IQ above 50

Your job is to find the model and equation for the emulsion process and then use the model to specify the levels of the controlling factors as well as their functional tolerances to obtain an image quality of 100 with a standard devation of 1.25.

The Emulsion simulator may also be found with other simulators on the Web at: www.crcpress.com by clicking on the link "Downloads and Updates" in the Electronic Products section.

BASIC Program to Generate Emulsion Example Responses

```
 70 RS=VAL(RIGHT$(TIME$,2))
 80 RANDOMIZE (RS)
 90 CLS
 95 DIM N$(9), X(9)
100 PRINT "ENTER THE VALUES OF YOUR EMULSION FORMULATION"
105 PRINT
110 FOR I=1 TO 9
115 READ N$(I)
120 DATA AGNO3,GELATINE,KBR, KI, TEMP
140 DATA RH,SPEED,MIXTIME,BATCHSIZE
150 NEXT I
170 FOR I=1 TO 9
180 PRINT N$(I);
185 INPUT X(I)
190 NEXT I
220 Y=-4*X(1)+X(3)*X(4)+10*X(5)-.05*X(5)*X(5)+10*X(7)-465
240 Y1=Y
260 TT=0
300 FOR J=1 TO 24:R=RND(1):TT=TT+R:NEXT J
320 RE=1*SQR(.5)*(TT-12)
380 Y=Y+RE
420 PRINT"IMAGE QUALITY=";Y
440 PRINT:PRINT "MORE? IF SO ENTER 1":INPUT M
450 PRINT
480 IF M=1 THEN 170
500 END
```

SAVE"EMUL"

If you enter the program in GWBASIC and save it as shown above, you will be able to generate image quality (IQ) responses for the experiment you design for Problem 6. The next page shows two sample runs of this program.

Use of Emulsion Program

LOAD"EMUL"
RUN

ENTER THE VALUES OF YOUR EMULSION FORMULATION

AGNO3? 7.5
GELATINE? 30
KBR? 7.5
KI? 2.5
TEMP? 105
RH? 65
SPEED? .5
MIXTIME? 12.5
BATCHSIZE? 5
IMAGE QUALITY= 27.48124

MORE? IF SO ENTER 1
1

AGNO3? 7.5
GELATINE? 30
KBR? 7.5
KI? 2.5
TEMP? 105
RH? 65
SPEED? .5
MIXTIME? 12.5
BATCHSIZE? 5
IMAGE QUALITY= 26.62939

MORE? IF SO ENTER 1
0

Notice that each time the program is run, you will get a different answer. This is random, residual error that is part of the program that simulates the real world!

APPENDIX 19

The KlatterMax experiment was presented to students at the Center for Quality and Applied Statistics at Rochester Institute of Technology in an advanced experimental design course as part of their M.S. program in Applied Statistics. They were shown several prototypes of the Max design and given the required performance characteristics. The class brainstormed the problem and created the list of factors that they subsequently used to elevate only seven of these factors to the first design.

The treatment combinations of this design were constructed using a 10-inch hollow ground table saw, drill press, and 10-inch disk sander. Appropriate finishing sanding was done to remove burrs from the wood. The wood used was dimensioned bass wood, which is a semisoft, straight-grained material.

Production of the KlatterMax using drill press equipment.

Response data were obtained and the class was given the assignment of analysis. Part of their recommendations was the second segment, fold-over design. This was constructed and the Maxes were built. Another group of students in the same class, but in a different quarter, was given the data from the fold-over design and assigned to do its analysis. Their recommendations led to the deconfounding runs for the interaction effects and the subsequent manufacturing specifications.

The author then did further work to build a regression model that will be used in Chapter 22 for the robustification process using the Taguchi concepts.

One of the Max configurations as it flies off.

PART V

Utilization of Empirical Equations

20

Robust Design

In the early 1980s when the quality of products made in the United States was considered subpar and the products from Japan were considered "world class," an engineer by the name of Genechi Taguchi (1) began to (and I quote) "share my ideas of quality control with my colleagues in the United States." This author was fortunate to have studied these ideas under Dr. Taguchi directly while employed in the R&D division of Xerox Corporation, which had engaged Taguchi as a consultant. Taguchi's ideas go far beyond the application of experimental design but they utilize experimental design in the implementation of "infused quality."

No book on experimental design would be complete without exploring Taguchi's ideas of quality and his approach to quality infusion.

The Concept of Quality

In the western world we usually have a difficult time defining that elusive concept of quality. Quality is that characteristic that we can feel and know when we have it, but we can't describe it in advance. Taguchi combines both business and engineering concepts in his definition.

Quality is the loss to society from the time the product is shipped. (20-1)

Now, we usually think of quality as a positive attribute and since the word "loss" is a negative concept, we may simply add the words "avoidance of" before the word loss to make the idea work in our society. So the westernization of Taguchi's definition of quality then becomes:

> *Quality is the <u>avoidance of</u> loss to society*
> *from the time the product is shipped.*   (20-2)

This modification in no way diminished his basic concept, which is to lower loss; with such lower losses, we have the type of products, processes, and services that delight our customers.

But Taguchi goes beyond the definition of quality in monetary terms. He has created the idea of a loss function that links monetary units with engineering characteristics. While loss functions are not new to the world of the MBA, **Raiffa and Schlaifer** (2), Taguchi introduced this concept to the world of engineering. In his consultating activities with a number of Japanese companies, he has developed a large number of market research based loss functions. All of these are proprietary and are not in the public domaine. However, there is a generic loss function that can cover a wide variety of situations and has the same form as found on page 188 of reference 2. This generic loss function defines the loss at a particular point (point loss function).

$$\mathbf{L(x) = k(\ x - m\ )^2} \qquad\qquad (20\text{-}3)$$

where $L(x)$ is the loss at the value of x

k is a constant for the monetary units

x is the point and

m is the aim point (or target) where no loss takes place

Expected Loss (for a Distribution)

Examples of the application of this point loss function are found in **Barker** (3). What is more important is the extension of this point loss function to the loss for a distribution that is named the "Expected Loss." Again, reference 3 shows the origin of the expected loss. The important aspect of the expected loss is the idea that there is a penalty for being off target and a penalty for variation.

$$\mathbf{EL = k\ [\ (\overline{X} - m)^2 + s^2\ ]} \qquad\qquad (20\text{-}4)$$

> where EL is the expected loss
> k is a constant for the monetary units
> $\overline{X}$ is the average of the distribution
> m is the aim point where no loss takes place (target)
> and s is the standard deviation of the distribution

Let's examine this very important component (20-4) of the quantification of quality in a bit more detail.

The expected loss (**EL**) is the average monetary value we expect to accumulate for the distribution under study (usually a normal distribution). The expected loss is a function of the location of the average of the distribution with respect to the targeted value (which the customer needs for proper operation). If the average is not the same as the target, then there is a penalty (increased loss) due to being off target. The second part of the expected loss is that due to variation. As the standard deviation increases the loss increases as its **square**. This is the penalty for having variation. The "k" is merely a scaling factor to put the loss in monetary terms.

> *It is interesting to observe that in the field of quality control a great deal of emphasis is given to the location ($\overline{X}$) while in practice the variation (s) is given more or less "lip service." QC practioners will strive for the right average output and often neglect the variation. The expected loss gives both of these components of quality equal emphasis. The application of the concepts of Robust Design enables the QC engineer to control **both** the location and the variation to produce the products, processes, and services that a quality conscious world expects today. Watch as we show that process.*

An Application of the Expected Loss Function

Example I shows a process that is a function of a variable that has an influence on both the location (average) and the variation (standard deviation). The customer needs an average value from the distribution of output that is 4.0 and can tolerate a variation no more than 0.5 (standard deviation units). Figure 20-1 shows two proposed configurations for the process designed to produce the component. The setting I of the parameter gives the correct average of 4, but the standard deviation is 1, which is twice the customer's tolerable variation. Setting II is 2 units lower than the specified average, but has a standard deviation of only 0.33, which is smaller than the customer's tolerable variation. There is a need for a trade-off.

However, before we go ahead and try a trade-off, let's compute the expected loss for these two situations. This way, we will be able to see just how much our process manipulations are worth – or if there is any merit in making changes to find a third setting. For simplicity, we will use a k value of 1.

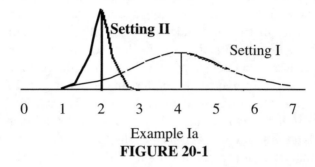

Example Ia
FIGURE 20-1

Expected Loss for Setting I:

$$EL = k \, [\, (\overline{X} - m)^2 + s^2 \,] \tag{20-5}$$

$$EL = \$1 \, [\, (4 - 4)^2 + 1^2 \,] = \$1 \tag{20-6}$$

Expected Loss for Setting II:

$$EL = k \, [\, (\overline{X} - m)^2 + s^2 \,] \tag{20-7}$$

$$EL = \$1 \, [\, (2 - 4)^2 + 0.33^2 \,] = \$4.11 \tag{20-8}$$

It is quite clear that the lower variation (but off-target) situation is the worse of the two cases. Many quality engineers will select setting II based on the lower variation and hope to change the average toward the target by some other means. That elusive "other means" often does not manifest itself in the process, and there are tremendous quality problems for that product.

By knowing the functional realtionship between the controlling parameter and the response, it is possible to find a set of conditions that better satisfy the cusiomer's needs and thus produce a lower loss. Figure 20-2 shows a new condition (Trade-off I) that does just that. When we compute the expected loss for this (see homework problem 1) we find a lower loss than either of the earlier settings. (For Trade-off I, the mean is 3.5 and the standard deviation is 0.5).

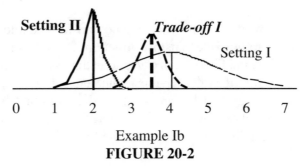

Example Ib
FIGURE 20-2

The Signal to Noise Transform – or Finding the Elusive Loss Function

In the previous example, we have seen the power of the expected loss in driving a process to deliver a response close to target with low variation. In this example, we assumed a k value of 1 for simplicity.

In most situations, the value of k is elusive, difficult to determine, and usually not a believable numerical value. If then, this is the most likely case, the power of the loss function is essentially lost. However, if we treat the loss function as a relative figure of demerit, it is possible to make relative, comparative assessments as we have done in the above example.

It is possible to transform the loss function into a utility function. In doing so, we lose the financial implications that the loss function holds (if we could ever establish viable financial implications in the first place). The utility function developed by Dr. Taguchi is called the "Signal to Noise Ratio (S/N)." The name stems from his work at Nippon Telephone & Telegraph (NTT) where he began his work on quality in the Electrical Communication Laboratories, the home of many electrical engineers who were acustomed to the metric S/N in their daily work. Taguchi wanted to use a familiar term and he adopted the name S/N for this purpose.

The units of S/N are measured in decibels (dB). The greater the value of S/N, the better the design. Think of the term S/N in your own daily life with regard to electronic appliances used to record and play music. If you recall the original phonograph records with their scratchy sound or even the hiss of recorder tape, you would agree that there is a rather large noise in these devices. Therefore the signal (the music you want to hear) is sometimes lost in the noise (the scratchiness or hiss). The S/N of such recorded music is quite low when compared with the sound recorded on a modern compact disk (CD).

Let's see how the loss function can be transformed into the S/N. We will use Example I to show the transform in action. Consider the fundamental concept that it is desirable to produce an output that is always on target. The target is, then, the **signal**. However, due to inherent variation, there will be a distribution of output around this target. This variation around the target is the **noise**. Using the symbols T for the target and S_T for the noise, the S/N is then defined as (since the S/N has units of dB, the Log_{10} is used):

$$S/N = 20 \ Log_{10} \ \frac{T}{S_T} \qquad\qquad (20\text{-}9)$$

The definition of the target (T) is easy. It is merely the customer-specified value that satisfies functional performance. The target is a constant for each process we investigate. The noise (S_T) is a bit more complex, but, given our new knowledge of the loss function, not too difficult to comprehend. Recall (expression 20-2) that deviations from the target (the target was symbolized as m in expression 20-2) cause losses. The definition of the expected loss (expression 20-4) pulls both of these principles together and shows that distributions of output that deviate from target and have variation will have larger average losses than processes that adhere to the target with little variation around this target. If we consider this in a mathematical expression which we call the "Target Deviation," we can create a utility function.

The target deviation (S_T) is defined as the average deviation from the target. There is a similarity in the S_T to the standard deviation (s) in as much as the target value (like the mean value) can be surrounded by the deviating values and the summation of these deviations could be zero (it is always zero in the case of the standard deviation since the mean is *always* in the center of the data). Therefore, as in the case of the standard deviation, we square the differences between the target and the values that deviate from this target. This gives the following expression for the target deviation:

$$S_T = \sqrt{\frac{\Sigma\,(x - T)^2}{n - 1}} \qquad\qquad (20\text{-}10)$$

S/N Compared with Expected Loss

Recall from Example I these conditions and the ensuing expected losses:

TABLE 20-1

	$\overline{X}$	s	Expected Loss	S_T	S/N
Setting I	4.0	1.0	$1.00	1.052	11.6dB
Setting II	2.0	0.33	$4.11	2.024	5.9dB
Trade-off I	3.5	0.5	(problem 1)	0.721	14.9dB

Using Minitab to create normal distributions of 1000 values for each of the above conditions, it is possible to then compute the target deviation and S/N as shown in Table 20-1 for each condition.

To make a linear comparison of the S/N differences to the expected loss differences, it is necessary to convert the S/N decibel (dB) values to linear values. To do so, we need to first divide by 10 ("deci" means 10) and then find the anti-log of this result. When this operation is completed, we find the following comparisons shown in Table 20-2.

TABLE 20-2

Comparison	Δ dB	Ratio Linear	Ratio Loss
Setting I - Setting II	5.7	3.7	4.1
Trade-off I - Setting I	3.3	2.1	2.0

In theory, the Ratio Linear should be exactly the same as the Ratio Loss. However, the simulation using 1000 normally distributed observations has a random error, and yet the results are quite close to the theoretical. If we were to do this simulation with more observations, the experimental results would converge more closely to the theoretical truth. In the next chapter on simulations, we will address this point.

This simulation has shown by example that the S/N is fulfilling the function of the Expected Loss. The k value in the underlying loss function behind the S/N is unity. While the S/N cannot be used to determine accounting objectives, it is a relative figure of *merit* that can help engineers determine optimum conditions that put a process on target with low variation.

Using the S/N in Optimization Experiments

Now that we have a figure of merit to maximize (remember the S/N is always maximized), let us see how we can use it to optimize product designs, processes, and any phenomena that are amenable to experimentation. Along the way we will discover other forms of the S/N that are based on the fundamental **Target Type** we have just developed which is commonly referred to as the **S/N$_T$**.

Remember, we need to give equal emphasis to *both* the location and the variation. The experiments we have done thus far have looked at the location. If we need to add the dimension of variation, we must study variation. If we need to study variation, differences in variation must occur in the experiment, just as differences in location must occur when we study the effect of our factors on the response's location.

We have already observed that variation does occur in our experimental data, and, in fact, we use analytical methods (like ANOVA and GLM) to characterize the amount of variation in comparison to the differences we observe as a result of the changes we make. The type of variation we have become accustomed to in our experiments is more of a plague than a variable we want to characterize. This is a stumbling block that often confuses traditional experimenters who encounter Taguchi's robust design approach. The confusion over random error and induced variation that is to be studied as well as the seemingly huge experiments that Taguchi proposes has resulted in the wholesale rejection of the Taguchi Approach by many prominent statisticians. This is an unfortunate rejection, for the concepts are a part of a dramatic new improvement in quality thinking.

The Parameter Design

The parameter design is the most important activity for product and process design engineers. We will outline the fundamental concept of parameter design and then show how the parameter design concept can be accomplished with far fewer experimental observations.

We begin by introducing the idea of inducing variation in the experiment. This induction of variation is purposely done so we may understand the functional relationship between variation of the response and the controlling factors. We have expanded our objective to look at location as well as variation. Therefore a new general objective looks like this:

> **Objective: To test the hypothesis that the response and its variation are functions of the controlling factors.**

Do we really need to induce variation? By now, you should have a lot of insight into *where* the variation comes from when we run and experiment. It's not just our inability to measure the response precisely. The variation is a function of our inability to set the conditions exactly on the prescribed conditions set forth in the experimental design matrix.

Take, for instance, the following small mechanical engineering experiment that looks at the relationship between the outside diameter (OD) of an aluminum tube and three machining factors. (The "data" shown throughout this example is generated via a simulation found on the Web at: www.crcpress.com by clicking on the link "Downloads and Updates" in the Electronic Products section.) A diagram of the device called a "Flo-former" is in Figure 20-3. All factor dimensions are in thousandths of an inch.

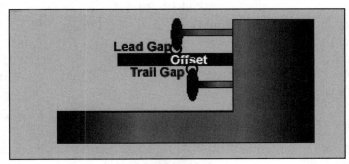

FIGURE 20-3

TABLE 20-3

tc	A Lead Gap	B Trail Gap	C LG - TG Offset	OD
(1)	65	45	60	2.48128"
a	80	45	60	2.49896"
b	65	60	60	2.49703"
ab	80	60	60	2.51471"
c	65	45	110	2.48128"
ac	80	45	110	2.49896"
bc	65	60	110	2.49703"
abc	80	60	110	2.51471"

In a "perfect world" the outside diameter (OD) response would be exactly as shown in Table 20-3 for each setting of the Lead Gap, Trail Gap, and Offset. But when we attempt to set a specific value of the factor, we can come close, but never get the exact setting. This inability to make the setting of the factor's level (say Lead Gap at 65 thousandths) creates variation in the (OD) response if there is a functional relationship between that response and that factor. Let's see what a change of minus one thousandth does to the first tc of the design in Table 20-3. Table 20-4 shows the original OD for tc (1) of 2.48128 inches and the modified run with the Lead Gap one thousandth lower at 2.48010 inches. This is a difference of 0.00118 inches. Now, you may say that such a difference of a little more than a thousandth of an inch is not very much, consider this amount of change in the light of a tolerance on the final product that has a standard deviation of 0.0075 inches. That "slightly more than a thousandth" is 15% of the tolerance!

TABLE 20-4

tc	Lead Gap	Trail Gap	LG - TG Offset		OD
(1)	65	45	60		2.48128"
LG modified	**64**	**45**	**60**		**2.48010"**
				$\Delta =$	**0.00118"**
All modified	*64*	*44*	*59*		*2.47905"*
				$\Delta =$	*0.00223"*

Looking further in Table 20-4, where *all* the factor settings have been reduced by a thousandth of an inch, we see that the difference is even larger. But what if the settings of the factors moved from the ideal, desired points in a more unpredictable manner and not always in one direction or the other, *and* in larger or smaller increments?

Process Capability Study Approach

What we have just speculated about is exactly what happens in production processes and causes these processes to produce variable output. A common method used by quality engineers to quantify such variation in the output of a process is called a "Process Capability Study." Figure 20-4 shows the schematic outline of a process capability study that could be applied to the three factor Flo-former process. Please note that the sampling points shown in Figure 20-4 do not represent the systematically chosen values from Table 20-4. The points are randomly selected from the distributions, for this is the manner in which a process capability study is conducted.

So in a process capability study, each factor has an underlying frequency distribution with a mean and standard deviation. Each time the

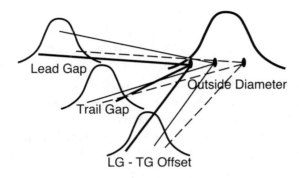

FIGURE 20-4
Schematic outline of a process capability study.

process is set up or a part is made, a random selection of a value of the setting for each factor is made from the distribution of that factor. Then together with the other factors' settings that were selected in the same random manner, an output is produced that depends on the functional relationship between the response and these factors.

Let's watch the beginning of a process capability study with the three illustrative setups for the Flo-former example shown in Figure 20-4. Only three setups have been chosen to make the diagram less cluttered. In reality, a process capability study requires a minimum of 30 setups.

Follow the thin solid lines that show samples from the middle of the lead gap factor's distribution; from the lower end of the trail gap factor's distribution; and a bit to the right of center of the offset factor's distribution. This produces a response about one standard deviation below the average of the outside diameter's distribution.

The dashed thin line has a lead gap setting above the mean of the factor's distribution; the trail gap setting is in the middle of the factor's distribution; and the offset setting is to the right of center of the factor's distribution. These produce an OD that is about one standard deviation above the average.

The final solid bold lines indicate samples from the low end of the lead gap factor's distribution; from the high end of the trail gap factor's distribution; and from the low end of the offset factor's distribution which produce an OD about two and one half standard deviations below the average.

Again, the sampling from each factor's distribution would take place until sufficient setups were made to begin to see the pattern of variation in the response's distribution and produce sufficient confidence intervals on the population parameters (μ, σ) of the response. This is normally about 30 setups for a process, or 30 parts for a piece part operation.

One procedural element that is very important in the execution of a process capability study is to make sure the observations studied are genuine replicates and not simply multiple observations of the same setup conditions. If multiple observations (which are faster and easier to do) are used, the variation will not represent the genuine setup induced variation. Such "easy to do" produced variation will be much lower than the genuine setup induced variation. This will give the process and quality engineers a false sense of security in the process, but that false security will come crashing down as soon as the process goes into full production.

Another false sense of security in the process based on a poorly executed process capability study emerges when the operators are aware of the process capability study and are "on their toes" and do a much better job of controlling the variation of the input factors. Thus the variation is never true and often underestimated. A remedy for this is to make the process capability study "blind" so the operators are not aware that something special is happening.

A Designed Approach

While the process capability study is a common practice in the field of quality engineering, there are (as we have cited) some possible operational aspects of this approach that can and often do give false information on the true capability of a process.

There is also one very fundamental, intrinsic aspect of a process capability study that is not very satisfying. If the process shows a poor capability, there is very little that can be done to analyze the data in order to rectify the deficiency. This is because the values of the settings of the controlling factors are usually never measured or recorded. And, if these settings were recorded, the random nature of the process capability study makes any analysis quite confounded.

Enter Statistical Experimental Design. If we were able to induce the variation (instead of letting it happen by chance) in a systematic manner, then we could do an analysis of the influence of the transmitted variation from the controlling factors to the response(s). This purposeful induction of variation would also overcome the problem with operators being "on their toes," since we would be prescribing specific changes for them to make. Plus, we would not need to be concerned with multiple observations of the same setup, since we would be making specific setups for each trial in the new *Designed Process Capability Study*." Let's see how this would work with the machining process example.

The first tc of the design matrix from Figure 20-3 is shown below:

tc	Lead Gap	Trail Gap	LG - TG Offset	
(1)	65	45	60	(20-11)

We will take the settings shown in excerpt 20-11 and systematically change them using the concept of "low cost tolerances" which in this case we will define numerically as a thousandth of an inch. In general a "low cost tolerance" is a component variation or process setting variation that is likely to be rather wide and subsequently inexpensive for production.

Beyond Just "Does it Work?"

The Parameter Design goes far beyond the single question a process capability study can answer, which is: "Does it Work?" The Parameter Design is a part of the engineering process and "Makes it Work!"

By inducing the changes in what is called the "Noise Matrix," we can see the effect of the of the low cost tolerances on the variation of the response and also link these changes in the variation to the nominal settings of each factor to find the optimal set points that will put the response on target with low variation (and if the low cost tolerances are sufficient) at low cost!

While this goal may sound like a "tall order," it can be accomplished with about the same effort that an ordinary process capability study requires. We will expend similar resources, but in the end, have more knowledge and process/product assurance by using Parameter Design.

TABLE 20-5 (NOISE MATRIX (1))

| | A | B | C$\approx$AB | |
tc	Lead Gap	Trail Gap	LG - TG Offset	OD
(c)	64	44	61	
a	66	44	59	
b	64	46	59	
ab(c)	66	46	61	

The noise matrix in Table 20-5 makes the purposeful changes to the three factors and utilizes a half fractional factorial design to do so. Is this efficient?

Certainly, the amount of work is minimal, but is the information we need available? Remember the criterion for efficiency is not just doing a minimal amount of work, but "**obtaining the required information** at the least expenditure of resources."

With this highly confounded experimental structure, do we keep the double edged sword of efficiency sharp on both sides? Yes, since the only purpose of the noise matrix (that's what we have in Table 20-5) is to create variation, and not to analyze for effects, we *may* use such highly confounded experimental configurations.

We will use the noise matrix *only* to calculate the average, the standard deviation, and the S/N. There will be no testing for significant single effects or interactions in the noise matrix. Its only purpose is to act like the process capability study. But the noise matrix is better than a process capability study since it is smaller (fewer runs) and automatically covers the full range of variation (unlike the process capability study where the operators are usually "on their toes" and allow less variation during the process capability study than will show up in the reality of production).

The noise matrix emulates the same level of variation that will show up in the reality of production! This alleviates the lament of the engineer, "It worked in the qualification test, but it failed in the field." Parameter Design does the job faster and with more engineering confidence.

Structure of the Parameter Design

Figure 20-6 shows the general structure of the Parameter Design. The Parameter Design consists of two experimental design structures. We have seen the noise matrix which creates the variation. The "design matrix" looks a lot like an ordinary experimental configuration we would use to study the influence on a response. Indeed it is! The big difference in a parameter design, is that this design matrix looks at two responses. It looks at the ordinary location characteristic (the mean) and also looks at the variation. Both of these responses are created in the noise matrix.

Think about an ordinary experiment that is focused only on the mean response. We get this mean by replicating the observations for a particular treatment combination. We have seen that the reasons for variation in the replicates is dependent on the ability to set up the runs and variation in the settings of the controlling factors causes the variation in the responses (plus any measurement error). So, if we purposely induce variation via the noise matrix, we are doing nothing more than emulating reality.

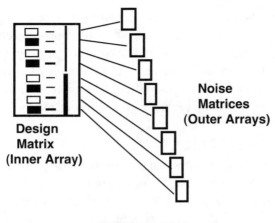

FIGURE 20-6

When Taguchi (4) first introduced his Parameter Design concept, he named the first part the Inner Array and the second parts the Outer Arrays. While these names are descriptive of location, a better set of names were developed by Kackar (5). These are the names we have been using. The original Taguchi names are referenced since many practitioners still refer to the design matrix as the inner array and the noise matrix as the outer array.

Let's see how much work is involved in the Flo-former process experiment when we complete the Parameter Design. Look at Figure 20-6 and use the first three columns (A,B,C) of the design matrix for the three factors (Lead Gap, Trail Gap, and Offset). There will be 8 tc's in this full factorial, 2-level design matrix. Each tc of the design matrix must have a set of runs that generate the systematically induced variation. These are the noise matrices. If we use the noise matrix suggested in Table 20-5, where there are four perturbing runs, there will be a total of 32 runs in this experiment.

If we had done just one process capability study (30 runs), we would have expended about the same amount of effort, but we would not have been given much direction after the process capability study had been completed and would have certainly needed to run another process capability study if the process had not proven capable. So the process capability study approach would have most likely required 60 observations.

Let's consider 60 observations as a more likely scenario in such a study and introduce another critical concept to the robustification process – the concept of "Outside Noise." While the process engineer has control over the settings in the flo-former device and can devise methods for setting and holding them on the correct targeted settings with the tolerable amount of variation in making these setups, there is often no control (except to pay high prices) over the incoming raw material (the tubes that are being machined). This is where the Parameter Design demonstrates its unique ability to accomplish the lofty ultimate goal of engineering, "To design a product, process, or service that is on **target** with **low variation** and at **low cost**."

Outside Noise Factors

An outside noise factor is beyond our engineering design authority. We can't specify the tolerance of and sometimes even the level of an outside noise factor. Environmental conditions are an extreme example of outside noise. If our output response is functionally related to the weather, we can't do much about that influence since we never know what the weather will be doing!

In the early days of electrostatic copy machines, the relative humidity played havoc with the image quality. The damp summer made smudged, dirty copies, and the dry winter had weak, light copies. Only when the materials used in the process were robustified against this outside noise of moisture did this problem go away.

The definition of an outside noise factor
An outside noise factor is beyond our engineering design authority. We can't specify the tolerance of and sometimes even the level of an outside noise factor once the product has left the source of production.

The idea is to make the product robust against the outside noise. To do so, we need to include the outside noise in the Parameter Design. Now, we will not make a study of this outside noise factor to find its best setting, since we may not select a setting value for an outside noise. Instead, we will look for the settings of the controlling factors where the influence of the outside noise is diminished.

The best way to see this is to continue the Flo-former example. The noise matrix now includes the outside noise factor, tube wall thickness. The vendor can supply tubes with wall thicknesses ranging from 78 to 100 thousandths of an inch. The low cost tolerance on thickness is five thousandths (one standard deviation). Table 20-6 shows the new design matrix and Table 20-7 is the first noise matrix for the Flo-former. We have included the outside noise factor since we need to find the level that is best suited to our requirements of an outside diameter of 2.5000 inches with a standard deviation of 0.0075 inches. The number of runs in this design has increased since we have added the noise, but also utilized a half fraction design (Template T8-4, Chapter 5 Appendix).

TABLE 20-6
Flo-former Design Matrix

tc	A Lead Gap	B Trail Gap	C Offset	D Tube	Response* Average OD	S/N_T
(1)	65	45	60	78		
a(d)	80	45	60	100		
b(d)	65	60	60	100		
ab	80	60	60	78		
c(d)	65	45	110	100		
ac	80	45	110	78		
bc	65	60	110	78		
abc(d)	80	60	110	100		

*Responses will come from noise matrices

We will show the noise matrix for treatment combination (1) of the design matrix and complete the remaining noise matrices to provide the average and type T S/N (since we are targeting an OD of 2.5000"). Table 20-7 shows the noise matrix for tc (1).

TABLE 20-7
Flo-former Noise Matrix

tc	A Lead Gap	B Trail Gap	C Offset	D Tube	Response Outside Diameter
(1)	64	44	59	73	2.47476
a(d)	66	44	59	83	2.48519
b(d)	64	46	59	83	2.48545
ab	66	46	59	73	2.47973
c(d)	64	44	61	83	2.48335
ac	66	44	61	73	2.47763
bc	64	46	61	73	2.47686
abc(d)	66	46	61	83	2.48729

Table 20-8 shows the results of the experimental effort done via the noise matrices. Notice that we have two response variables. We will find the functional relationship between the average outside diameter (OD) and the four controlling factors. This will give us a handle on the effect of these factors on the location. We will also find the functional relationship between the S/N_T and the controlling factors.

While the S/N_T will have the ability to guide us to the factor level selection, the Average OD is an additional help if the physical function includes a factor that has an influence on the mean, but little if any influence on the variation. This type of factor (has no variation influence, but has location influence) is called a "signal factor."

TABLE 20-8
Flo-former Design Matrix with Responses

tc	A Lead Gap	B Trail Gap	C Offset	D Tube	Response Average OD	S/N_T
(1)	65	45	60	78	2.48128	41.7
a(d)	80	45	60	100	2.49976	66.4
b(d)	65	60	60	100	2.51479	43.6
ab	80	60	60	78	2.51471	44.0
c(d)	65	45	110	100	2.49904	54.7
ac	80	45	110	78	2.49896	61.7
bc	65	60	110	78	2.49703	53.0
abc(d)	80	60	110	100	2.51551	43.5

It is very important to understand the difference between the outside noise as it applies to the "field" and to the development activities associated with Parameter Design. During the design-development activities, the outside noises are controllable by the engineering team. Environmental conditions are often emulated in environmental chambers. Special materials can be ordered from the vendor or made in the model shop. Engineers are accustomed to these concepts and often practice such "stress testing" to see if a design will work under extreme conditions.

The advantage of the Parameter Design over ordinary stress tests is the ability to use the stresses to find the set points of the controlling factors that make the overall design robust to the outside noises. Parameter Design does not just have a goal to see if something works, but has a goal to *make* the process, product, or service work under any noise that is "thrown against it!"

In this way, the Parameter Design is analogous to the medical practice of developing a vaccine for a disease. The vaccine is usually a form of the dead virus. The human body creates the anti-virus upon injection of the vaccine and builds an immunity to the disease.

Parameter Design creates a vaccine (knowledge of the right settings) to ward off the bacteria of variation. To create the vaccine, the Parameter Design must become infected with the virus (the noises) which takes place in the noise matrices. The analysis of the experiment gives us the information on how to set the controlling factors to prevent the virus (variation) from infecting our product, process, or service. We experiment, study, and characterize variation in the Parameter Design to ward off its effect.

Formulating the Vaccine from the Noise Matrices

The eight observations were made in each of the eight noise matrices following the pattern established in Table 20-7. It is with these noise matrices that we create the variation (combined with the location) as the S/N_T and also separately as the location, the Average OD (outside Diameter) which is summarized in Table 20-8.

While we might be tempted to simply "pick the winner" with the highest S/N (which is tc a(d)), we know that a systematic analysis which allows us to find trends is a far better way to understand our process and fulfill the goal by accomplishing the objective. (See Problem 6 for the writing of the Goal and Objective.)

A simple contrast analysis of the S/N_T response reveals the following shown in Table 20-9. The bold values are the best settings for the factors. The average S/N_T are shown for each setting with the italic indicating the larger S/N_T.

TABLE 20-9

Lead Gap		Trail Gap		Offset		Tube Thickness	
65	48.3dB	**45**	*56.1dB*	60	48.9dB	78	50.1dB
80	*53.9dB*	60	46.0dB	**110**	*53.2dB*	**100**	*52.1dB*

Upon trying these settings, we come very close to the required diameter of 2.5000 with 2.4998. This shows the simplicity and the effectiveness of such a design/analysis approach. However, we will refine this analysis using Minitab and show another approach that is more traditional in its foundations.

Using Minitab for Parameter Design

We will now create the parameter design for the Flo-Former using Minitab software which has a "Taguchi" option in the "DOE" section. As usual, we begin with Stat>DOE> but now go to "Taguchi" and "Create Taguchi Design" as shown in Figure 20-7.

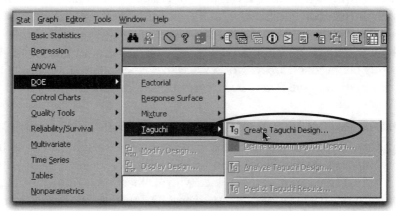

FIGURE 20-7

There are many types of "Taguchi Designs" available in the Minitab catalog. Click on the button "Display Available Designs" from the main dialog box (shown in Figure 20-9) to show the table of available designs that is reproduced in Figure 20-8. Appendix 20-1 describes selected 2 and 3 level design configurations and shows the column assignments as well as the "equivalent" columns from traditional, ("classical") design configurations that we have developed in Chapters 4, 5, and 7. The 4 level and 5 level designs are hybrids that should not be used, due to undefined confounding patterns.

For our example, we need a 2-level design with 4 factors that we

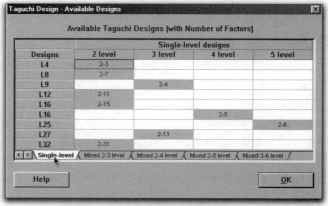

FIGURE 20-8

select as shown in Figure 20-9. Next, click on "Designs" where we choose the L_8 by highlighting the line as shown in Figure 20-10.

FIGURE 20-9

Since we have chosen only 4 factors, the L_8 is shown as a 2^4 (note that Minitab uses the notation "**" to indicate exponentiation). The L8 is capable of holding up to 7 factors (see Appendix 20-1).

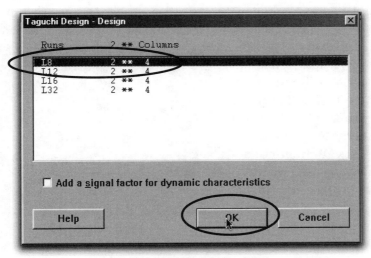

FIGURE 20-10

From the main dialog box, we choose "Factors" (which had been grayed out before we picked the design) and here is where we get to name the factors, add the levels, and select the columns from the design we want to use. The default names are simply letters of the alphabet and the default levels are 1 and 2. In Figure 20-11, we have given the proper physical names and levels

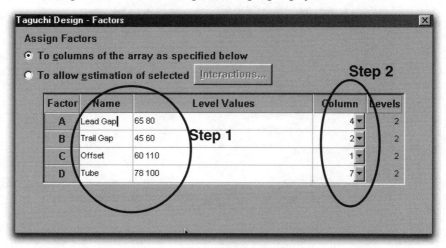

FIGURE 20-11

to the four factors we are investigating in this design matrix (Step 1). More importantly, we have scrolled through the column designator (Step 2) and chosen the correct columns (the ones that match the ones chosen in the previous "by hand" design we did earlier [Table 20-6] in this chapter). The L_8 design shown in Appendix 20-1 is very helpful in translating the columns of the Orthogonal Array (OA) into the equivalent columns of a classical 2^{k-p} experiment.

Click **OK** from the **Factors** dialog box and we return to the main dialog box (Figure 20-12). The only option (which is already clicked as default) is

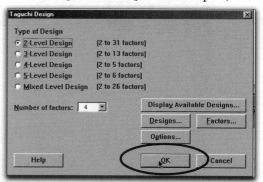

FIGURE 20-12

to store the design in the worksheet. Click **OK** and the design is placed into the worksheet (Figure 20-13). It is prudent to save the project at this time to avoid losing the work that has been completed.

Worksheet 1 ***						
C1	C2	C3	C4	C5	C6	C7
Lead Gap	Trail Gap	Offset	Tube			
1	65	45	60	78		
2	80	45	60	100		
3	65	60	60	100		
4	80	60	60	78		
5	65	45	110	100		
6	80	45	110	78		
7	65	60	110	78		
8	80	60	110	100		
9						

FIGURE 20-13

Figure 20-13 has the same configuration as Table 20-6. We now add the responses from the noise matrices. Notice that these responses are added as row extensions for each treatment combination of the design matrix.

Worksheet 1 ***											
C3	C4	C5-T	C6	C7	C8	C9	C10	C11	C12	C13
Offset	Tube	tc ID								
2	60	100	a(d)	2.49821	2.49867	2.50119	2.50092	2.49909	2.49882	2.50031	2.50077
3	60	100	b(d)	2.50939	2.51756	2.52008	2.51210	2.51798	2.51000	2.51149	2.51966
4	60	78	ab	2.51204	2.51476	2.51501	2.51701	2.51291	2.51491	2.51414	2.51686
5	110	100	c(d)	2.49364	2.50187	2.50433	2.49635	2.50223	2.49425	2.49574	2.50391
6	110	78	ac	2.49629	2.49901	2.49926	2.50126	2.49716	2.49916	2.49839	2.50111
7	110	78	bc	2.49050	2.50093	2.50119	2.49547	2.49909	2.49337	2.49260	2.50303
8	110	100	abc(d)	2.51396	2.51442	2.51694	2.51667	2.51484	2.51457	2.51606	2.51652

FIGURE 20-14

This is different from the ordinary manner in which Minitab adds "replicates" to a design when done in the traditional way we have learned earlier. The ordinary manner of adding replicates in Minitab is to repeat the row of the treatment combination and have a single column for the response. While we could use random replicates for the responses in the parameter design, we realize that such a strategy would require far more replicates than is affordable. This point of affordability was discussed under the heading "Process Capability Study Approach" earlier in this chapter. Using the systematic noise matrix approach makes the induction of variation an affordable option in our quest for quality, which we realize has both the dimension of location and variation as equal partners in its make up.

Analysis Procedure in Minitab

With the design and the responses, we are now ready to go beyond the simple contrast analysis done previously. Figure 20-15 shows the path to the **Analyze Taguchi Design** from the dropdown.

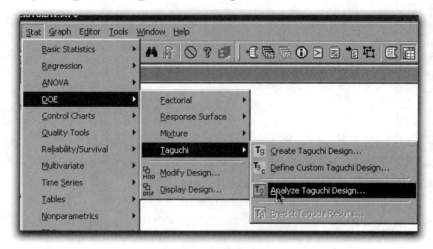

FIGURE 20-15

The first thing we do is to choose the responses from the main dialog box as shown in Figure 20-16. Place your cursor in the **Response** box, highlight, and select all the columns with responses that sends them there.

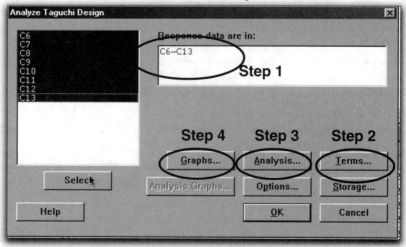

FIGURE 20-16

Next click **Terms** in the main dialog box as shown in Figure 20-16 (Step 2). In the **Terms** dialog box (Figure 20-17) we will see the single effects in the **Selected Terms** box by default.

Add the interactions that you believe to be possible. If you don't know the interactions, you will need to add all the interactions and observe which are statistically significant.

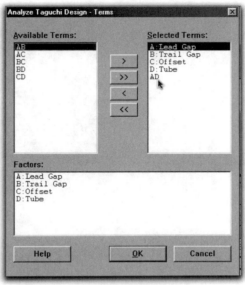

From previous experience, the AD (Lead Gap x Tube Thickness) interaction has been active and it is included in the model.

The Taguchi module in Minitab preforms limited statistical significance testing. A more extensive analysis may be done separately by compiling the summarized responses and doing an "ordinary" DOE analysis in Minitab.

FIGURE 20-17

Now we go to step 3 in the Main dialog box and click on **Analysis**. This (Figure 20-18) is where we select the summarized responses we shall investigate. For this example, we will select all of the possibilities.

Step 4 (shown in Figure 20-16) is choosing the content for the graphs. Followers of Taguchi are more prone to just look at graphs rather than determine if the effects are statistically significant. While this may seem to be a departure from good scientific thinking, they argue that the factor would not have been included in the investigation if it had not been influential.

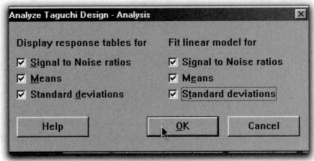

FIGURE 20-18

In Figure 20-19 we specify the plotted responses. The interaction plots will look much like those we have experienced in the ordinary DOE analysis. It is possible to plot the interactions on separate sheets. We'll go with the matrix plot option.

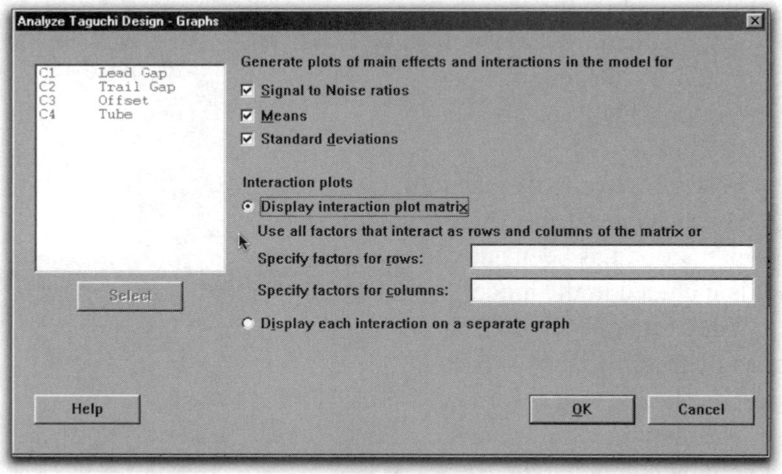

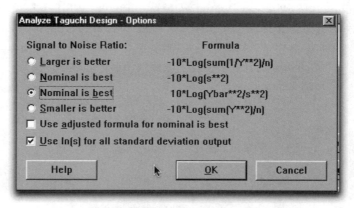

FIGURE 20-19

Again we go back to the main dialog box and click on **Options** (Figure 20-20). This is where we select the Signal to Noise that we wish to use in this analysis. Notice that the Type T S/N is not available. Taguchites sneer at the Type T S/N insisting that the S/N must be independent of adjustment (6). The closest S/N to the S/N_T is the Nominal is best with the average in the numerator. Since there is a very small amount of deviation from the target value in this

FIGURE 20-20

experiment, the Type N S/N will be sufficient for our purposes. The last visit to the main dialog box is for storage. In Figure 20-21 we store the summarized values in the worksheet. Problem 6 asks the student to do the statistical analysis of the function in the Flo-Form and test the statistical significance of the terms. We store the S/N, mean, and standard deviations to allow this exercise. These values are found in Figure 20-22.

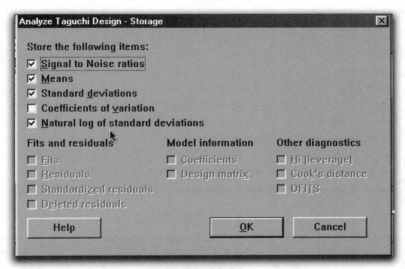

FIGURE 20-21

The more formal analysis now unfolds in the form of graphs found on the next pages (Figures 20-22, 23, 24). Both the single effects plots as well as the interaction plot shows the same conclusions we had found in the simple contrast analysis, but now we have a greater insight into the reasons for these settings.

	C1	C2	C3	C4	C5-T	C14	C15	C16	C17
	Lead Gap	Trail Gap	Offset	Tube	tc ID	SNRA1	LSTD1	STDE1	MEAN1
2	80	45	60	100	a(d)	66.5739	-6.74841	0.0011727	2.49975
3	65	60	60	100	b(d)	55.0070	-5.41073	0.0044683	2.51478
4	80	60	60	78	ab	63.2948	-6.36492	0.0017209	2.51470
5	65	45	110	100	c(d)	54.9420	-5.40952	0.0044737	2.49904
6	80	45	110	78	ac	63.2402	-6.36492	0.0017209	2.49895
7	65	60	110	78	bc	54.6136	-5.37252	0.0046424	2.49702
8	80	60	110	100	abc(d)	66.6274	-6.74829	0.0011729	2.51550

FIGURE 20-22

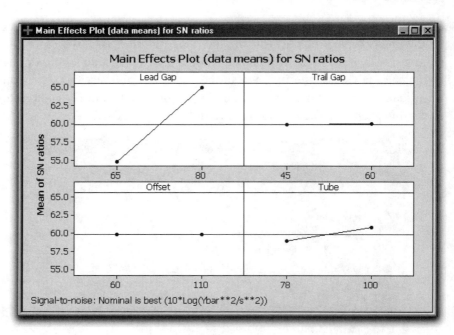

FIGURE 20-23

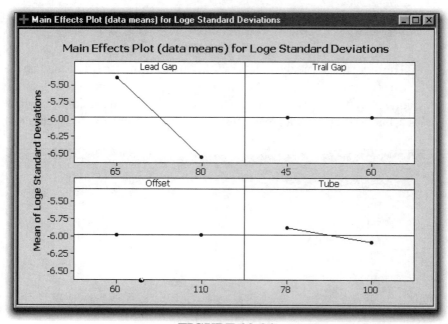

FIGURE 20-24

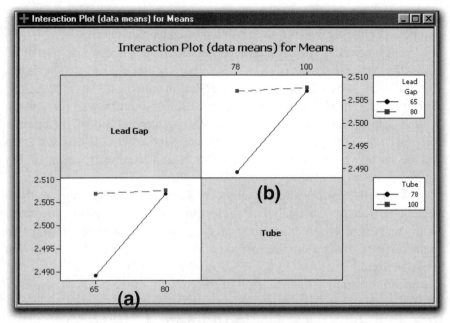

FIGURE 20-25

Interpretation of the Plots

We will begin by looking at the interaction plot of the average diameter. In Figure 20-25 (a) the average diameter is plotted against the lead gap while isolating the tube thickness at its two levels. At the thicker (100) tube, the change in the diameter as a function of lead gap is practically nonexistent. Further examination of the interaction plotted with the tube thickness on the x axis and lead gap isolated (Figure 20-25 (b)) shows that with the lead gap at 100, the changes induced by tube thickness are again nonexistent.

Since tube thickness is a factor somewhat outside our design authority (we can choose a nominal level, but the variation is a function of cost), this interaction is very helpful in deciding on the levels of lead gap and tube thickness that will be unaffected (robust, immune) to changes in these settings. So, to have a greater immunity to tube thickness changes, use the 80 lead gap setting. To have a greater immunity to lead gap variation, use the 100 tube wall thickness.

Let's see how the S/N and Log s functions plotted on the previous page can lead to this same conclusion. First note that the Log s plots are a direct inverse of the S/N plots. If you consider the calculations done to create these metrics, you will understand immediately why there is an inverse relationship between these two summarized responses (see Problem #7).

In Figure 20-23, we see that the maximum S/N is at a lead gap of 80 and a tube wall thickness of 100. The other factors are inert. The Log s (Figure 20-24) shows the same recommendation with the lowest values (Log s is minimized) at 80 for the lead gap and 100 for the tube wall thickness.

These are the same recommendations our interpretation of the interaction plot of the average lead us to! So by using the S/N or the Log s, we can find the settings of the factors that will minimize variation. But what about achieving the target? (See problem #9.)

With all the very exacting set ups required in the noise matrices there is the question of practicality. While by applying the principle of the parameter design we are able to get the requirements of low variation and also hit the target, there is a simpler method of doing the work called the Internal Stress Method (3) or the Combined Array Method (7). We will look at that approach now that we have a grasp of the concept of the study of noise.

Practical Parameter Design – Internal Stress Method

We have seen that the main departure of parameter design from ordinary location-based experimental investigations is the inclusion of the study of variation. This concept is of course very important. However the amount of effort to set up and execute the multitude of experimental runs required to accomplish parameter design by means of the Design Matrix/Noise Matrix approach is often overwhelming.

Many engineers have abandoned parameter design because of the immense amount of work required to accomplish the task. However, as we begin to understand the underlying engineering reasons behind the parameter design function, we can look for and find alternative approaches that can accomplish the effect of the parameter design without the added work.

One such approach is referred to by Taguchi in an obscure appendix to a chapter in one of his books (4). This author further developed and tested that approach and named it "The Internal Stress Method (3)." Other authors have also written about this abbreviated approach to parameter design and have named it "The Combined Array (7)."

No matter what the method is called (Taguchi did not give it a name), the concept is simple and elegant. It also utilizes traditional experimental design and analysis concepts. We will now continue the Flo-Form example using the Internal Stress method.

The Flo-Form process involves a targeting situation. We need to have a diameter of 2.5000 inches with a standard deviation of 0.0075 inches. Finding the correct settings to place the response on target with low standard deviation is a difficult activity especially with a centered target. If we choose the levels in our experiment that bracket the target, then there will be an optimal level and any deviation from this optimal in either the positive or negative direction will give a sub-optimal solution.

This is where we need an experimental configuration that allows the building of a quadratic function. The smallest such design comes in the class of 3^{k-p} designs. Taguchi is famous for his 3-level Orthogonal Arrays, and the Minitab software includes many of these designs. Appendix 20-1 includes two of the more useful 3-level designs, the L_9 and the L_{27}.

Since we have found from our previous work with the Flo-Former that there were only three active factors (Lead Gap, Trail Gap, and Tube Wall Thickness), we will be able to use the smallest 3-level design, the L_9. We need to take care with this design, since we had discovered an interaction between the Lead Gap and the Tube Wall Thickness.

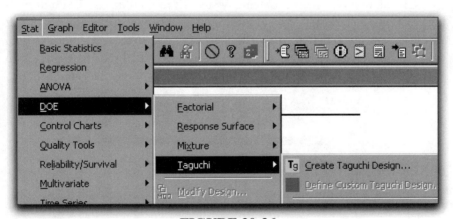

FIGURE 20-26

We begin to build the experimental structure in the usual way, by dropping down from **Stat=>DOE=>Taguchi=>Create Taguchi Design** as shown in Figure 20-26.

We reach the main dialog box where we click on **Three level designs** (Step 1) as shown in Figure 20-27. We scroll down to select 3 factors (Step 2) and then click on **Designs** (Step 3). This brings up the sub dialog box (Figure 20-28) where we choose the L_9 design, which will hold up to four factors given that there are **NO** interactions.

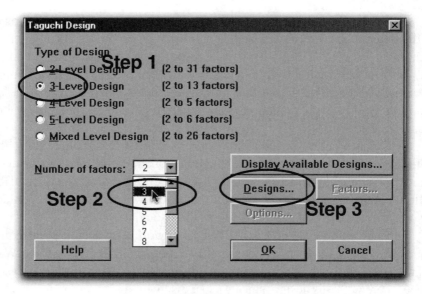

FIGURE 20-27

If interactions are present and we use all four columns, the confound-ing between the interactions and the single effects will distort the results. It would be necessary in that circumstance to move up to the next larger design, the L_{27}. In this example, we are "stretching our luck" by using the L_9 with three factors where we know there is an interaction between two of them.

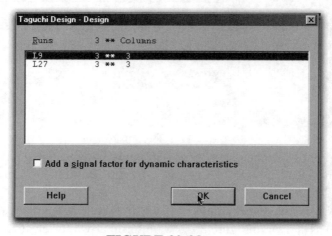

FIGURE 20-28

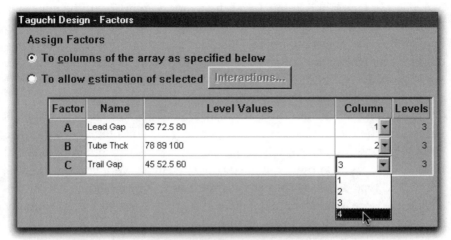

FIGURE 20-29

We click **OK** and return to the main dialog box where we now may click on **Factors** (it was grayed out before since we had not chosen the design). In the Factors sub dialog box,(Figure 20-29) we enter the names and the levels of the factors we are selecting. Since we had already told Minitab we had three factors, that's all that show up.

We assign the Lead Gap and Tube Thickness to columns 1 and 2, but we carefully avoid using column 3 for the Trail Gap, since we know there is an interaction between Lead Gap and Tube Thickness. The linear interaction

↓	C1	C2	C3	C4
	Lead Gap	Tube Thck	Trail Gap	Diameter
1	65.0	78	45.0	
2	65.0	89	52.5	
3	65.0	100	60.0	
4	72.5	78	60.0	
5	72.5	89	45.0	
6	72.5	100	52.5	
7	80.0	78	52.5	
8	80.0	89	60.0	
9	80.0	100	45.0	
10				

Worksheet 1 ×××

FIGURE 20-30

between the factor in column 1 and the factor column 2 will be appear in column 3 (see Chapter 7 to review this principle). So we place the Trail Gap into column 4 which contains the quadratic interaction between the factor in column 1 and the factor column 2. Since we expect a linear interaction, we are reasonably safe in using this design with all three factors. Note that a large amount of prior knowledge is necessary to make such assignments.

Often, we do not have such information and a larger design is necessary (the L_{27} in this case). This simply reinforces the idea that, "The more you know about the process, the less work (experimentation) you need to do. The less you know about the process, the more work you need to do."

The type T S/N is the correct utility function we should use in this situation. Since Minitab does not include the type T, we need to use another analysis method to solve this type of problem.

We will now extract the experimental design with the responses from Minitab and export them to Excel where a macro will compute the S/N's (all of them including type T) and plot the results for interpretation.

We also need to add a column with the name "Row" to the front of the design matrix. To do so, highlight the first column in the design matrix, then go to the **Editor** dropdown in Minitab and click on **Insert Columns** as shown in Figure 20-31. Then simply type in the numbers of the rows (from 1 to 9 in this case). Highlight the five columns and nine rows (including the column headings) and copy the cells as shown in Figure 20-32.

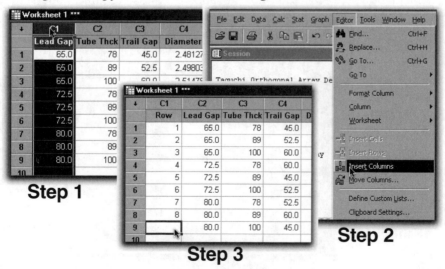

FIGURE 20-31

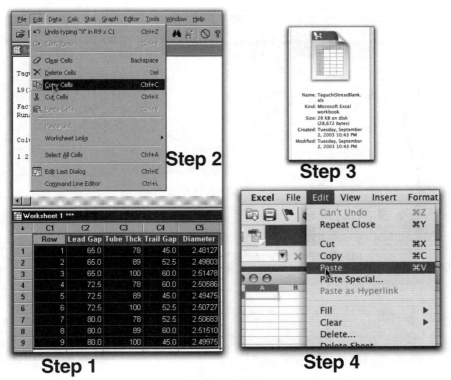

FIGURE 20-32

Following step 3 in Figure 20-32, open the Excel macro (obtainable from www.crcpress.com by clicking on the link "Downloads and Updates" in the Electronic Products section). Place the cursor in the "A-1" position and paste as shown in step 4 of Figure 20-32.

Now go to the Excel dropdown **Tools** and move down to **Macros...** as shown in Figure 20-33. This opens a dialog box (Figure 20-34) where we will see the macro named **Stress**. Click on **Run**. A question box (Figure 20-35) will ask for the target value. For this problem enter 2.5000.

For proper operation of this macro, you must be sure that it is loaded into Excel as a "clean" (no data has been entered into it before) load. To preserve the original, do a **Save As** once the macro has executed. This will preserve the original "shell" that can then be used for other problems. Also note that this macro works with 3-level designs and is only capable of doing the Internal Stress Method. Use Minitab for other parameter design methods such as the Design Matrix/Noise Matrix approach that we have previously illustrated in this chapter.

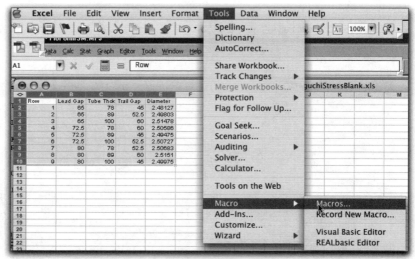

FIGURE 20-33

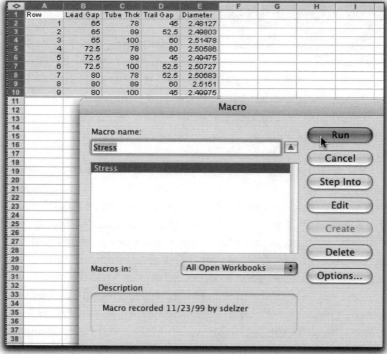

FIGURE 20-34

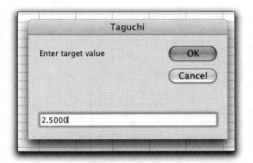

FIGURE 20-35

Interpretation of Results

Figure 20-36 from Excel shows plots of functional relationships between the average outside diameter and the three single effects as well as the type T S/N and these three factors.

In a targeting situation like this one, it is best to use the S/N_T to make the initial selection of the level of the factor. If necessary, make any adjustments to reach the exact target by choosing a "semi-signal factor" to tweak the process to that target. Also, realize while the quadratic plots in this figure are plotted point-to-point, the interpolated optimum should be used since we can safely assume we have continuous functions.

So, for the Lead Gap, we find an optimum by fitting the points to a quadratic function, taking the derivative and setting it equal to zero. The result is shown in Table 20-10. This is also done for the Trail Gap. We do not choose an optimum based on this criterion for the Tube Wall Thickness, since this is a part that we need to order from a supplier. Such parts come in standard

TABLE 20-10

Lead Gap	65	43.4	Quadratic function:
Lead Gap	72.5	50.4	$S/N_T = -469.67 + 14.13*LG - 0.096*LG^2$
Lead Gap	80	46.6	Optimum: Set Lead Gap at 73.6

Trail Gap	45	45.2	Quadratic function:
Trail Gap	52.5	50.8	$S/N_T = -246.7 + 11.41*TG - 0.1093*TG^2$
Trail Gap	60	44.1	Optimum: Set Trail Gap at 52.2

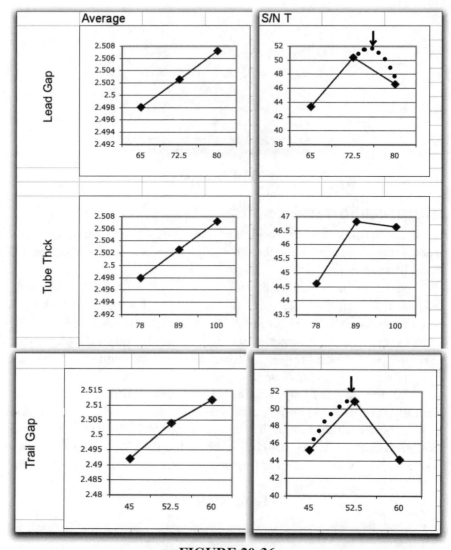

FIGURE 20-36

dimensions. Special-order parts would be much more costly than off-the-shelf commodities, so we will use the 100 thousandths tube thickness (a standard dimension).

Proof of the Pudding

Using the optimal settings of 73.6 thousandths (round to 74) for the Lead Gap and 52.2 thousandths (round to 52) for the Trail Gap with the 100 thousandths for the tube wall thickness, we make a confirmatory trial which produces an outside diameter of 2.5068 inches. This result is 68 ten thousandths too large. We need to bring the outside diameter down. We have only two signal factors (recall a signal factor influences the average without influencing the variation) since the tube wall thickness is beyond our (economic) control to select a non-standard level.

Of the two factors that we do have design authority over, neither is a true signal factor since they both influence the S/N. However, the Trail Gap shows a greater amount of influence on the Outside Diameter than the Lead Gap. (The Excel plots have changed the scales for the factors, so you must fit an equation to find this sensitivity.) This gives the Trail Gap the ability to change the average of the response without too much disruption in the variation especially in the region of interest.

To find the setting to put the outside diameter on the target of 2.5000 inches, we find the linear function between outside diameter and Trail Gap.

Outside Diameter $= 2.43266167 + 0.00133267*TG$ (20-11)

Expression 20-11 shows this linear relationship which tells us that for an increase in each thousandth of Trail Gap, we gain 13 ten thousandths in outside diameter. So to lose 68 ten thousandths, we would need to lower the Lead Gap by 5.2 thousandths. This gives a new setting of 46.8 for the Trail Gap. This produces an Outside Diameter of 2.5013 inches, so another thousandth must be shaved off the Trail Gap and with a "more convenient" setting of 45.5 thousandths, we obtain an OD of exactly 2.5000.

We have found the criteria for obtaining the exact outside diameter. However, even though we reached this point with a minimal number of runs (12 runs including the confirmatory trials), and very little mental gymnastics in balancing the choice of levels had we used the traditional average output, the process still did not converge to the optimal immediately. There were two tweaking runs that were needed. This tweaking was done with knowledge of the functionality and was not the typical "shot in the dark" that is often used.

So why was any tweaking needed at all? The root of this is in the size of the parameter design's Orthogonal Array. The L_9 had a confounded interaction which lessened the discrimination of the experiment. However, the next size 3-level design (L27) would have required 3 times the effort. So the extra two confirmatory runs is a low price to pay compared with the extra 18 runs that the L_{27} would have required. (See Problem 10 for more on this.)

TABLE 20-11

Factor	Setting	Tolerance (1 standard deviation)
Lead Gap	74 thousandths	1 thousandth
Trail Gap	45.5 thousandths	1 thousandth
Tube Thick.	100 thousandths	5 thousandths

Table 20-11 shows the settings and the low cost tolerances for this Flo-Former device. There are other factors involved with the machine, but they do not influence the outside diameter, so their setting may be based on other criteria such as thruput or other productivity measures. We will visit the Flo-Former problem in the next chapter where we study tolerance analysis.

Considerations for the Internal Stress Approach

The internal stress method (ISM) saves a considerable amount of resources while still achieving the requirements of on target (after tweaking) and low variation transmission from the controlling factors to the responses.

Since, we never get something for nothing, there must be constraints for this method. These constraints are not difficult to adhere to. Here they are:

1) The experimental design configuration (OA) must be balanced.

2) There must be at least 4 *active* factors in the investigation.

3) Noise is put into the design matrix, but no optimal level is selected for noise.

These three criteria are easy to fulfill. All Orthogonal Arrays are balanced. However, the CCD is not balanced and cannot be used in the ISM. The reason for the balance is the way the variation is induced into the experiment. If the variation comes in due to the way the experimental configuration is set up, then the noise is not being induced by the noise factors (or pseudo noises), but by the way the runs are made. This is a false type of noise.

There needs to be sufficient perturbation in the design, so that is the reason for at least four active factors. Interactions and quadratic physical effects are most instrumental in producing "sweet spots" where the variation is transmitted with less vengeance to the response.

The final criterion for the ISM being able to work is fundamental to it's whole idea. Parameter design has the avowed purpose to optimize both location *and* variation. Good designs will be robust to outer noise. Instead of testing a design configuration in the pre-production stage to see if it can withstand outside noises, we include these outside noises during the prototype development process. This is the way we build quality into the design. So, those outside noises will be flexing their influence and leading us to the settings of the controlling factors that make our design immune to the ravages of the outside noise. *If we ignore noise in the design process, we pay the penalty in the redesign phase. If we exploit the noises in the design process, there is no need for a redesign!*

Theory of Parameter Design

We have learned that the basic idea that makes an engineering design robust (on target with low variation) is the statistical experimental design that investigates the functionality between the responses of location *and* variation with the controlling (under our design authority) factors and the noises (outside our design authority). We have learned to include in our goal the concept of low variation as well as reaching target. Our objective has been expanded to include not just the functional relationship between the response and the factors, but also the *variation* in the response and these factors.

So Parameter Design is an expansion of ordinary experimentation to include the element of *the transmission of variation*. There is no magic in this process. It is completely dependent on the basic functionality of the process being investigated. Figure 20-37 illustrates this concept. The function in **a** is a steep linear. Any change in the factor on the X axis will induce a large change in the response on the Y axis. However, the function in **b** has a much shallower slope and there is less variation transmitted from the factor on the X axis to the response. The *interaction* in **c** has either a high transmission or a low transmission, which depends on the choice of the isolated factor's level. The solid level is lower than the dashed level. The same differential slope phenomenon is seen in the curve. The slope change

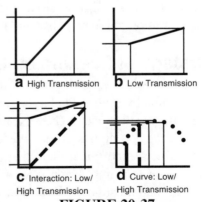

a High Transmission   **b** Low Transmission

c Interaction: Low/ High Transmission   **d** Curve: Low/ High Transmission

FIGURE 20-37

depends on the level of the factor in the quadratic. When the factor on the X axis is low to medium, the slope is very severe and transmits variation with a vengeance. However, when the factor hovers around the mid values, there is almost no transmission of variation to the response. In the case of this symmetrical quadratic, the transmission of variation also becomes severe when we have a range of values from mid to high.

In the quadratic it is as if the factor is interacting with itself! In fact it is. If we think of the fundamental concept of an interaction, we realize it merely means the non additive influence of two or more factors on the response. The interacting factors have differential slopes which depend on the level of the other factor(s). In a sense, the quadratic is doing the same thing! The only difference is with the quadratic, the factor is "interacting" with itself. The slope of the function depends on the level of the factor. So, as the X factor in Figure 20-36d increases, the slope gets lower and lower, until it becomes zero and then as X increases, the slope becomes larger and larger.

The "theory" behind Parameter Design is to find the minimal slope where the transmission of change in the X factor to the Y response is a minimum. Figure 20-36 shows the likely functions that will do this very thing for us. The power of statistical experimental design allows us to find these functions and take advantage of them. The original approach to Parameter Design using the noise matrices to induce error simply allows us to find the effect of the slopes (or differential slopes in the form of interactions or quadratics) on the response.

The Internal Stress Method simply by-passes the external generation of variation and allows the inherent functions to transmit this variation. Where there is a shallower slope, the S/N is larger. Steeper slopes transmit more variation to the response and garner smaller S/N's.

Strategy of Seeking an Optimum

When the goal is either larger is better or smaller is better, the optimum is easily found by finding the larger or smaller response that has the largest S/N. In most of these cases, the S/N and the average will track synchronously.

If the optimum is beyond the levels of the experiment we will need to set up a new experiment with the new levels projected from the direction given from the first design. As in all experimentation, extrapolation is incorrect and must be avoided. Management sometimes thinks experimentation has "failed" if the goal sought is not found in the first experiment. Think about this attitude and the harm it may do to good experimentation.

In situations where the target is the goal, the type T S/N will give direction to the optimum settings if the levels of the experiment bracket that optimum. Remember in the targeting situation, we are seeking both the target and the low variation. However, the S/N_T will respond more strongly to the target portion of the metric if we are nowhere near that target. This is where the traditional "Taguchites" recommend the use of the S/N_N to find the point of lowest variation and then use an adjustment factor (called a "signal" factor) to bring the average on target. This is a reasonable strategy, but it does rely on a factor that has the properties of being able to move the average without disrupting the variation. Such signal factors are not often present in most physical situations. Therefore the use of S/N_T is a more global approach and does not rely on a 2-step optimization procedure.

Summary

We have explored ways to measure quality via the loss function and its transform, the signal to noise ratio. Using the S/N, we can measure both location and variation in a single metric. This metric becomes the focus of our efforts to systematically investigate the functional relationships that cause location shifts and variation spreads.

Using Parameter Design, in conjunction with the S/N, we are able to find the settings of the controlling factors that put our product, process, or service on target with low variation while producing a low cost product, process, or service.

Using the Internal Stress Method (ISM), we are able to simulate the induction of variation without using external arrays (noise matrices). This ISM allows us to do the Parameter Design work with as little as 10% of the effort needed in the more basic, traditional Design Matrix/Noise Matrix approach.

What's Next?

If the Parameter Design is unable to obtain the settings for the required low variation, then we will need to find the contributors to the variation. Often not all of the controlling factors are contributors. Tolerance Design finds the "Quality Sensitive Components" and gives us a direct focus on those factors that need adjusting to meet the variation specification. We will investigate Tolerance Design in Chapter 22.

Problems for Chapter 20

1. Compute and compare the expected loss for the "Trade-off I" results found in Figure 20-2 to the expected losses for "Setting I" and "Setting II."

2. Try to visualize the function that causes the average and the standard deviation to behave in the way that it does for the above process.
    a. Is this an interaction? If so, what does it look like?
    b. Is there a curve? If so, what does it look like?

3. Consider and discuss the similarities between the standard deviation and the target deviation.

4. Explain why the multiplier of 20 in the S/N shown in expression 20-9 is correct given that the units of the S/N are decibels (deci means 10). Hint: work with the target variance (square of target deviation) and the target squared in defining the S/N_T with a multiplier of 10.

5. Confirm the calculations in Table 20-2 using the expected loss you found in problem 1.

6. Write a Goal and Objective for the Floform problem used in the parameter design as shown in Table 20-6.

7. Use the summarized responses (mean, S/N, log s) found in Figure 20-21 to test the significance of the single effects and interactions in the Flo-Former experiment.

8. Consider the calculations done to create the S/N_N used in the Flo-Former and compare this with the Log s to understand and explain why there is an inverse relationship between these two summary metrics.

9. Using the Single effects plot shown on the next page, determine the settings for obtaining both low variation and a target of 2.5000 for the Flo-Former problem. (Hint: Use the "Robust" settings already found for the Lead Gap and Tube Thickness, then find a setting of the Trail Gap to attain the target.) Use the data generator at the web site to confirm your recommendations. Comment on how this two-step approach compares in ease with the Internal Stress approach shown in this chapter.

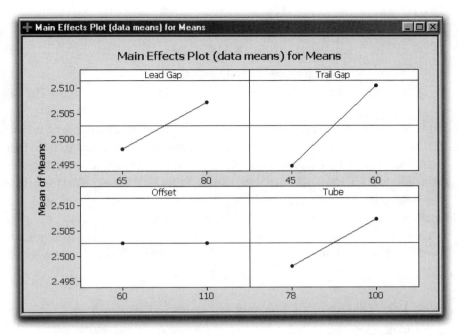

10. Re-do the Flo-Form device using an L_{27} and in the Internal Stress mode. Put the Lead Gap in column 1 of the L_{27}; Trail Gap in column 2 and Tube Wall Thickness in column 5. Use the Excel macro and compare the results with those reported in this chapter. If you see any differences, think about the reasons for these differences and explain the differences. You may need to revisit Chapter 7 to "brush up" on 3^{k-p} fractional factorial designs.

11. Choose one of the simulation problems found on the Web (www.crcpress.com; click on the link "Downloads and Updates" in the Electronic Products section). Use the Internal Stress method to do the parameter design. Write a report on your findings. (Note: This can be assigned as a team project with teams of three students working together.)

12. Explain why shallow slopes transmit less variation to the response.

13. Comment on: Management sometimes thinks experimentation has "failed" if the goal sought is not found in the first experiment.

APPENDIX 20-1
OA'S
Two Level Orthogonal Arrays
(These designs were generated in Minitab. Also see (3) Chapter 10)

Taguchi has created a catalog of 2- and 3-level designs that are based on the 2^{k-p} and 3^{k-p} structures. Due to his association with electrical engineers, and their Boolean counting conventions, the order in his Orthogonal Arrays (as he calls them) is "backwards" from the usual YATES order we are accustomed to. The **bold** column labels are the Classical YATES labeling. Also note that a negative generator is used in some cases to create the added factor columns. This assures that in one of the rows all factors are at the low level. This was an important consideration during the era these designs were developed (late 1940s) where a "Base Case" (all low level) was an integral part of the design.

L_4 Maximum of 3 Factors at 2 levels in 4 runs

Col:	1	2	3	
	A	B	C	"Taguchi"
	B	**A**	**AB**	**"Classical"**
	1	1	1	
	1	2	2	
	2	1	2	
	2	2	1	

L_8 Maximum of 7 Factors at 2 levels in 8 runs

Col:	1	2	3	4	5	6	7	
	A	B	C	D	E	F	G	"Taguchi"
	C	**B**	**-BC**	**A**	**-AC**	**-AB**	**ABC**	**"Classical"**
	1	1	1	1	1	1	1	
	1	1	1	2	2	2	2	
	1	2	2	1	1	2	2	
	1	2	2	2	2	1	1	
	2	1	2	1	2	1	2	
	2	1	2	2	1	2	1	
	2	2	1	1	2	2	1	
	2	2	1	2	1	1	2	

L$_{16}$ Maximum of 15 Factors at 2 levels in 16 runs

Col:	1	2	3	4	5	6	7	8	9	10	11	12	13	14	15	
	A	B	C	D	E	F	G	H	J	K	L	M	N	O	P	"Taguchi"
	D	C	-CD	B	-BD	-BC	BCD	A	-AD	-AC	ACD	-AB	ABD	ABC	-ABCD	"Classical"
	1	1	1	1	1	1	1	1	1	1	1	1	1	1	1	
	1	1	1	1	1	1	1	2	2	2	2	2	2	2	2	
	1	1	1	2	2	2	2	1	1	1	1	2	2	2	2	
	1	1	1	2	2	2	2	2	2	2	2	1	1	1	1	
	1	2	2	1	1	2	2	1	1	2	2	1	1	2	2	
	1	2	2	1	1	2	2	2	2	1	1	2	2	1	1	
	1	2	2	2	2	1	1	1	1	2	2	2	2	1	1	
	1	2	2	2	2	1	1	2	2	1	1	1	1	2	2	
	2	1	2	1	2	1	2	1	2	1	2	1	2	1	2	
	2	1	2	1	2	1	2	2	1	2	1	2	1	2	1	
	2	1	2	2	1	2	1	1	2	1	2	2	1	2	1	
	2	1	2	2	1	2	1	2	1	2	1	1	2	1	2	
	2	2	1	1	2	2	1	1	2	2	1	1	2	2	1	
	2	2	1	1	2	2	1	2	1	1	2	2	1	1	2	
	2	2	1	2	1	1	2	1	2	2	1	2	1	1	2	
	2	2	1	2	1	1	2	2	1	1	2	1	2	2	1	

Three Level Orthogonal Arrays

(These designs were generated in Minitab)

L_9 Maximum of 4 Factors at 3 levels in 9 runs

Col:	1	2	3	4	
	A	B	C	D	"Taguchi"
	B	**A**	**AB**	**A^{2}B**	**"Classical"**
	1	1	1	1	
	1	2	2	2	
	1	3	3	3	
	2	1	2	3	
	2	2	3	1	
	2	3	1	2	
	3	1	3	2	
	3	2	1	3	
	3	3	2	1	

Linear Graph for L_9:

The Use of Linear Graphs

A linear graph is a pictorial method of showing some of the confounding in an Orthogonal Array. The columns at the end of the line belong to factors that may be assigned to study a single effect. The column numbers on the line are columns involved with the computation of the interaction of the factors on the ends of the line. So, in the L_9 columns 1 and 2 may be used to study factors A and B, but the AB interaction will be contained in columns 3 and 4. Therefore if a physical interaction exists between factors A and B, we cannot assign any factors to the 3rd and 4th columns. If factors are assigned to the 3rd and 4th columns in the presence of interactions, all the effects are confounded and the interpretation of the results is questionable.

L_{27} Maximum of 13 Factors at 3 levels in 27 runs

1	2	3	4	5	6	7	8	9	10	11	12	13
A	B	C	D	E	F	G	H	J	K	L	M	N
1	1	1	1	1	1	1	1	1	1	1	1	1
1	1	1	1	2	2	2	2	2	2	2	2	2
1	1	1	1	3	3	3	3	3	3	3	3	3
1	2	2	2	1	1	1	2	2	2	3	3	3
1	2	2	2	2	2	2	3	3	3	1	1	1
1	2	2	2	3	3	3	1	1	1	2	2	2
1	3	3	3	1	1	1	3	3	3	2	2	2
1	3	3	3	2	2	2	1	1	1	3	3	3
1	3	3	3	3	3	3	2	2	2	1	1	1
2	1	2	3	1	2	3	1	2	3	1	2	3
2	1	2	3	2	3	1	2	3	1	2	3	1
2	1	2	3	3	1	2	3	1	2	3	1	2
2	2	3	1	1	2	3	2	3	1	3	1	2
2	2	3	1	2	3	1	3	1	2	1	2	3
2	2	3	1	3	1	2	1	2	3	2	3	1
2	3	1	2	1	2	3	3	1	2	2	3	1
2	3	1	2	2	3	1	1	2	3	3	1	2
2	3	1	2	3	1	2	2	3	1	1	2	3
3	1	3	2	1	3	2	1	3	2	1	3	2
3	1	3	2	2	1	3	2	1	3	2	1	3
3	1	3	2	3	2	1	3	2	1	3	2	1
3	2	1	3	1	3	2	2	1	3	3	2	1
3	2	1	3	2	1	3	3	2	1	1	3	2
3	2	1	3	3	2	1	1	3	2	2	1	3
3	3	2	1	1	3	2	3	2	1	2	1	3
3	3	2	1	2	1	3	1	3	2	3	2	1
3	3	2	1	3	2	1	2	1	3	1	3	2

Linear Graph for L_{27}:

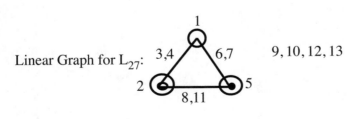

References

1. **How to Make It Right the First Time,** *Business Week*, June 8, 1987, pp. 142-143.

2. **Applied Statistical Decision Theory**, Raiffa, Howard and Schlaifer, Robert. Harvard Business School Press, Boston, MA, 1961.

3. **Engineering Quality by Design, Interpreting the Taguchi Approach**, Barker, T.B. Marcel Dekker, New York, 1990, Chapter 7.

4. **Introduction to Quality Engineering**, Taguchi, G. Asian Productivity Organization, Tokyo, 1986, p. 144.

5. **Off-line Quality Control,Parameter Design, and the Taguchi Method**, Kackar, R. N., *J. Quality Technology* 17(4), 1985.

6. **Engineering Methods for Robust Product Design**, Fowlkes, William J, and Creveling, C. Addison Wesley, Reading, MA, 1995.

7. **Economical Experimentation Methods for Robust Design**, Shoemaker, Anne C., Tsui, Kwok-Leung, and Wu, C.F. Jeff., *Technometrics* 33 (4): 415-427, 1991.

21

Monte Carlo Simulation and Tolerance Design

Monte Carlo is a famous gambling commune in the small, 368-acre country of Monaco. The broad field of statistics displays some of its underpinnings by naming a probabilistic technique after this city where the laws of chance are a way of life and a way of living.

Monte Carlo simulation is a technique that in its simplest form allows us to look at a portion or a whole of a real situation and ask the question "what if?" Since even simple problems have many "what ifs" or branch points, we would like to see the most likely (expected value) results as a consequence of trying all the combinations of branches. Besides the expected value, we would also like to know how much variation is involved in our predictions. This information can be used to build confidence in our prediction (with low variation) or "unconfidence" (for large variation). The probabilistic part of Monte Carlo simulation establishes the degree of variability which produces the level of confidence in our prediction.

Simulation

The simulation part of the Monte Carlo technique is done using the equation we have derived via designed experiments. Simulation itself is a common daily practice in many of the things we tackle and accomplish. To illustrate, think of a trip to the grocery store. We usually have a shopping *list*. The list is the model of what we *intend* to buy at the store. If we know the store, we may even have

613

the list arranged in a specific order so we go through the store in a systematic manner. This system might have the forethought and consideration to move through the store in such a way as to fill the shopping basket with the frozen and meat items last to avoid the possibility of thawing or spoilage.

Another illustration of mapping a plan of action via simulation is the effort we invest when planning a vacation. The extent and degree of planning depends on the goal of the vacation. For a camping trip to the mountains, our goal is to get there as fast as possible with the least traffic and small towns along the way. If we are touring the country or a section of the country, we *want* to move more slowly to savor the climate and the surroundings.

In both cases, we search out a model (which is a road atlas) and go over alternate routes to accomplish our goal. We might even write an itinerary for the longer trip which includes places to visit and where to stay each night. There are many examples of simulation that we take for granted, but are a large part of our daily lives.

Combining Simulations and Probabilities

Let's go back to the grocery store illustration and build it into a Monte Carlo example. We want to know how much money to bring to the store to finance a party. We would like to buy specific brand names, but we have at least one alternative brand as a back-up. Further, we know the prices of the items and the chances of the items being on the shelf. Our **model** is simply a map of the store with the prices. The **chances** are our beliefs in a product's availability. Table 21-1 has a listing of the products, prices, and chances. If we were able to obtain all the primary items, our grocery bill would appear as follows:

4	@	1.19
		4.79
3	@	1.98
		5.94
4	@	1.19
		4.76
4	@	1.19
		4.76
8	@	2.89
		23.12
TOTAL		$43.34

TABLE 21-1

Product	Price	Quality	Chances(p)	Primary (P) /Backup (B)
Lite Potato Chips	1.19	4	.75	P
Potato Frills	1.39	4	.85	B1
Corn Chips	1.69	4	.65	B2
Clam Dip	1.98	3	.45	P
Onion Dip	1.59	3	.70	B1
Vetgetable Dip	.98	3	.70	B1
Canada Dry Ginger Ale	1.19	4	.95	P
Generic Ginger Ale	.99	4	.80	B1
Schweppes	1.59	4	.50	B0
7-UP	1.19	4	.90	P
Sprite	1.29	4	.80	B1
Mellow Yellow	1.20	4	.60	B2
Miller	2.89	8	.75	P
Bud	2.99	8	.85	B1
Schlitz	2.99	8	.75	B2

If we were to go through the entire set of possible purchases, then the most costly bill would come to $48.14; the least costly would be $39.54. We can prepare for the worst and take $50.00 or hope for the least costly and take $40.00. This approach, of course, has *not* taken the probabilities into account. We would like to know what the expected cost will be. To do so, we would have go to the store a great number of times to *experience* the shopping. We would also have to give a lot of parties to do this. So instead of actually going to the store, we *simulate* a great number of shopping trips. A computer provides a convenient mechanism to act out this simulation. The BASIC program (Table 21-2) on the next page "goes to the store" for us and "buys" our party supplies as many times as we tell it, and finds the average expenditure as well as the standard deviation of our spending. The program is really no more complicated than a series of instructions to buy a primary item and if it is not available, to buy the first back-up or second back-up. The page after the program code shows the program in action.

The simulation of the shopping trip found in Table 21-3 has given us an average cost of about $42.50. Notice that while the mean value is fairly stable over many such simulations (trials), the variation does fluctuate. We would expect less fluctuation in both the mean value and the variation as the sample size increases. However, as the sample size goes up, the computing time increases. A typical 10,000 trial run would take 40 minutes. There are faster computers than the turtle I own, and their existence has been justified by extensive Monte Carlo simulations.

TABLE 21-2

```
50 REM Simulation of a Shopping Trip: "SIMSHOP"
55 CLEAR 300
85 RS=VAL(RIGHT$(TIME$,2))
95 RANDOMIZE (RS)
110 REM The "DIM" sets the space up for our variables
115 REM I=Item Price; P=Probability; Q=Quantity;N$=Names
120 DIM I(5,3),P(5,3),Q(5),N$(5,3)
125 DATA 1.19,1.39,1.69,1.98,1.59,.98,1.19,.99,1.59,1.19,1.29,1.29,2.89
126 DATA 2.99,2.99,.75,.85,.65,.45,.70,.90,.95,.80,.50,.90,.80,.60,.75,.85,.75
127 DATA 4,3,4,4,8
128 DATA Chips,Frills,Corn Chips,Clam Dip,Onion Dip,Vegetable Dip,Canada Dry
129 DATA Generic,Schweppes,7-UP,Sprite,Mellow Yellow,Millers,Bud,Schlitz
130 FOR J=1 TO 5:FOR K=1 TO 3
132 READ I(J,K):NEXT K:NEXT J
135 FOR J=1 TO 5:FOR K=1 TO 3
137 READ P(J,K):NEXT K:NEXT J
140 FOR J=1 TO 5:READ Q(J):NEXT J
143 FOR J=1 TO 5:FOR K= 1 TO 3
145 READ N$(J,K):NEXT K:NEXT J
150 PRINT"How Many Shopping Trips";
151 INPUT N
152 PRINT "Do You Want to See Each Trip's Result printed <Y/N>";
153 INPUT Z$
157 S=0:S1=0
160 FOR I1=1 TO N
165 C=0
166 IF Z$="N" GOTO 170
167 PRINT "Buy ";
170 FOR J=1 TO 5
180 FOR K=1 TO 3
190 R=RND(1)
200 IF R<=P(J,K) GOTO 215
210 NEXT K
212 IF Z$="Y" THEN PRINT "Nothing";
213 GOTO 240
215 IF Z$="N" GOTO 230
220 PRINT N$(J,K);" ";
```

```
230 C=C+I(J,K)*Q(J)
240 NEXT J
242 IF Z$="N" GOTO 250
245 PRINT
250 S=C+S
260 S1=C*C+S1
270 NEXT I1
280 A=S/N
290 S2=SQR((S1-(S*S)/N)/(N-1))
300 PRINT "Average Cost= ";A;" Standard Deviation= ";S2
310 END
```

TABLE 21-3

How Many Shopping Trips? 10
Do You Want to See Each Trip's Result printed <Y/N>? Y
Buy Chips Onion Dip Canada Dry 7-UP Millers
Buy Chips Clam Dip Canada Dry 7-UP Millers
Buy Frills Vegetable Dip Canada Dry 7-UP Millers
Buy Chips Clam Dip Canada Dry 7-UP Millers
Buy Chips Onion Dip Canada Dry Sprite Bud
Buy Chips Onion Dip Canada Dry 7-UP Millers
Buy Chips Onion Dip Canada Dry 7-UP Millers
Buy Chips Onion Dip Canada Dry 7-UP Bud
Buy Frills Vegetable Dip Canada Dry 7-UP Millers
Buy Chips Onion Dip Generic 7-UP Millers
Average Cost = 42.32 Standard Deviation = .90

How Many Shopping Trips? 1000
Do You Want to See Each Trip's Result printed <Y/N>? N **first 1000 run**
Average Cost = 42.45 Standard Deviation = 2.18   **trial**

How Many Shopping Trips? 1000
Do You Want to See Each Trip's Result printed <Y/N>? N **second 1000**
Average Cost = 42.41 Standard Deviation = 2.74   **run trial**

Now, why should we bother with Monte Carlo when in many cases (including our shopping spree example) it is possible to make a direct calculation based on probability of the expected cost? Table 21-4 does an *exact* calculation by combining probabilities and gives an expected cost of $42.41. This expected cost is what we had observed in the second of the 1000 trial runs in Table 21-3 using Monte Carlo. The other 1000 Run Trial (1) comes within 4¢ of the truth and as the sample size increases, the chances of the Monte Carlo hitting the exact value

Table 21-4

Direct Calculation

Type	Prob.	Cost	Kind	Number	Expected Cost
Chips	.75	1.19	(Lite)	4	= 4.92
	.2125	1.39	(Frills)		
	.024375	1.69	(Corn)		
	.013125	0.00			
Dip	.45	1.98	(Clam)	3	= 4.95
	.385	1.59	(Onion)		
	.1485	.98	(Vege.)		
	.0165	0.00			
Ginger Ale	.95	1.19	(C.D.)	4	= 4.76
	.04	.80	(GEN)		
	.005	1.59	(Schweppes)		
	.005	0.00			
Lemon Drink	.90	1.19	(7-Up)	4	= 4.76
	.08	1.29	(Sprite)		
	.012	1.29	(M.Y.)		
	.008	0.00			
Beer	.75	2.89	(Mill)	8	= 23.10
	.2125	2.99	(Bud)		
	.028125	2.99	(Sch)		
	.009375	0.00			
Total Expected Cost					$42.41

The Direct Calculation Method works like this: Our first choice for chips has a 75% chance of getting into our basket. If it is not available (which is 25% of the time) we will look for Frills which have an 85% chance of being bought. So the P of looking for Frills (.25) and the P of getting Frills is .25 x .85 = .2125. If Lite and Frills are both not available (.25 x .15), then we purchase corn chips if they are available (.25 x .15x .65=.024375). There is the probability that there will be no chip item and we spend nothing (and have a poor chipless party).

increases. So, while the Monte Carlo takes a "brute force" whack at finding the expected value of the response, it gets there. The direct calculation method has the advantage of always getting the expected value. There is an advantage of Monte Carlo, however, that is neither statistical nor mathematical in nature. It is a psychological advantage. What we do in the Monte Carlo Method is exactly what happens in real life. Therefore, we can explain what we have done more easily to the end users of the analysis who may not understand probabilities. Another prime reason for Monte Carlo is the fact that sometimes we neither have the proper distribution function for the probabilities or the function is too complex to determine how to combine the various functions in a reasonable time frame.

It is for these reasons that we will tend to use Monte Carlo Method rather than exact calculation methods. The results speak for themselves and the method, although tedious, simulates the situation in a clear, straightforward manner.

Application to More Complex Situations

Our example on buying party items at a store serves to lay the groundwork for more complex and useful applications of the Monte Carlo method. In the application of simulation to an empirical equation, we are usually interested in obtaining an expected (average) value and a variation about this expected value. We will follow the same general procedure for a Monte Carlo simulation with an empirical equation as we followed for the shopping example. With the equation our model **is** the function and all we need to do is run our equation with various input levels of the factors in this equation.

To illustrate, let's use a very simple additive equation that says the height of an assembly is equal to piece A plus part B.

$$Y_{Height} = A + B \qquad (21\text{-}1)$$

Where
$$A = 2 \qquad s_A = 0.1$$
$$B = 3 \qquad s_B = 0.2$$

Again, this example is simple enough to solve in closed form. The average height of Y will simply be A + B or 5 units. The standard deviation is the square root of the sum of the variances (s^2). So we find this to be:

$$s_Y = \sqrt{0.1^2 + 0.2^2} = \qquad 0.22361 \qquad (21\text{-}2)$$

Using Monte Carlo methods we will exercise our equation N times for samples from the A population and the B population. When we solved the problem in 21-1 and 21-2 using the closed form method we did not assume any distribution function. With Monte Carlo, we must decide on the distribution of the components in the equation. We will look at two distributions: a uniform (all values have an equal chance) and a normal distribution (extreme values have a lower chance). We will also look at a "true" random Monte Carlo and a discrete simulation.

We will compare the outputs of each distribution method to see if any differences exist. We already understand how the "true" Monte Carlo method reaches into each distribution at random and picks a piece and then makes the assembly. The discrete method, on the other hand, picks a specific combination of pieces from the two populations in a factorial design manner. Then the expectation of the event (assembly) is weighted by the probability of the pieces that were selected. In this way, the discrete method combines the closed form concept with simulation techniques. Let's look at the various outputs for the four methods/distributions.

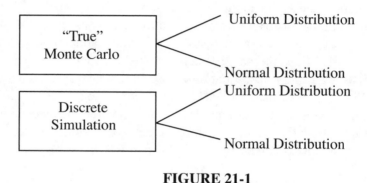

FIGURE 21-1

TABLE 21-5

Method	Distribution	Code	Mean	Std. Dev.	Trials
True Monte Carlo	Uniform	TU	5.017	0.38956	1000
True Monte Carlo	Normal	TN	5.005	0.22714	1000
Discrete	Uniform	DU	5.000	0.45185	49
Discrete	Normal	DN	5.000	0.22314	49

A visualization for each of the four simulation procedures is shown in Figures 21-2 and 21-3. Note that the distribution shapes are the same despite the change in simulation techniques. The only difference we see is how the sampling is done. In the Monte Carlo methods, we have a continuous distribution and all positions in the distribution have a chance of being selected for the assembly. In the discrete method, only seven values may be chosen.

The results in Table 21-5 show that the uniform distribution has a broader variation then the normal. This is expected since the resultant distribution of two uniforms is broader than the combination of two normals. The normal distribution gives standard deviation values very close to the closed form solution to this simple problem, and there is a closer agreement between the random and discrete methods using a normal than when we use a uniform.

Uniform Distribution with a random continuous selection

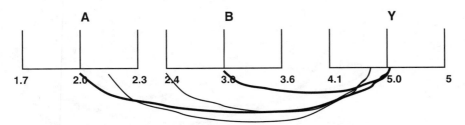

Uniform Distribution with a specific discrete selection

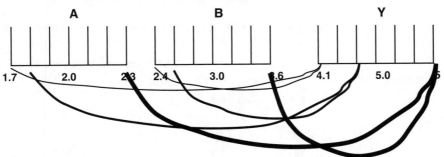

FIGURE 21-2

Since the discrete method using a normal assumption produces an equivalent result with the random method with fewer trials, we conclude that for small numbers of variables (under 5), the discrete method has a computational advantage over the random method. However, with a 7-level discrete simulation, the number of combinations grows rapidly beyond 5 factors ($7^5 = 16807$). We can get around this problem by using a 3-level fractional factorial. To approximate a

Normal Distribution with a random continuous selection

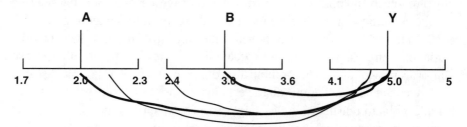

Normal Distribution with a specific discrete selection

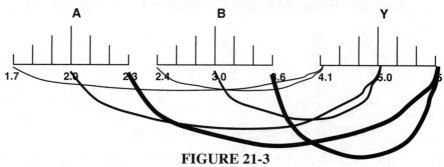

FIGURE 21-3

normal distribution based variance for the resulting distribution of Y, we transform the standard deviations of the input distribution by a factor of $s \times \sqrt{3/2}$ (see Appendix 21B for the source of this transform). Table 21-6 shows the nine calculations that give a standard deviation very close to the normal distribution value.

TABLE 21-6

A	B	Y	
1.878	2.756	4.634	
2.000	2.756	4.756	A: $\overline{X} = 2$
2.122	2.756	4.878	$s = 0.1$
1.878	3.000	4.878	$\sqrt{3/2}\, s = 0.122$
2.000	3.000	5.000	
2.122	3.00	5.122	B: $\overline{X} = 3$
1.878	3.244	5.122	$s = 0.2$
2.00	3.244	5.122	$\sqrt{3/2}\, s = 0.244$
2.122	3.244	5.366	

$$Y = 5.000$$
$$s_Y = 0.23625$$

The frequency distribution pattern of Y is, however, *not* normal in shape. The interesting aspect of this 3-level design application is the fact that 3-level designs can be easily fractionalized so that a discrete simulation can be accomplished with very few runs.

We will next apply (in Chapter 22) these simulation methods to the equation for the photographic process that we derived in Chapter 19.

Random Numbers

The following is a method for computing random normal numbers from a set of uniform random numbers from Box and Muller (1).

$$\text{Random normal} = \bar{X} + \sqrt{-2\text{Log}_e(\text{RU})} \times \text{COS}(2\pi \times \text{RU}) \times s$$

Where: $\bar{X}$ is the average
s is the standard deviation
RU is the random uniform (on a 0 to 1 interval)
COS is in radians

TABLE 21-7
Random Normal Numbers with mean = 0, and s = 1

```
 0.464  0.137  2.455 -0.323 -0.068  0.296 -0.288  1.298  0.241-0.957
 0.060 -2.526 -0.531 -0.194  0.543 -1.558  0.187 -1.190  0.022 0.525
 1.486 -0.354 -0.634  0.697  0.926  1.375  0.785 -0.963 -0.853-1.865
 1.022 -0.472  1.279  3.521  0.571 -1.851  0.194  1.192 -0.501-0.273
 1.394 -0.555  0.046  0.321  2.945  1.974 -0.258  0.412  0.439-0.035
 0.906 -0.513 -0.525  0.595  0.881 -0.934  1.579  0.161 -1.885 0.371
 1.179 -1.055  0.007  0.769  0.971  0.712  1.090 -0.631 -0.255-0.702
-1.501 -0.488 -0.162 -0.136  1.033  0.203  0.448  0.748 -0.423-0.432
-0.690  0.756 -1.618 -0.345 -0.511 -2.051 -0.457 -0.218  0.857-0.465
 1.372  0.225  0.378  0.761  0.181 -0.736  0.960 -1.530 -0.260 0.120
-0.482  1.678 -0.057 -1.229 -0.486  0.856 -0.491 -1.983 -2.830-0.238
-1.376 -0.150  1.356 -0.561 -0.256 -0.212  0.219  0.779  0.953-0.869
-1.010  0.598 -0.918  1.598  0.065  0.415 -0.169  0.313 -0.973-1.016
-0.005 -0.899  0.012 -0.725  1.147 -0.121  1.096  0.481 -1.691 0.417
 1.393 -1.163 -0.911  1.231 -0.199 -0.246  1.239 -2.574 -0.558 0.056
```

Transmission of Variation

As we have observed, there is a variety of different methods focused on the fundamental concept of "transmission of variation." Figure 21-4 shows how in three different functional relationships the variation of the controlling factor influences the variation of the response.

Each of the three graphs has been set up so that the scales are identical on both the X (controlling factor) and the Y (response). This is an important consideration when having a computer automatically plot your functions. The automatic plotting routine will often scale the plots to give the "best resolution" in preference to plots that show the relative impact of the factors on the response. Let's see how each of these functions influence the variation in the response. To do so, the same distance on the controlling factor axis has been selected as the tolerance range for all three functions.

With the A factor, the relatively shallow slope has a much lower impact on the variation (spread – shown as a black bar) of the response than the B factor which has a much steeper slope and twice the spread. The C factor has practically no slope in the region selected and has only half the spread of that which is transmitted by the A factor.

For both the A and B factors, the transmission of variation is not dependent on the nominal level (designated with "**N**") since the function is

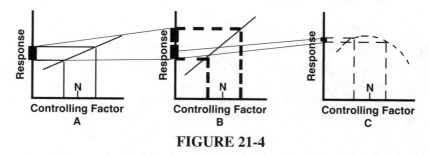

FIGURE 21-4

constant over the range of interest, so we may say that the variation transmission is independent of adjustment. In other words, we may change the overall level of the response (by changing the set-point of the controlling factor) without changing the variation in the response.

However, the selection of the set-point for the C factor will have a large influence on the transmission of the variation. In Figure 21-5 the set-point has been changed in an upward direction for each of the factors. The new settings are designated "**N2**." The new transmitted variation is shown in a slanted line rectangle while the original transmitted variation is shown as a solid rectangle. While the transmission of variation has not changed for the A and B factors, the C factor's influence has been increased by a factor of five!

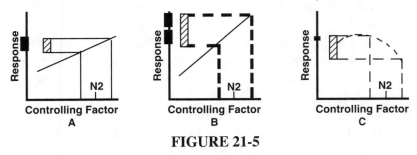

FIGURE 21-5

The variation transmitted to the response by the C factor is **not** independent of adjustment. The same may be said about interaction functions. Consider the form of an interaction (a cross product between two or more factors $(X_1 * X_2)$) and a quadratic like factor C (a cross product of the factor itself $(X_1 * X_1)$). We will leave the further development of this stream of thought for problem 4 which asks you to extend the logic developed about the quadratic function to the interaction function.

Adding a Frequency Distribution

It is important to understand the concept of the transmission of variation before we plunge in with the use of the methods described earlier in this chapter. Given the concepts illustrated in Figure 21-4, we now need to add one more important ingredient. That ingredient is the *distribution* of the variation in controlling factor.

Figure 21-6 shows how the points of the distribution of the controlling factor are picked up through the function and reproduced as a distribution of the response. With the steep slope of the B factor, that reproduction is amplified and the reproduction of the input distribution of the A factor is diminished. The need for the distribution is driven by the fact that almost all

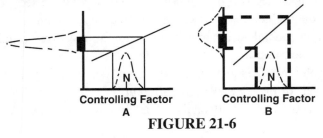

FIGURE 21-6

inputs (controlling factors) have a frequency distribution pattern. Many times this pattern (or distribution shape) follows the Normal Model. In fact, the name "Normal" was given to this distribution shape since so many items found in nature followed this shape. Problem 5 will ask you to verify this phenomenon.

The need to find a way to transmit the variation from the controlling factor to the response given that we have a functional relationship in the form of an equation was one of the reasons for the development of Monte Carlo methods. Monte Carlo, as we have seen, takes points from the input distribution(s), processes these points through a function to obtain a response, and builds a distribution of these output responses by repeating this process multiple (often tens of thousands) times.

Tolerance Design

While Monte Carlo will always do a fine job of finding the correct distribution of the output response, it does require sufficient computer power and when it gives the final distribution, it is unable to identify those components that contribute more or less to the overall variation.

The "Tolerance Design" that was developed by Dr. Taguchi provides solutions to the problems of both the many thousands of computations and the sensitivity analysis shortfalls of the traditional Monte Carlo approach. (The company, Decisioneering, produces a program called "Crystal Ball" that does a sensitivity analysis making this software non-traditional. Their Web address is: www.crystalball.com.)

We have seen a small example of the tolerance design in Table 21-6. The underlying concept of the Tolerance Design is to use fixed levels covering the expected range of variation of the controlling factor in an orthogonal design to systematically transmit the variation from the controlling factors to the response. The key difference between the Monte Carlo approach and the Tolerance Design approach is the fixed levels used in the Tolerance Design in contrast to the randomly selected values used in the Monte Carlo approach. Because the levels in the Tolerance Design are fixed and put into an experimental design matrix, there are far fewer runs required due to this systematic sampling.

Since the Tolerance Design uses an orthogonal design, it is possible to complete an analysis of the sources of variation utilizing ordinary ANOVA calculations. This is the "sensitivity analysis" lacking in an ordinary Monte Carlo simulation.

Making Uniform Distributions Look Like Normals

There is one slight problem in the utilization of the designed experiment in Tolerance Design. A designed experiment has a uniform distribution of the levels. In Figure 21-7, we can see that each factor (in this 3-level design) has the same frequency for each level. This uniform distribution of the levels is necessary to assure the balance in the design which leads to the orthogonality

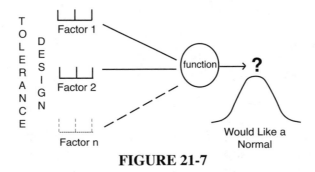

FIGURE 21-7

which is so important in being able to make independent decisions on each factor. But this uniform distribution will not produce a realistic, normal distribution of the out put response that we would expect in most systems (see problem 5 concerning the expectation of normality).

By making a slight adjustment in the levels of the design, we can overcome this problem and produce output statistics (mean and standard deviation) that appear to have been produced by controlling factor input distributions that were normally distributed (even though they are still very uniform!).

We have seen this transform in the example (shown in Table 21-6) which amounts to multiplying the one standard deviation value of the tolerance on the factor by 1.2247. Let's see what is behind this transform. Figure 21-8 shows the uniformly distributed three levels of a 3-level experimental design configuration. We will be applying a set point with a population parameter mean of μ and an upper level and lower level **a** units away from this center point. The variance is defined for a population as:

FIGURE 21-8

$$\sigma^2 = \frac{\Sigma (x - \mu)^2}{n} \qquad 21\text{-}1$$

We now insert the three points from the uniform distribution of the experimental design structure.

$$\sigma^2 = \frac{((\mu-a)-\mu)^2 + (\mu-\mu)^2 + ((\mu+a)-\mu)^2}{3} \qquad 21\text{-}2$$

When we gather up the terms, all the μ's cancel and we are left with:

$$\sigma^2 = \frac{(-a)^2 + (a)^2}{3} \qquad 21\text{-}3$$

This reduces to:
$$\sqrt{\frac{3}{2}}\,\sigma = a \qquad 21\text{-}4$$

$$a = 1.2247 * \sigma \qquad 21\text{-}5$$

So how does this transform give the uniformly distributed levels of the experimental structure any hope for normality? If we examine the area of the expanded uniform and compare that area to a $\pm$ 3σ normal, we find equality. Figure 21-9 illustrates visually how the "extra area" in the expanded uniform above the 3σ normal fits into the "extra area" of the 3σ normal beyond uniform. Since area is probability, we conclude that the probability density of the expanded uniform is the same as the probability density of a 3σ normal and therefore the expanded uniform produces the same parameters (which we estimate with sample statistics) as the 3σ normal. We have seen the result of the "1.2247 * s" transform, courtesy of the example illustrated in Table 21-6. This author has done extensive comparisons over many other examples and has found no reason to believe the transform will fail to produce the desired result of normal distribution-like statistics. However the frequency distribution of the results in the Tolerance Design will not *look* like a normal.

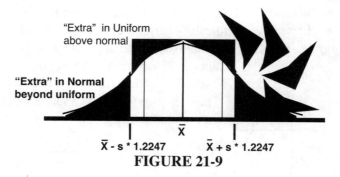

FIGURE 21-9

Finding the Quality Sensitive Components
- A Sensitivity Analysis

We will look at a more extensive example of Tolerance Design and show how by using a designed experiment, we may identify and quantify the "Quality Sensitive Components" by using ANOVA. We will use Minitab to build the design and also complete the analysis. This example is an investigation of a manufacturing process called "Friction Welding" for joining two pieces of steel.

This process is done on a device that has components that resemble a lathe and an arbor press (Figure 21-10). One of the parts to be joined with the other part is placed in the chuck of the lathe and the other part is held in the chuck of the press. The lathe is spun to a high speed while the part in the chuck of the press is held stationary. The press is activated to move the stationary part in contact with the spinning part.

There are two stages of contact at different pressures. During the 3.5 second heating pressure, the interface heats up, but a weld is not completed. At the nearly doubled upset pressure which is held for 3.6 seconds, the pieces are welded. Tensile strength is the response. Table 21-8 shows the going in or "low cost" tolerances for the 6 components of the friction welding process. The goal for the tensile strength variation is a standard deviation of 20 psi.

FRICTION WELDING

FIGURE 21-10

TABLE 21-8

The current low cost tolerances are (stated as 1 standard deviation):

Speed	100RPM	Length	1 thousandth
Heat Pressure	500psi	Heat time	.5 sec
Upset Pressure	1200psi	Upset time	.5 sec

Factor	Set Point	s * 1.2247
SPEED	1200 +/- 122.47	
HT PRES	4800 +/- 612.35	
UP PRES	9500 +/- 1469.64	
LENGTH	0 +/- 1.2247	
HT TIME	3.5 +/- .61235	
UP TIME	3.6 +/- .61235	

Before we begin to build the experimental design, we need to comment on the format for the specification of tolerances. There can be considerable confusion on just what the ± tolerance values mean. Many engineers assume a normal distribution and when they specify a tolerance range. To them the ± is for a ±3 sigma range.

But not all think statistically (although the **6σ** initiative has corrected much in the way of statistical thinking) and the ± may not have anything to do with a frequency distribution at all. The most unambiguous way to specify a tolerance is to state the allowable standard deviation. This has been done in Table 21-8. Yet there could still be confusion based on industry-specific practices (see problem 7).

Using Minitab for Tolerance Design

We will build a L_{27} tolerance design for the friction welder. Begin by following the dropdown from **Stat=>DoE=>Taguchi=>Create Taguchi Design** as shown in Figure 21-11. This brings up the main dialog box (shown in Figure 21-12) where we select **3 Level Designs**. We can look at the available designs for 3-Levels or simply scroll down to the number of factors (6) and then go to the **Designs** radio button which shows us in Figure 21-13 that there is only one available design for our purposes which is the L27. Click **OK** to return to the main dialog box where **Factors** is no longer grayed out, so we click on **Factors** which brings up the sub dialog box where we are able to name the factors and insert their levels. This is shown in Figure 21-14.

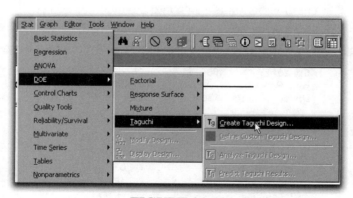

FIGURE 21-11

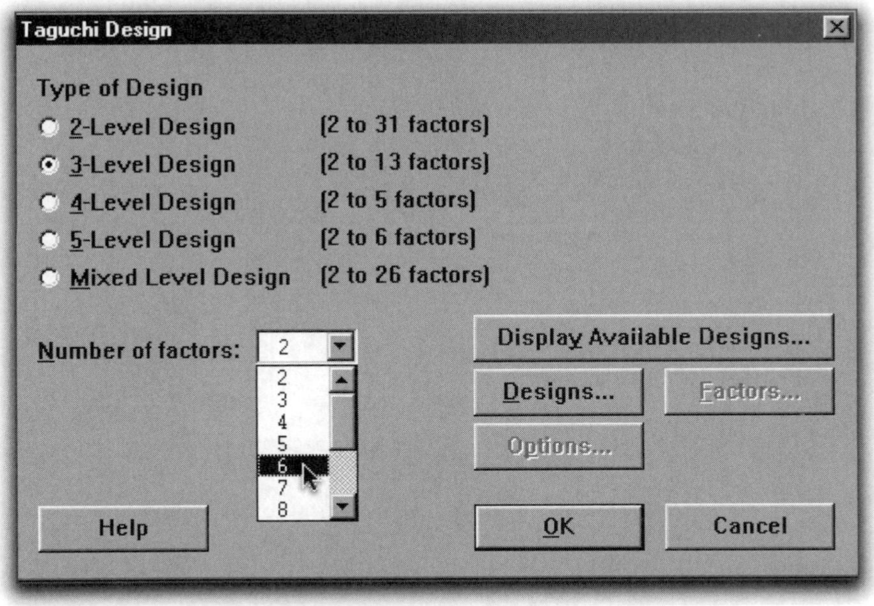

FIGURE 21-12

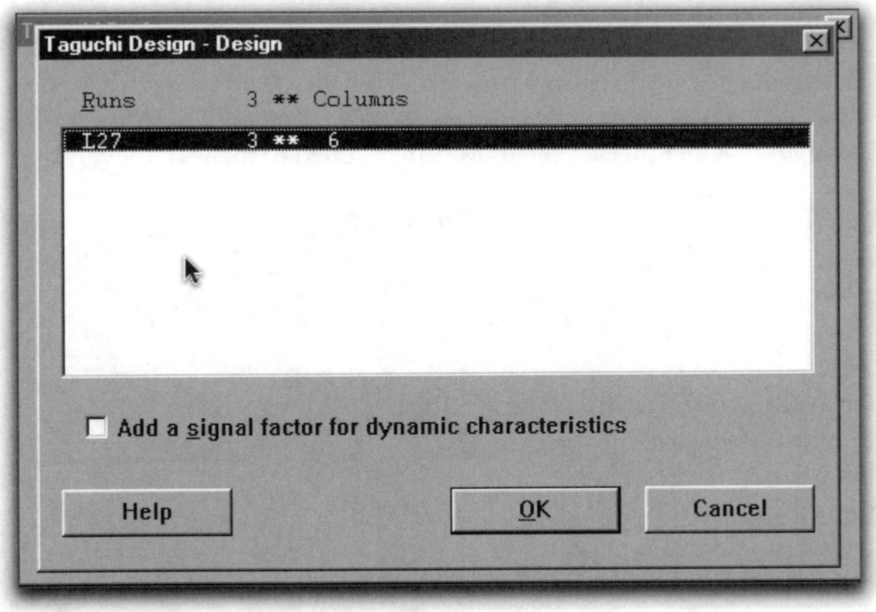

FIGURE 21-13

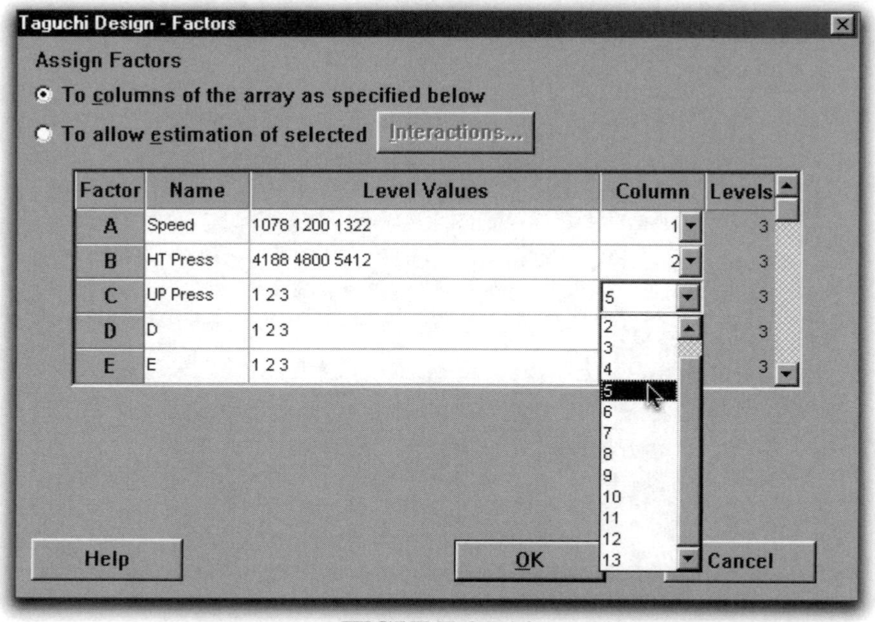

FIGURE 21-14

Enter the names and levels of the factors, as shown in Figure 21-14. Simply leave a space between levels. Be sure to select the proper columns to avoid confounding as we discussed in Chapter 20 with regard to the linear graphs associated with the Orthogonal Arrays. The levels of this design are outlined in Table 21-8 and include the transform required to make a 3-level design

produce statistics that appear to have been generated from a normal distribution input. We click **OK** which returns us to the main dialog box. The option radio button has but one option which is defaulted to store the design in the worksheet. Click **OK** to produce the design

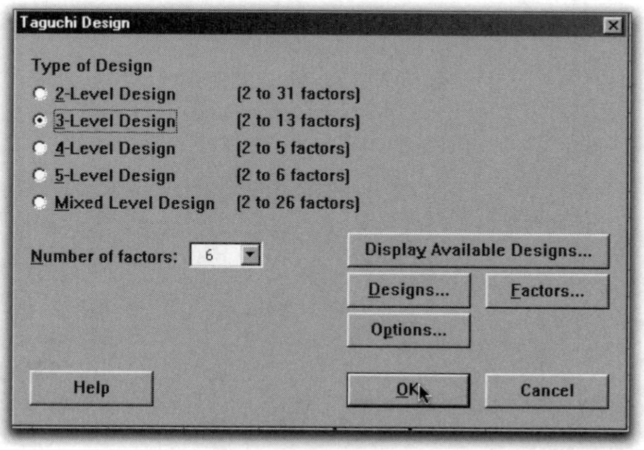

FIGURE 21-15

TABLE 21-9

Speed	HTPress	UPPress	Length	HTTime	UPTime	TenslStr
1078	4188	8030	-1.22	2.89	2.99	102.4
1078	4188	9500	0.00	3.50	3.60	156.1
1078	4188	10970	1.22	4.11	4.21	162.4
1078	4800	8030	0.00	3.50	4.21	125.4
1078	4800	9500	-1.22	4.11	2.99	172.2
1078	4800	10970	-1.22	2.89	3.60	186.3
1078	5412	8030	1.22	4.11	3.60	205.6
1078	5412	9500	-1.22	2.89	4.21	138.8
1078	5412	10970	0.00	3.50	2.99	202.3
1200	4188	8030	0.00	4.11	3.60	134.0
1200	4188	9500	1.22	2.89	4.21	134.8
1200	4188	10970	-1.22	3.50	2.99	164.2
1200	4800	8030	1.22	2.89	2.99	111.0
1200	4800	9500	-1.22	3.50	3.60	180.9
1200	4800	10970	0.00	4.11	4.21	204.0
1200	5412	8030	-1.22	3.50	4.21	150.2
1200	5412	9500	0.00	4.11	2.99	213.7
1200	5412	10970	1.22	2.89	3.60	194.9
1322	4188	8030	1.22	3.50	4.21	104.6
1322	4188	9500	-1.22	4.11	2.99	134.0
1322	4188	10970	0.00	2.89	3.60	182.3
1322	4800	8030	1.22	4.11	3.60	167.5
1322	4800	9500	0.00	2.89	4.21	134.8
1322	4800	10970	1.22	3.50	2.99	181.5
1322	5412	8030	0.00	2.89	2.99	111.0
1322	5412	9500	1.22	3.50	3.60	198.2
1322	5412	10970	-1.22	4.11	4.21	237.4

shown in Table 21-9. We will use this design to either simulate the experimentation via a mathematical model or actually set up a physical experiment. This is one of the advantages of the Tolerance Design approach in as much as the number of runs is not prohibitive and actual experimentation can be done to produce the responses. In this case, the responses were produced by solving a model for each trial in the L27 array. Since this is a deterministic solution, there is no need to randomize the run order.

With the design and the responses, we may now complete the analysis. There are many different approaches to

FIGURE 21-16

this in Minitab. We will utilize the GLM, which we select from the dropdown in Figure 21-16. This brings up the main dialog box shown in Figure 21-17. We select the response by highlighting "C7 TenslStr" and clicking on the "**Select**" radio button. Move the cursor to the "**Model**" entry box and highlight the 6 factors (Speed through UPTime) and click on "**Select**."

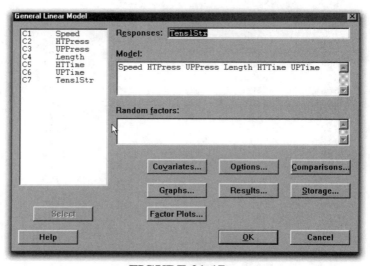

FIGURE 21-17

Since we are only interested in an ANOVA table, we need not open any of the options, graphs, storage, or factor plots sub dialog boxes. However, to make a neat ANOVA, we will open the "Results" sub dialog box and change

from the default to sim-ply the "Analysis of variance table" as shown in Figure 21-18. Now, Minitab will not give one piece of informa-tion needed to identify and quantify the qual-ity sensitive compo-nents. So, Table 21-10 has been modified by hand to include the "% of SS" column.

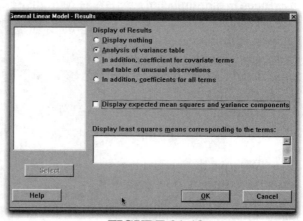

FIGURE 21-18

TABLE 21-10

Analysis of Variance for TenslStr, using Adjusted SS for Tests

```
Source    DF   Seq SS    Adj SS    Adj MS     F      % Of SS
Speed     2      97.6      97.6     48.8     0.22     0.0%
HTPress   2    7908.6    7908.6   3954.3    17.60    22.5%
UPPress   2   14089.5   14089.5   7044.8    31.35    40.2%
Length    1       0.0       0.0      0.0     0.0      0.0%
HTTime    2    6216.1    6216.1   3108.1    13.83    17.7%
UPTime    2    3374.9    3374.9   1687.4     7.51     9.6%
Error     15   3370.8    3370.8    224.7              9.6%
Total     26  35057.5                               99.6%
```

To add this column, take each Sum of Squares (SS) and divide it by the Total. So, the Speed factor has a SS of 97.6 which when divided by the Total SS of 35057 gives a tiny percent contribution which rounds to zero. The Heat Pressure (HTPress) has a more substantial contribution of 22.5% which is the SS of HTPress (7908) divided by the Total SS (35057). The % Of SS does not add up to 100% due to negligible rounding in these calculations.

Using the % Contribution to Rationally Apply Tolerances

The real power of the Tolerance Design lies in its ability to use the proportional contributions of the factors to determine the allocation of final tolerances in a design.

This is done by setting up a modification equation. The first element in this equation is the level *to* which we must modify the overall variation in the system. This is determined by taking the variance we need (400 for this friction welder) and dividing this variance by the variance we currently are experiencing. We may calculate this current variance by taking the responses from the designed experiment and calculating their variance or by using the ANOVA's Total SS divided by the Total df which is exactly what is done in the variance calculation. For this example that current variance is:

$$\frac{35057}{26} = 1348.3 \qquad\qquad 21\text{-}6$$

The general form of the modification equation is as follows. There is the reduced to term on the left with a set of terms that must sum to the value of the left side. We will use trial values of the Variation Modification Factor (VMF) in solving the equation. The VMF is squared since the contribution component (which weights the impact of the factor) is the ratio of the sum of squares of that item divided by the total sum of squares. To maintain the units properly we need to then square the modification factor.

Fraction Reduction = Σ { (VMF$_{Factor}$)^2 $\star$ CONTRIBUTION$_{Factor}$ } 21 -7

Note that this modification function is a single equation with multiple unknowns which means there is no single solution. The trial values of the VMF will depend on the cost and feasibility of attaining the reductions. Now let's see how we may set new tolerances for the friction weld example.

$$\frac{400}{1348.3} = \left\{ \overset{\text{Heat Pressure}}{(VMF_{HP})^2 0.225} + \overset{\text{Upset Pressure}}{(VMF_{UP})^2 0.402} + \overset{\text{Heat time}}{(VMF_{HT})^2 0.177} + \right.$$

$$\left. \overset{\text{Upset time}}{(VMF_{UT})^2 0.096} + \overset{\text{Residual}}{0.096} \right\} \qquad \text{21 -8}$$

Since the required standard deviation is 20 (p629), and we must work with the additive element of variation, the variance, we use $20^2 = 400$ in the numerator of the fraction reduction left side of the equation, indicating the value which we must attain to "keep the customer happy." We will need to reduce from a variance of 1348.3 to a variance of 400. This sets up the complete left side of our modification equation.

While there were six factors in the Tolerance Design investigation, two of them were inert and contributed nothing to the components of variance. Therefore there are only four quality sensitive components in this tolerance reduction exercise. We can simplify the tolerance modification even more by considering that there is a common pressure sensor and common timer. So, we combine the contributions of the pressures and timers and produce an equation with two unknowns. Before inserting trial VMF's for these two components, we must first remove the unknown (residual) from both sides of the equation. What if the residual were larger than the fraction reduction term on the left side of the equation? (It is not in this case.)

Since the residual is essentially the unknown, having a residual that is larger than our desired reduction means that it is impossible to achieve our requirement! Having a large amount of unknown variation means we cannot control the process. Here is a case where the math of the solution makes perfect sense. "The less you know about your process, the less you can do."

• Simplify for one pressure, one timer

$$0.297 = \left\{ (VMF_P)^2\, 0.627 + (VMF_T)^2\, 0.273 + 0.096 \right\} \qquad 21\text{-}9$$

• Remove Residual

$$0.201 = \left\{ (VMF_P)^2\, 0.627 + (VMF_T)^2\, 0.2737 \right\} \qquad 21\text{-}10$$

We begin the modification with the pressure which has the largest contribution to the variation. Our vendor, Nucor Corporation, has quoted us an "up cost" of $15,000 for new controllers on pressure plus the cost of $25,000 for a computer to read and feed back to the friction welder's hydraulic press. This will reduce the variation to one half of the current value. We square the VMF of 0.5 and multiply the result by the contribution of the pressure factor. This result (0.15675) is subtracted from both sides of the equation. Now we have one equation in one unknown and we solve for the VMF for the timer which is 0.4. So, given the 0.5 reduction in the pressure and 0.4 reduction in the timer, we have the new tolerance (stated as standard deviations) for the friction welding process shown in Table 21-11.

$$\begin{aligned}
-\,0.15675 \qquad &\quad -\,0.15675 \\
\cancel{0.201} = &\left\{ \cancel{(.5_P)^2\, 0.627} + (VMF_T)^2\, 0.2737 \right\} \qquad 21\text{-}11 \\
0.04425 = &\left\{ (VMF_T)^2\, 0.2737 \right\} \\
\frac{0.04425}{0.2737} = &\ (VMF_T)^2 \\
VMF_T = &\ 0.4
\end{aligned}$$

TABLE 21-11

New Tolerances on Friction Welder

Speed	100RPM (No change)
Heat Pressure	250psi (Reduced to 0.5 of original)
Upset Pressure	600psi (Reduced to 0.5 of original)
Length	1 thousandth (No change)
Heat time	0.2 sec (Reduced to 0.4 of original)
Upset time	0.2 sec (Reduced to 0.4 of original)

The advantage of the designed experiment approach to tolerancing lies in the fact that it is possible to try different solutions without having to re-run any simulations or physical experiments. We can "what if" the process simply and easily by trying different VMF's and comparing the sets of new, proposed tolerances for cost effectiveness.

Problems for Chapter 21

1. Write all the detailed steps for preparing a breakfast of bacon, eggs, buttered toast, orange juice, and coffee.

2. With your computer, generate 100 normally distributed random numbers with mean zero and a standard deviation of one. Plot a frequency distribution of the 100 values.

3. Use the table you generated in Problem 2 or, if you do not have a computer, use Table 20-7 to run a Monte Carlo simulation of the following equation. Use 10 trials and do this problem manually.
$$Y = A + B + C$$

Factor	Distribution	$\overline{X}$	s
A	Normal	2.5	0.15
B	Normal	1.0	0.05
C	Normal	0.5	0.01

4. Continue and extend the logic of the transmission of variation with a quadratic function as explained on pp. 624 and 625 to an interaction function. Use an interaction plot to visually support your arguments.

5. Given your prior knowledge of basic statistics including the consequences of the Central Limit Theorem, and any other information you can find in references, explain why so many phenomena have distributions that are good approximations to the Normal Distribution Model.

6. Create a frequency distribution of the responses from the Tolerance Design shown in Table 21-9. Comment on the shape of this frequency distribution.

7. Consider why using ± 1 sigma to specify the tolerance value could be less confusing or more confusing in your field. What educational initiatives would need to be implemented to make this statistical thinking approach (± 1 sigma) viable?

8. Comment on why the RPM and length did not contribute to the variation in the friction weld Tolerance Design. (Hint: a Parameter Design was completed before the Tolerance Design was done.)

9. Complete the variation analysis and propose a rational modification of variance given the % contribution and the going in low cost tolerances.

Factors	Set Point	Low Cost s	Contribution
Temperature	200°C	10°C	30.2%
Pressure	20 psi	1 psi	38.5%
Reaction Time	20 min	0.5 min	10.2%
Catalyst Concentration	0.5%	0.05%	8.8%
Residual (unknown)			12.2%

The current standard deviation is 40,000 and the required standard deviation is 20,000.

Costs to make variance reductions are:

Factor	3/4	1/2	1/4
Temperature	$500	$2000	$10,000
Pressure	$100	$500	$2000
Reaction Time	$10	$50	$100
Catalyst Concentration	$500	$2000	$10,000

APPENDIX 21A

Summary of Monte Carlo Methods

Method	How	Advantages	Disadvantages
Random Selection "True Monte Carlo"	Reach into the Population and grab samples. Assemble the samples using the equation. Repeat many times (1,000 to 10,000).	Realistic. Frequency distributions look "right."	Slow. Requires massive computer power.
Discrete Selection (Full Factorial)	Pick only certain(by design) values. Assemble in a factorial fashion using the equation.	Fast for small number of factors. Gives proper $\bar{Y}$ and s.	Frequency distribution of result (Y) does not look right.
Discrete Selection (3-level fractional factorial design) "Tolerance Design" from Taguchi	Use a 3-level factorial to pick values. For normal distribution statistics, use transform of factors' variation: s x $\sqrt{2/3}$ which is added and subtracted from the nominal level.	Very fast for even large number of factors. Direct analysis of the sources of variation (Quality Sensitive Components) via ANOVA.	Frequency distribution of the result (Y) does not look right.

References

1. Box, G.E.P., and Muller, M.E., A note on the generation of random normal deviates, *Ann. Math. Stat.*, Vol. 29 (1958), p. 610.

2. Barker, T.B., **Engineering Quality by Design, Interpreting the Taguchi Approach**, Dekker, New York, 1990.

3. Taguchi, G., **Introduction to Quality Engineering**, Asian Productivity Organization, Tokyo, 1986.

4. D'Errico, J.R., and Zaino, N.A., Statistical tolerancing using a modification of Taguchi's methods. *Proceedings QED 87*, Rochester Institute of Technology, Rochester, NY, 1988.

## 22

Case History Completed:
The Utilization of the Equation

If we were to simply derive the equation for a process, pat ourselves on the back for a good, efficient job and then quit, we would have done only half the job of infusing quality into our efforts. Once the equation is developed and verified, we need to exercise it. This is less a job for experimental design and more an effort for the engineer and therefore has been left out of most books on experimental design. However, it is probably the most important aspect of the "proof of the pudding" phase of the experimental effort. In Chapter 19 we derived an equation describing a photographic process. In Chapter 20 we learned that any equation could be put through its paces via simulation. We shall now take that photographic processing solution equation and see what makes it tick.

We have already solved the equation to determine the best midpoint operating conditions for the significant influencing factors. Recall that the equation was:

$$IQ = -1685.11 + 43.3517 \times \text{Temp.} - 0.268167 \times \text{Temp.}^2 + 1.317 \times \text{Metol} + 1.109 \times \text{HQ} - 1.08 \times \text{Speed}$$

Midpoint values are:	Temperature	80.82° F
	Metol	16 g/l
	HQ	10 g/l
	Speed	10 ips

When we used the above values, the IQ (image quality) of the process was computed at 100, which was our goal. However, the ultimate goal of quality is to achieve the target value and do so consistently. We realize that it would be impossible to hold a temperature at 80.82 all the time. To do so would require very expensive feedback control systems. To be able to weigh the chemicals to their exact specifications would also be impossible and would impose a restriction on the production process that is too costly. A speed of exactly 10 ips may be accomplished with another system, but again, this is an imposition that a production process can't afford. What we need is a toleranced specification. That is a midpoint with an acceptable degree of slop around it. This is a common practice in most industry. However, the practice of deriving the degree of variation printed on the specification is not often practiced in a scientific or even economic manner.

First we must realize that the tolerance on the components must be related to the customer's requirements. Quality products must meet customer needs in a consistent manner. In other words, a quality product is a product that is on target with little variation. We must first determine the target and then understand the amount of variation tolerable. In the photographic processing process, the target value is determined by market research. We must know what level of image quality is required by the customer. We must also know what level of deviation from the target will cause the customer dissatisfaction.

We will use the following market research information to set the product "specs" on the processing example.

The Image Quality should not fall below 95 IQ units.

Further, the distribution shall follow a normal density function.

The above specification could be interpreted to mean (incorrectly) that the mean level shall be at 95. But if there is any variation in the process in the downward direction, the spec would be violated since we would produce a product below the 95 lower limit. We already know how to get to the IQ of 100, but we don't know how much variation in each of the four critical components is tolerable. This is where simulation techniques come in handy. We will make systematic changes in the levels of the critical components and study their effect on the overall IQ of the process. We will try two approaches to this simulation. One will be the traditional Monte Carlo random method and the other will be to make changes according to a pattern based on an orthogonal experimental design.

Random Method

While there are many ways of accomplishing the same goals using Monte Carlo concepts, we will in this instance use the very simple method that was the basis for Problem 20-2. Instead of using a manual selection of the random normal numbers, we will use a self-contained computer program. The program is shown in the Appendix of this chapter. It is written in GWBASIC and should run without modification on most computers. It is set up to allow data to be generated for control chart plotting. If you are not familiar with the concept of a control chart, the idea is to study a series of outputs and observe if the averages of selected groups stay within the "natural" variation of the process. We will not use the subgroup feature of this simulation program for this exercise, but will use only the final mean and standard deviation.

```
ENTER CONTROL PARAMETER # 1 MEAN,STD. DEV.? 80.82,2
ENTER CONTROL PARAMETER # 2 MEAN,STD. DEV.? 16,1
ENTER CONTROL PARAMETER # 3 MEAN,STD. DEV.? 20,1
ENTER CONTROL PARAMETER # 4 MEAN,STD. DEV.? 10,.5
```

SUBGROUP # 1	101.266	98.830	98.391	98.549	97.392
SUBGROUP # 2	98.513	97.725	94.678	98.329	94.581
SUBGROUP # 3	97.640	97.411	96.395	97.255	97.938
SUBGROUP # 4	97.268	99.960	96.495	99.248	96.700
SUBGROUP # 5	95.881	97.598	98.048	98.267	95.979
SUBGROUP # 6	100.379	98.390	97.742	99.943	99.759
SUBGROUP # 7	97.239	97.502	100.821	93.555	97.505
SUBGROUP # 8	98.459	98.500	99.406	95.074	99.933
SUBGROUP # 9	97.823	100.157	98.831	96.619	97.573
SUBGROUP # 10	98.391	97.622	101.788	98.422	98.995
SUBGROUP # 11	98.640	99.307	101.535	99.898	99.797
SUBGROUP # 12	96.802	97.156	98.917	99.045	98.120
SUBGROUP # 13	100.238	95.728	100.337	97.749	99.234
SUBGROUP # 14	96.156	92.417	99.772	99.975	93.870
SUBGROUP # 15	97.128	99.629	100.278	97.783	98.388

```
AVERAGE= 98.11598  STD. DEV.= 1.825356
```

The above output shows a run of 15 subgroups with 5 observations per subgroup. For a reliable Monte Carlo analysis, we will need more total observations, so we will eliminate the printing of each subgroup data set and concentrate only on the bottom line summary statistics.

```
RUN
HOW MANY CONTROL PARAMETERS IN YOUR PROCESS? 4
HOW MANY SUBGROUPS FOR THE CONTROL CHART? 100
HOW MANY ITEMS PER  SUBGROUP? 5
ENTER CONTROL PARAMETER #  1 MEAN,STD. DEV.? 80.82,2
ENTER CONTROL PARAMETER #  2 MEAN,STD. DEV.? 16,1
ENTER CONTROL PARAMETER #  3 MEAN,STD. DEV.? 20,1
ENTER CONTROL PARAMETER #  4 MEAN,STD. DEV.? 10,.5

AVERAGE= 98.1  STD. DEV.=  2.26


RUN
HOW MANY CONTROL PARAMETERS IN YOUR PROCESS? 4
HOW MANY SUBGROUPS FOR THE CONTROL CHART? 100
HOW MANY ITEMS PER  SUBGROUP? 5
ENTER CONTROL PARAMETER #  1 MEAN,STD. DEV.? 80.82,1
ENTER CONTROL PARAMETER #  2 MEAN,STD. DEV.? 16,.5
ENTER CONTROL PARAMETER #  3 MEAN,STD. DEV.? 20,.5
ENTER CONTROL PARAMETER #  4 MEAN,STD. DEV.? 10,.25

AVERAGE= 99.0  STD. DEV.=  1.01

HOW MANY CONTROL PARAMETERS IN YOUR PROCESS? 4
HOW MANY SUBGROUPS FOR THE CONTROL CHART? 100
HOW MANY ITEMS PER  SUBGROUP? 5

ENTER CONTROL PARAMETER #  1 MEAN,STD. DEV.? 80.82,.75
ENTER CONTROL PARAMETER #  2 MEAN,STD. DEV.? 16,.75
ENTER CONTROL PARAMETER #  3 MEAN,STD. DEV.? 20,.75
ENTER CONTROL PARAMETER #  4 MEAN,STD. DEV.? 10,.25

AVERAGE= 99.0  STD. DEV.=  1.37
```

In the program, lines 70-160 hold the random normal values from a distribution with a mean of zero and a standard deviation of one. We will "reach into" this population and sample a value for each of our 4 control parameters. The actual mean and standard will depend upon the mean and standard deviation of the parameter and the random value picked. This is exactly like the homework problem, only we have automated the process. After all 4 parameter values have been selected, the equation is activated and a response is generated. We will continue this process of selecting the random values and apply them to the parameters and then solve the equation until we have satisfied the required number of trials. This is exactly what would have happened in running the photographic process, except we do it with an equation rather than the real chemicals and process settings. Table 22-1 shows a summary of the simulation work and the final settings that accomplish the goal of IQ greater than 95.

TABLE 22-1

Temp. Mean/SD	Metol Mean/SD	HQ Mean/SD	Speed Mean/SD	Sample Size	IQ Mean/SD
80.82	16	20	10	150	98.1
2.0	1.0	1.0	0.5		2.26
80.82	16	20	10	150	99.0
1.0	0.5	0.5	0.25		1.01
80	16	20	10	150	99.0
0.5	0.75	0.75	0.25		1.37

The last trial with a mean of 99 and a standard deviation of 1.37 produces a lower 3 sigma level of 94.9, which is very close to the specification of 95. The first Monte Carlo trial represented low cost tolerances and the selection of the tolerances was made based on the prior knowledge of the photographic process. Such knowledge is very important in making the first set of trial runs in such a simulation. The educated guessing of the subsequent tolerances is based on the knowledge and an observation of the previous run. In the second trial, we cut the tolerances in half and got a greater reduction in the overall IQ standard deviation than we may have expected. We had "overcorrected" and the last run brought us back to the right set of conditions. Had we tried to do this in a production environment, we would have had to run nearly 500 rolls of film through the process with exact control of the conditions to observe the results that we obtained with only a small amount of computer time and a little bit of programming effort. The combined use of simulation techniques and the results of empirical equations derived from statistically designed experiments pays genuine dividends in shortening schedules, giving us the quality products the customer desires

Another Approach

While the Monte Carlo approach will give us the answers we desire, there is a certain amount of guesswork involved in using the random process in simulating the parameter set points and especially the variation around these set points. The problem is recognized by the experts in the field of simulation and there are far more sophisticated computer routines for doing the job than the simple one we just used. Some of these programs have "sensitivity" analysis built in to point to the parameters that have the greatest influence on the output variable. Of course, as the simulation programs become more complex, they take longer to run and require bigger and bigger computer resources. What is worse is the involvement in the programming and the learning required to run these monster programs. No matter how complex the simulation program, we must remember that it is only a tool to help us make a good decision. I have seen some projects that have become so involved with the simulation tool that the users forget about the real problem.

As experimental designers, we should have realized that we have always had the most powerful method for simulation and *rational* reduction of variance in the product or process design. The purpose of a Monte Carlo simulation is to make purposeful changes in the input parameters to see how these changes effect the output. This is exactly the same purpose that a designed experiment has! While the Monte Carlo method will give us the good news or the bad news of meeting or not meeting the desired goal, it can't do much more. It does not tell us which of the parameters has an influence on the variation. The Monte Carlo method works in the same way a one-factor-at-a-time test works. It also lacks a method of systematic analysis.

On the other hand, a designed experiment is capable of being analyzed. We can determine which parameters have an influence on the output. We can determine the magnitude as well as the identity of these factors. In our second look at the photo process simulation, we will not use random samples of the parameter settings, but we will use set levels. We will use an orthogonal experimental design to generate variation and then analyze the results of this design to determine how much each factor contributes to the variation in the response variable (1).

In Chapter 7 we introduced the idea of 3-level orthogonal designs. We said that they would become useful in later applications. The time has arrived to put these designs to work in a simulation exercise. While other designs will do an adequate job, the 3-level design is particularly well-suited because it provides a nominal (midpoint) and an excursion on both sides of this nominal. We can (as we indicated in Chapter 19) even simulate a normal distribution from such a uniform pattern by a clever transform of the factor's standard deviation.

To expedite the use of designed experiments in such simulation situations, computers can be programmed to construct the design structure and create the responses (2). For our example we will use a 3^{4-1} design. With this design, we find that the confounding is minimal. Also, there are no interactions in our model, so we need not be concerned about anything except single effects.

TABLE 22-2

Temp.	Metol	HQ	Speed	IQ
78.37	14.78	18.78	9.39	95.46
78.37	14.78	20.00	10.00	96.16
78.37	14.78	21.22	10.61	96.85
78.37	16.00	18.78	10.00	96.41
78.37	16.00	20.00	10.61	97.11
78.37	16.00	21.22	9.39	99.79
78.37	17.22	18.78	10.61	97.37
78.37	17.22	20.00	9.39	100.05
78.37	17.22	21.22	10.00	100.74
80.82	14.78	18.78	10.00	96.42
80.82	14.78	20.00	10.61	97.12
80.82	14.78	21.22	9.39	99.80
80.82	16.00	18.78	10.61	97.37
80.82	16.00	20.00	9.39	100.06
80.82	16.00	21.22	10.00	100.75
80.82	17.22	18.78	9.39	100.30
80.82	17.22	20.00	10.00	101.01
80.82	17.22	21.22	10.61	101.70
83.26	14.78	18.78	10.61	94.17
83.26	14.78	20.00	9.39	96.85
83.26	14.78	21.22	10.00	97.54
83.26	16.00	18.78	9.39	97.10
83.26	16.00	20.00	10.00	97.80
83.26	16.00	21.22	10.61	98.49
83.26	17.22	18.78	10.00	98.05
83.26	17.22	20.00	10.61	98.75
83.26	17.22	21.22	9.39	101.43

ANOVA of the Results

Table 22-2 shows the systematic changes in the four controlling factors and the solution of the photo process equation after each treatment combination. We now take each of the 27 responses and do a simple ANOVA, which is shown in Table 22-3. The results in this table will help us appreciate the power of a fixed experimental design over the random Monte Carlo approach.

TABLE 22-3

ANOVA:

Source	ΣSq	df	% Contribution
Temperature	15.53	2	15.01
Metol	46.83	2	45.27
HQ	33.20	2	32.10
Speed	7.87	2	7.61
Total	103.43	26	

First, look at the bottom line, Total Sum of Squares. If we divide this by 26 df (the total number of df in this design), we obtain the variance of the 27 observations. The square root of this variance is 2.00, which is the standard deviation of the 27 observations. This is not surprising, but very important. If the total is made up of the parts, we should be able to dissect the total to find which of the parameters have contributed to the overall variation. Remember, the variation induced into the response (the IQ) came only from the variation in the purposely induced variation of the four control parameters. There is no experimental error since we are solving an equation. If we add up the individual sums of squares, we should get the total of 103.43 (it does add up!).

The important result of the above observation is the fact that we can now determine the fractional contribution of each control parameter to the total. With this knowledge, we can determine the amount of reduction required to meet our goal and set the tolerances on the control parameters from a rational, numerical basis.

Now let's look at our specification on IQ. If we can attain a value of 100 and the specified lower limit for 3 standard deviations is 95, we need a standard deviation of 5/3 or 1.67 to accomplish our goal. Our current standard deviation is 2.00. Since variances are additive (standard deviations are not), we will square these values and find the amount of reduction required to attain the goal.

We will take the required variance of $1.67^2 = 2.7889$ and divide it by the current variance $2.00^2 = 4.00$. This gives a .70 reduction factor which is nearly three quarters of the variation. How can we get the current variance down to 70% of its value? We will have to cut the variation in each of the factors. This is what we did in the random Monte Carlo case, but there we did not know which were the major contributors. We could plot a Pareto chart of the situation to communicate, but the real power comes from the following concept.

Sum reduced variance = Sum weighted fraction reductions

$$0.70 = [VMF_T{}^2 \times 0.155 + VMF_M{}^2 \times 0.468 + VMF_{HQ}{}^2 \times 0.332 + VMF_S{}^2 \times 0.079]$$

Since the weights are for variances, we must square the variation modification factor (VMF), which is then multiplied by the original standard deviation of the factor's setting to obtain the new allowable variation.

We now have an expression that we can exercise in a flexible way to determine the most economic allocation of the tolerances on the control parameters. There is no single solution to the above expression. But we can see that the speed has a small contribution to the total, so it could be left at its present tolerance. We also know that the control of the temperature is more difficult than control of the weight of the chemicals. We will try the following solution to the equation. Leave T and S at the present level of variation which makes the values of T and S equal to 1 (no modification of these gives a VMF of 1). The equation is now reduced to the following:

$$0.466 = [VMF_M{}^2 \times 0.468 + VMF_{HQ}{}^2 \times 0.332]$$

If we cut the current variation of Metol in half, then the VMF_M is 0.5. When we square 0.5 and multiply this result by the weight of the Metol (0.468), then subtract this from the left side of the equation, we are left with a remainder of 0.349 for the HQ. Now we have one equation in a single unknown and we may solve for the VMF for HQ.

$$0.349 = [VMF_{HQ}{}^2 \times 0.332]$$

$$VMF_{HQ} = \sqrt{\frac{0.349}{0.332}} = 1.03$$

This says that we may leave the variation of the HQ alone and need not reduce it. This might be considered a cost saving, *but* the weighing device we need to buy to improve the precision of the Metol tolerance is the same device we will be using for the HQ! So instead of imposing a tighter specification on the Metol and a looser one on the HQ, we will combine the contributions of Metol and HQ (0.4527+0.321=0.774) and re-express the tolerance reduction equation as follows:

$$0.70 = [VMF_T^2 \times 0.155 + VMF_{WP}^2 \times 0.774 + VMF_S^2 \times 0.079]$$

We will again leave T and S at the present level of variation which makes the values of VMF_T and VMF_S equal to 1 (no modification of these gives a VMF of 1). The equation is now reduced to the following:

$$0.466 = [VMF_{WP}^2 \times 0.774]$$

It is easy to solve for the VMF_{WP}, since it is one equation in one unknown.

$$VMF_{WP} = \sqrt{\frac{0.466}{0.774}} = 0.776$$

With a VMF for the weighing process of .776, the new standard deviation of the weighing process is the original standard deviation (1.0) times this VMF. We will round the tolerance to 0.75 for simplicity. Now we have the new toleranced specifications for the developer process.

> Temperature set point: 80.82, standard deviation: 2.0
> Metol   set point: 16, standard deviation: 0.75
> HQ   set point: 20, standard deviation: 0.75
> Speed   set point: 10, standard deviation: 0.5

A confirmation of this set of conditions shows that we meet our target. The systematic appraoch has produced a different set of operating tolerances then we obtained from the *random* Monte Carlo method. These conditions are more economical than the previous set. Notice that the temperature and speed have very little effect and do not have to be controlled as tightly as the Monte Carlo method trial suggested. Of course, if the tolerances were too costly, we would have gone back and tried different settings. But if there had been no complaint, then we would have imposed specifications that were too tight for no real reason!

Monte Carlo methods are good as simulations of the situation. However, because they only tell if a process will work but not *why* it works, they are not sufficient. The experimental design approach to systematically finding the sensitivity of parameter tolerances on the process response is a more structured and engineering oriented approach to rational specifications.

References

1. Barker, T. B. (1990). **Engineering Quality by Design – Interpreting the Taguchi Approach.** Marcel Dekker, Inc., New York.

2. Barker, T. B. (1993). **QED Programs for MSBASIC**. Rochester Institute of Technology, Rochester, New York.

Problems for Chapter 22

1. Using the following equations that were derived from the KlatterMax experiment, determine the sensitivity to tolerances on the settings of the factors. Use starting standard deviations = ±20%.

Time = 35.4 - 6.81xLength (A) + 6.72xInside Length (E) - 80.0xSlot Width (F)
    - 3.71xSlot LengthxInside Length (DE)

Hang-up = 2.73 - 3.49xLength (A) + 39.3xSlot Length (D)
        + 14.9xInside Length (E) - 20.1xSlot Width (F)
        - 50.7 Slot LengthxInside Length (DE)

Fly-off = 84.9 - 28.1xLength (A) - 3.06xThick (B) - 46.7xWidth (C)
        - 9.24xHole (G) + 16.9 LengthxWidth (AC)

2. Recommend the tolerances needed to achieve a robust KlatterMax based on your above sensitivity analysis.

3. After obtaining an equation for the emulsion problem in Chapter 19, determine the best set points and tolerances for this process.

APPENDIX 22

Control Chart Monte Carlo Computer Program

```
5 CLS:CLEAR 300
10 INPUT "HOW MANY CONTROL PARAMETERS IN YOUR PROCESS";N5
20 INPUT "HOW MANY SUBGROUPS FOR THE CONTROL CHART";T1
30 INPUT "HOW MANY ITEMS PER SUBGROUP";I1
40 DIM P(N5,2),ND(100),X(N5)
50 CLS:PRINT "PLEASE WAIT"
55 RS=VAL(RIGHT$(TIME$,2)):RANDOMIZE RS
60 FOR I=1 TO 100:READ ND(I):NEXT
70 DATA .906,-.513,-.525,.595,.881,-.934,1.579,.161,-1.885,.371
80 DATA 1.179,-1.055,.007,.769,.971,.712,1.09,-.631,-.255,-.702
90 DATA -1.501,-.488,-.162,-.136,1.033,.203,.448,.748,-.423,-.432
100 DATA -.69,.756,-1.618,-.345,-.511,-2.051,-.457,-.218,.857,-.465
110 DATA 1.372,.225,.378,.761,.181,-.736,.96,-1.53,-.26,.12
120 DATA -1.787,-.261,1.237,1.046,-.508,-1.63,-.146,-.392,-.627,.561
130 DATA -.105,-.357,-1.384,.36,-.992,-.116,-1.698,-2.832,-1.108,-2.357
140 DATA -1.339,1.827,-.959,.424,.969,-1.141,-1.041,.362,-1.726,1.956
150 DATA 1.041,.535,.731,1.377,.983,-1.33,1.62,-1.04,.524,-.281
160 DATA .279,-2.056,.717,-.0872,-1.096,-1.396,1.047,.089,-.573,.932
170 FOR I=1 TO N5:PRINT "ENTER CONTROL PARAMETER # ";I;:INPUT "
MEAN,STD. DEV.";P(I,1),P(I,2):NEXT
180 FOR I=1 TO T1:LPRINT "SUBGROUP #";I;
190 FOR J=1 TO I1
200 FOR K=1 TO N5:R=RND(1)*100:IF R=0 THEN R=1
210 X(K)=P(K,1)+(ND(R)*P(K,2)):NEXT K
220 REM Insert equations here (LINE 280)
230 REM of the form Y=b0 + b1*X(i),etc.
240 REM be sure to use Y for the response and X(I) for the
250 REM control parameter where I=1,2,3,...N5(the max #)
270 REM PLACE AN "LPRINT Y;" STATEMENT AFTER EACH EQUATION
280 Y=-1685.11+43.3517*X(1)-.268167*X(1)*X(1)+1.317*X(2)
290 Y=Y+1.109*X(3)-1.08**X(4)
300 REM X(1)=TEMP;X(2)=METOL;X(3)=HQ;X(4)=SPEED
305 LPRINT  Y;
307 S=S+Y:SS=SS+Y*Y
400 NEXT J
405 LPRINT
410 NEXT I
412 N=T1*I1
415 XB=S/N:SD=SQR((SS-(S*S/N))/(N-1))
420 LPRINT "AVERAGE=";XB;" STD. DEV.= ";SD
430 END
```

Statistical Tables

TABLE 1
Areas under the Normal Curve

z	0.00	0.01	0.02	0.03	0.04	0.05	0.06	0.07	0.08	0.09
−3.4	0.0003	0.0003	0.0003	0.0003	0.0003	0.0003	0.0003	0.0003	0.0003	0.0002
−3.3	0.0005	0.0005	0.0005	0.0004	0.0004	0.0004	0.0004	0.0004	0.0004	0.0003
−3.2	0.0007	0.0007	0.0006	0.0006	0.0006	0.0006	0.0006	0.0005	0.0005	0.0005
−3.1	0.0010	0.0009	0.0009	0.0009	0.0008	0.0008	0.0008	0.0008	0.0007	0.0007
−3.0	0.0013	0.0013	0.0013	0.0012	0.0012	0.0011	0.0011	0.0011	0.0010	0.0010
−2.9	0.0019	0.0018	0.0017	0.0017	0.0016	0.0016	0.0015	0.0015	0.0014	0.0014
−2.8	0.0026	0.0025	0.0024	0.0023	0.0023	0.0022	0.0021	0.0021	0.0020	0.0019
−2.7	0.0035	0.0034	0.0033	0.0032	0.0031	0.0030	0.0029	0.0028	0.0027	0.0026
−2.6	0.0047	0.0045	0.0044	0.0043	0.0041	0.0040	0.0039	0.0038	0.0037	0.0036
−2.5	0.0062	0.0060	0.0059	0.0057	0.0055	0.0054	0.0052	0.0051	0.0049	0.0048
−2.4	0.0082	0.0080	0.0078	0.0075	0.0073	0.0071	0.0069	0.0068	0.0066	0.0064
−2.3	0.0107	0.0104	0.0102	0.0099	0.0096	0.0094	0.0091	0.0089	0.0087	0.0084
−2.2	0.0139	0.0136	0.0132	0.0129	0.0125	0.0122	0.0119	0.0116	0.0113	0.0110
−2.1	0.0179	0.0174	0.0170	0.0166	0.0162	0.0158	0.0154	0.0150	0.0146	0.0143
−2.0	0.0228	0.0222	0.0217	0.0212	0.0207	0.0202	0.0197	0.0192	0.0188	0.0183
−1.9	0.0287	0.0281	0.0274	0.0268	0.0262	0.0256	0.0250	0.0244	0.0239	0.0233
−1.8	0.0359	0.0352	0.0344	0.0336	0.0329	0.0322	0.0314	0.0307	0.0301	0.0294
−1.7	0.0446	0.0436	0.0427	0.0418	0.0409	0.0401	0.0392	0.0384	0.0375	0.0367
−1.6	0.0548	0.0537	0.0526	0.0516	0.0505	0.0495	0.0485	0.0475	0.0465	0.0455
−1.5	0.0668	0.0655	0.0643	0.0630	0.0618	0.0606	0.0594	0.0582	0.0571	0.0559
−1.4	0.0808	0.0793	0.0778	0.0764	0.0749	0.0735	0.0722	0.0708	0.0694	0.0681
−1.3	0.0968	0.0951	0.0934	0.0918	0.0901	0.0885	0.0869	0.0853	0.0838	0.0823
−1.2	0.1151	0.1131	0.1112	0.1093	0.1075	0.1056	0.1038	0.1020	0.1003	0.0985
−1.1	0.1357	0.1335	0.1314	0.1292	0.1271	0.1251	0.1230	0.1210	0.1190	0.1170
−1.0	0.1587	0.1562	0.1539	0.1515	0.1492	0.1469	0.1446	0.1423	0.1401	0.1379
−0.9	0.1841	0.1814	0.1788	0.1762	0.1736	0.1711	0.1685	0.1660	0.1635	0.1611
−0.8	0.2119	0.2090	0.2061	0.2033	0.2005	0.1977	0.1949	0.1922	0.1894	0.1867
−0.7	0.2420	0.2389	0.2358	0.2327	0.2296	0.2266	0.2236	0.2206	0.2177	0.2148
−0.6	0.2743	0.2709	0.2676	0.2643	0.2611	0.2578	0.2546	0.2514	0.2483	0.2451
−0.5	0.3085	0.3050	0.3015	0.2981	0.2946	0.2912	0.2877	0.2843	0.2810	0.2776
−0.4	0.3446	0.3409	0.3372	0.3336	0.3300	0.3264	0.3228	0.3192	0.3156	0.3121
−0.3	0.3821	0.3783	0.3745	0.3707	0.3669	0.3632	0.3594	0.3557	0.3520	0.3483
−0.2	0.4207	0.4168	0.4129	0.4090	0.4052	0.4013	0.3974	0.3936	0.3897	0.3859
−0.1	0.4602	0.4562	0.4522	0.4483	0.4443	0.4404	0.4364	0.4325	0.4286	0.4247
−0.0	0.5000	0.4960	0.4920	0.4880	0.4840	0.4801	0.4761	0.4721	0.4681	0.4641
0.0	0.5000	0.5040	0.5080	0.5120	0.5160	0.5199	0.5239	0.5279	0.5319	0.5359
0.1	0.5398	0.5438	0.5478	0.5517	0.5557	0.5596	0.5636	0.5675	0.5714	0.5753
0.2	0.5793	0.5832	0.5871	0.5910	0.5948	0.5987	0.6026	0.6064	0.6103	0.6141
0.3	0.6179	0.6217	0.6255	0.6293	0.6331	0.6368	0.6406	0.6443	0.6480	0.6517
0.4	0.6554	0.6591	0.6628	0.6664	0.6700	0.6736	0.6772	0.6808	0.6844	0.6879
0.5	0.6915	0.6950	0.6985	0.7019	0.7054	0.7088	0.7123	0.7157	0.7190	0.7224
0.6	0.7257	0.7291	0.7324	0.7357	0.7389	0.7422	0.7454	0.7486	0.7517	0.7549
0.7	0.7580	0.7611	0.7642	0.7673	0.7704	0.7734	0.7764	0.7794	0.7823	0.7852
0.8	0.7881	0.7910	0.7939	0.7967	0.7995	0.8023	0.8051	0.8078	0.8106	0.8133
0.9	0.8159	0.8186	0.8212	0.8238	0.8264	0.8289	0.8315	0.8340	0.8365	0.8389
1.0	0.8413	0.8438	0.8461	0.8485	0.8508	0.8531	0.8554	0.8577	0.8599	0.8621
1.1	0.8643	0.8665	0.8686	0.8708	0.8729	0.8749	0.8770	0.8790	0.8810	0.8830
1.2	0.8849	0.8869	0.8888	0.8907	0.8925	0.8944	0.8962	0.8980	0.8997	0.9015
1.3	0.9032	0.9049	0.9066	0.9082	0.9099	0.9115	0.9131	0.9147	0.9162	0.9177
1.4	0.9192	0.9207	0.9222	0.9236	0.9251	0.9265	0.9278	0.9292	0.9306	0.9319
1.5	0.9332	0.9345	0.9357	0.9370	0.9382	0.9394	0.9406	0.9418	0.9429	0.9441
1.6	0.9452	0.9463	0.9474	0.9484	0.9495	0.9505	0.9515	0.9525	0.9535	0.9545
1.7	0.9554	0.9564	0.9573	0.9582	0.9591	0.9599	0.9608	0.9616	0.9625	0.9633
1.8	0.9641	0.9649	0.9656	0.9664	0.9671	0.9678	0.9686	0.9693	0.9699	0.9706
1.9	0.9713	0.9719	0.9726	0.9732	0.9738	0.9744	0.9750	0.9756	0.9761	0.9767
2.0	0.9772	0.9778	0.9783	0.9788	0.9793	0.9798	0.9803	0.9808	0.9812	0.9817
2.1	0.9821	0.9826	0.9830	0.9834	0.9838	0.9842	0.9846	0.9850	0.9854	0.9857
2.2	0.9861	0.9864	0.9868	0.9871	0.9875	0.9878	0.9881	0.9884	0.9887	0.9890
2.3	0.9893	0.9896	0.9898	0.9901	0.9904	0.9906	0.9909	0.9911	0.9913	0.9916
2.4	0.9918	0.9920	0.9922	0.9925	0.9927	0.9929	0.9931	0.9932	0.9934	0.9936
2.5	0.9938	0.9940	0.9941	0.9943	0.9945	0.9946	0.9948	0.9949	0.9951	0.9952
2.6	0.9953	0.9955	0.9956	0.9957	0.9959	0.9960	0.9961	0.9962	0.9963	0.9964
2.7	0.9965	0.9966	0.9967	0.9968	0.9969	0.9970	0.9971	0.9972	0.9973	0.9974
2.8	0.9974	0.9975	0.9976	0.9977	0.9977	0.9978	0.9979	0.9979	0.9980	0.9981
2.9	0.9981	0.9982	0.9982	0.9983	0.9984	0.9984	0.9985	0.9985	0.9986	0.9986
3.0	0.9987	0.9987	0.9987	0.9988	0.9988	0.9989	0.9989	0.9989	0.9990	0.9990
3.1	0.9990	0.9991	0.9991	0.9991	0.9992	0.9992	0.9992	0.9992	0.9993	0.9993
3.2	0.9993	0.9993	0.9994	0.9994	0.9994	0.9994	0.9994	0.9995	0.9995	0.9995
3.3	0.9995	0.9995	0.9995	0.9996	0.9996	0.9996	0.9996	0.9996	0.9996	0.9997
3.4	0.9997	0.9997	0.9997	0.9997	0.9997	0.9997	0.9997	0.9997	0.9997	0.9998

NORMAL TABLE

Source: *Xerox Corporation SPC Training Program.*

TABLE 2
Critical Values of Student's t Distribution

$\overset{\dot{\alpha}}{\diagdown}$ ν	IIb – Two Tail Critical Values						
	0.50	0.25	0.10	0.05	0.025	0.01	0.005
1	1.00000	2.4142	6.3138	12.706	25.452	63.657	127.32
2	0.81650	1.6036	2.9200	4.3027	6.2053	9.9248	14.089
3	0.76489	1.4226	2.3534	3.1825	4.1765	5.8409	7.4533
4	0.74070	1.3444	2.1318	2.7764	3.4954	4.6041	5.5976
5	0.72669	1.3009	2.0150	2.5706	3.1634	4.0321	4.7733
6	0.71756	1.2733	1.9432	2.4469	2.9687	3.7074	4.3168
7	0.71114	1.2543	1.8946	2.3646	2.8412	3.4995	4.0293
8	0.70639	1.2403	1.8595	2.3060	2.7515	3.3554	3.8325
9	0.70272	1.2297	1.8331	2.2622	2.6850	3.2498	3.6897
10	0.69981	1.2213	1.8125	2.2281	2.6338	3.1693	3.5814
11	0.69745	1.2145	1.7959	2.2010	2.5931	3.1058	3.4966
12	0.69548	1.2089	1.7823	2.1788	2.5600	3.0545	3.4284
13	0.69384	1.2041	1.7709	2.1604	2.5326	3.0123	3.3725
14	0.69242	1.2001	1.7613	2.1448	2.5096	2.9768	3.3257
15	0.69120	1.1967	1.7530	2.1315	2.4899	2.9467	3.2860
16	0.69013	1.1937	1.7459	2.1199	2.4729	2.9208	3.2520
17	0.68919	1.1910	1.7396	2.1098	2.4581	2.8982	3.2225
18	0.68837	1.1887	1.7341	2.1009	2.4450	2.8784	3.1966
19	0.68763	1.1866	1.7291	2.0930	2.4334	2.8609	3.1737
20	0.68696	1.1848	1.7247	2.0860	2.4231	2.8453	3.1534
21	0.68635	1.1831	1.7207	2.0796	2.4138	2.8314	3.1352
22	0.68580	1.1816	1.7171	2.0739	2.4055	2.8188	3.1188
23	0.68531	1.1802	1.7139	2.0687	2.3979	2.8073	3.1040
24	0.68485	1.1789	1.7109	2.0639	2.3910	2.7969	3.0905
25	0.68443	1.1777	1.7081	2.0595	2.3846	2.7874	3.0782
26	0.68405	1.1766	1.7056	2.0555	2.3788	2.7787	3.0669
27	0.68370	1.1757	1.7033	2.0518	2.3734	2.7707	3.0565
28	0.68335	1.1748	1.7011	2.0484	2.3685	2.7633	3.0469
29	0.68304	1.1739	1.6991	2.0452	2.3638	2.7564	3.0380
30	0.68276	1.1731	1.6973	2.0423	2.3596	2.7500	3.0298
40	0.68066	1.1673	1.6839	2.0211	2.3289	2.7045	2.9712
60	0.67862	1.1616	1.6707	2.0003	2.2991	2.6603	2.9146
120	0.67656	1.1559	1.6577	1.9799	2.2699	2.6174	2.8599
∞	0.67449	1.1503	1.6449	1.9600	2.2414	2.5758	2.8070
$\overset{\nu}{\diagup}$ α	0.25	0.125	0.05	0.025	0.0125	0.005	0.0025
	IIa – One Tail Critical Values						

Source: *E.S. Pearson, **Critical Values of Student's t Distribution**, Biometrika, vol. 32, pp 168-181, 1941. Used with permission of the Biometrika Trust.*

TABLE 3

$H_0: \mu = \mu_0$

NUMBER OF OBSERVATIONS FOR t-TEST OF MEAN

The entries in this table show the numbers of observations needed in a t-test of the significance of a mean in order to control the probabilities of errors of the first and second kinds at α and β respectively.

Value of D = δ/σ	Level of t-test																				Value of D = δ/σ
Single-sided test / Double-sided test	**0·01**: a=0·005 / a=0·01					**0·02**: a=0·01 / a=0·02					**0·05**: a=0·025 / a=0·05					**0·1**: a=0·05 / a=0·1					
β =	0·01	0·05	0·1	0·2	0·5	0·01	0·05	0·1	0·2	0·5	0·01	0·05	0·1	0·2	0·5	0·01	0·05	0·1	0·2	0·5	
0·05																					0·05
0·10																					0·10
0·15																				122	0·15
0·20										139					99					70	0·20
0·25					110					90				128	64			139	101	45	0·25
0·30				134	78				115	63			119	90	45		122	97	71	32	0·30
0·35			125	99	58			109	85	47		109	88	67	34		90	72	52	24	0·35
0·40		115	97	77	45		101	85	66	37	117	84	68	51	26	101	70	55	40	19	0·40
0·45		92	77	62	37	110	81	68	53	30	93	67	54	41	21	80	55	44	33	15	0·45
0·50	100	75	63	51	30	90	66	55	43	25	76	54	44	34	18	65	45	36	27	13	0·50
0·55	83	63	53	42	26	75	55	46	36	21	63	45	37	28	15	54	38	30	22	11	0·55
0·60	71	53	45	36	22	63	47	39	31	18	53	38	32	24	13	46	32	26	19	9	0·60
0·65	61	46	39	31	20	55	41	34	27	16	46	33	27	21	12	39	28	22	17	8	0·65
0·70	53	40	34	28	17	47	35	30	24	14	40	29	24	19	10	34	24	19	15	8	0·70
0·75	47	36	30	25	16	42	31	27	21	13	35	26	21	16	9	30	21	17	13	7	0·75
0·80	41	32	27	22	14	37	28	24	19	12	31	22	19	15	9	27	19	15	12	6	0·80
0·85	37	29	24	20	13	33	25	21	17	11	28	21	17	13	8	24	17	14	11	6	0·85
0·90	34	26	22	18	12	29	23	19	16	10	25	19	16	12	7	21	15	13	10	5	0·90
0·95	31	24	20	17	11	27	21	18	14	9	23	17	14	11	7	19	14	11	9	5	0·95
1·00	28	22	19	16	10	25	19	16	13	9	21	16	13	10	6	18	13	11	8	5	1·00
1·1	24	19	16	14	9	21	16	14	12	8	18	13	11	9	6	15	11	9	7		1·1
1·2	21	16	14	12	8	18	14	12	10	7	15	12	10	8	5	13	10	8	6		1·2
1·3	18	15	13	11	8	16	13	11	9	6	14	10	9	7		11	8	7	6		1·3
1·4	16	13	12	10	7	14	11	10	9	6	12	9	8	7		10	8	7	5		1·4
1·5	15	12	11	9	7	13	10	9	8	6	11	8	7	6		9	7	6			1·5
1·6	13	11	10	8	6	12	10	9	7	5	10	8	7	6		8	6	6			1·6
1·7	12	10	9	8	6	11	9	8	7		9	7	6	5		8	6	5			1·7
1·8	12	10	9	8	6	10	8	7	7		8	7	6			7	6				1·8
1·9	11	9	8	7	6	10	8	7	6		8	6	6			7	5				1·9
2·0	10	8	8	7	5	9	7	7	6		7	6	5			6					2·0
2·1	10	8	7	7		8	7	6	6		7	6				6					2·1
2·2	9	8	7	6		8	7	6	5		7	6				6					2·2
2·3	9	7	7	6		8	6	6			6	5				5					2·3
2·4	8	7	7	6		7	6	6			6										2·4
2·5	8	7	6	6		7	6	6			6										2·5
3·0	7	6	6	5		6	5	5			5										3·0
3·5	6	5	5			5															3·5
4·0	6																				4·0

Use last entry in an "incomplete" column.

Source: *O. L. Davies, **Design and Analysis of Industrial Experiments**, Oliver and Boyd, Ltd. 1956. Used with permission.*

TABLE 4

$$H_0: \mu_A = \mu_B$$

NUMBER OF OBSERVATIONS FOR t-TEST OF DIFFERENCE BETWEEN TWO MEANS

The entries in this table show the number of observations needed in a t-test of the significance of the difference between two means in order to control the probabilities of the errors of the first and second kinds at α and β respectively.

	Level of t-test																				
	0·01					0·02					0·05					0·1					
Single-sided test	a = 0·005					a = 0·01					a = 0·025					a = 0·05					
Double-sided test	a = 0·01					a = 0·02					a = 0·05					a = 0·1					
$\beta =$	0·01	0·05	0·1	0·2	0·5	0·01	0·05	0·1	0·2	0·5	0·01	0·05	0·1	0·2	0·5	0·01	0·05	0·1	0·2	0·5	$D = \dfrac{\delta}{\sigma}$
0·05																					0·05
0·10																					0·10
0·15																					0·15
0·20																				137	0·20
0·25															124					88	0·25
0·30										123					87					61	0·30
0·35					110					90					64				102	45	0·35
0·40					85					70				100	50			108	78	35	0·40
0·45				118	68				101	55			105	79	39		108	86	62	28	0·45
0·50				96	55			106	82	45		106	86	64	32		88	70	51	23	0·50
0·55			101	79	46		106	88	68	38		87	71	53	27	112	73	58	42	19	0·55
0·60		101	85	67	39		90	74	58	32	104	74	60	45	23	89	61	49	36	16	0·60
0·65		87	73	57	34	104	77	64	49	27	88	63	51	39	20	76	52	42	30	14	0·65
0·70	100	75	63	50	29	90	66	55	43	24	76	55	44	34	17	66	45	36	26	12	0·70
0·75	88	66	55	44	26	79	58	48	38	21	67	48	39	29	15	57	40	32	23	11	0·75
0·80	77	58	49	39	23	70	51	43	33	19	59	42	34	26	14	50	35	28	21	10	0·80
0·85	69	51	43	35	21	62	46	38	30	17	52	37	31	23	12	45	31	25	18	9	0·85
0·90	62	46	39	31	19	55	41	34	27	15	47	34	27	21	11	40	28	22	16	8	0·90
0·95	55	42	35	28	17	50	37	31	24	14	42	30	25	19	10	36	25	20	15	7	0·95
1·00	50	38	32	26	15	45	33	28	22	13	38	27	23	17	9	33	23	18	14	7	1·00
1·1	42	32	27	22	13	38	28	23	19	11	32	23	19	14	8	27	19	15	12	6	1·1
1·2	36	27	23	18	11	32	24	20	16	9	27	20	16	12	7	23	16	13	10	5	1·2
1·3	31	23	20	16	10	28	21	17	14	8	23	17	14	11	6	20	14	11	9	5	1·3
1·4	27	20	17	14	9	24	18	15	12	8	20	15	12	10	6	17	12	10	8	4	1·4
1·5	24	18	15	13	8	21	16	14	11	7	18	13	11	9	5	15	11	9	7	4	1·5
1·6	21	16	14	11	7	19	14	12	10	6	16	12	10	8	5	14	10	8	6	4	1·6
1·7	19	15	13	10	7	17	13	11	9	6	14	11	9	7	4	12	9	7	6	3	1·7
1·8	17	13	11	10	6	15	12	10	8	5	13	10	8	6	4	11	8	7	5		1·8
1·9	16	12	11	9	6	14	11	9	8	5	12	9	7	6	4	10	7	6	5		1·9
2·0	14	11	10	8	6	13	10	9	7	5	11	8	7	6	4	9	7	6	4		2·0
2·1	13	10	9	8	5	12	9	8	7	5	10	8	6	5	3	8	6	5	4		2·1
2·2	12	10	8	7	5	11	9	7	6	4	9	7	6	5		8	6	5	4		2·2
2·3	11	9	8	7	5	10	8	7	6	4	9	7	6	5		7	5	5	4		2·3
2·4	11	9	8	6	5	10	8	7	6	4	8	6	5	4		7	5	4	4		2·4
2·5	10	8	7	6	4	9	7	6	5	4	8	6	5	4		6	5	4	3		2·5
3·0	8	6	6	5	4	7	6	5	4	3	6	5	4	4		5	4	3			3·0
3·5	6	5	5	4	3	6	5	4	4		5	4	4	3		4	3				3·5
4·0	6	5	4	4		5	4	4	3		4	4	3			4					4·0

Use last entry in an "incomplete" column.

Source: *O. L. Davies, **Design and Analysis of Industrial Experiments**, Oliver and Boyd, Ltd. 1956. Used with permission.*

TABLE 5
Critical Values of the F Distribution

α = 0.10

v_2 \ v_1	1	2	3	4	5	6	7	8	9	10	12	15	20	24	30	40	60	120	∞
1	39.864	49.500	53.593	55.833	57.241	58.204	58.906	59.439	59.858	60.195	60.705	61.220	61.740	62.002	62.265	62.529	62.794	63.061	63.328
2	8.5263	9.0000	9.1618	9.2434	9.2926	9.3255	9.3491	9.3668	9.3805	9.3916	9.4081	9.4247	9.4413	9.4496	9.4579	9.4663	9.4746	9.4829	9.4913
3	5.5383	5.4624	5.3908	5.3427	5.3092	5.2847	5.2662	5.2517	5.2400	5.2304	5.2156	5.2003	5.1845	5.1764	5.1681	5.1597	5.1512	5.1425	5.1337
4	4.5448	4.3246	4.1908	4.1073	4.0506	4.0098	3.9790	3.9549	3.9357	3.9199	3.8955	3.8689	3.8443	3.8310	3.8174	3.8036	3.7896	3.7753	3.7607
5	4.0604	3.7797	3.6195	3.5202	3.4530	3.4045	3.3679	3.3393	3.3163	3.2974	3.2682	3.2380	3.2067	3.1905	3.1741	3.1573	3.1402	3.1228	3.1050
6	3.7760	3.4633	3.2888	3.1808	3.1075	3.0546	3.0145	2.9830	2.9577	2.9369	2.9047	2.8712	2.8363	2.8183	2.8000	2.7812	2.7620	2.7423	2.7222
7	3.5894	3.2574	3.0741	2.9605	2.8833	2.8274	2.7849	2.7516	2.7247	2.7025	2.6681	2.6322	2.5947	2.5753	2.5555	2.5351	2.5142	2.4928	2.4708
8	3.4579	3.1131	2.9238	2.8064	2.7265	2.6683	2.6241	2.5893	2.5612	2.5380	2.5020	2.4642	2.4246	2.4041	2.3830	2.3614	2.3391	2.3162	2.2926
9	3.3603	3.0065	2.8129	2.6927	2.6106	2.5509	2.5053	2.4694	2.4403	2.4163	2.3789	2.3396	2.2983	2.2768	2.2547	2.2320	2.2085	2.1843	2.1592
10	3.2850	2.9245	2.7277	2.6053	2.5216	2.4606	2.4140	2.3772	2.3473	2.3226	2.2841	2.2435	2.2007	2.1784	2.1554	2.1317	2.1072	2.0818	2.0554
11	3.2252	2.8595	2.6602	2.5362	2.4512	2.3891	2.3416	2.3040	2.2735	2.2482	2.2087	2.1671	2.1230	2.1000	2.0762	2.0516	2.0261	1.9997	1.9721
12	3.1765	2.8068	2.6055	2.4801	2.3940	2.3310	2.2828	2.2446	2.2135	2.1878	2.1474	2.1049	2.0597	2.0360	2.0115	1.9861	1.9597	1.9323	1.9036
13	3.1362	2.7632	2.5603	2.4337	2.3467	2.2830	2.2341	2.1953	2.1638	2.1376	2.0966	2.0532	2.0070	1.9827	1.9576	1.9315	1.9043	1.8759	1.8462
14	3.1022	2.7265	2.5222	2.3947	2.3069	2.2426	2.1931	2.1539	2.1220	2.0954	2.0537	2.0095	1.9625	1.9377	1.9119	1.8852	1.8572	1.8280	1.7973
15	3.0732	2.6952	2.4898	2.3614	2.2730	2.2081	2.1582	2.1185	2.0862	2.0593	2.0171	1.9722	1.9243	1.8990	1.8728	1.8454	1.8168	1.7867	1.7551
16	3.0481	2.6682	2.4618	2.3327	2.2438	2.1783	2.1280	2.0880	2.0553	2.0281	1.9854	1.9399	1.8913	1.8656	1.8388	1.8108	1.7816	1.7507	1.7182
17	3.0262	2.6446	2.4374	2.3077	2.2183	2.1524	2.1017	2.0613	2.0284	2.0009	1.9577	1.9117	1.8624	1.8362	1.8090	1.7805	1.7506	1.7191	1.6856
18	3.0070	2.6239	2.4160	2.2858	2.1958	2.1296	2.0785	2.0379	2.0047	1.9770	1.9333	1.8868	1.8368	1.8103	1.7827	1.7537	1.7232	1.6910	1.6567
19	2.9899	2.6056	2.3970	2.2663	2.1760	2.1094	2.0580	2.0171	1.9836	1.9557	1.9117	1.8647	1.8142	1.7873	1.7592	1.7298	1.6988	1.6659	1.6308
20	2.9747	2.5893	2.3801	2.2489	2.1582	2.0913	2.0397	1.9985	1.9649	1.9367	1.8924	1.8449	1.7938	1.7667	1.7382	1.7083	1.6768	1.6433	1.6074
21	2.9609	2.5746	2.3649	2.2333	2.1423	2.0751	2.0232	1.9819	1.9480	1.9197	1.8750	1.8272	1.7756	1.7481	1.7193	1.6890	1.6569	1.6228	1.5862
22	2.9486	2.5613	2.3512	2.2193	2.1279	2.0605	2.0084	1.9668	1.9327	1.9043	1.8593	1.8111	1.7590	1.7312	1.7021	1.6714	1.6389	1.6042	1.5668
23	2.9374	2.5493	2.3387	2.2065	2.1149	2.0472	1.9949	1.9531	1.9189	1.8903	1.8450	1.7964	1.7439	1.7159	1.6864	1.6554	1.6224	1.5871	1.5490
24	2.9271	2.5383	2.3274	2.1949	2.1030	2.0351	1.9826	1.9407	1.9063	1.8775	1.8319	1.7831	1.7302	1.7019	1.6721	1.6407	1.6073	1.5715	1.5327
25	2.9177	2.5283	2.3170	2.1843	2.0922	2.0241	1.9714	1.9292	1.8947	1.8658	1.8200	1.7708	1.7175	1.6890	1.6589	1.6272	1.5934	1.5570	1.5176
26	2.9091	2.5191	2.3075	2.1745	2.0822	2.0139	1.9610	1.9188	1.8841	1.8550	1.8090	1.7596	1.7059	1.6771	1.6468	1.6147	1.5805	1.5437	1.5036
27	2.9012	2.5106	2.2987	2.1655	2.0730	2.0045	1.9515	1.9091	1.8743	1.8451	1.7989	1.7492	1.6951	1.6662	1.6356	1.6032	1.5686	1.5313	1.4906
28	2.8939	2.5028	2.2906	2.1571	2.0645	1.9959	1.9427	1.9001	1.8652	1.8359	1.7895	1.7395	1.6852	1.6560	1.6252	1.5925	1.5575	1.5198	1.4784
29	2.8871	2.4955	2.2831	2.1494	2.0566	1.9878	1.9345	1.8918	1.8568	1.8274	1.7808	1.7306	1.6759	1.6465	1.6155	1.5825	1.5472	1.5090	1.4670
30	2.8807	2.4887	2.2761	2.1422	2.0492	1.9803	1.9269	1.8841	1.8490	1.8195	1.7727	1.7223	1.6673	1.6377	1.6065	1.5732	1.5376	1.4989	1.4564
40	2.8354	2.4404	2.2261	2.0909	1.9968	1.9269	1.8725	1.8289	1.7929	1.7627	1.7146	1.6624	1.6052	1.5741	1.5411	1.5056	1.4672	1.4248	1.3769
60	2.7914	2.3932	2.1774	2.0410	1.9457	1.8747	1.8194	1.7748	1.7380	1.7070	1.6574	1.6034	1.5435	1.5107	1.4755	1.4373	1.3952	1.3476	1.2915
120	2.7478	2.3473	2.1300	1.9923	1.8959	1.8238	1.7675	1.7220	1.6843	1.6524	1.6012	1.5450	1.4821	1.4472	1.4094	1.3676	1.3203	1.2646	1.1926
∞	2.7055	2.3026	2.0838	1.9449	1.8473	1.7741	1.7167	1.6702	1.6315	1.5987	1.5458	1.4871	1.4206	1.3832	1.3419	1.2951	1.2400	1.1686	1.0000

numerator df

denominator df

Source: E.S. Pearson, *Tables of Percentage Points of the Inverted Beta (F) Distribution*, Biometrika, vol. 32, pp. 73-88, 1943 Used with permission of the Biometrika Trust.

TABLE 5 (CONTINUED)
Critical Values of the F Distribution

$\alpha = 0.05$

numerator df

v_2 \ v_1	1	2	3	4	5	6	7	8	9	10	12	15	20	24	30	40	60	120	∞
1	161.45	199.50	215.71	224.58	230.16	233.99	236.77	238.88	240.54	241.88	243.91	245.95	248.01	249.05	250.09	251.14	252.20	253.25	254.32
2	18.513	19.000	19.164	19.247	19.296	19.330	19.353	19.371	19.385	19.396	19.413	19.429	19.446	19.454	19.462	19.471	19.479	19.487	19.496
3	10.128	9.5521	9.2766	9.1172	9.0135	8.9406	8.8868	8.8452	8.8123	8.7855	8.7446	8.7029	8.6602	8.6385	8.6166	8.5944	8.5720	8.5494	8.5265
4	7.7086	6.9443	6.5914	6.3883	6.2560	6.1631	6.0942	6.0410	5.9988	5.9644	5.9117	5.8578	5.8025	5.7744	5.7459	5.7170	5.6878	5.6581	5.6281
5	6.6079	5.7861	5.4095	5.1922	5.0503	4.9503	4.8759	4.8183	4.7725	4.7351	4.6777	4.6188	4.5581	4.5272	4.4957	4.4638	4.4314	4.3984	4.3650
6	5.9874	5.1433	4.7571	4.5337	4.3874	4.2839	4.2066	4.1468	4.0990	4.0600	3.9999	3.9381	3.8742	3.8415	3.8082	3.7743	3.7398	3.7047	3.6688
7	5.5914	4.7374	4.3468	4.1203	3.9715	3.8660	3.7870	3.7257	3.6767	3.6365	3.5747	3.5108	3.4445	3.4105	3.3758	3.3404	3.3043	3.2674	3.2298
8	5.3177	4.4590	4.0662	3.8378	3.6875	3.5806	3.5005	3.4381	3.3881	3.3472	3.2840	3.2184	3.1503	3.1152	3.0794	3.0428	3.0053	2.9669	2.9276
9	5.1174	4.2565	3.8626	3.6331	3.4817	3.3738	3.2927	3.2296	3.1789	3.1373	3.0729	3.0061	2.9365	2.9005	2.8637	2.8259	2.7872	2.7475	2.7067
10	4.9646	4.1028	3.7083	3.4780	3.3258	3.2172	3.1355	3.0717	3.0204	2.9782	2.9130	2.8450	2.7740	2.7372	2.6996	2.6609	2.6211	2.5801	2.5379
11	4.8443	3.9823	3.5874	3.3567	3.2039	3.0946	3.0123	2.9480	2.8962	2.8536	2.7876	2.7186	2.6464	2.6090	2.5705	2.5309	2.4901	2.4480	2.4045
12	4.7472	3.8853	3.4903	3.2592	3.1059	2.9961	2.9134	2.8486	2.7964	2.7534	2.6866	2.6169	2.5436	2.5055	2.4663	2.4259	2.3842	2.3410	2.2962
13	4.6672	3.8056	3.4105	3.1791	3.0254	2.9153	2.8321	2.7669	2.7144	2.6710	2.6037	2.5331	2.4589	2.4202	2.3803	2.3392	2.2966	2.2524	2.2064
14	4.6001	3.7389	3.3439	3.1122	2.9582	2.8477	2.7642	2.6987	2.6458	2.6021	2.5342	2.4630	2.3879	2.3487	2.3082	2.2664	2.2230	2.1778	2.1307
15	4.5431	3.6823	3.2874	3.0556	2.9013	2.7905	2.7066	2.6408	2.5876	2.5437	2.4753	2.4035	2.3275	2.2878	2.2468	2.2043	2.1601	2.1141	2.0658
16	4.4940	3.6337	3.2389	3.0069	2.8524	2.7413	2.6572	2.5911	2.5377	2.4935	2.4247	2.3522	2.2756	2.2354	2.1938	2.1507	2.1058	2.0589	2.0096
17	4.4513	3.5915	3.1968	2.9647	2.8100	2.6987	2.6143	2.5480	2.4943	2.4499	2.3807	2.3077	2.2304	2.1898	2.1477	2.1040	2.0584	2.0107	1.9604
18	4.4139	3.5546	3.1599	2.9277	2.7729	2.6613	2.5767	2.5102	2.4563	2.4117	2.3421	2.2686	2.1906	2.1497	2.1071	2.0629	2.0166	1.9681	1.9168
19	4.3808	3.5219	3.1274	2.8951	2.7401	2.6283	2.5435	2.4768	2.4227	2.3779	2.3080	2.2341	2.1555	2.1141	2.0712	2.0264	1.9796	1.9302	1.8780
20	4.3513	3.4928	3.0984	2.8661	2.7109	2.5990	2.5140	2.4471	2.3928	2.3479	2.2776	2.2033	2.1242	2.0825	2.0391	1.9938	1.9464	1.8963	1.8432
21	4.3248	3.4668	3.0725	2.8401	2.6848	2.5727	2.4876	2.4205	2.3661	2.3210	2.2504	2.1757	2.0960	2.0540	2.0102	1.9645	1.9165	1.8657	1.8117
22	4.3009	3.4434	3.0491	2.8167	2.6613	2.5491	2.4638	2.3965	2.3419	2.2967	2.2258	2.1508	2.0707	2.0283	1.9842	1.9380	1.8895	1.8380	1.7831
23	4.2793	3.4221	3.0280	2.7955	2.6400	2.5277	2.4422	2.3748	2.3201	2.2747	2.2036	2.1282	2.0476	2.0050	1.9605	1.9139	1.8649	1.8128	1.7570
24	4.2597	3.4028	3.0088	2.7763	2.6207	2.5082	2.4226	2.3551	2.3002	2.2547	2.1834	2.1077	2.0267	1.9838	1.9390	1.8920	1.8424	1.7897	1.7331
25	4.2417	3.3852	2.9912	2.7587	2.6030	2.4904	2.4047	2.3371	2.2821	2.2365	2.1649	2.0889	2.0075	1.9643	1.9192	1.8718	1.8217	1.7684	1.7110
26	4.2252	3.3690	2.9751	2.7426	2.5868	2.4741	2.3883	2.3205	2.2655	2.2197	2.1479	2.0716	1.9898	1.9464	1.9010	1.8533	1.8027	1.7488	1.6906
27	4.2100	3.3541	2.9604	2.7278	2.5719	2.4591	2.3732	2.3053	2.2501	2.2043	2.1323	2.0558	1.9736	1.9299	1.8842	1.8361	1.7851	1.7307	1.6717
28	4.1960	3.3404	2.9467	2.7141	2.5581	2.4453	2.3593	2.2913	2.2360	2.1900	2.1179	2.0411	1.9586	1.9147	1.8687	1.8203	1.7689	1.7138	1.6541
29	4.1830	3.3277	2.9340	2.7014	2.5454	2.4324	2.3463	2.2782	2.2229	2.1768	2.1045	2.0275	1.9446	1.9005	1.8543	1.8055	1.7537	1.6981	1.6377
30	4.1709	3.3158	2.9223	2.6896	2.5336	2.4205	2.3343	2.2662	2.2107	2.1646	2.0921	2.0148	1.9317	1.8874	1.8409	1.7918	1.7396	1.6835	1.6223
40	4.0848	3.2317	2.8387	2.6060	2.4495	2.3359	2.2490	2.1802	2.1240	2.0772	2.0035	1.9245	1.8389	1.7929	1.7444	1.6928	1.6373	1.5766	1.5089
60	4.0012	3.1504	2.7581	2.5252	2.3683	2.2540	2.1665	2.0970	2.0401	1.9926	1.9174	1.8364	1.7480	1.7001	1.6491	1.5943	1.5343	1.4673	1.3893
120	3.9201	3.0718	2.6802	2.4472	2.2900	2.1750	2.0867	2.0164	1.9588	1.9105	1.8337	1.7505	1.6587	1.6084	1.5543	1.4952	1.4290	1.3519	1.2539
∞	3.8415	2.9957	2.6049	2.3719	2.2141	2.0986	2.0096	1.9384	1.8799	1.8307	1.7522	1.6664	1.5705	1.5173	1.4591	1.3940	1.3180	1.2214	1.0000

denominator df

TABLE 5 (CONTINUED)
Critical Values of the F Distribution

α = 0.025

numerator df

denominator df ν_2 \ ν_1	1	2	3	4	5	6	7	8	9	10	12	15	20	24	30	40	60	120	∞
1	647.79	799.50	864.16	899.58	921.85	937.11	948.22	956.66	963.28	968.63	976.71	984.87	993.10	997.25	1001.4	1005.6	1009.8	1014.0	1018.3
2	38.506	39.000	39.165	39.248	39.298	39.331	39.355	39.373	39.387	39.398	39.415	39.431	39.448	39.456	39.465	39.473	39.481	39.490	39.498
3	17.443	16.044	15.439	15.101	14.885	14.735	14.624	14.540	14.473	14.419	14.337	14.253	14.167	14.124	14.081	14.037	13.992	13.947	13.902
4	12.218	10.649	9.9792	9.6045	9.3645	9.1973	9.0741	8.9796	8.9047	8.8439	8.7512	8.6565	8.5599	8.5109	8.4613	8.4111	8.3604	8.3092	8.2573
5	10.007	8.4336	7.7636	7.3879	7.1464	6.9777	6.8531	6.7572	6.6810	6.6192	6.5246	6.4277	6.3285	6.2780	6.2269	6.1751	6.1225	6.0693	6.0153
6	8.8131	7.2598	6.5988	6.2272	5.9876	5.8197	5.6955	5.5996	5.5234	5.4613	5.3662	5.2687	5.1684	5.1172	5.0652	5.0125	4.9589	4.9045	4.8491
7	8.0727	6.5415	5.8898	5.5226	5.2852	5.1186	4.9949	4.8994	4.8232	4.7611	4.6658	4.5678	4.4667	4.4150	4.3624	4.3089	4.2544	4.1989	4.1423
8	7.5709	6.0595	5.4160	5.0526	4.8173	4.6517	4.5286	4.4332	4.3572	4.2951	4.1997	4.1012	3.9995	3.9472	3.8940	3.8398	3.7844	3.7279	3.6702
9	7.2093	5.7147	5.0781	4.7181	4.4844	4.3197	4.1971	4.1020	4.0260	3.9639	3.8682	3.7694	3.6669	3.6142	3.5604	3.5055	3.4493	3.3918	3.3329
10	6.9367	5.4564	4.8256	4.4683	4.2361	4.0721	3.9498	3.8549	3.7790	3.7168	3.6209	3.5217	3.4186	3.3654	3.3110	3.2554	3.1984	3.1399	3.0798
11	6.7241	5.2559	4.6300	4.2751	4.0440	3.8807	3.7586	3.6638	3.5879	3.5257	3.4296	3.3299	3.2261	3.1725	3.1176	3.0613	3.0035	2.9441	2.8828
12	6.5538	5.0959	4.4742	4.1212	3.8911	3.7283	3.6065	3.5118	3.4358	3.3736	3.2773	3.1772	3.0728	3.0187	2.9633	2.9063	2.8478	2.7874	2.7249
13	6.4143	4.9653	4.3472	3.9959	3.7667	3.6043	3.4827	3.3880	3.3120	3.2497	3.1532	3.0527	2.9477	2.8932	2.8373	2.7797	2.7204	2.6590	2.5955
14	6.2979	4.8567	4.2417	3.8919	3.6634	3.5014	3.3799	3.2853	3.2093	3.1469	3.0501	2.9493	2.8437	2.7888	2.7324	2.6742	2.6142	2.5519	2.4872
15	6.1995	4.7650	4.1528	3.8043	3.5764	3.4147	3.2934	3.1987	3.1227	3.0602	2.9633	2.8621	2.7559	2.7006	2.6437	2.5850	2.5242	2.4611	2.3953
16	6.1151	4.6867	4.0768	3.7294	3.5021	3.3406	3.2194	3.1248	3.0488	2.9862	2.8890	2.7875	2.6808	2.6252	2.5678	2.5085	2.4471	2.3831	2.3163
17	6.0420	4.6189	4.0112	3.6648	3.4379	3.2767	3.1556	3.0610	2.9849	2.9222	2.8249	2.7230	2.6158	2.5598	2.5021	2.4422	2.3801	2.3153	2.2474
18	5.9781	4.5597	3.9539	3.6083	3.3820	3.2209	3.0999	3.0053	2.9291	2.8664	2.7689	2.6667	2.5590	2.5027	2.4445	2.3842	2.3214	2.2558	2.1869
19	5.9216	4.5075	3.9034	3.5587	3.3327	3.1718	3.0509	2.9563	2.8800	2.8173	2.7196	2.6171	2.5089	2.4523	2.3937	2.3329	2.2695	2.2032	2.1333
20	5.8715	4.4613	3.8587	3.5147	3.2891	3.1283	3.0074	2.9128	2.8365	2.7737	2.6758	2.5731	2.4645	2.4076	2.3486	2.2873	2.2234	2.1562	2.0853
21	5.8266	4.4199	3.8188	3.4754	3.2501	3.0895	2.9686	2.8740	2.7977	2.7348	2.6368	2.5338	2.4247	2.3675	2.3082	2.2465	2.1819	2.1141	2.0422
22	5.7863	4.3828	3.7829	3.4401	3.2151	3.0546	2.9338	2.8392	2.7628	2.6998	2.6017	2.4984	2.3890	2.3315	2.2718	2.2097	2.1446	2.0760	2.0032
23	5.7498	4.3492	3.7505	3.4083	3.1835	3.0232	2.9024	2.8077	2.7313	2.6682	2.5699	2.4665	2.3567	2.2989	2.2389	2.1763	2.1107	2.0415	1.9677
24	5.7167	4.3187	3.7211	3.3794	3.1548	2.9946	2.8738	2.7791	2.7027	2.6396	2.5412	2.4374	2.3273	2.2693	2.2090	2.1460	2.0799	2.0099	1.9353
25	5.6864	4.2909	3.6943	3.3530	3.1287	2.9685	2.8478	2.7531	2.6766	2.6135	2.5149	2.4110	2.3005	2.2422	2.1816	2.1183	2.0517	1.9811	1.9055
26	5.6586	4.2655	3.6697	3.3289	3.1048	2.9447	2.8240	2.7293	2.6528	2.5895	2.4909	2.3867	2.2759	2.2174	2.1565	2.0928	2.0257	1.9545	1.8781
27	5.6331	4.2421	3.6472	3.3067	3.0828	2.9228	2.8021	2.7074	2.6309	2.5676	2.4688	2.3644	2.2533	2.1946	2.1334	2.0693	2.0018	1.9299	1.8527
28	5.6096	4.2205	3.6264	3.2863	3.0625	2.9027	2.7820	2.6872	2.6106	2.5473	2.4484	2.3438	2.2324	2.1735	2.1121	2.0477	1.9796	1.9072	1.8291
29	5.5878	4.2006	3.6072	3.2674	3.0438	2.8840	2.7633	2.6686	2.5919	2.5286	2.4295	2.3248	2.2131	2.1540	2.0923	2.0276	1.9591	1.8861	1.8072
30	5.5675	4.1821	3.5894	3.2499	3.0265	2.8667	2.7460	2.6513	2.5746	2.5112	2.4120	2.3072	2.1952	2.1359	2.0739	2.0089	1.9400	1.8664	1.7867
40	5.4239	4.0510	3.4633	3.1261	2.9037	2.7444	2.6238	2.5289	2.4519	2.3882	2.2882	2.1819	2.0677	2.0069	1.9429	1.8752	1.8028	1.7242	1.6371
60	5.2857	3.9253	3.3425	3.0077	2.7863	2.6274	2.5068	2.4117	2.3344	2.2702	2.1692	2.0613	1.9445	1.8817	1.8152	1.7440	1.6668	1.5810	1.4822
120	5.1524	3.8046	3.2270	2.8943	2.6740	2.5154	2.3948	2.2994	2.2217	2.1570	2.0548	1.9450	1.8249	1.7597	1.6899	1.6141	1.5299	1.4327	1.3104
∞	5.0239	3.6889	3.1161	2.7858	2.5665	2.4082	2.2875	2.1918	2.1136	2.0483	1.9447	1.8326	1.7085	1.6402	1.5660	1.4835	1.3883	1.2684	1.0000

TABLE 5 (CONTINUED)

Critical Values of the F Distribution

$\alpha = 0.01$

v_2 \ v_1	1	2	3	4	5	6	7	8	9	10	12	15	20	24	30	40	60	120	∞
1	4052.2	4999.5	5403.3	5624.6	5763.7	5859.0	5928.3	5981.6	6022.5	6055.8	6106.3	6157.3	6208.7	6234.6	6260.7	6286.8	6313.0	6339.4	6366.0
2	98.503	99.000	99.166	99.249	99.299	99.332	99.356	99.374	99.388	99.399	99.416	99.432	99.449	99.458	99.466	99.474	99.483	99.491	99.501
3	34.116	30.817	29.457	28.710	28.237	27.911	27.672	27.489	27.345	27.229	27.052	26.872	26.690	26.598	26.505	26.411	26.316	26.221	26.125
4	21.198	18.000	16.694	15.977	15.522	15.207	14.976	14.799	14.659	14.546	14.374	14.198	14.020	13.929	13.838	13.745	13.652	13.558	13.463
5	16.258	13.274	12.060	11.392	10.967	10.672	10.456	10.289	10.158	10.051	9.8883	9.7222	9.5527	9.4665	9.3793	9.2912	9.2020	9.1118	9.0204
6	13.745	10.925	9.7795	9.1483	8.7459	8.4661	8.2600	8.1016	7.9761	7.8741	7.7183	7.5590	7.3958	7.3127	7.2285	7.1432	7.0568	6.9690	6.8801
7	12.246	9.5466	8.4513	7.8467	7.4604	7.1914	6.9928	6.8401	6.7188	6.6201	6.4691	6.3143	6.1554	6.0743	5.9921	5.9084	5.8236	5.7372	5.6495
8	11.259	8.6491	7.5910	7.0060	6.6318	6.3707	6.1776	6.0289	5.9106	5.8143	5.6668	5.5151	5.3591	5.2793	5.1981	5.1156	5.0316	4.9460	4.8588
9	10.561	8.0215	6.9919	6.4221	6.0569	5.8018	5.6129	5.4671	5.3511	5.2565	5.1114	4.9621	4.8080	4.7290	4.6486	4.5667	4.4831	4.3978	4.3105
10	10.044	7.5594	6.5523	5.9943	5.6363	5.3858	5.2001	5.0567	4.9424	4.8492	4.7059	4.5582	4.4054	4.3269	4.2469	4.1653	4.0819	3.9965	3.9090
11	9.6460	7.2057	6.2167	5.6683	5.3160	5.0692	4.8861	4.7445	4.6315	4.5393	4.3974	4.2509	4.0990	4.0209	3.9411	3.8596	3.7761	3.6904	3.6025
12	9.3302	6.9266	5.9526	5.4119	5.0643	4.8206	4.6395	4.4994	4.3875	4.2961	4.1553	4.0096	3.8584	3.7805	3.7008	3.6192	3.5355	3.4494	3.3608
13	9.0738	6.7010	5.7394	5.2053	4.8616	4.6204	4.4410	4.3021	4.1911	4.1003	3.9603	3.8154	3.6646	3.5868	3.5070	3.4253	3.3413	3.2548	3.1654
14	8.8616	6.5149	5.5639	5.0354	4.6950	4.4558	4.2779	4.1399	4.0297	3.9394	3.8001	3.6557	3.5052	3.4274	3.3476	3.2656	3.1813	3.0942	3.0040
15	8.6831	6.3589	5.4170	4.8932	4.5556	4.3183	4.1415	4.0045	3.8948	3.8049	3.6662	3.5222	3.3719	3.2940	3.2141	3.1319	3.0471	2.9595	2.8684
16	8.5310	6.2262	5.2922	4.7726	4.4374	4.2016	4.0259	3.8896	3.7804	3.6909	3.5527	3.4089	3.2588	3.1808	3.1007	3.0182	2.9330	2.8447	2.7528
17	8.3997	6.1121	5.1850	4.6690	4.3359	4.1015	3.9267	3.7910	3.6822	3.5931	3.4552	3.3117	3.1615	3.0835	3.0032	2.9205	2.8348	2.7459	2.6530
18	8.2854	6.0129	5.0919	4.5790	4.2479	4.0146	3.8406	3.7054	3.5971	3.5082	3.3706	3.2273	3.0771	2.9990	2.9185	2.8354	2.7493	2.6597	2.5660
19	8.1850	5.9259	5.0103	4.5003	4.1708	3.9386	3.7653	3.6305	3.5225	3.4338	3.2965	3.1533	3.0031	2.9249	2.8442	2.7608	2.6742	2.5839	2.4893
20	8.0960	5.8489	4.9382	4.4307	4.1027	3.8714	3.6987	3.5644	3.4567	3.3682	3.2311	3.0880	2.9377	2.8594	2.7785	2.6947	2.6077	2.5168	2.4212
21	8.0166	5.7804	4.8740	4.3688	4.0421	3.8117	3.6396	3.5056	3.3981	3.3098	3.1729	3.0299	2.8796	2.8011	2.7200	2.6359	2.5484	2.4568	2.3603
22	7.9454	5.7190	4.8166	4.3134	3.9880	3.7583	3.5867	3.4530	3.3458	3.2576	3.1209	2.9780	2.8274	2.7488	2.6675	2.5831	2.4951	2.4029	2.3055
23	7.8811	5.6637	4.7649	4.2635	3.9392	3.7102	3.5390	3.4057	3.2986	3.2106	3.0740	2.9311	2.7805	2.7017	2.6202	2.5355	2.4471	2.3542	2.2559
24	7.8229	5.6136	4.7181	4.2184	3.8951	3.6667	3.4959	3.3629	3.2560	3.1681	3.0316	2.8887	2.7380	2.6591	2.5773	2.4923	2.4035	2.3099	2.2107
25	7.7698	5.5680	4.6755	4.1774	3.8550	3.6272	3.4568	3.3239	3.2172	3.1294	2.9931	2.8502	2.6993	2.6203	2.5383	2.4530	2.3637	2.2695	2.1694
26	7.7213	5.5263	4.6366	4.1400	3.8183	3.5911	3.4210	3.2884	3.1818	3.0941	2.9579	2.8150	2.6640	2.5848	2.5026	2.4170	2.3273	2.2325	2.1315
27	7.6767	5.4881	4.6009	4.1056	3.7848	3.5580	3.3882	3.2558	3.1494	3.0618	2.9256	2.7827	2.6316	2.5522	2.4699	2.3840	2.2938	2.1984	2.0965
28	7.6356	5.4529	4.5681	4.0740	3.7539	3.5276	3.3581	3.2259	3.1195	3.0320	2.8959	2.7530	2.6017	2.5223	2.4397	2.3535	2.2629	2.1670	2.0642
29	7.5976	5.4205	4.5378	4.0449	3.7254	3.4995	3.3302	3.1982	3.0920	3.0045	2.8685	2.7256	2.5742	2.4946	2.4118	2.3253	2.2344	2.1378	2.0342
30	7.5625	5.3904	4.5097	4.0179	3.6990	3.4735	3.3045	3.1726	3.0665	2.9791	2.8431	2.7002	2.5487	2.4689	2.3860	2.2992	2.2079	2.1107	2.0062
40	7.3141	5.1785	4.3126	3.8283	3.5138	3.2910	3.1238	2.9930	2.8876	2.8005	2.6648	2.5216	2.3689	2.2880	2.2034	2.1142	2.0194	1.9172	1.8047
60	7.0771	4.9774	4.1259	3.6491	3.3389	3.1187	2.9530	2.8233	2.7185	2.6318	2.4961	2.3523	2.1978	2.1154	2.0285	1.9360	1.8363	1.7263	1.6006
120	6.8510	4.7865	3.9493	3.4796	3.1735	2.9559	2.7918	2.6629	2.5586	2.4721	2.3363	2.1915	2.0346	1.9500	1.8600	1.7628	1.6557	1.5330	1.3805
∞	6.6349	4.6052	3.7816	3.3192	3.0173	2.8020	2.6393	2.5113	2.4073	2.3209	2.1848	2.0385	1.8783	1.7908	1.6964	1.5923	1.4730	1.3246	1.0000

numerator df

denominator df

TABLE 6

Upper 5-Percent Points of Studentized Range q*

n_2	2	3	4	5	6	7	8	9	10	11	12	13	14	15	16	17	18	19	20
1	18.0	26.7	32.8	37.2	40.5	43.1	45.4	47.3	49.1	50.6	51.9	53.2	54.3	55.4	56.3	57.2	58.0	58.8	59.6
2	6.09	8.28	9.80	10.89	11.73	12.43	13.03	13.54	13.99	14.39	14.75	15.08	15.38	15.65	15.91	16.14	16.36	16.57	16.77
3	4.50	5.88	6.83	7.51	8.04	8.47	8.85	9.18	9.46	9.72	9.95	10.16	10.35	10.52	10.69	10.84	10.98	11.12	11.24
4	3.93	5.00	5.76	6.31	6.73	7.06	7.35	7.60	7.83	8.03	8.21	8.37	8.52	8.67	8.80	8.92	9.03	9.14	9.24
5	3.61	4.54	5.18	5.64	5.99	6.28	6.52	6.74	6.93	7.10	7.25	7.39	7.52	7.64	7.75	7.86	7.95	8.04	8.13
6	3.46	4.34	4.90	5.31	5.63	5.89	6.12	6.32	6.49	6.65	6.79	6.92	7.04	7.14	7.24	7.34	7.43	7.51	7.59
7	3.34	4.16	4.68	5.06	5.35	5.59	5.80	5.99	6.15	6.29	6.42	6.54	6.65	6.75	6.84	6.93	7.01	7.08	7.16
8	3.26	4.04	4.53	4.89	5.17	5.40	5.60	5.77	5.92	6.05	6.18	6.29	6.39	6.48	6.57	6.65	6.73	6.80	6.87
9	3.20	3.95	4.42	4.76	5.02	5.24	5.43	5.60	5.74	5.87	5.98	6.09	6.19	6.28	6.36	6.44	6.51	6.58	6.65
10	3.15	3.88	4.33	4.66	4.91	5.12	5.30	5.46	5.60	5.72	5.83	5.93	6.03	6.12	6.20	6.27	6.34	6.41	6.47
11	3.11	3.82	4.26	4.58	4.82	5.03	5.20	5.35	5.49	5.61	5.71	5.81	5.90	5.98	6.06	6.14	6.20	6.27	6.33
12	3.08	3.77	4.20	4.51	4.75	4.95	5.12	5.27	5.40	5.51	5.61	5.71	5.80	5.88	5.95	6.02	6.09	6.15	6.21
13	3.06	3.73	4.15	4.46	4.69	4.88	5.05	5.19	5.32	5.43	5.53	5.63	5.71	5.79	5.86	5.93	6.00	6.06	6.11
14	3.03	3.70	4.11	4.41	4.64	4.83	4.99	5.13	5.25	5.36	5.46	5.56	5.64	5.72	5.79	5.86	5.92	5.98	6.03
15	3.01	3.67	4.08	4.37	4.59	4.78	4.94	5.08	5.20	5.31	5.40	5.49	5.57	5.65	5.72	5.79	5.85	5.91	5.96
16	3.00	3.65	4.05	4.34	4.56	4.74	4.90	5.03	5.15	5.26	5.35	5.44	5.52	5.59	5.66	5.73	5.79	5.84	5.90
17	2.98	3.62	4.02	4.31	4.52	4.70	4.86	4.99	5.11	5.21	5.31	5.39	5.47	5.55	5.61	5.68	5.74	5.79	5.84
18	2.97	3.61	4.00	4.28	4.49	4.67	4.83	4.96	5.07	5.17	5.27	5.35	5.43	5.50	5.57	5.63	5.69	5.74	5.79
19	2.96	3.59	3.98	4.26	4.47	4.64	4.79	4.92	5.04	5.14	5.23	5.32	5.39	5.46	5.53	5.59	5.65	5.70	5.75
20	2.95	3.58	3.96	4.24	4.45	4.62	4.77	4.90	5.01	5.11	5.20	5.28	5.36	5.43	5.50	5.56	5.61	5.66	5.71
24	2.92	3.53	3.90	4.17	4.37	4.54	4.68	4.81	4.92	5.01	5.10	5.18	5.25	5.32	5.38	5.44	5.50	5.55	5.59
30	2.89	3.48	3.84	4.11	4.30	4.46	4.60	4.72	4.83	4.92	5.00	5.08	5.15	5.21	5.27	5.33	5.38	5.43	5.48
40	2.86	3.44	3.79	4.04	4.23	4.39	4.52	4.63	4.74	4.82	4.90	4.98	5.05	5.11	5.17	5.22	5.27	5.32	5.36
60	2.83	3.40	3.74	3.98	4.16	4.31	4.44	4.55	4.65	4.73	4.81	4.88	4.94	5.00	5.06	5.11	5.15	5.20	5.24
120	2.80	3.36	3.69	3.92	4.10	4.24	4.36	4.47	4.56	4.64	4.71	4.78	4.84	4.90	4.95	5.00	5.04	5.09	5.13
∞	2.77	3.32	3.63	3.86	4.03	4.17	4.29	4.39	4.47	4.55	4.62	4.68	4.74	4.80	4.84	4.89	4.93	4.97	5.01

* From J. M. May, "Extended and Corrected Tables of the Upper Percentage Points of the Studentized Range," *Biometrika*, vol. 39 (1952), pp. 192-193. Reproduced by permission of the trustees of *Biometrika*.

** p is the number of quantities (for example, means) whose range is involved. n_2 is the degrees of freedom in the error estimate.

Upper 1-Percent Points of Studentized Range q

n_2	2	3	4	5	6	7	8	9	10	11	12	13	14	15	16	17	18	19	20
1	90.0	135	164	186	202	216	227	237	246	253	260	266	272	227	282	286	290	294	298
2	14.0	19.0	22.3	24.7	26.6	28.2	29.5	30.7	31.7	32.6	33.4	34.1	34.8	35.4	36.0	36.5	37.0	37.5	37.9
3	8.26	10.6	12.2	13.3	14.2	15.0	15.6	16.2	16.7	17.1	17.5	17.9	18.2	18.5	18.8	19.1	19.3	19.5	19.8
4	6.51	8.12	9.17	9.96	10.6	11.1	11.5	11.9	12.3	12.6	12.8	13.1	13.3	13.5	13.7	13.9	14.1	14.2	14.4
5	5.70	6.97	7.80	8.42	8.91	9.32	9.67	9.97	10.24	10.48	10.70	10.89	11.08	11.24	11.40	11.55	11.68	11.81	11.93
6	5.24	6.33	7.03	7.56	7.97	8.32	8.61	8.87	9.10	9.30	9.49	9.65	9.81	9.95	10.08	10.21	10.32	10.43	10.54
7	4.95	5.92	6.54	7.01	7.37	7.68	7.94	8.17	8.37	8.55	8.71	8.86	9.00	9.12	9.24	9.35	9.46	9.55	9.65
8	4.74	5.63	6.20	6.63	6.96	7.24	7.47	7.68	7.87	8.03	8.18	8.31	8.44	8.55	8.66	8.76	8.85	8.94	9.03
9	4.60	5.43	5.96	6.35	6.66	6.91	7.13	7.32	7.49	7.65	7.78	7.91	8.03	8.13	8.23	8.32	8.41	8.49	8.57
10	4.48	5.27	5.77	6.14	6.43	6.67	6.87	7.05	7.21	7.36	7.48	7.60	7.71	7.81	7.91	7.99	8.07	8.15	8.22
11	4.39	5.14	5.62	5.97	6.25	6.48	6.67	6.84	6.99	7.13	7.25	7.36	7.46	7.56	7.65	7.73	7.81	7.88	7.95
12	4.32	5.04	5.50	5.84	6.10	6.32	6.51	6.67	6.81	6.94	7.06	7.17	7.26	7.36	7.44	7.52	7.59	7.66	7.73
13	4.26	4.96	5.40	5.73	5.98	6.19	6.37	6.53	6.67	6.79	6.90	7.01	7.10	7.19	7.27	7.34	7.42	7.48	7.55
14	4.21	4.89	5.32	5.63	5.88	6.08	6.26	6.41	6.54	6.66	6.77	6.87	6.96	7.05	7.12	7.20	7.27	7.33	7.39
15	4.17	4.83	5.25	5.56	5.80	5.99	6.16	6.31	6.44	6.55	6.66	6.76	6.84	6.93	7.00	7.07	7.14	7.20	7.26
16	4.13	4.78	5.19	5.49	5.72	5.92	6.08	6.22	6.35	6.46	6.56	6.66	6.74	6.82	6.90	6.97	7.03	7.09	7.15
17	4.10	4.74	5.14	5.43	5.66	5.85	6.01	6.15	6.27	6.38	6.48	6.57	6.66	6.73	6.80	6.87	6.94	7.00	7.05
18	4.07	4.70	5.09	5.38	5.60	5.79	5.94	6.08	6.20	6.31	6.41	6.50	6.58	6.65	6.72	6.79	6.85	6.91	6.96
19	4.05	4.67	5.05	5.33	5.55	5.73	5.89	6.02	6.14	6.25	6.34	6.43	6.51	6.58	6.65	6.72	6.78	6.84	6.89
20	4.02	4.64	5.02	5.29	5.51	5.69	5.84	5.97	6.09	6.19	6.29	6.37	6.45	6.52	6.59	6.65	6.71	6.76	6.82
24	3.96	4.54	4.91	5.17	5.37	5.54	5.69	5.81	5.92	6.02	6.11	6.19	6.26	6.33	6.39	6.45	6.51	6.56	6.61
30	3.89	4.45	4.80	5.05	5.24	5.40	5.54	5.65	5.76	5.85	5.93	6.01	6.08	6.14	6.20	6.26	6.31	6.36	6.41
40	3.82	4.37	4.70	4.93	5.11	5.27	5.39	5.50	5.60	5.69	5.77	5.84	5.90	5.96	6.02	6.07	6.12	6.17	6.21
60	3.76	4.28	4.60	4.82	4.99	5.13	5.25	5.36	5.45	5.53	5.60	5.67	5.73	5.79	5.84	5.89	5.93	5.98	6.02
120	3.70	4.20	4.50	4.71	4.87	5.01	5.12	5.21	5.30	5.38	5.44	5.51	5.56	5.61	5.66	5.71	5.75	5.79	5.83
∞	3.64	4.12	4.40	4.60	4.76	4.88	4.99	5.08	5.16	5.23	5.29	5.35	5.40	5.45	5.49	5.54	5.57	5.61	5.65

* p is the number of quantities (for example, means) whose range is involved. n_2 is the degrees of freedom in the error estimate.

TABLE 7
Coefficients of Orthogonal Polynomials

n	Polynomial	X=1	2	3	4	5	6	7	8	9	10	Σz^2	λ
2	Linear	-1	1									2	2
3	Linear	-1	0	1								2	1
	Quadratic	1	-2	1								6	3
4	Linear	-3	-1	1	3							20	2
	Quadratic	1	-1	-1	1							4	1
	Cubic	-1	3	-3	1							20	10/3
5	Linear	-2	-1	0	1	2						10	1
	Quadratic	2	-1	-2	-1	2						14	1
	Cubic	-1	2	0	-2	1						10	5/6
	Quartic	1	-4	6	-4	1						70	35/12
6	Linear	-5	-3	-1	1	3	5					70	2
	Quadratic	5	-1	-4	-4	-1	5					84	3/2
	Cubic	-5	7	4	-4	-7	5					180	5/3
	Quartic	1	-3	2	2	-3	1					28	7/12
7	Linear	-3	-2	-1	0	1	2	3				28	1
	Quadratic	5	0	-3	-4	-3	0	5				84	1
	Cubic	-1	1	1	0	-1	-1	1				6	1/6
	Quartic	3	-7	1	6	1	-7	3				154	7/12
8	Linear	-7	-5	-3	-1	1	3	5	7			168	2
	Quadratic	7	1	-3	-5	-5	-3	1	7			168	1
	Cubic	-7	5	7	3	-3	-7	-5	7			264	2/3
	Quartic	7	-13	-3	9	9	-3	-13	7			616	7/12
	Quintic	-7	23	-17	-15	15	17	-23	7			2184	7/10
9	Linear	-4	-3	-2	-1	0	1	2	3	4		60	1
	Quadratic	28	7	-8	-17	-20	-17	-8	7	28		2772	3
	Cubic	-14	7	13	9	0	-9	-13	-7	14		990	5/6
	Quartic	14	-21	-11	9	18	9	-11	-21	14		2002	7/12
	Quintic	-4	11	-4	-9	0	9	4	-11	4		468	3/20
10	Linear	-9	-7	-5	-3	-1	1	3	5	7	9	330	2
	Quadratic	6	2	-1	-3	-4	-4	-3	-1	2	6	132	1/2
	Cubic	-42	14	35	31	12	-12	-31	-35	-14	42	8580	5/3
	Quartic	18	-22	-17	3	18	18	3	-17	-22	18	2860	5/12
	Quintic	-6	14	-1	-11	-6	6	11	1	-14	6	780	1/10

TABLE 8

PERCENTILE POINTS FOR Q-TEST, FOR EQUAL DEGREES OF FREEDOM ν, AND FOR p SAMPLES

p	$\nu = 1$		$\nu = 2$		$\nu = 3$		$\nu = 4$	
	.99	.999	.99	.999	.99	.999	.99	.999
3	*	*	.863	*	.757	.919	.684	.828
4	.920	*	.720	.898	.605	.754	.549	.675
5	.828	*	.608	.773	.512	.644	.443	.552
6	.744	.949	.539	.690	.430	.546	.369	.461
7	.671	.865	.469	.606	.372	.471	.318	.394
8	.609	.793	.412	.537	.325	.411	.276	.342
9	.576	.750	.371	.481	.287	.363	.244	.300
10	.528	.694	.333	.433	.257	.324	.218	.267
12	.448	.598	.276	.358	.211	.265	.179	.217
14	.391	.522	.234	.303	.178	.222	.151	.181
15	.365	.490	.217	.280	.165	.205	.140	.167
16	.343	.460	.202	.261	.154	.190	.130	.155
18	.304	.409	.178	.228	.135	.165	.114	.135
20	.273	.367	.158	.202	.120	.146	.101	.119
22	.246	.332	.142	.180	.108	.130	.090	.106
24	.224	.302	.129	.162	.098	.117	.082	.096
26	.206	.276	.118	.148	.090	.107	.075	.087
28	.190	.254	.108	.135	.082	.098	.069	.080
30	.176	.234	.100	.124	.075	.090	.064	.074
32	.163	.218	.093	.115	.070	.083	.060	.068
36	.143	.189	.082	.100	.062	.072	.052	.060
40	.127	.167	.072	.088	.055	.064	.047	.053
45	.111	.145	.063	.076	.048	.055	.041	.046
50	.098	.127	.056	.067	.043	.049	.037	.041
60	.080	.102	.045	.053	.035	.039	.030	.033
64	.074	.094	.042	.049	.033	.037	.028	.031

*These entries exceeded 1 using the approximating distribution. Since $Q \geq 1$, they are omitted.

Source: Anderson and McLean, **Design of Experiments: A Realistic Approach**, Marcel Dekker, Inc., New York, 1974. Used with permission.

TABLE 8 (CONTINUED)

p	$\nu = 5$ .99	$\nu = 5$ .999	$\nu = 6$ .99	$\nu = 6$ .999	$\nu = 8$ .99	$\nu = 8$ .999	$\nu = 10$ .99	$\nu = 10$ .999
3	.631	.760	.593	.708	.539	.633	.512	.596
4	.498	.608	.461	.558	.413	.490	.383	.446
5	.399	.490	.368	.446	.328	.388	.303	.351
6	.334	.407	.307	.368	.271	.318	.250	.288
7	.284	.345	.261	.311	.230	.268	.212	.242
8	.246	.298	.226	.268	.199	.231	.184	.209
9	.217	.261	.199	.235	.176	.202	.162	.183
10	.194	.232	.178	.208	.157	.179	.145	.163
15	.123	.145	.113	.131	.101	.113	.094	.103
20	.090	.104	.083	.094	.074	.082	.069	.075
30	.058	.065	.053	.059	.048	.052	.045	.048
40	.042	.047	.039	.043	.035	.038	.033	.035
50	.033	.036	.031	.033	.028	.030	.026	.028
60	.027	.029	.025	.027	.023	.024	.022	.023

p	$\nu = 12$ .99	$\nu = 12$ .999	$\nu = 14$ .99	$\nu = 14$ .999	$\nu = 16$ .99	$\nu = 16$ .999	$\nu = 20$ .99	$\nu = 20$ .999
3	.486	.558	.466	.530	.451	.508	.429	.476
4	.362	.415	.347	.393	.335	.375	.319	.351
5	.287	.326	.275	.308	.265	.295	.252	.276
6	.236	.267	.227	.253	.219	.242	.209	.226
7	.201	.225	.192	.213	.186	.204	.178	.191
8	.174	.194	.167	.184	.162	.176	.154	.166
9	.154	.170	.148	.162	.143	.155	.136	.146
10	.137	.152	.132	.144	.128	.138	.122	.130
15	.089	.097	.086	.092	.083	.089	.080	.084
20	.066	.070	.063	.067	.062	.065	.059	.062
30	.043	.045	.042	.043	.040	.042	.039	.040
40	.032	.033	.031	.032	.030	.031	.029	.030
50	.025	.026	.024	.025	.024	.025	.023	.024
60	.021	.022	.020	.021	0.20	.020	.019	.020

For $\nu > 60$, calculate $p\nu(pq-1)$ and compare with χ^2 with $(p-1)$ degrees of freedom in Appendix 7.

TABLE 9

Coefficients $\{a_{n-i+1}\}$ for the W test for normality,

for $n = 2(1)50$

i \ n	2	3	4	5	6	7	8	9	10
1	0.7071	0.7071	0.6872	0.6646	0.6431	0.6233	0.6052	0.5888	0.5739
2		.0000	.1677	.2413	.2806	.3031	.3164	.3244	.3291
3				.0000	.0875	.1401	.1743	.1976	.2141
4						.0000	.0561	.0947	.1224
5								.0000	.0399

i \ n	11	12	13	14	15	16	17	18	19	20
1	0.5601	0.5475	0.5359	0.5251	0.5150	0.5056	0.4968	0.4886	0.4808	0.4734
2	.3315	.3325	.3325	.3318	.3306	.3290	.3273	.3253	.3232	.3211
3	.2260	.2347	.2412	.2460	.2495	.2521	.2540	.2553	.2561	.2565
4	.1429	.1586	.1707	.1802	.1878	.1939	.1988	.2027	.2059	.2085
5	.0695	.0922	.1099	.1240	.1353	.1447	.1524	.1587	.1641	.1686
6	0.0000	0.0303	0.0539	0.0727	0.0880	0.1005	0.1109	0.1197	0.1271	0.1334
7			.0000	.0240	.0433	.0593	.0725	.0837	.0932	.1013
8					.0000	.0196	.0359	.0496	.0612	.0711
9							.0000	.0163	.0303	.0422
10									.0000	.0410

i \ n	21	22	23	24	25	26	27	28	29	30
1	0.4643	0.4590	0.4542	0.4493	0.4450	0.4407	0.4366	0.4328	0.4291	0.4254
2	.3185	.3156	.3126	.3098	.3069	.3043	.3018	.2992	.2968	.2944
3	.2578	.2571	.2563	.2554	.2543	.2533	.2522	.2510	.2499	.2487
4	.2119	.2131	.2139	.2145	.2148	.2151	.2152	.2151	.2150	.2148
5	.1736	.1764	.1787	.1807	.1822	.1836	.1848	.1857	.1864	.1870
6	0.1399	0.1443	0.1480	0.1512	0.1539	0.1563	0.1584	0.1601	0.1616	0.1630
7	.1092	.1150	.1201	.1245	.1283	.1316	.1346	.1372	.1395	.1415
8	.0804	.0878	.0941	.0997	.1046	.1089	.1128	.1162	.1192	.1219
9	.0530	.0618	.0696	.0764	.0823	.0876	.0923	.0965	.1002	.1036
10	.0263	.0368	.0459	.0539	.0610	.0672	.0728	.0778	.0822	.0862
11	0.0000	0.0122	0.0228	0.0321	0.0403	0.0476	0.0540	0.0598	0.0650	0.0697
12			.0000	.0107	.0200	.0284	.0358	.0424	.0483	.0537
13					.0000	.0094	.0178	.0253	.0320	.0381
14							.0000	.0084	.0159	.0227
15									.0000	.0076

Source: *Anderson and McLean, **Design of Experiments: A Realistic Approach**, Marcel Dekker, Inc., New York, 1974. Used with permission.*

TABLE 9 (CONTINUED)

Coefficients $\{a_{n-i+1}\}$ for the W test for normality,

for n = 2(1)50 (cont.)

n i	31	32	33	34	35	36	37	38	39	40
1	0.4220	0.4188	0.4156	0.4127	0.4096	0.4068	0.4040	0.4015	0.3989	0.3964
2	.2921	.2898	.2876	.2854	.2834	.2813	.2794	.2774	.2755	.2737
3	.2475	.2463	.2451	.2439	.2427	.2415	.2403	.2391	.2380	.2368
4	.2145	.2141	.2137	.2132	.2127	.2121	.2116	.2110	.2104	.2098
5	.1874	.1878	.1880	.1882	.1883	.1883	.1883	.1881	.1880	.1878
6	0.1641	0.1651	0.1660	0.1667	0.1673	0.1678	0.1683	0.1686	0.1689	0.1691
7	.1433	.1449	.1463	.1475	.1487	.1496	.1505	.1513	.1520	.1526
8	.1243	.1265	.1284	.1301	.1317	.1331	.1344	.1356	.1366	.1376
9	.1066	.1093	.1118	.1140	.1160	.1179	.1196	.1211	.1225	.1237
10	.0899	.0931	.0961	.0988	.1013	.1036	.1056	.1075	.1092	.1108
11	0.0739	0.0777	0.0812	0.0844	0.0873	0.0900	0.0924	0.0947	0.0967	0.0986
12	.0585	.0629	.0669	.0706	.0739	.0770	.0798	.0824	.0848	.0870
13	.0435	.0485	.0530	.0572	.0610	.0645	.0677	.0706	.0733	.0759
14	.0289	.0344	.0395	.0441	.0484	.0523	.0559	.0592	.0622	.0651
15	.0144	.0206	.0262	.0314	.0361	.0404	.0444	.0481	.0515	.0546
16	0.0000	0.0068	0.0131	0.0187	0.0239	0.0287	0.0331	0.0372	0.0409	0.0444
17		.0000	.0062	.0119	.0172	.0220	.0264	.0305	.0343
18				.0000	.0057	.0110	.0158	.0203	.0244
19						.0000	.0053	.0101	.0146
20								.0000	.0049

n i	41	42	43	44	45	46	47	48	49	50
1	0.3940	0.3917	0.3894	0.3872	0.3850	0.3830	0.3808	0.3789	0.3770	0.3751
2	.2719	.2701	.2684	.2667	.2651	.2635	.2620	.2604	.2589	.2574
3	.2357	.2345	.2334	.2323	.2313	.2302	.2291	.2281	.2271	.2260
4	.2091	.2085	.2078	.2072	.2065	.2058	.2052	.2045	.2038	.2032
5	.1876	.1874	.1871	.1868	.1865	.1862	.1859	.1855	.1851	.1847
6	0.1693	0.1694	0.1695	0.1695	0.1695	0.1695	0.1695	0.1693	0.1692	0.1691
7	.1531	.1535	.1539	.1542	.1545	.1548	.1550	.1551	.1553	.1554
8	.1384	.1392	.1398	.1405	.1410	.1415	.1420	.1423	.1427	.1430
9	.1249	.1259	.1269	.1278	.1286	.1293	.1300	.1306	.1312	.1317
10	.1123	.1136	.1149	.1160	.1170	.1180	.1189	.1197	.1205	.1212
11	0.1004	0.1020	0.1035	0.1049	0.1062	0.1073	0.1085	0.1095	0.1105	0.1113
12	.0891	.0909	.0927	.0943	.0959	.0972	.0986	.0998	.1010	.1020
13	.0782	.0804	.0824	.0842	.0860	.0876	.0892	.0906	.0919	.0932
14	.0677	.0701	.0724	.0745	.0765	.0783	.0801	.0817	.0832	.0846
15	.0575	.0602	.0628	.0651	.0673	.0694	.0713	.0731	.0748	.0764
16	0.0476	0.0506	0.0534	0.0560	0.0584	0.0607	0.0628	0.0648	0.0667	0.0685
17	.0379	.0411	.0442	.0471	.0497	.0522	.0546	.0568	.0588	.0608
18	.0283	.0318	.0352	.0383	.0412	.0439	.0465	.0489	.0511	.0532
19	.0188	.0227	.0263	.0296	.0328	.0357	.0385	.0411	.0436	.0459
20	.0094	.0136	.0175	.0211	.0245	.0277	.0307	.0335	.0361	.0386
21	0.0000	0.0045	0.0087	0.0126	0.0163	0.0197	0.0229	0.0259	0.0288	0.0314
22		.0000	.0042	.0081	.0118	.0153	.0185	.0215	.0244
23				.0000	.0039	.0076	.0111	.0143	.0174
24						.0000	.0037	.0071	.0104
25								.0000	.0035

TABLE 10

Percentage points of the W test for n = 3(1)50

					Level				
n	0.01	0.02	0.05	0.10	0.50	0.90	0.95	0.98	0.99
3	0.753	0.756	0.767	0.789	0.959	0.998	0.999	1.000	1.000
4	.687	.707	.748	.792	.935	.987	.992	.996	.997
5	.686	.715	.762	.806	.927	.979	.986	.991	.993
6	0.713	0.743	0.788	0.826	0.927	0.974	0.981	0.986	0.989
7	.730	.760	.803	.838	.928	.972	.979	.985	.988
8	.749	.778	.818	.851	.932	.972	.978	.984	.987
9	.764	.791	.829	.859	.935	.972	.978	.984	.986
10	.781	.806	.842	.869	.938	.972	.978	.983	.986
11	0.792	0.817	0.850	0.876	0.940	0.973	0.979	0.984	0.986
12	.805	.828	.859	.883	.943	.973	.979	.984	.986
13	.814	.837	.866	.889	.945	.974	.979	.984	.986
14	.825	.846	.874	.895	.947	.975	.980	.984	.986
15	.835	.855	.881	.901	.950	.975	.980	.984	.987
16	0.844	0.863	0.887	0.906	0.952	0.976	0.981	0.985	0.987
17	.851	.869	.892	.910	.954	.977	.981	.985	.987
18	.858	.874	.897	.914	.956	.978	.982	.986	.988
19	.863	.879	.901	.917	.957	.978	.982	.986	.988
20	.868	.884	.905	.920	.959	.979	.983	.986	.988
21	0.873	0.888	0.908	0.923	0.960	0.980	0.983	0.987	0.989
22	.878	.892	.911	.926	.961	.980	.984	.987	.989
23	.881	.895	.914	.928	.962	.981	.984	.987	.989
24	.884	.898	.916	.930	.963	.981	.984	.987	.989
25	.888	.901	.918	.931	.964	.981	.985	.988	.989
26	0.891	0.904	0.920	0.933	0.965	0.982	0.985	0.988	0.989
27	.894	.906	.923	.935	.965	.982	.985	.988	.990
28	.896	.908	.924	.936	.966	.982	.985	.988	.990
29	.898	.910	.926	.937	.966	.982	.985	.988	.990
30	.900	.912	.927	.939	.967	.983	.985	.988	.900
31	0.902	0.914	0.929	0.940	0.967	0.983	0.986	0.988	0.990
32	.904	.915	.930	.941	.968	.983	.986	.988	.990
33	.906	.917	.931	.942	.968	.983	.986	.989	.990
34	.908	.919	.933	.943	.969	.983	.986	.989	.990
35	.910	.920	.934	.944	.969	.984	.986	.989	.990
36	0.912	0.922	0.935	0.945	0.970	0.984	0.986	0.989	0.990
37	.914	.924	.936	.946	.970	.984	.987	.989	.990
38	.916	.925	.938	.947	.971	.984	.987	.989	.990
39	.917	.927	.939	.948	.971	.984	.987	.989	.991
40	.919	.928	.940	.949	.972	.985	.987	.989	.991
41	0.920	0.929	0.941	0.950	0.972	0.985	0.987	0.989	0.991
42	.922	.930	.942	.951	.972	.985	.987	.989	.991
43	.923	.932	.943	.951	.973	.985	.987	.990	.991
44	.924	.933	.944	.952	.973	.985	.987	.990	.991
45	.926	.934	.945	.953	.973	.985	.988	.990	.991
46	0.927	0.935	0.945	0.953	0.974	0.985	0.988	0.990	0.991
47	.928	.936	.946	.954	.974	.985	.988	.990	.991
48	.929	.937	.947	.954	.974	.985	.988	.990	.991
49	.929	.937	.947	.955	.974	.985	.988	.990	.991
50	.930	.938	.947	.955	.974	.985	.988	.990	.991

Source: *Anderson and McLean, **Design of Experiments: A Realistic Approach**, Marcel Dekker, Inc., New York, 1974. Used with permission.*

TABLE 11
Generators for 2-level Fractional Factorial Designs

Number of Factors k	Fractionalization Element p	Number of Runs	Generator Components	
			Single Effect	Interaction
3	1	4	C	AB (12)
4	1	8	D	ABC (123)
5	1	16	E	ABCD (1234)
5	2	8	D	AB (12)
			E	AC (13)
6	1	32	F	ABCDE (12345)
6	2	16	E	ABC (123)
6	3	8	D	AB (12)
			E	AC (13)
			F	BC (23)
7	1	64	G	ABCDEF (123456)
7	2	32	F	ABCD (1234)
			G	ABDE (1245)

TABLE 11 (CONTINUED)
Generators for 2-level Fractional Factorial Designs

Number of Factors k	Fractionalization Element p	Number of Runs	Generator Components	
			Single Effect	Interaction
7	3	16	E	ABC (123)
			F	BCD (234)
			G	ACD (134)
7	4	8	D	AB (12)
			E	AC (13)
			F	BC(23)
			G	ABC (123)
8	1	128	H	ABCDEFG(1234567)
8	2	64	G	ABCD (1234)
			H	ABEF (1256)
8	3	32	F	ABC (123)
			G	ABD (124)
			H	BCDE (2345)
8	4	16	E	BCD (234)
			F	ACD (134)
			G	ABC (123)
			H	ABD (124)
9	2	128	H	ACDFG (13467)
			J	BCEFG (23567)

TABLE 11 (CONTINUED)
Generators for 2-level Fractional Factorial Designs

Number of Factors k	Fractionalization Element p	Number of Runs	Generator Components	
			Single Effect	Interaction
9	3	64	G	ABCD (1234)
			H	ACEF (1356)
			J	CDEF (3456)
9	4	32	F	BCDE (2345)
			G	ACDE (1345)
			H	ABDE (1245)
			J	ABCE (1235)
9	5	16	E	ABC (123)
			F	BCD (234)
			G	ACD (134)
			H	ABD (124)
			J	ABCD (1234)
10	3	128	H	ABCG (1237)
			J	BCDE (2345)
			K	ACDF (1346)
10	4	64	G	BCDF (2346)
			H	ACDF (1346)
			J	ABDE (1245)
			K	ABCE (1235)

TABLE 11 (CONTINUED)
Generators for 2-level Fractional Factorial Designs

Number of Factors k	Fractionalization Element p	Number of Runs	Generator Components	
			Single Effect	Interaction
10	6	16	E	ABC (123)
			F	BCD (234)
			G	ACD (134)
			H	ABD (124)
			J	ABCD (1234)
			K	AB (12)
11	5	64	G	CDE (345)
			H	ABCD (1234)
			J	ABF (126)
			K	BDEF (2456)
			L	ADEF (1456)
11	6	32	F	ABC (123)
			G	BCD (234)
			H	CDE (345)
			J	ACD (134)
			K	ADE (145)
			L	BDE (245)

Index